Boundary Element Methods in Engineering and Sciences

Computational and Experimental Methods in Structures

Series Editor: Ferri M. H. Aliabadi *(Imperial College London, UK)*

Vol. 1 Buckling and Postbuckling Structures: Experimental, Analytical and Numerical Studies
edited by B. G. Falzon and M. H. Aliabadi (Imperial College London, UK)

Vol. 2 Advances in Multiphysics Simulation and Experimental Testing of MEMS
edited by A. Frangi, C. Cercignani (Politecnico di Milano, Italy), S. Mukherjee (Cornell University, USA) and N. Aluru (University of Illinois at Urbana Champaign, USA)

Vol. 3 Multiscale Modeling in Solid Mechanics: Computational Approaches
edited by U. Galvanetto and M. H. Aliabadi (Imperial College London, UK)

Vol. 4 Boundary Element Methods in Engineering and Sciences
by M. H. Aliabadi (Imperial College, UK) and P. Wen (Queen Mary University of London, UK)

Computational and Experimental Methods in Structures – Vol. 4

Boundary Element Methods in Engineering and Sciences

M H Aliabadi
Imperial College, UK

P H Wen
Queen Mary University of London, UK

ICP
Imperial College Press

Published by

Imperial College Press
57 Shelton Street
Covent Garden
London WC2H 9HE

Distributed by

World Scientific Publishing Co. Pte. Ltd.
5 Toh Tuck Link, Singapore 596224
USA office: 27 Warren Street, Suite 401-402, Hackensack, NJ 07601
UK office: 57 Shelton Street, Covent Garden, London WC2H 9HE

British Library Cataloguing-in-Publication Data
A catalogue record for this book is available from the British Library.

BOUNDARY ELEMENT METHODS IN ENGINEERING AND SCIENCES
Computational and Experimental Methods in Structures — Vol. 4

ISBN-13 978-1-84816-579-3
ISBN-10 1-84816-579-X

Typeset by Stallion Press
Email: enquiries@stallionpress.com

Printed in Singapore by World Scientific Printers.

PREFACE

The Boundary Element Method (BEM) has in recent years attained a level of development that has made it a numerical tool of choice in areas such as fracture mechanics and acoustics. This book brings together contribution from leading experts in the field to give a comprehensive descrption of recently developed techniques. These include boundary element method for modelling viscous flow, boundary element method for free surface flows, simulation of cavitating and free surface flows, boundary element method for geometrically nonlinear analysis of plates and shells, modelling of plates and shells by meshless local boundary element method, fast hierarchical boundary element method for large scale 3D elastic problems, time-domain BEM techniques, Green's function evaluation for three dimensional exponentially graded elasticity; boundary element method for the fracture analysis of the general piezoelectric solids; using BEM and condition numbers and local errors in the boundary element method.

We are grateful to all contributors who have displayed great enthusiasm for this book.

M H Aliabadi
Imperical College London

P H Wen
Queen Mary, University of London

CONTENTS

Chapter 1

THE BOUNDARY ELEMENT METHOD FOR GEOMETRICALLY NON-LINEAR ANALYSES OF PLATES AND SHELLS

P.H. Wen
Department of Engineering, Queen Mary, University of London
Mile End Road, London E1 4NS, United Kingdom
p.h.wen@qmul.ac.uk

T. Dirgantara
Faculty of Mechanical and Aerospace Engineering, Institut Teknologi Bandung, Jl Ganesha 10 Bandung 40132, Indonesia
tdirgantara@ae.itb.ac.id

P.M. Baiz and M.H. Aliabadi
Department of Aeronautics, Imperial College London
South Kensington Campus, London, SW7 2AZ, United Kingdom
p.m.baiz@imperial.ac.uk, m.h.aliabadi@imperial.ac.uk

This chapter describes recent developments on the non-linear analysis of thin-walled structures by the Boundary Element Method (BEM), with particular emphasis on geometrically non-linear behaviour (large deflection). Non-linear boundary integral equations are obtained by coupling two-dimensional plane stress elasticity with boundary element formulation of shear deformable (Reissner) plate bending. Several examples of flat and curved plates with different geometries and boundary conditions are presented. Results are compared with other numerical solutions and good agreement is obtained.

1.1. Introduction

Only a relatively small number of problems can be solved by means of the exact field equations of elasticity theory; therefore the structural engineer is forced to make a number of simplifying assumptions. These assumptions depend on the relative dimensions of the structural element and its boundary conditions in the three-dimensional space. Plate and shell structures are one of such simplifications widely used in engineering applications, such as aircrafts, cars, pressure vessels, building slabs, etc.

There are two widely used plate theories. The first plate theory was developed by Kirchhoff[17] and is commonly referred to as the classical theory. The other was developed by Reissner,[31] and is known as the shear deformable theory. The classical theory is adequate for analysing many engineering applications; however, for problems involving stress concentrations and cracks, the theory has been shown not to be in agreement with experimental measurements. In the shear deformable plate theory, the problem is represented in terms of three independent degrees of freedom, involving generalised displacements (i.e. two rotations and deflection) and generalised tractions (i.e. moments and transverse shear force).

The first direct boundary element formulation for analysis of Reissner type plates was proposed by Van der Weeën.[42] Later, Karam and Telles[15] extended the formulation to account for infinite regions and also reported that Reissners plate model is suitable for both thin and thick plates. Several years later Barcellos and Silva[4] presented a formulation based on Mindlin theory, differing only with Van der Weeën[42] in the shear factor constant. Since these early works, further extensions of the BEM to shear deformable plate theories have been made.

Boundary element formulations for analysis of shallow shells are very limited, particularly in the case of shear deformable shallow shells theory. Applications in this area are relatively difficult because of the high order differential equations governing the problem. Newton and Tottenham,[26] and Tottenham[41] presented a BEM formulation of shallow shell problems, by decomposition of the fourth-order governing equations into second-order equations. The fundamental solutions used in the direct boundary element formulation of shell problems tend to be complicated,[19,22] which means that further extensions will be cumbersome.[5] It is for this reason that boundary element formulations based on the coupling of boundary element formulation for plate bending and the two-dimensional plane stress elasticity have been proposed. The main disadvantage of this technique is that domain integrals appear in the formulations. Some works based on these kinds of formulations are those of Forbes and Robinson,[11] Zhang and Atluri[52] and Beskos.[5] In the case of shear deformable shallow shell theories, the first work was presented by Dirgantara and Aliabadi[7] in 1999. Since then, further works and extensions have been presented by Aliabadi[1] and co-workers.

Boundary element formulations for geometrically non-linear analysis of plates were first presented in the early eighties. Kamiya and Sawaki[12] introduced a formulation based on the Berger equation. This formulation was later extended[14] to include post-buckling behaviour under thermal loading. Tanaka[37] presented a formulation based on the Von Karman's equation, obtaining boundary and domain integrals in terms of stress and displacement. Other works dealing with large deflection and post-buckling of Kirchhoff's plates are those of Ye and Liu,[50] O'Donoghue and Atluri,[27] Kawabe,[16] Qin and Huang,[29] Atluri and Pipkins,[3] Tanaka, Matsumoto and Zheng,[38,39] Elzein and Singellakis[10] and Wang, Ji and Tanaka,[44] among others.

Boundary element formulations for geometrically non-linear analysis of Reissner plates were first introduced in the early nineties. Lei, Huang and Wang[18,19] presented an integral formulation of the Von Karman type non-linear governing equations. Sun, He and Qin[36] derived exact integral equations based on a variational principle. More recent works on shear deformable plates are those presented by Wen, Aliabadi and Young, describing large deflection[47] and post-buckling.[48] Supriyono and Aliabadi[35] presented a formulation for combined geometrical and material non-linear behaviour.

Geometrically non-linear formulations for shallow shells by BEM were first introduced in the early eighties. Kamiya and Sawaki[13] presented an approach for geometrically non-linear shallow shells with arbitrary contours. Tosaka and Miyake[40] developed a theoretical foundation for the large deflection analysis of shells, proposing three types of integral equations. Zhang and Atluri[53] presented a non-linear boundary element formulation for snap-through phenomena in thin shallow shells using the arc length method. Other works dealing with large deflection and post-buckling analysis of classical shallow shells are those of Ye[51] and Lin and Long.[20] Boundary element formulations for shear deformable shallow shells are scarce. Dirgantara and Aliabadi[9] introduced the first work on large deflection while Aliabadi and Baiz[2] presented a formulation for post-buckling analysis.

This chapter reviews boundary element formulations for geometrical non-linear analysis of shear deformable plates and shallow shells as developed by Wen, Aliabadi and Young[47] and Dirgantara and Aliabadi,[9] respectively. The formulation is derived for geometrically non-linear shallow shells[9] by coupling boundary element formulations of shear deformable plate bending and two-dimensional plane stress elasticity. This general

BEM formulation can be reduced to the geometrically non-linear plate[47] integral equations by neglecting the curvature terms, or to the linear plate and shallow shell expressions by neglecting the non-linear terms. Domain integrals are transformed into boundary integrals using the dual reciprocity technique. At the end, several examples are shown to demonstrate the accuracy and efficiency of the present BEM formulation.

1.2. Basic Definitions of Shear Deformable Plates and Shallow Shells

Plate and shell structures can be seen as bodies bounded by two surfaces that are separated by a distance known as the thickness, h, which are capable of resisting bending and which transfer membrane forces from one of its edges to another. The present work considers isotropic plates or shallow shells of uniform thickness h, Young's modulus E, Poisson's ratio ν and a quadratic middle surface defined only by $k_{11} = 1/R_1$ and $k_{22} = 1/R_2$, which are principal curvatures of the shell in the x_1^- and x_2^- directions respectively, while $k_{12} = k_{21} = 0$ (see Fig. 1.1).

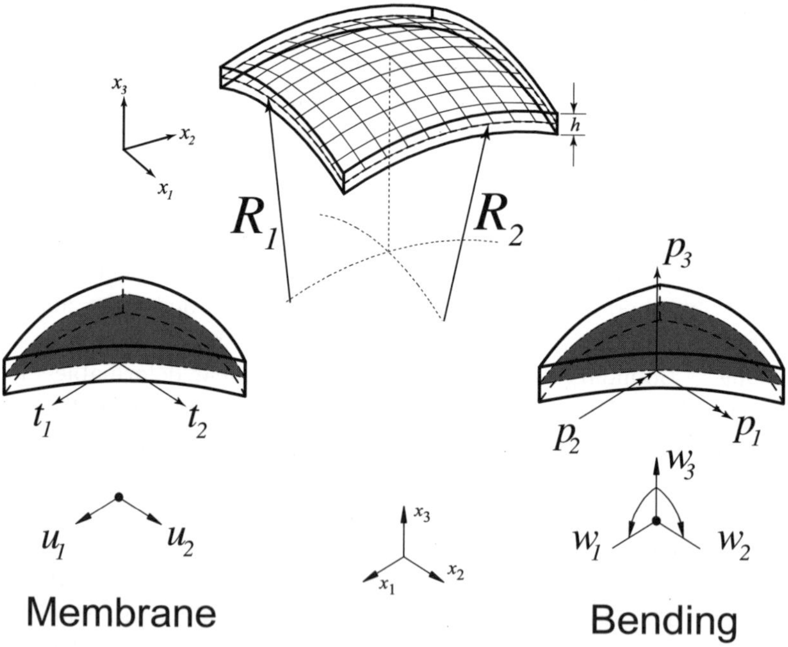

Fig. 1.1. Degrees of freedom and curved plate geometry.

Classical linear elastic shell theory, based on the hypothesis of Kirchhoff,[17] was presented by Love.[21] The basic assumptions are the following:

(1) The shell thickness h should be at least an order of magnitude smaller than the least radius of curvature R of the middle surface.
(2) The strains and displacements are small enough for changes in geometry to be negligible.
(3) The component of stress normal to the middle surface is small compared to the other components of stress.
(4) Straight lines normal to the undeformed middle surface would remain inextensible, straight and normal to the middle surface when the shell is in a deformed state (transverse shear strains are neglected).

This last assumption represents the main difference between the classical and shear deformable theories. In Kirchhoff's hypothesis, the rotations of the cross section are approximated with the first derivatives of the deflection, while in the shear deformable theories, the straight lines normal to the undeformed middle surface would still remain inextensible and straight when the shell is in a deformed state, but would not necessarily remain normal to the middle surface. In other words, in shear deformable theories, rotations of the cross section are considered totally independent degrees of freedom.

Shear deformable plate theories have been proposed by Reissner[30] and Mindlin,[23] and the difference between them is in the shear factor (5/6 for Reissner and $\pi^2/12$ for Mindlin). Shear deformable theories for thin elastic isotropic shells have been derived by Reissner[33] and Naghdi.[24] All theories have been successfully applied to both thick and thin plates and shells.

Simplification of the general equations of shells can be valid in many situations. A shallow shell (also known as a curved plate) is the one whose rise with regard to any cord is small; this means that all points on the middle surface have to be sufficiently close to a plane. According to Vlasov,[43] a shell can be considered shallow if the ratio of the rise to the shorter side (for a shell of rectangular plan) or to the diameter (for a shell of circular plan) is less than 1/5. Figure 1.1 shows the generalised displacements w_i and u_α, where w_α denotes rotations of the middle surface, w_3 denotes the out-of-plane displacement and u_α denotes in-plane displacements. The corresponding generalised tractions are denoted as p_i and t_α, where p_α denotes tractions due to the stress couples, p_3 denotes the traction due to shear stress resultant and t_α denotes tractions due to membrane stress resultants. At this point, it is good to mention that the indicial notation used throughout this chapter is as follows: the Greek indices (α, β, γ) will vary from 1 to 2 and Roman indices (i, j, k) from 1 to 3.

1.2.1. *Forces and moments for stress resultants and stress couples*

Consider an element from a shell (see Fig. 1.2). As in beam theory, in shell theory it is more convenient to transform the conventional stresses shown in Fig. 1.2 into a statically equivalent system of forces and bending moments shown in Fig. 1.3.

The relation between the conventional stress state and the shell stress resultants system can be obtained from the total force acting on each of

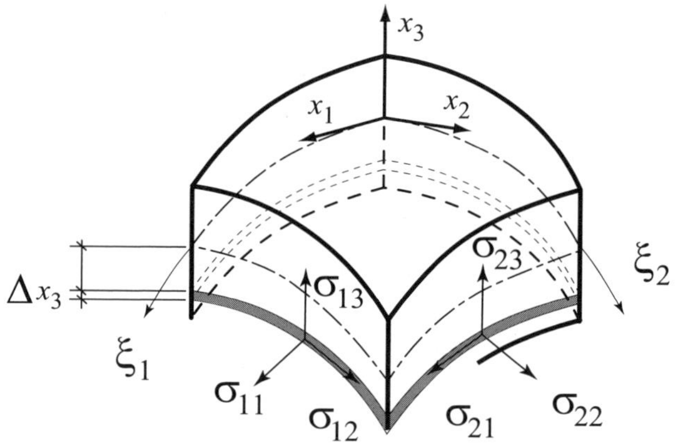

Fig. 1.2. General stress components.

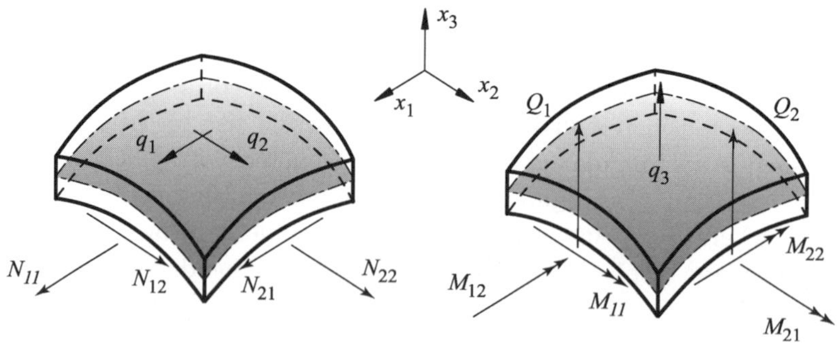

Fig. 1.3. Shell stress resultants.

the faces. On the face with stresses σ_{11}, σ_{12} and σ_{13}, the total forces (as a resultant of the stresses) acting on the entire surface of this face is given by:

$$F_1 = R_2 d\theta \int_{-h/2}^{+h/2} \left(1 + \frac{x_3}{R_2}\right) \sigma_{11} dx_3 \tag{1.1}$$

$$F_2 = R_2 d\theta \int_{-h/2}^{+h/2} \left(1 + \frac{x_3}{R_2}\right) \sigma_{12} dx_3 \tag{1.2}$$

$$F_3 = R_2 d\theta \int_{-h/2}^{+h/2} \left(1 + \frac{x_3}{R_2}\right) \sigma_{13} dx_3 \tag{1.3}$$

Therefore, the membrane $N_{\alpha\beta}$ and shearing Q_α stress resultants (defined as forces per unit length along dx_2) on this face are given as:

$$N_{11} = \frac{F_1}{R_2 d\theta} = \int_{-h/2}^{+h/2} \left(1 + \frac{x_3}{R_2}\right) \sigma_{11} dx_3 \tag{1.4}$$

$$N_{12} = \frac{F_2}{R_2 d\theta} = \int_{-h/2}^{+h/2} \left(1 + \frac{x_3}{R_2}\right) \sigma_{12} dx_3 \tag{1.5}$$

$$Q_1 = \frac{F_3}{R_2 d\theta} = \int_{-h/2}^{+h/2} \left(1 + \frac{x_3}{R_2}\right) \sigma_{13} dx_3 \tag{1.6}$$

while the bending stress couples $M_{\alpha\beta}$ (defined as the bending and twisting moments per unit length along dx_2) are:

$$M_{11} = \int_{-h/2}^{+h/2} \left(1 + \frac{x_3}{R_2}\right) x_3 \sigma_{11} dx_3 \tag{1.7}$$

$$M_{12} = \int_{-h/2}^{+h/2} \left(1 + \frac{x_3}{R_2}\right) x_3 \sigma_{12} dx_3 \tag{1.8}$$

Similarly, stress resultants and stress couples on the other face of the shell element (perpendicular to the x_2^- axis):

$$N_{22} = \int_{-h/2}^{+h/2} \left(1 + \frac{x_3}{R_1}\right) \sigma_{22} dx_3 \tag{1.9}$$

$$N_{21} = \int_{-h/2}^{+h/2} \left(1 + \frac{x_3}{R_1}\right) \sigma_{21} dx_3 \tag{1.10}$$

$$Q_2 = \int_{-h/2}^{+h/2} \left(1 + \frac{x_3}{R_1}\right) \sigma_{23} dx_3 \tag{1.11}$$

$$M_{22} = \int_{-h/2}^{+h/2} \left(1 + \frac{x_3}{R_1}\right) x_3 \sigma_{22} dx_3 \tag{1.12}$$

$$M_{21} = \int_{-h/2}^{+h/2} \left(1 + \frac{x_3}{R_1}\right) x_3 \sigma_{21} dx_3 \tag{1.13}$$

The behaviour of shell stresses through the thickness is assumed as follows;[24,33] those due to membrane forces are uniform, those due to bending and twisting moments vary linearly and the transverse shear stresses vary parabolically over the thickness:

$$\left(1 + \frac{x_3}{R_\gamma}\right) \sigma_{\alpha\beta} = \frac{1}{h} N_{\alpha\beta} + \frac{12 x_3}{h^3} M_{\alpha\beta}; \quad \begin{cases} \gamma = \beta & \text{if } \alpha \neq \beta \\ \gamma \neq \beta & \text{if } \alpha = \beta \end{cases} \tag{1.14}$$

and

$$\left(1 + \frac{x_3}{R_\gamma}\right) \sigma_{\alpha 3} = \frac{3}{2h} \left[1 - \left(\frac{2x_3}{h}\right)^2\right] Q_\alpha; \ \gamma \neq \alpha \tag{1.15}$$

Relation between stress resultants and the generalised tractions at the boundary are written as:

$$p_\alpha = M_{\alpha\beta} n_\beta \tag{1.16}$$

$$p_3 = Q_\alpha n_\alpha \tag{1.17}$$

$$t_\alpha = N_{\alpha\beta} n_\beta \tag{1.18}$$

where n_β are the components of the outward normal vector to the projected boundary on the $x_1^-\ x_2^-$ plane.

1.2.2. *Linear kinematic equations*

Strain tensors in shear deformable linear elastic theories can be derived from the deformation pattern of a differential element. The membrane strain resultant tensor can be expressed as follows:

$$\varepsilon_{\alpha\beta}^{(l)} = \frac{1}{2} \left(u_{\alpha,\beta} + u_{\beta,\alpha}\right) + w_3 k_{\alpha\beta} \tag{1.19}$$

where $(\ )_{,\beta} = \partial(\ )/\partial x_\beta$ and the superscript (l) refers to the linear part. The transverse shear strain resultant can be expressed as:

$$\gamma_{\alpha 3} = w_\alpha + w_{3,\alpha} \tag{1.20}$$

and finally, the flexural strain resultant can be written as:

$$\kappa_{\alpha\beta} = 2\chi_{\alpha\beta} = w_{\alpha,\beta} + w_{\beta,\alpha} \tag{1.21}$$

1.2.3. *Linear constitutive equations*

The relationships between stress resultants and strains were derived by Naghdi,[24] using Reissner's variational theorem of elasticity,[32] and are given by:

$$M_{\alpha\beta} = D\frac{1-\nu}{2}\left(2\chi_{\alpha\beta} + \frac{2\nu}{1-\nu}\chi_{\gamma\gamma}\delta_{\alpha\beta}\right) \tag{1.22}$$

$$Q_\alpha = C\gamma_{\alpha 3} \tag{1.23}$$

$$N_{\alpha\beta} = B\frac{1-\nu}{2}\left(2\varepsilon_{\alpha\beta} + \frac{2\nu}{1-\nu}\varepsilon_{\gamma\gamma}\delta_{\alpha\beta}\right) \tag{1.24}$$

where $\delta_{\alpha\beta}$ is the Kronecker delta function; $B(= Eh/\left(1-\nu^2\right))$ is known as the tension stiffness; $D(= Eh^3/\left[12\left(1-\nu^2\right)\right])$ is the bending stiffness; $C(=\left[D\left(1-\nu\right)\lambda_s^2\right]/2)$ is the shear stiffness and $\lambda_s(=\sqrt{10}/h)$ is called the shear factor.

Stress-displacement relationships can be obtained after Eqs. (1.19–1.21) are substituted into Eqs. (1.22–1.24):

$$M_{\alpha\beta} = D\frac{1-\nu}{2}\left(w_{\alpha,\beta} + w_{\beta,\alpha} + \frac{2\nu}{1-\nu}w_{\gamma,\gamma}\delta_{\alpha\beta}\right) \tag{1.25}$$

$$Q_\alpha = C(w_\alpha + w_{3,\alpha}) \tag{1.26}$$

and

$$\begin{aligned} N_{\alpha\beta}^{(i)} &= B\frac{1-\nu}{2}\left(u_{\alpha,\beta} + u_{\beta,\alpha} + \frac{2\nu}{1-\nu}u_{\gamma,\gamma}\delta_{\alpha\beta}\right) \\ N_{\alpha\beta}^{(ii)} &= B\left[(1-\nu)k_{\alpha\beta} + \nu\delta_{\alpha\beta}k_{\phi\phi}\right]w_3 \\ N_{\alpha\beta}^{(l)} &= N_{\alpha\beta}^{(i)} + N_{\alpha\beta}^{(ii)} \end{aligned} \tag{1.27}$$

To make the representation more convenient, the linear membrane stress $N_{\alpha\beta}^{(l)}$ is separated into $N_{\alpha\beta}^{(i)}$ which is due to in-plane displacements and $N_{\alpha\beta}^{(ii)}$ which is due to curvature and out-of-plane displacement. This notation facilitates the derivation of the membrane boundary integral equations.

1.2.4. *Linear equilibrium equations*

Equilibrium equations for shear deformable plate bending (Reissner-Mindlin) and 2D elasticity can be written in indicial notation as follows[1]:

$$M_{\alpha\beta,\beta} - Q_\alpha = 0; \tag{1.28}$$

$$Q_{\alpha,\alpha} + f_3^b = 0; \tag{1.29}$$

$$N_{\alpha\beta,\beta}^{(i)} + f_\alpha^m = 0 \tag{1.30}$$

where f_3^b, f_α^m are body forces that can be expressed as:

$$f_3^b = q_3 - k_{\alpha\beta} N_{\alpha\beta} \tag{1.31}$$

$$f_\alpha^m = q_\alpha + N_{\alpha\beta,\beta}^{(ii)} \tag{1.32}$$

where q_i are applied body forces and $k_{\alpha\beta} N_{\alpha\beta}$ and $N_{\alpha\beta,\beta}^{(ii)}$ are terms due to the curvature of the shallow shell.

1.3. Large Deflection Theory

For geometrical non-linear theory (large deflection) of plates and shells, the membrane and bending deflections are coupled in the membrane strain equation (out-of-plane displacement contributions are included in strain-displacement relationships). The new assumption can be stated as follows: Deflections (w_3) are large (several times greater than the thickness, h); but strains are still small compared to unity.

Based on this, the strain tensors can be obtained by considering the geometrical changes as the element is stretched and deformed,

$$\varepsilon_{\alpha\beta} = \frac{1}{2}(u_{\alpha,\beta} + u_{\beta,\alpha}) + w_3 k_{\alpha\beta} + \left[\frac{1}{2} w_{3,\alpha} w_{3,\beta}\right] \tag{1.33}$$

from which the following expressions can be written (see Eq. (1.19)):

$$\varepsilon_{\alpha\beta}^{(l)} = \frac{1}{2}(u_{\alpha,\beta} + u_{\beta,\alpha}) + w_3 k_{\alpha\beta}; \quad \varepsilon_{\alpha\beta}^{(n)} = \frac{1}{2} w_{3,\alpha} w_{3,\beta}$$

Equilibrium equations are also affected by the membrane and bending coupling. In a geometrically non-linear analysis, the pseudo body force Eqs. (1.31–1.32) are now given as:

$$f_3^b = q_3 - k_{\alpha\beta} N_{\alpha\beta} + \left[\left(N_{\alpha\beta} w_{3,\beta} \right)_{,\alpha} \right] \tag{1.34}$$

and

$$f_\alpha^m = q_\alpha + N_{\alpha\beta,\beta}^{(ii)} + \left[N_{\alpha\beta,\beta}^{(n)} \right] \tag{1.35}$$

where,

$$N_{\alpha\beta} = N_{\alpha\beta}^{(i)} + N_{\alpha\beta}^{(ii)} + N_{\alpha\beta}^{(n)} \tag{1.36}$$

$$N_{\alpha\beta}^{(n)} = B \frac{1-\nu}{2} \left(w_{3,\alpha} w_{3,\beta} + \frac{\nu}{1-\nu} w_{3,\gamma} w_{3,\gamma} \delta_{\alpha\beta} \right)$$

Constitutive Eqs. (1.22–1.24) remain the same in a pure geometrically non-linear analysis. Therefore, Eq. (1.36) is obtained after introducing Eq. (1.33) into Eq. (1.24). Non-linear terms in the kinematic Eq. (1.33) and pseudo body force (equilibrium) Eqs. (1.34–1.35) are given by the terms in square brackets.

Equilibrium equations can also be written in term of displacements as follows:

$$L_{ik}^b w_k + f_i^b = 0 \tag{1.37}$$

and

$$L_{\alpha\beta}^m u_\beta + f_\alpha^m = 0 \tag{1.38}$$

where L_{ik}^b is the Navier differential operator for shear deformable plate bending problems:

$$L_{\alpha\beta}^b = \frac{D(1-\nu)}{2} \left[\left(\nabla^2 - \lambda^2 \right) \delta_{\alpha\beta} + \frac{(1+\nu)}{(1-\nu)} \frac{\partial}{\partial x_\alpha} \frac{\partial}{\partial x_\beta} \right] \tag{1.39}$$

$$L_{\alpha 3}^b = -\frac{(1-\nu) D}{2} \lambda^2 \frac{\partial}{\partial x_\alpha} \tag{1.40}$$

$$L_{3\alpha}^b = -L_{\alpha 3}^b \tag{1.41}$$

$$L_{33}^b = \frac{(1-\nu) D}{2} \lambda^2 \nabla^2 \tag{1.42}$$

while $L^m_{\alpha\beta}$ is the Navier differential operator for two-dimensional plane stress problems:

$$L^m_{\alpha\beta} = \frac{B(1-\nu)}{2}\left[\nabla^2\delta_{\alpha\beta} + \frac{(1+\nu)}{(1-\nu)}\frac{\partial}{\partial x_\alpha}\frac{\partial}{\partial x_\beta}\right] \tag{1.43}$$

Shallow shells bear the same relation to plates as curved beams to straight beams. Therefore, in the literature, including the present work, shallow shells are also known as curved plates. It follows from this that flat plate equations can be easily obtained by considering an infinite radius ($k_{\alpha\beta} = 0$) in the governing equations presented in this work.

1.4. Integral Representations

Displacement integral equations for geometrically and material non-linear shear deformable shallow shell/plate bending problems can be derived by considering the following integral identities representations of the governing equilibrium Eqs. (1.28–1.30)[7,8]:

$$\int_\Omega \left[(M_{\alpha\beta,\beta} - Q_\alpha)\, W^*_\alpha + (Q_{\alpha,\alpha} + f^b_3)\, W^*_3\right] d\Omega = 0 \tag{1.44}$$

and

$$\int_\Omega (N^{(i)}_{\alpha\beta,\beta} + f^m_\alpha)\, U^*_\alpha d\Omega = 0 \tag{1.45}$$

where U^*_α and W^*_i are weighting functions and Ω represents the projected domain of a shallow shell/plate on the x_1–x_2 plane. Equation (1.44) is the integral representation for bending and transverse shear stress resultants and Eq. (1.45) is the integral representation for in-plane stress resultants. The terms f^m_α and f^b_i are given by Eqs. (1.34–1.35).

It can be seen that integral representations (1.44–1.45) have the form of shear deformable plate bending and two dimensional plane stress, for which boundary integral formulation procedures and fundamental solutions have already been developed.[1]

1.4.1. *Rotations and out-of-plane integral representations*

Integrating by parts, Eq. (1.44) will result in:

$$\int_\Gamma M_{\alpha\beta} n_\beta W^*_\alpha d\Gamma - \int_\Omega M_{\alpha\beta} W^*_{\alpha,\beta} d\Omega - \int_\Omega Q_\alpha W^*_\alpha d\Omega$$
$$+ \int_\Gamma Q_\alpha n_\alpha W^*_3 d\Gamma - \int_\Omega Q_\alpha W^*_{3,\alpha} d\Omega + \int_\Omega f^b_3 W^*_3 d\Omega = 0 \tag{1.46}$$

Let's now consider the relationships between stresses and tractions given in Eqs. (1.16–1.17):

$$\int_\Gamma p_j W_j^* d\Gamma - \int_\Omega M_{\alpha\beta} W_{\alpha,\beta}^* d\Omega - \int_\Omega Q_\alpha (W_\alpha^* + W_{3,\alpha}^*) d\Omega + \int_\Omega f_3^b W_3^* d\Omega = 0 \tag{1.47}$$

Bending and shear stress resultants ($M_{\alpha\beta}$ and Q_α) in Eq. (1.47) can now be replaced with the stress displacements relationships in Eqs. (1.25–1.26), resulting in the following equation:

$$\int_\Gamma p_j W_j^* d\Gamma - \int_\Omega W_{\alpha,\beta}^* \left\{ \frac{D(1-\nu)}{2} \left(w_{\alpha,\beta} + w_{\beta,\alpha} + \frac{2\nu}{(1-\nu)} \delta_{\alpha\beta} w_{\gamma,\gamma} \right) \right\} d\Omega - \int_\Omega C(w_\alpha + w_{3,\alpha})(W_\alpha^* + W_{3,\alpha}^*) d\Omega + \int_\Omega f_3^b W_3^* d\Omega = 0 \tag{1.48}$$

Integrating the second integral by parts and rearranging the third integral in Eq. (1.48):

$$\begin{aligned} &\int_\Gamma p_j W_j^* d\Gamma - \int_\Gamma W_{\alpha,\beta}^* \\ &\quad \times \left\{ \frac{D(1-\nu)}{2} \left(w_\alpha n_\beta + w_\beta n_\alpha + \frac{2\nu}{(1-\nu)} \delta_{\alpha\beta} w_\gamma n_\gamma \right) \right\} d\Gamma \\ &\quad + \int_\Omega \frac{D(1-\nu)}{2} \left(w_\alpha W_{\alpha,\beta\beta}^* + w_\beta W_{\alpha,\beta\alpha}^* + \frac{2\nu}{(1-\nu)} w_\gamma W_{\alpha,\beta\gamma}^* \delta_{\alpha\beta} \right) d\Omega \\ &\quad - \int_\Omega Q_\alpha^* (w_\alpha + w_{3,\alpha}) d\Omega + \int_\Omega f_3^b W_3^* d\Omega = 0 \end{aligned} \tag{1.49}$$

Using integration by parts, stress-displacement relationships, tractions-stress relationships and some useful properties of the Kronecker delta function, it can be shown that[8]:

$$\int_\Gamma W_{\alpha,\beta}^* \left\{ \frac{D(1-\nu)}{2} \left(w_\alpha n_\beta + w_\beta n_\alpha + \frac{2\nu}{(1-\nu)} \delta_{\alpha\beta} w_\gamma n_\gamma \right) \right\} d\Gamma = \int_\Gamma w_\alpha P_\alpha^* d\Gamma \tag{1.50}$$

$$\int_\Omega \frac{D(1-\nu)}{2} \left(w_\alpha W_{\alpha,\beta\beta}^* + w_\beta W_{\alpha,\beta\alpha}^* + \frac{2\nu}{(1-\nu)} w_\gamma W_{\alpha,\beta\gamma}^* \delta_{\alpha\beta} \right) d\Omega = \int_\Omega w_\alpha M_{\alpha\beta,\beta}^* d\Omega \tag{1.51}$$

$$\int_{\Omega} Q_{\alpha}^{*}(w_{\alpha}+w_{3,\alpha})d\Omega = \int_{\Gamma} P_{3}^{*}w_{3}d\Gamma + \int_{\Omega}(Q_{\alpha}^{*}w_{\alpha} - Q_{\alpha,\alpha}^{*}w_{3})d\Omega \tag{1.52}$$

Now, substituting Eqs. (1.50–1.52) into Eq. (1.49) and grouping:

$$\int_{\Gamma} W_{j}^{*}p_{j}d\Gamma - \int_{\Gamma} P_{j}^{*}w_{j}d\Gamma + \int_{\Omega}[(M_{\alpha\beta,\beta}^{*} - Q_{\alpha}^{*})w_{\alpha} + Q_{\alpha,\alpha}^{*}w_{3}]d\Omega + \int_{\Omega} f_{3}^{b}W_{3}^{*}d\Omega = 0 \tag{1.53}$$

The weighting functions $(\cdot)^{*}$ can be chosen arbitrarily. In the present boundary element formulation, they are defined from a state of concentrated generalised loads: two bending moments and one concentrated shear force at an arbitrary source point $\mathbf{X}' \in \Omega$ as shown in Fig. 1.4.

Based on this fundamental state, the loading direction can be introduced in Eq. (1.53) as follows:

$$\int_{\Gamma} W_{ij}^{*}p_{j}d\Gamma - \int_{\Gamma} P_{ij}^{*}w_{j}d\Gamma + \int_{\Omega}[(M_{i\alpha\beta,\beta}^{*} - Q_{i\alpha}^{*})w_{\alpha} + Q_{i\alpha,\alpha}^{*}w_{3}]d\Omega + \int_{\Omega} W_{i3}^{*}f_{3}^{b}d\Omega = 0 \tag{1.54}$$

The fundamental state $(\cdot)^{*}$ is chosen such that:

$$M_{i\alpha\beta,\beta}^{*}(\mathbf{X}',\mathbf{X}) - Q_{i\alpha}^{*}(\mathbf{X}',\mathbf{X}) + \delta(\mathbf{X}',\mathbf{X})\delta_{i\alpha} = 0 \tag{1.55}$$

$$Q_{i\alpha,\alpha}^{*}(\mathbf{X}',\mathbf{X}) + \delta(\mathbf{X}',\mathbf{X})\delta_{i3} = 0 \tag{1.56}$$

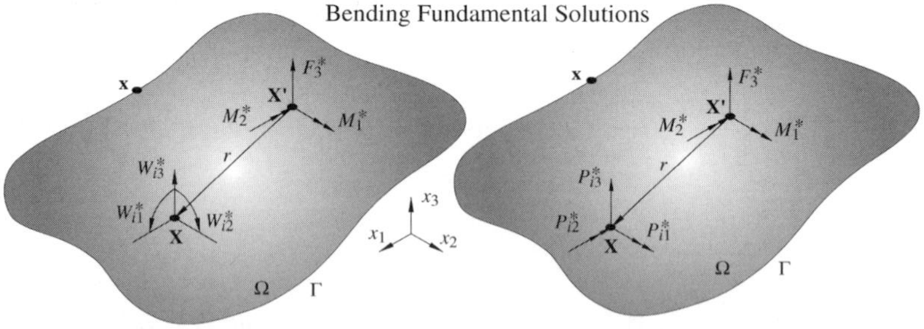

Fig. 1.4. Fundamental state of displacements and tractions due to concentrated shear force and bending moments.

where $\delta(\mathbf{X}', \mathbf{X})$ is the Dirac Delta function, which has the following property:

$$\int_\Omega \delta(\mathbf{X}', \mathbf{X}) w_i(\mathbf{X}) d\Omega = w_i(\mathbf{X}') \tag{1.57}$$

Now, after making use of the property in Eq. (1.57) with Eqs. (1.55–1.56), Eq. (1.54) can be rewritten for an internal source point $\mathbf{X}'$ as follows:

$$\begin{aligned} & w_i(\mathbf{X}') - \int_\Gamma W_{ij}^*(\mathbf{X}', \mathbf{x}) p_j(\mathbf{x})\, d\Gamma(\mathbf{x}) + \int_\Gamma P_{ij}^*(\mathbf{X}', \mathbf{x}) w_j(\mathbf{x})\, d\Gamma(\mathbf{x}) \\ & \quad - \int_\Omega W_{i3}^*(\mathbf{X}', \mathbf{X}) f_3^b(\mathbf{X})\, d\Omega(\mathbf{X}) = 0 \end{aligned} \tag{1.58}$$

$W_{ij}^*(\mathbf{X}', \mathbf{x})$ and $P_{ij}^*(\mathbf{X}', \mathbf{x})$ are the fundamental solutions for rotations and out-of-plane displacements, and bending and shear tractions, respectively, derived by Van der Weeen[42] and can be found in Aliabadi.[1] They represent the displacements or the tractions at the point $\mathbf{x}$ or $\mathbf{X}$ in the direction j due to unit point load applied at $\mathbf{X}'$ in the direction i.

After introducing $f_i^b(\mathbf{X})$ from Eq. (1.34) in Eq. (1.58):

$$\begin{aligned} & w_i(\mathbf{X}') + \int_\Gamma P_{ij}^*(\mathbf{X}', \mathbf{x}) w_j(\mathbf{x})\, d\Gamma(\mathbf{x}) \\ & \quad = \int_\Gamma W_{ij}^*(\mathbf{X}', \mathbf{x}) p_j(\mathbf{x})\, d\Gamma(\mathbf{x}) \\ & \quad - \int_\Omega k_{\alpha\beta} B \frac{1-\nu}{2} \left(u_{\alpha,\beta}(\mathbf{X}) + u_{\beta,\alpha}(\mathbf{X}) + \frac{2\nu}{1-\nu} u_{\phi,\phi}(\mathbf{X})\, \delta_{\alpha\beta} \right) \\ & \quad \times W_{i3}^*(\mathbf{X}', \mathbf{X})\, d\Omega(\mathbf{X}) \\ & \quad - \int_\Omega k_{\alpha\beta} B \left[(1-\nu) k_{\alpha\beta} + \nu \delta_{\alpha\beta} k_{\gamma\gamma} \right] w_3(\mathbf{X})\, W_{i3}^*(\mathbf{X}', \mathbf{X})\, d\Omega(\mathbf{X}) \\ & \quad - \int_\Omega W_{i3}^*(\mathbf{X}', \mathbf{X}) k_{\alpha\beta} N_{\alpha\beta}^{(n)}(\mathbf{X}) d\Omega(\mathbf{X}) + \int_\Omega W_{i3}^*(\mathbf{X}', \mathbf{X}) q_3(\mathbf{X}) d\Omega(\mathbf{X}) \\ & \quad + \int_\Omega W_{i3}^*(\mathbf{X}', \mathbf{X}) \left(N_{\alpha\beta}(\mathbf{X}) w_{3,\beta}(\mathbf{X}) \right)_{,\alpha} d\Omega(\mathbf{X}) \end{aligned} \tag{1.59}$$

Equation (1.59) is the geometrically non-linear integral representation for rotations and out-of-plane displacement of shear deformable shallow shell theory.

1.4.2. *In-plane integral representations*

Membrane integral representations can also be obtained in a similar way to the bending and out-of-plane integral representations. Integrating Eq. (1.45) by parts gives:

$$\int_\Gamma N^{(i)}_{\alpha\beta} n_\beta U^*_\alpha d\Gamma - \int_\Omega N^{(i)}_{\alpha\beta} U^*_{\alpha,\beta} d\Omega + \int_\Omega f^m_\alpha U^*_\alpha d\Omega = 0 \tag{1.60}$$

Using Eq. (1.18) and Eq. (1.27), Eq. (1.60) can be written as follows:

$$\int_\Gamma t^{(i)}_\alpha U^*_\alpha d\Gamma - \int_\Omega B\frac{1-\nu}{2}\left(u_{\alpha,\beta} + u_{\beta,\alpha} + \frac{2\nu}{1-\nu} u_{\gamma,\gamma}\delta_{\alpha\beta}\right) U^*_{\alpha,\beta} d\Omega + \int_\Omega f^m_\alpha U^*_\alpha d\Omega = 0 \tag{1.61}$$

Integrating the second integral of Eq. (1.61) by parts:

$$\int_\Gamma t^{(i)}_\alpha U^*_\alpha d\Gamma - \int_\Gamma B\frac{1-\nu}{2}\left(u_\alpha n_\beta + u_\beta n_\alpha + \frac{2\nu}{1-\nu} u_\gamma n_\gamma \delta_{\alpha\beta}\right) U^*_{\alpha,\beta} d\Gamma + \int_\Omega B\frac{1-\nu}{2}\left(u_\alpha U^*_{\alpha,\beta\beta} + u_\beta U^*_{\alpha,\beta\alpha} + \frac{2\nu}{1-\nu}\delta_{\alpha\beta} u_\gamma U^*_{\alpha,\beta\gamma}\right) d\Omega + \int_\Omega f^m_\alpha U^*_\alpha d\Omega = 0 \tag{1.62}$$

Using again the stress-displacement Eq. (1.27) and some useful properties of the Kronecker delta functions, it can be shown that [8]:

$$\int_\Gamma B\frac{1-\nu}{2}\left(u_\alpha n_\beta + u_\beta n_\alpha + \frac{2\nu}{1-\nu} u_\gamma n_\gamma \delta_{\alpha\beta}\right) U^*_{\alpha,\beta} d\Gamma = \int_\Gamma u_\alpha T^{(i)*}_\alpha d\Gamma \tag{1.63}$$

$$\int_\Omega B\frac{1-\nu}{2}\left(u_\alpha U^*_{\alpha,\beta\beta} + u_\beta U^*_{\alpha,\beta\alpha} + \frac{2\nu}{1-\nu}\delta_{\alpha\beta} u_\gamma U^*_{\alpha,\beta\gamma}\right) d\Omega = \int_\Omega u_\alpha N^{(i)*}_{\alpha\beta,\beta} d\Omega \tag{1.64}$$

Now, introducing Eqs. (1.63–1.64) into Eq. (1.62) and grouping, gives:

$$\int_\Gamma t^{(i)}_\alpha U^*_\alpha d\Gamma - \int_\Gamma u_\alpha T^{(i)*}_\alpha d\Gamma + \int_\Omega u_\alpha N^{(i)*}_{\alpha\beta,\beta} d\Omega + \int_\Omega f^m_\alpha U^*_\alpha d\Omega = 0 \tag{1.65}$$

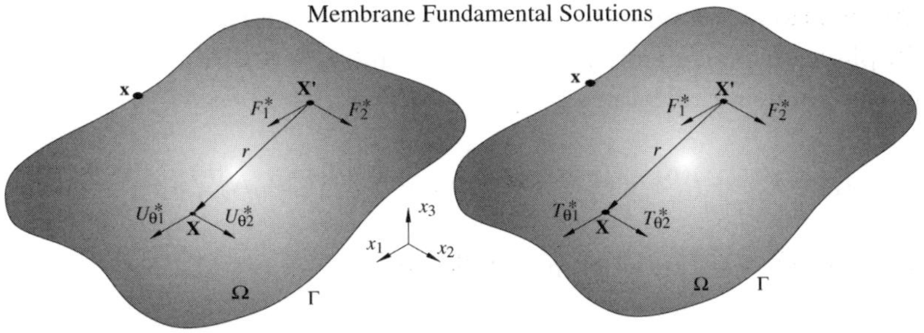

Fig. 1.5. Fundamental state of displacements and tractions due to concentrated membrane forces.

Again, the fundamental state $(\cdot)^*$ is defined for concentrated generalised loads; in this case, two concentrated membrane forces at an arbitrary point $\mathbf{X}' \in \Omega$, as shown in Fig. 1.5.

After introducing the direction of the load θ in Eq. (1.65):

$$0 = \int_\Gamma U^*_{\theta\alpha}(\mathbf{X}',\mathbf{x})\, t^{(i)}_\alpha(\mathbf{x})\, d\Gamma(\mathbf{x}) - \int_\Gamma T^{(i)*}_{\theta\alpha}(\mathbf{X}',\mathbf{x})\, u_\alpha(\mathbf{x})\, d\Gamma(\mathbf{x}) + \int_\Omega N^{(i)*}_{\theta\alpha\beta,\beta}(\mathbf{X}',\mathbf{X})\, u_\alpha(\mathbf{X})\, d\Omega(\mathbf{X}) + \int_\Omega U^*_{\theta\alpha}(\mathbf{X}',\mathbf{X})\, f^m_\alpha(\mathbf{X})\, d\Omega(\mathbf{X}) \tag{1.66}$$

The fundamental state $(\cdot)^*$ is chosen, such that:

$$N^{(i)*}_{\theta\alpha\beta,\beta}(\mathbf{X}',\mathbf{X}) + \delta(\mathbf{X}',\mathbf{X})\delta_{\theta\alpha} = 0 \tag{1.67}$$

Using the property of the Dirac Delta function (1.57) and Eq. (1.67), Eq. (1.66) can be rewritten for an internal source point $\mathbf{X}'$ as follows:

$$u_\theta(\mathbf{X}') + \int_\Gamma T^{(i)*}_{\theta\alpha}(\mathbf{X}',\mathbf{x})\, u_\alpha(\mathbf{x})\, d\Gamma(\mathbf{x}) = \int_\Gamma U^*_{\theta\alpha}(\mathbf{X}',\mathbf{x})\, t^{(i)}_\alpha(\mathbf{x})\, d\Gamma(\mathbf{x}) + \int_\Omega U^*_{\theta\alpha}(\mathbf{X}',\mathbf{X})\, f^m_\alpha(\mathbf{X})\, d\Omega(\mathbf{X}) \tag{1.68}$$

where $U^*_{\theta\alpha}(\mathbf{X}',\mathbf{x})$ and $T^{(i)*}_{\theta\alpha}(\mathbf{X}',\mathbf{x})$ are the fundamental solutions for in-plane displacements and membrane tractions, respectively, and are given in Aliabadi.[1] They represent displacements or tractions at the field point $\mathbf{x}$ or $\mathbf{X}$ in the direction α due to unit load applied at the source point $\mathbf{X}'$ in the direction θ.

As mentioned previously, the upper index (i) on the tractions refers to the in-plane displacement. After introducing $f_\alpha^m(\mathbf{X})$ from Eq. (1.35) and using the following relation:

$$t_\alpha = t_\alpha^{(i)} + t_\alpha^{(ii)} + t_\alpha^{(n)} = N_{\alpha\beta}^{(i)} n_\beta + N_{\alpha\beta}^{(ii)} n_\beta + N_{\alpha\beta}^{(n)} n_\beta \tag{1.69}$$

Equation (1.68) can be written as follows:

$$\begin{aligned}
u_\theta(\mathbf{X}') &+ \int_\Gamma T_{\theta\alpha}^{(i)*}(\mathbf{X}',\mathbf{x})\, u_\alpha(\mathbf{x})\, d\Gamma(\mathbf{x}) \\
&+ \int_\Gamma U_{\theta\alpha}^*(\mathbf{X}',\mathbf{x}) B\left[k_{\alpha\beta}(1-\nu) + \nu\delta_{\alpha\beta}k_{\phi\phi}\right] w_3(\mathbf{x}) n_\beta(\mathbf{x}) d\Gamma(\mathbf{x}) \\
&- \int_\Omega U_{\theta\alpha}^*(\mathbf{X}',\mathbf{X}) B\left[k_{\alpha\beta}(1-\nu) + \nu\delta_{\alpha\beta}k_{\phi\phi}\right] w_{3,\beta}(\mathbf{X}) d\Omega(\mathbf{X}) \\
&- \int_\Omega U_{\theta\alpha}^*(\mathbf{X}',\mathbf{X}) N_{\alpha\beta,\beta}^{(n)}(\mathbf{X}) d\Omega(\mathbf{X}) \\
&+ \int_\Gamma U_{\theta\alpha}^*(\mathbf{X}',\mathbf{x})\, N_{\alpha\beta}^{(n)} n_\beta(\mathbf{x})\, d\Gamma(\mathbf{x}) \\
= &\int_\Gamma U_{\theta\alpha}^*(\mathbf{X}',\mathbf{x})\, t_\alpha(\mathbf{x})\, d\Gamma(\mathbf{x}) \\
&+ \int_\Omega U_{\theta\alpha}^*(\mathbf{X}',\mathbf{X})\, q_\alpha(\mathbf{X})\, d\Omega(\mathbf{X})
\end{aligned} \tag{1.70}$$

Equation (1.70) is the geometrically non-linear integral representation for in-plane displacement of shear deformable shallow shell theory.

1.5. Boundary Integral Equations

In the case where the collocation point is taken to the boundary, $\mathbf{X}' \to \mathbf{x}' \in \Gamma$; and the integration process is performed, there will be a boundary region where the distance, r, tends to zero. Assuming that the displacements w_i and u_θ satisfy the Hölder continuity,

$$|w_i(\mathbf{x}) - w_i(\mathbf{x}')| < Ar^\alpha; \quad A: \text{constant, and } 0 < \alpha \leqslant 1 \tag{1.71}$$

and analysing the limit of a semi-circular domain with boundary Γ_ε^* and radius ε centred at the source point $\mathbf{x}'$, as shown in Fig. 1.6, rotations and

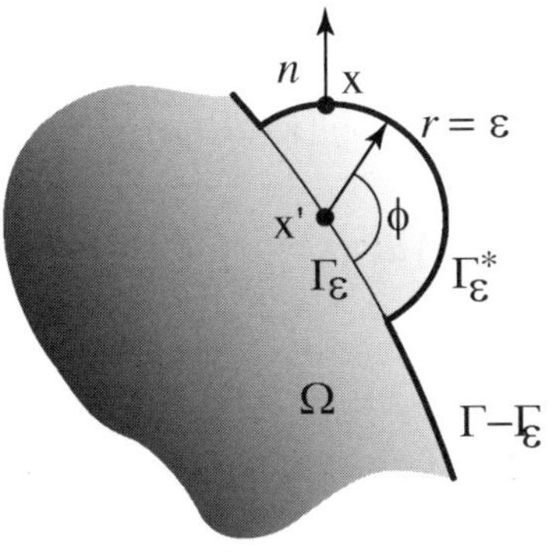

Fig. 1.6. Semi-circular region around the source point on the boundary.

out-of-plane displacement integral Eq. (1.58) can be written as follows:

$$w_i(\mathbf{x}') + \lim_{\varepsilon \to 0} \int_{\Gamma - \Gamma_\varepsilon + \Gamma_\varepsilon^*} P_{ij}^*(\mathbf{x}', \mathbf{x}) w_j(\mathbf{x}) \, d\Gamma(\mathbf{x})$$
$$= \lim_{\varepsilon \to 0} \int_{\Gamma - \Gamma_\varepsilon + \Gamma_\varepsilon^*} W_{ij}^*(\mathbf{x}', \mathbf{x}) p_j(\mathbf{x}) \, d\Gamma(\mathbf{x})$$
$$+ \int_\Omega W_{i3}^*(\mathbf{x}', \mathbf{X}) f_3^b(\mathbf{X}) d\Omega(\mathbf{X}) \tag{1.72}$$

and the in-plane Eq. (1.68):

$$u_\theta(\mathbf{x}') + \lim_{\varepsilon \to 0} \int_{\Gamma - \Gamma_\varepsilon + \Gamma_\varepsilon^*} T_{\theta\alpha}^{(i)*}(\mathbf{x}', \mathbf{x}) \, u_\alpha(\mathbf{x}) \, d\Gamma(\mathbf{x})$$
$$= \lim_{\varepsilon \to 0} \int_{\Gamma - \Gamma_\varepsilon + \Gamma_\varepsilon^*} U_{\theta\alpha}^*(\mathbf{x}', \mathbf{x}) \, t_\alpha^{(i)}(\mathbf{x}) \, d\Gamma(\mathbf{x})$$
$$+ \int_\Omega U_{\theta\alpha}^*(\mathbf{x}', \mathbf{X}) \, f_\alpha^m d\Omega(\mathbf{X}) \tag{1.73}$$

Boundary and domain integrals involving the displacement fundamental solutions ($U_{\theta\alpha}^*$ and W_{ij}^*) contain weakly singular or regular kernels that lead to no jump terms and therefore exist in the limit in the normal sense. Boundary integrals involving the traction fundamental solutions ($T_{\theta\alpha}^{(i)*}$ and P_{ij}^*) contain a strong singularity of $O(1/r)$ and therefore will lead to a jump term and will exist in the limit as a Cauchy principal value integral. After performing all the limiting processes,[8] Eqs. (1.72–1.73) can be written as follows:

$$c_{ij}(\mathbf{x}')w_j(\mathbf{x}') + \fint_{\Gamma} P^*_{ij}(\mathbf{x}',\mathbf{x})w_j\,(\mathbf{x})\,d\Gamma\,(\mathbf{x})$$
$$= \int_{\Gamma} W^*_{ij}(\mathbf{x}',\mathbf{x})p_j\,(\mathbf{x})\,d\Gamma\,(\mathbf{x}) + \int_{\Omega} W^*_{i3}(\mathbf{x}',\mathbf{X})f^b_3(\mathbf{X})d\Omega(\mathbf{X}) \quad (1.74)$$

and

$$c_{\theta\alpha}\,(\mathbf{x}')\,u_\alpha(\mathbf{x}') + \fint_{\Gamma} T^{*(i)}_{\theta\alpha}(\mathbf{x}',\mathbf{x})u_\alpha(\mathbf{x})d\Gamma(\mathbf{x})$$
$$= \int_{\Gamma} U^*_{\theta\alpha}(\mathbf{x}',\mathbf{x})t^{(i)}_\alpha(\mathbf{x})d\Gamma(\mathbf{x}) + \int_{\Omega} U^*_{\theta\alpha}(\mathbf{x}',\mathbf{X})f^m_\alpha(\mathbf{X})d\Omega(\mathbf{X}) \quad (1.75)$$

where $\fint$ denotes a Cauchy principal value integral and $\mathbf{x}', \mathbf{x} \in \Gamma$ are source and field points, respectively. $c_{ij}(\mathbf{x}')$ represent the jump terms, whose value is equal to $\frac{1}{2}\delta_{ij}$ when $\mathbf{x}'$ is located on a smooth boundary.

Using Eqs. (1.34–1.35) and (1.69), Eqs. (1.74–1.75) can be rewritten as:

$$c_{ij}(\mathbf{x}')w_j(\mathbf{x}') + \fint_{\Gamma} P^*_{ij}(\mathbf{x}',\mathbf{x})w_j\,(\mathbf{x})\,d\Gamma\,(\mathbf{x})$$
$$= \int_{\Gamma} W^*_{ij}(\mathbf{x}',\mathbf{x})p_j\,(\mathbf{x})\,d\Gamma\,(\mathbf{x})$$
$$- \int_{\Omega} k_{\alpha\beta}B\frac{1-\nu}{2}\left(u_{\alpha,\beta}\,(\mathbf{X}) + u_{\beta,\alpha}\,(\mathbf{X}) + \frac{2\nu}{1-\nu}u_{\phi,\phi}\,(\mathbf{X})\,\delta_{\alpha\beta}\right)$$
$$\times W^*_{i3}\,(\mathbf{x}',\mathbf{X})\,d\Omega\,(\mathbf{X})$$
$$- \int_{\Omega} k_{\alpha\beta}B\left[(1-\nu)k_{\alpha\beta} + \nu\delta_{\alpha\beta}k_{\gamma\gamma}\right]w_3\,(\mathbf{X})\,W^*_{i3}\,(\mathbf{x}',\mathbf{X})\,d\Omega\,(\mathbf{X})$$
$$- \int_{\Omega} W^*_{i3}(\mathbf{x}',\mathbf{X})k_{\alpha\beta}N^{(n)}_{\alpha\beta}(\mathbf{X})d\Omega\,(\mathbf{X}) + \int_{\Omega} W^*_{i3}(\mathbf{x}',\mathbf{X})q_3(\mathbf{X})d\Omega(\mathbf{X})$$
$$+ \int_{\Omega} W^*_{i3}(\mathbf{x}',\mathbf{X})\left(N_{\alpha\beta}(\mathbf{X})w_{3,\beta}(\mathbf{X})\right)_{,\alpha}\,d\Omega\,(\mathbf{X}) \quad (1.76)$$

and

$$c_{\theta\alpha}\,(\mathbf{x}')\,u_\alpha\,(\mathbf{x}') + \fint_{\Gamma} T^{(i)*}_{\theta\alpha}\,(\mathbf{x}',\mathbf{x})\,u_\alpha\,(\mathbf{x})\,d\Gamma\,(\mathbf{x})$$
$$+ \int_{\Gamma} U^*_{\theta\alpha}(\mathbf{x}',\mathbf{x})B\left[k_{\alpha\beta}\,(1-\nu) + \nu\delta_{\alpha\beta}k_{\phi\phi}\right]w_3(\mathbf{x})n_\beta(\mathbf{x})d\Gamma(\mathbf{x})$$

$$
\begin{aligned}
&- \int_{\Omega} U^*_{\theta\alpha}(\mathbf{x}', \mathbf{X}) B \left[k_{\alpha\beta}(1-\nu) + \nu \delta_{\alpha\beta} k_{\phi\phi}\right] w_{3,\beta}(\mathbf{X}) d\Omega(\mathbf{X}) \\
&- \int_{\Omega} U^*_{\theta\alpha}(\mathbf{x}', \mathbf{X}) N^{(n)}_{\alpha\beta,\beta}(\mathbf{X}) d\Omega(\mathbf{X}) + \int_{\Gamma} U^*_{\theta\alpha}(\mathbf{x}', \mathbf{x}) N^{(n)}_{\alpha\beta} n_\beta(\mathbf{x}) d\Gamma(\mathbf{x}) \\
&= \int_{\Gamma} U^*_{\theta\alpha}(\mathbf{x}', \mathbf{x}) t_\alpha(\mathbf{x}) d\Gamma(\mathbf{x}) + \int_{\Omega} U^*_{\theta\alpha}(\mathbf{x}', \mathbf{X}) q_\alpha(\mathbf{X}) d\Omega(\mathbf{X}) \qquad (1.77)
\end{aligned}
$$

Equations (1.76–1.77) are the boundary geometrically non-linear integral representation for rotations and out-of-plane displacement and in-plane displacements, respectively. It is important to mention that, due to the curvature terms (containing $k_{\alpha\beta}$) or non-linear terms, Eqs. (1.76–1.77) have to be solved simultaneously and not only for collocation at boundary points $\mathbf{x}$ but also at domain points $\mathbf{X}$.

By applying the divergence theorem, the last domain integral in Eq. (1.76) can be transferred to the boundary, in the case of a uniform domain load ($q_3 = constant$):

$$
\int_{\Omega} W^*_{i3}(\mathbf{x}', \mathbf{X}) q_3(\mathbf{X}) d\Omega(\mathbf{X}) = q_3 \int_{\Gamma} V^*_{i,\alpha}(\mathbf{x}', \mathbf{x}) n_\alpha(\mathbf{x}) d\Gamma(\mathbf{x}) \qquad (1.78)
$$

where V^*_i are the particular solutions of the Eq. $V^*_{i,\theta\theta} = W^*_{i3}$. For not uniform q_3, the transformation of domain integrals can be carried out by using the dual reciprocity technique.

1.6. Stresses in Shear Deformable Plates/Shallow Shells

Once the values of displacements and tractions are known, displacements and stresses at any point in the structure can be obtained as a post processing procedure. In the case of internal stresses, evaluation can be directly performed by derivation of the displacement integral equation and substitution in the displacement-strain (kinematics) and strain-stress (constitutive)relationships.

In the case of boundary stresses, two different methods can be used. The indirect approach relies on the use of boundary tractions and displacements obtained from the BEM solution in a local coordinate system where tangential strains are calculated by differentiation of the shape functions. Later, these strains are used to obtain stresses by means of Hooke's law. In the direct approach, stresses are obtained directly from the integral equations for stresses. The direct approach is mathematically very cumbersome because of the singularities involved in the boundary integrals and therefore is only recommended in special cases. The indirect approach, although economical and therefore more popular, is prone to inaccuracies when a coarse mesh is used.

1.6.1. *Internal stress resultants*

Evaluation of stresses at any domain point in the structure can be obtained when derivatives of Eqs. (1.58) and (1.68), with respect to the source point $\mathbf{X}'$ are introduced in Eqs. (1.25–1.26) and (1.36)[9,47]:

$$M_{\alpha\beta}(\mathbf{X}') = \int_\Gamma W^*_{\alpha\beta k}(\mathbf{X}',\mathbf{x})p_k(\mathbf{x})d\Gamma(\mathbf{x}) - \int_\Gamma P^*_{\alpha\beta k}(\mathbf{X}',\mathbf{x})w_k(\mathbf{x})d\Gamma(\mathbf{x}) + \int_\Omega W^*_{\alpha\beta 3}(\mathbf{X}',\mathbf{X})f^b_3(\mathbf{X})d\Omega(\mathbf{X}) \tag{1.79}$$

$$Q_{\beta}(\mathbf{X}') = \int_\Gamma W^*_{3\beta k}(\mathbf{X}',\mathbf{x})p_k(\mathbf{x})d\Gamma(\mathbf{x}) - \int_\Gamma P^*_{3\beta k}(\mathbf{X}',\mathbf{x})w_k(\mathbf{x})d\Gamma(\mathbf{x}) + \int_\Omega W^*_{3\beta 3}(\mathbf{X}',\mathbf{X})f^b_3(\mathbf{X})d\Omega(\mathbf{X}) \tag{1.80}$$

and

$$\begin{aligned} N_{\alpha\beta}(\mathbf{X}') &= \int_\Gamma U^*_{\alpha\beta\gamma}(\mathbf{X}',\mathbf{x})t^{(i)}_\gamma(\mathbf{x})d\Gamma(\mathbf{x}) - \int_\Gamma T^{(i)*}_{\alpha\beta\gamma}(\mathbf{X}',\mathbf{x})u_\gamma(\mathbf{x})d\Gamma(\mathbf{x}) \\ &+ \int_\Omega U^*_{\alpha\beta\gamma}(\mathbf{X}',\mathbf{X})f^m_\gamma(\mathbf{X})d\Omega(\mathbf{X}) + B\left[(1-\nu)k_{\alpha\beta} + \nu\delta_{\alpha\beta}k_{\phi\phi}\right]w_3(\mathbf{X}') \\ &+ B\frac{1-\nu}{2}\left(w_{3,\alpha}(\mathbf{X}')\,w_{3,\beta}(\mathbf{X}') + \frac{\nu}{1-\nu}w_{3,\gamma}(\mathbf{X}')\,w_{3,\gamma}(\mathbf{X}')\,\delta_{\alpha\beta}\right) \end{aligned} \tag{1.81}$$

The body force terms f^b_3 and f^m_α are given by Eqs. (1.34–1.35) and $t^{(i)}_\alpha$ by Eq. (1.69). The kernels $W^*_{i\beta k}$, $P^*_{i\beta k}$, $U^*_{\alpha\beta\gamma}$, $T^*_{\alpha\beta\gamma}$ and $Q^*_{i\beta}$ are linear combinations of the first derivatives of W^*_{ij}, P^*_{ij}, $U^*_{\alpha\beta}$, $T^*_{\alpha\beta}$ and $V^*_{i,\beta}$, respectively. The expression for all these kernels was derived by Van der Weeën[42] and are given in Aliabadi.[1]

1.6.2. *Boundary stress resultants: indirect approach*

Let's consider a local coordinate system at a boundary source point $\mathbf{x}'$, as shown in Fig. 1.7. Generalised local displacements (w^o_i, u^o_θ) and tractions

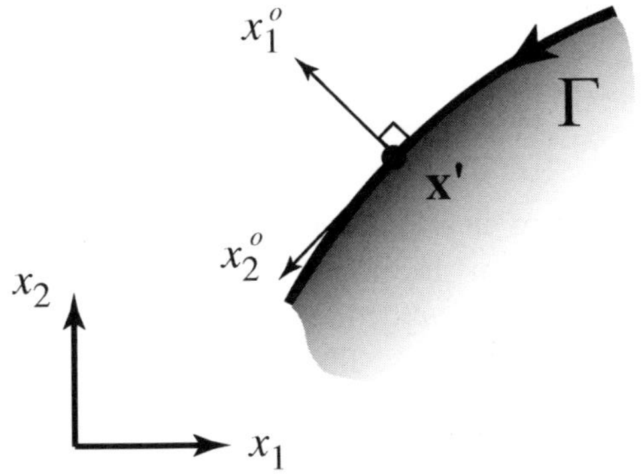

Fig. 1.7. Local and global coordinate system at $\mathbf{x}'$.

$(p_i^o,\ t_\theta^o)$ at $\mathbf{x}'$ are related to global displacements $(w_i,\ u_\theta)$ and tractions (p_i, t_θ), as follows:

$$\begin{aligned} w_i^o &= e_{ij}^o w_j \quad \text{and} \quad u_\theta^o = e_{\theta\gamma}^o u_\gamma \\ p_i^o &= e_{ij}^o p_j \quad \text{and} \quad t_\theta^o = e_{\theta\gamma}^o t_\gamma \end{aligned} \tag{1.82}$$

where e_{ij}^o is the rotation matrix and can be written in terms of the known normal components as:

$$e_{ij}^o = \begin{bmatrix} n_1 & n_2 & 0 \\ -n_2 & n_1 & 0 \\ 0 & 0 & 1 \end{bmatrix} \tag{1.83}$$

Stress resultants at the local coordinate system can be obtained from equilibrium considerations at the boundary as follows:

$$\begin{aligned} M_{1\alpha}^o &= p_\alpha^o \\ Q_1^o &= p_3^o \\ N_{1\alpha}^o &= t_\alpha^o \end{aligned} \tag{1.84}$$

The remaining stress resultants on the local coordinate system can be obtained with Eqs. (1.25–1.26) and (1.36):

$$\begin{aligned} M_{22}^o &= \nu p_1^o + D\left(1-\nu^2\right) w_{2,2}^o \\ Q_2^o &= \frac{D(1-\nu)\lambda^2}{2}(w_2^o + w_{3,2}^o) \end{aligned}$$

$$N_{22}^{o} = \nu t_{1}^{o} + B\left(1-\nu^{2}\right)u_{2,2}^{o} + B\left(1-\nu^{2}\right)k_{22}^{o}w_{3}^{o} + \left[\frac{B\left(1-\nu^{2}\right)}{2}\left(w_{3,2}^{o}\right)^{2}\right] \tag{1.85}$$

The non-linear term in Eq. (1.85) is given by the term in square brackets. As can be seen from Eq. (1.85), derivatives of the local displacements ($w_{2,2}^{o}$, $w_{3,2}^{o}$ and $u_{2,2}^{o}$) have to be computed in order to obtain some of the local stress resultants. This can be achieved by first expressing the local displacements as a product of the shape functions Φ^{m} and the nodal displacements as follows:

$$\begin{aligned} w_{j}^{o} &= \sum_{m=1}^{3} \Phi^{m} w_{j}^{om} \\ u_{\theta}^{o} &= \sum_{m=1}^{3} \Phi^{m} u_{\theta}^{om} \end{aligned} \tag{1.86}$$

Now, the displacement derivatives in Eq. (1.85) can be obtained from:

$$\begin{aligned} w_{j,2}^{o} &= \sum_{m=1}^{3} \Phi_{,2}^{m} w_{j}^{om} = \sum_{m=1}^{3} \frac{\partial \Phi^{m}}{\partial \xi} e_{ji}^{o} w_{i}^{m} \frac{\partial \xi}{\partial x_{2}^{o}(\mathbf{x}')} \\ u_{\theta,2}^{o} &= \sum_{m=1}^{3} \Phi_{,2}^{m} u_{\theta}^{om} = \sum_{m=1}^{3} \frac{\partial \Phi^{m}}{\partial \xi} e_{\theta\gamma}^{o} u_{\gamma}^{m} \frac{\partial \xi}{\partial x_{2}^{o}(\mathbf{x}')} \end{aligned} \tag{1.87}$$

where $\frac{\partial \xi}{\partial x_{2}^{o}(\mathbf{x}')}$ is given by the inverse of the Jacobian[1]:

$$\frac{\partial \xi}{\partial x_{2}^{o}(\mathbf{x}')} = \frac{1}{J(\mathbf{x}')} \tag{1.88}$$

The local curvature term k_{22}^{o} can be obtained following a similar strategy[8]:

$$k_{22}^{o} = n_{2}^{2} k_{11} + n_{1}^{2} k_{22} \tag{1.89}$$

Global stress resultants can now be obtained using the following transformation:

$$\begin{aligned} M_{\alpha\beta} &= e_{\theta\alpha}^{o} e_{\gamma\beta}^{o} M_{\theta\gamma}^{o} \\ Q_{\alpha} &= e_{\beta\alpha}^{o} Q_{\beta}^{o} \\ N_{\alpha\beta} &= e_{\theta\alpha}^{o} e_{\gamma\beta}^{o} N_{\theta\gamma}^{o} \end{aligned} \tag{1.90}$$

1.6.3. *Boundary stress resultants: direct approach*

The direct approach for boundary stress resultants is based on Eqs. (1.79–1.81) when $\mathbf{X}'$ approaches the boundary point $\mathbf{x}'$. The same procedure used for the displacement integral equations is adopted. Considering the limit of a semi-circular domain with boundary Γ_ε^* and radius ε centered at the source point $\mathbf{x}'$, as shown in Fig. 1.6, Eqs. (1.79–1.81) can be rewritten as follows[45]:

$$\begin{aligned} M_{\alpha\beta}(\mathbf{x}') &+ \lim_{\varepsilon\to 0}\int_{\Gamma-\Gamma_\varepsilon+\Gamma_\varepsilon^*} P^*_{\alpha\beta\gamma}(\mathbf{x}',\mathbf{x})w_\gamma(\mathbf{x})d\Gamma(\mathbf{x}) \\ &+ \lim_{\varepsilon\to 0}\int_{\Gamma-\Gamma_\varepsilon+\Gamma_\varepsilon^*} P^*_{\alpha\beta 3}(\mathbf{x}',\mathbf{x})w_3(\mathbf{x})d\Gamma(\mathbf{x}) \\ &= \lim_{\varepsilon\to 0}\int_{\Gamma-\Gamma_\varepsilon+\Gamma_\varepsilon^*} W^*_{\alpha\beta\gamma}(\mathbf{x}',\mathbf{x})p_\gamma(\mathbf{x})d\Gamma(\mathbf{x}) \\ &+ \lim_{\varepsilon\to 0}\int_{\Gamma-\Gamma_\varepsilon+\Gamma_\varepsilon^*} W^*_{\alpha\beta 3}(\mathbf{x}',\mathbf{x})p_3(\mathbf{x})d\Gamma(\mathbf{x}) \\ &+ \lim_{\varepsilon\to 0}\int_\Omega W^*_{\alpha\beta 3}(\mathbf{x}',\mathbf{X})f_3^b(\mathbf{X})d\Omega(\mathbf{X}); \end{aligned} \tag{1.91}$$

$$\begin{aligned} Q_{\beta}(\mathbf{x}') &+ \lim_{\varepsilon\to 0}\int_{\Gamma-\Gamma_\varepsilon+\Gamma_\varepsilon^*} P^*_{3\beta\gamma}(\mathbf{x}',\mathbf{x})w_\gamma(\mathbf{x})d\Gamma(\mathbf{x}) \\ &+ \lim_{\varepsilon\to 0}\int_{\Gamma-\Gamma_\varepsilon+\Gamma_\varepsilon^*} P^*_{3\beta 3}(\mathbf{x}',\mathbf{x})w_3(\mathbf{x})d\Gamma(\mathbf{x}) \\ &= \lim_{\varepsilon\to 0}\int_{\Gamma-\Gamma_\varepsilon+\Gamma_\varepsilon^*} W^*_{3\beta\gamma}(\mathbf{x}',\mathbf{x})p_\gamma(\mathbf{x})d\Gamma(\mathbf{x}) \\ &+ \lim_{\varepsilon\to 0}\int_{\Gamma-\Gamma_\varepsilon+\Gamma_\varepsilon^*} W^*_{3\beta 3}(\mathbf{x}',\mathbf{x})p_3(\mathbf{x})d\Gamma(\mathbf{x}) \\ &+ \lim_{\varepsilon\to 0}\int_\Omega W^*_{3\beta 3}(\mathbf{x}',\mathbf{X})f_3^b(\mathbf{X})d\Omega(\mathbf{X}) \end{aligned} \tag{1.92}$$

and,

$$\begin{aligned} N_{\alpha\beta}(\mathbf{x}') &+ \lim_{\varepsilon\to 0}\int_{\Gamma-\Gamma_\varepsilon+\Gamma_\varepsilon^*} T^{(i)*}_{\alpha\beta\gamma}(\mathbf{x}',\mathbf{x})u_\gamma(\mathbf{x})d\Gamma(\mathbf{x}) \\ = \lim_{\varepsilon\to 0}&\int_{\Gamma-\Gamma_\varepsilon+\Gamma_\varepsilon^*} U^*_{\alpha\beta\gamma}(\mathbf{x}',\mathbf{x})t^{(i)}_\gamma(\mathbf{x})d\Gamma(\mathbf{x}) \\ + \lim_{\varepsilon\to 0}&\int_\Omega U^*_{\alpha\beta\gamma}(\mathbf{x}',\mathbf{X})f^m_\alpha(\mathbf{X})d\Omega(\mathbf{X}) \end{aligned}$$

$$+B\left[(1-\nu)k_{\alpha\beta}+\nu\delta_{\alpha\beta}k_{\phi\phi}\right]w_3\left(\mathbf{x}'\right)$$
$$+B\frac{1-\nu}{2}\left(w_{3,\alpha}\left(\mathbf{x}'\right)w_{3,\beta}\left(\mathbf{x}'\right)+\frac{\nu}{1-\nu}w_{3,\gamma}\left(\mathbf{x}'\right)w_{3,\gamma}\left(\mathbf{x}'\right)\delta_{\alpha\beta}\right) \tag{1.93}$$

Several integrals in Eqs. (1.91–1.92) were split due to the different levels of their singularity. A detailed description of the treatment of these singularities can be found in Dirgantara,[8] including the derivation of jump terms that arise from the limiting process of some integrals. Equations (1.91–1.92) represent the bending and shear stress resultant boundary integral equations, respectively, while Eq. (1.93) represents the membrane stress resultant boundary integral equation.

Due to the singularity level in some of these integrals, the generalised displacements w_i and u_α are required to be $C^{1,\alpha}(0 < \alpha < 1)$ and the generalised tractions p_i and t_α are required to be $C^{0,\alpha}(0 < \alpha < 1)$ for the principal value integrals to exist. In order to satisfy this requirement, the point $\mathbf{x}'$ is assumed to be on a smooth boundary, and this is achieved by using totally discontinuous elements.

Taking into account all the limits and the jump terms as $\varepsilon \to 0$ for a source point on a smooth boundary, stress resultant integral equations are obtained as follows:

$$\frac{1}{2}M_{\alpha\beta}(\mathbf{x}')+\ \substack{=\\ \int_\Gamma}\ P^*_{\alpha\beta\gamma}(\mathbf{x}',\mathbf{x})w_\gamma(\mathbf{x})d\Gamma(\mathbf{x})+\ \substack{-\\ \int_\Gamma}\ P^*_{\alpha\beta3}(\mathbf{x}',\mathbf{x})w_3(\mathbf{x})d\Gamma(\mathbf{x})$$
$$=\ \substack{-\\ \int_\Gamma}\ W^*_{\alpha\beta\gamma}(\mathbf{x}',\mathbf{x})p_\gamma(\mathbf{x})d\Gamma(\mathbf{x})+\int_\Gamma W^*_{\alpha\beta3}(\mathbf{x}',\mathbf{x})p_3(\mathbf{x})d\Gamma(\mathbf{x})$$
$$+\int_\Omega W^*_{\alpha\beta3}(\mathbf{x}',\mathbf{X})f^b_3(\mathbf{X})d\Omega(\mathbf{X}) \tag{1.94}$$

$$\frac{1}{2}Q_{\beta}(\mathbf{x}')+\ \substack{-\\ \int_\Gamma}\ P^*_{3\beta\gamma}(\mathbf{x}',\mathbf{x})w_\gamma(\mathbf{x})d\Gamma(\mathbf{x})+\ \substack{=\\ \int_\Gamma}\ P^*_{3\beta3}(\mathbf{x}',\mathbf{x})w_3(\mathbf{x})d\Gamma(\mathbf{x})$$
$$=\int_\Gamma W^*_{3\beta\gamma}(\mathbf{x}',\mathbf{x})p_\gamma(\mathbf{x})d\Gamma(\mathbf{x})+\ \substack{-\\ \int_\Gamma}\ W^*_{3\beta3}(\mathbf{x}',\mathbf{x})p_3(\mathbf{x})d\Gamma(\mathbf{x})$$
$$+\int_\Omega W^*_{3\beta3}(\mathbf{x}',\mathbf{X})f^b_3(\mathbf{X})d\Omega(\mathbf{X}) \tag{1.95}$$

$$\frac{1}{2}N_{\alpha\beta}(\mathbf{x}')+\ \substack{=\\ \int_\Gamma}\ T^{(i)*}_{\alpha\beta\gamma}(\mathbf{x}',\mathbf{x})u_\gamma(\mathbf{x})d\Gamma(\mathbf{x})$$
$$=\ \substack{-\\ \int_\Gamma}\ U^*_{\alpha\beta\gamma}(\mathbf{x}',\mathbf{x})t^{(i)}_\gamma(\mathbf{x})d\Gamma(\mathbf{x})$$

$$+ \int_{\Omega} U^{*}_{\alpha\beta\gamma}(\mathbf{x}', \mathbf{X}) f^{m}_{\alpha}(\mathbf{X}) d\Omega(\mathbf{X}) + \frac{1}{2} B \left[(1-\nu) k_{\alpha\beta} + \nu \delta_{\alpha\beta} k_{\phi\phi}\right] w_3(\mathbf{x}')$$
$$+ B \frac{1-\nu}{2} \left(w_{3,\alpha}(\mathbf{x}') \, w_{3,\beta}(\mathbf{x}') + \frac{\nu}{1-\nu} w_{3,\gamma}(\mathbf{x}') \, w_{3,\gamma}(\mathbf{x}') \, \delta_{\alpha\beta} \right) \tag{1.96}$$

where $⨎$ denotes a Hadamard principal value integral. Equations (1.94–1.96) represent eight stress resultant integral equations for boundary point $\mathbf{x}'$ on a smooth boundary Γ.

1.7. Non-linear Terms

1.7.1. *Boundary non-linear terms*

After a closer look at the non-linear system of Eqs. (1.76–1.77), a limitation can be found due to the presence of the following non-linear term at the boundary:

$$\int_{\Gamma} U^{*}_{\theta\alpha}(\mathbf{X}', \mathbf{x}) N^{(n)}_{\alpha\beta}(\mathbf{x}) \, n_{\beta}(\mathbf{x}) d\Gamma(\mathbf{x})$$

The limitation arises because the non-linear membrane stress $N^{(n)}_{\alpha\beta}$ depends on derivatives of the out-of-plane deflection as shown by Eq. (1.36). The boundary integral equation for $w_{\alpha}(\mathbf{x}')$ will lead to an equations similar to (1.94–1.96). As mentioned before, Eqs. (1.94–1.96) are mathematically very cumbersome because of the singularities involved in the boundary integrals and therefore are only recommended for use in special cases. In the large deflection analysis proposed by Dirgantara and Aliabadi,[9] this limitation was overcome by just considering linear in-plane tractions in the integral representations. Let's first recall the following relationship for the membrane tractions:

$$t_{\alpha}(\mathbf{x}) = N^{(i)}_{\alpha\beta}(\mathbf{x}) \, n_{\beta}(\mathbf{x}) + N^{(ii)}_{\alpha\beta}(\mathbf{x}) \, n_{\beta}(\mathbf{x}) + N^{(n)}_{\alpha\beta}(\mathbf{x}) \, n_{\beta}(\mathbf{x})$$

Therefore, non-linear and total membrane tractions can be written as,

$$t^{(l)}_{\alpha}(\mathbf{x}) = t^{(i)}_{\alpha}(\mathbf{x}) + t^{(ii)}_{\alpha}(\mathbf{x}) = t_{\alpha}(\mathbf{x}) - t^{(n)}_{\alpha} \tag{1.97}$$

It is possible to write Eq. (1.77) as follows:

$$c_{\theta\alpha}(\mathbf{x}') \, u_{\alpha}(\mathbf{x}') + ⨍_{\Gamma} T^{*(i)}_{\theta\alpha}(\mathbf{x}', \mathbf{x}) u_{\alpha}(\mathbf{x}) d\Gamma(\mathbf{x})$$
$$+ \int_{\Gamma} U^{*}_{\theta\alpha}(\mathbf{x}', \mathbf{x}) B \left[k_{\alpha\beta}(1-\nu) + \nu \delta_{\alpha\beta} k_{\phi\phi}\right] w_3(\mathbf{x}) n_{\beta}(\mathbf{x}) d\Gamma(\mathbf{x})$$

$$
\begin{aligned}
&-\int_{\Omega} U^{*}_{\theta\alpha}(\mathbf{x}',\mathbf{X})N^{(n)}_{\alpha\beta,\beta}(\mathbf{X})d\Omega(\mathbf{X})\\
&-\int_{\Omega} U^{*}_{\theta\alpha}(\mathbf{x}',\mathbf{X})B\left[k_{\alpha\beta}\left(1-\nu\right)+\nu\delta_{\alpha\beta}k_{\phi\phi}\right]w_{3,\beta}(\mathbf{X})d\Omega(\mathbf{X})\\
&=\int_{\Gamma} U^{*}_{\theta\alpha}(\mathbf{x}',\mathbf{x})t^{(l)}_{\alpha}(\mathbf{x})d\Gamma(\mathbf{x})+\int_{\Omega} U^{*}_{\theta\alpha}(\mathbf{x}',\mathbf{X})q_{\alpha}(\mathbf{X})d\Omega(\mathbf{X}) \qquad (1.98)
\end{aligned}
$$

The use of linear membrane tractions $t^{(l)}_{\alpha}$ in the integral Eq. (1.98), does not reduce the general performance of the formulation, but it limits the type of boundary conditions that can be analysed. In other words, as long as total membrane tractions t_{α} are not required to be known (input or output data), the formulation can be used.

1.7.2. *Domain non-linear terms*

Non-linear terms in Eqs. (1.76) and (1.98) include primarily derivatives of the out-of-plane displacement $w_{3,\beta}$ and total membrane stress resultants $N_{\alpha\beta}$. These non-linear terms are needed in order to solve the non-linear system of equations. Dirgantara and Aliabadi[9] derived these terms from the displacement integral Eqs. (1.59) and (1.70) as follows:

$$
\begin{aligned}
&w_{3,\theta}(\mathbf{X}')+\int_{\Gamma} P^{*}_{3j,\theta}(\mathbf{X}',\mathbf{x})w_{j}\left(\mathbf{x}\right)d\Gamma\left(\mathbf{x}\right)\\
&\quad=\int_{\Gamma} W^{*}_{3j,\theta}(\mathbf{X}',\mathbf{x})p_{j}\left(\mathbf{x}\right)d\Gamma\left(\mathbf{x}\right)\\
&\quad-\int_{\Omega} W^{*}_{33,\theta}\left(\mathbf{X}',\mathbf{X}\right)k_{\alpha\beta}B\frac{1-\nu}{2}\\
&\quad\times\left(u_{\alpha,\beta}\left(\mathbf{X}\right)+u_{\beta,\alpha}\left(\mathbf{X}\right)+\frac{2\nu}{1-\nu}u_{\phi,\phi}\left(\mathbf{X}\right)\delta_{\alpha\beta}\right)d\Omega\left(\mathbf{X}\right)\\
&\quad-\int_{\Omega} W^{*}_{33,\theta}\left(\mathbf{X}',\mathbf{X}\right)k_{\alpha\beta}B\left((1-\nu)k_{\alpha\beta}+\nu\delta_{\alpha\beta}k_{\phi\phi}\right)w_{3}\left(\mathbf{X}\right)d\Omega\left(\mathbf{X}\right)\\
&\quad-\int_{\Omega} W^{*}_{33,\theta}\left(\mathbf{X}',\mathbf{X}\right)k_{\alpha\beta}N^{(n)}_{\alpha\beta}\left(\mathbf{X}\right)d\Omega\left(\mathbf{X}\right)\\
&\quad+\int_{\Omega} W^{*}_{33,\theta}\left(\mathbf{X}',\mathbf{X}\right)\left(N_{\alpha\beta}\left(\mathbf{X}\right)w_{3,\beta}\left(\mathbf{X}\right)\right)_{,\alpha}d\Omega\left(\mathbf{X}\right)\\
&\quad+\int_{\Omega} W^{*}_{33,\theta}\left(\mathbf{X}',\mathbf{X}\right)q_{3}(\mathbf{X})d\Omega(\mathbf{X}) \qquad (1.99)
\end{aligned}
$$

And for linear membrane stress resultants, Eq. (1.81) can be rewritten as,

$$\begin{aligned}
&N_{\alpha\beta}^{(l)}(\mathbf{X}') + \int_\Gamma T_{\alpha\beta\gamma}^{(i)*}(\mathbf{X}',\mathbf{x})u_\gamma(\mathbf{x})d\Gamma(\mathbf{x}) \\
&= \int_\Gamma U_{\alpha\beta\gamma}^*(\mathbf{X}',\mathbf{x})t_\gamma^{(l)}(\mathbf{x})d\Gamma(\mathbf{x}) \\
&- \int_\Gamma U_{\alpha\beta\gamma}^*(\mathbf{X}',\mathbf{x})B\left[k_{\gamma\theta}(1-\nu)+\nu\delta_{\gamma\theta}k_{\phi\phi}\right]w_3(\mathbf{x})n_\theta(\mathbf{x})d\Gamma(\mathbf{x}) \\
&+ \int_\Omega U_{\alpha\beta\gamma}^*(\mathbf{X}',\mathbf{X})\,B\left[k_{\gamma\theta}(1-\nu)+\nu\delta_{\gamma\theta}k_{\phi\phi}\right]w_{3,\theta}(\mathbf{X})d\Omega(\mathbf{X}) \\
&+ \int_\Omega U_{\alpha\beta\gamma}^*(\mathbf{X}',\mathbf{X})N_{\gamma\theta,\theta}^{(n)}(\mathbf{X})d\Omega(\mathbf{X}) + \int_\Omega U_{\alpha\beta\gamma}^*(\mathbf{X}',\mathbf{X})q_\gamma(\mathbf{X})d\Omega(\mathbf{X}) \\
&+ B\left[(1-\nu)k_{\alpha\beta}+\nu\delta_{\alpha\beta}k_{\phi\phi}\right]w_3(\mathbf{X}')
\end{aligned} \tag{1.100}$$

Recalling Eq. (1.36), the total in-plane membrane stress resultant is given by:

$$N_{\alpha\beta}(\mathbf{X}') = N_{\alpha\beta}^{(l)}(\mathbf{X}') + N_{\alpha\beta}^{(n)}(\mathbf{X}')$$

where:

$$N_{\alpha\beta}^{(n)}(\mathbf{X}') = B\frac{1-\nu}{2}\left(w_{3,\alpha}(\mathbf{X}')w_{3,\beta}(\mathbf{X}') + \frac{\nu}{1-\nu}w_{3,\gamma}(\mathbf{X}')w_{3,\gamma}(\mathbf{X}')\delta_{\alpha\beta}\right)$$

1.7.3. *Derivatives of non-linear terms*[47]

Non-linear terms can be approximated by

$$S(x_1,x_2) = \sum_{m=1}^{M} f(r)\alpha^m \tag{1.101}$$

where S can be either $N_{\alpha\beta}^{(n)}(\mathbf{X})$ or $(N_{\alpha\beta}(\mathbf{X})\,w_{3,\beta}(\mathbf{X}))$. α^m are coefficients which can be determined by the values at selected domain points, and the approximating function is given as

$$f(r) = \sqrt{c^2+r^2} \tag{1.102}$$

where c is a constant and $r = \sqrt{(x_1-x_1^m)^2+(x_2-x_2^m)^2}$.

By substituting Eq. (1.102) into Eq. (1.101), the derivatives of non-linear terms can be obtained as follows:

$$S_{,\alpha}(x_1, x_2) = \sum_{m=1}^{M} \frac{(x_\alpha - x_\alpha^m)\alpha^m}{\sqrt{c^2 + r^2}} \tag{1.103}$$

Using Eq. (1.103), the derivatives of non-linear terms in the domain Ω can be determined, namely $N_{\alpha\beta}^{(n)}(\mathbf{X})_{,\alpha}$ and $\left(N_{\alpha\beta}(\mathbf{X})\, w_{3,\beta}(\mathbf{X})\right)_{,\alpha}$.

1.8. Transformation of Domain Integrals (The Dual Reciprocity Method)

There are several techniques for the treatment of domain integrals in BEM. Some of the most important are the analytical integration of the domain integrals, the multiple reciprocity method and the dual reciprocity method.[25] Out of all of these methods, the Dual Reciprocity Method (DRM) constitutes the most general technique other than domain cell integration.

The DRM is based on the use of a series of particular solutions, $\widehat{u}_m$, instead of a single function, u, of the differential equation. The number of $\widehat{u}_m$ used is equal to the total number of selected nodes in the problem. If there are N boundary nodes and L domain points, there could be $N + L$ values of $\widehat{u}_m$.

It is assumed that the domain terms in the non-linear system of integral equations presented in the previous sections can be approximated by,

$$b = \sum_{m=1}^{M} f^m(r)\alpha^m \tag{1.104}$$

where $f^m(r)$ are approximating functions, α^m are a set of initially unknown coefficients, r denotes the distance between the calculating point and the selected point in the domain Ω, and M is the total number of selected points. To obtain α^m, Eq. (1.104) may be inverted to give

$$\alpha = \mathbf{F}^{-1}\mathbf{b} \tag{1.105}$$

where $\mathbf{F}$ consists of a vector $\mathbf{f}^m$ determined from Eq. (1.104). A complete description of the DRM can be found in the book by Partridge *et al.*[28]

The domain integrals in Eqs. (1.76) and (1.98) are:

$$I_{1i}^D = \int_\Omega W_{i3}^* w_3 d\Omega, \quad I_{2i}^D = \int_\Omega W_{i3}^* \frac{\partial u_1}{\partial x_1} d\Omega,$$

$$I^D_{3i} = \int_\Omega W^*_{i3}\frac{\partial u_2}{\partial x_2}d\Omega, \quad I^D_{4i} = \int_\Omega W^*_{i3}q^*_3 d\Omega$$

$$I^D_{5\alpha} = \int_\Omega U^*_{\alpha 1}\frac{\partial w_3}{\partial x_1}d\Omega, \quad I^D_{6\alpha} = \int_\Omega U^*_{\alpha 2}\frac{\partial w_3}{\partial x_2}d\Omega$$

$$I^D_{7\alpha} = \int_\Omega U^*_{\alpha\beta}q^*_\beta d\Omega \tag{1.106}$$

where,

$$q^*_3 = q_3 - k_{\alpha\beta}N^{(n)}_{\alpha\beta} + \left(N_{\alpha\beta}w_{3,\beta}\right)_{,\alpha}$$

and

$$q^*_\beta = q_\beta + N^{(n)}_{\beta\alpha,\alpha}$$

The above domain integrals are transferred to the boundary by employing the dual reciprocity technique as described by Wen *et al.*[45,46] Using the particular solution $\hat{w}^3_{mk}$, which satisfies the differential equation,

$$L^{b,adj}_{\alpha k}\hat{w}^3_{mk} = 0 \quad \text{and} \quad L^{b,adj}_{3k}\hat{w}^3_{mk} = f^m(r)$$

where $L^{b,adj}_{\alpha k}$ and $L^{b,adj}_{3k}$ are adjoint operators of the original differential operator L^b_{ik} for the plate bending problem in Eq. (1.37), the boundary integral equations for plate bending problem becomes

$$c_{ik}(\mathbf{x}')\hat{w}^3_{mk}(\mathbf{x}') = \int_\Gamma W^*_{ik}(\mathbf{x}',\mathbf{x})\hat{p}^3_{mk}(\mathbf{x})\,d\Gamma(\mathbf{x}) - \fint_\Gamma P^*_{ik}(\mathbf{x}',\mathbf{x})\hat{w}^3_{mk}(\mathbf{x})\,d\Gamma(\mathbf{x})$$
$$+ \int_\Omega W^*_{i3}(\mathbf{x}',\mathbf{X})f^m(r)d\Omega(\mathbf{X}) \tag{1.107}$$

which implies that

$$I^D_{1i} = \sum_{m=1}^{M_T}\left[c_{ik}(\mathbf{x}')\hat{w}^3_{mk}(\mathbf{x}') - \int_\Gamma W^*_{ik}(\mathbf{x}',\mathbf{x})\hat{p}^3_{mk}(\mathbf{x})\,d\Gamma(\mathbf{x})\right.$$
$$\left. + \fint_\Gamma P^*_{ik}(\mathbf{x}',\mathbf{x})\hat{w}^3_{mk}(\mathbf{x})\,d\Gamma(\mathbf{x})\right]\mathbf{F}^{-1}w_3 \tag{1.108}$$

Similar to the previous procedure, the following plate bending differential equations are considered for the case of I^D_{2i} and I^D_{3i}:

$$L^{b,adj}_{\alpha k}\hat{w}^\gamma_{mk} = 0 \quad \text{and} \quad L^{b,adj}_{3k}\hat{w}^\gamma_{mk} = \frac{\partial F_m(r)}{\partial x_\gamma} = \frac{x_\gamma}{r}$$

This leads to the following expressions:

$$I_{2i}^{D} = \sum_{m=1}^{M_T} \left[c_{ik}(\mathbf{x}')\hat{w}_{mk}^{1}(\mathbf{x}') - \int_{\Gamma} W_{ik}^{*}(\mathbf{x}',\mathbf{x})\hat{p}_{mk}^{1}(\mathbf{x})\, d\Gamma(\mathbf{x}) \right.$$

$$\left. + \int_{\Gamma} P_{ik}^{*}(\mathbf{x}',\mathbf{x})\hat{w}_{mk}^{1}(\mathbf{x})\, d\Gamma(\mathbf{x}) \right] \mathbf{F}^{-1}u_{1} \tag{1.109}$$

and

$$I_{3i}^{D} = \sum_{m=1}^{M_T} \left[c_{ik}(\mathbf{x}')\hat{w}_{mk}^{2}(\mathbf{x}') - \int_{\Gamma} W_{ik}^{*}(\mathbf{x}',\mathbf{x})\hat{p}_{mk}^{2}(\mathbf{x})\, d\Gamma(\mathbf{x}) \right.$$

$$\left. + \int_{\Gamma} P_{ik}^{*}(\mathbf{x}',\mathbf{x})\hat{w}_{mk}^{2}(\mathbf{x})\, d\Gamma(\mathbf{x}) \right] \mathbf{F}^{-1}u_{2} \tag{1.110}$$

In the particular case of the pseudo transversal body force, the domain integral I_{4i}^{D} can be obtained with the same particular solutions of Eq. (1.108):

$$I_{4i}^{D} = \sum_{m=1}^{M} \left[\hat{w}_{mi}^{3}(\mathbf{X}') - \int_{\Gamma} W_{ik}^{*}(\mathbf{X}',\mathbf{x})\hat{p}_{mk}^{3}(\mathbf{x})d\Gamma(\mathbf{x}) \right.$$

$$\left. + \int_{\Gamma} P_{ik}^{*}(\mathbf{X}',\mathbf{x})\hat{w}_{mk}^{3}(\mathbf{x})d\Gamma(\mathbf{x}) \right] \mathbf{F}^{-1}q_{3}^{*} \tag{1.111}$$

$I_{5\alpha}^{D}$ can be evaluated from the particular solution $\hat{u}_{m\alpha}^{1}$ for two-dimensional plane stress elasticity problems, that satisfy the differential equation

$$L_{1\alpha}^{m,adj}\hat{u}_{m\alpha}^{1} = \frac{\partial F_m(r)}{\partial x_1} = \frac{x_1}{r} \quad \text{and} \quad L_{2\alpha}^{m,adj}\hat{u}_{m\alpha}^{1} = 0$$

to give

$$I_{5\alpha}^{D} = \sum_{m=1}^{M_T} \left[c_{\alpha\beta}(\mathbf{x}')\hat{u}_{m\beta}^{1}(\mathbf{x}') - \int_{\Gamma} U_{\alpha\beta}^{*}(\mathbf{x}',\mathbf{x})\hat{t}_{m\beta}^{1}(\mathbf{x})\, d\Gamma(\mathbf{x}) \right.$$

$$\left. + \int_{\Gamma} T_{\alpha\beta}^{*}(\mathbf{x}',\mathbf{x})\hat{u}_{m\beta}^{1}(\mathbf{x})\, d\Gamma(\mathbf{x}) \right] \mathbf{F}^{-1}w_{3} \tag{1.112}$$

The domain integral $I_{6\alpha}^{D}$ can be obtained from the particular solution $\hat{u}_{m\alpha}^{2}$ for two-dimensional plane stress elasticity problems, that satisfy the differential equation

$$L_{1\alpha}^{m,adj}\hat{u}_{m\alpha}^{2} = 0 \quad \text{and} \quad L_{2\alpha}^{m,adj}\hat{u}_{m\alpha}^{2} = \frac{\partial F_m(r)}{\partial x_2} = \frac{x_2}{r}$$

to give

$$I_{6\alpha}^{D} = \sum_{m=1}^{M_T}\left[c_{\alpha\beta}(\mathbf{x}')\hat{u}_{m\beta}^{2}(\mathbf{x}') - \int_{\Gamma} U_{\alpha\beta}^{*}(\mathbf{x}',\mathbf{x})\hat{t}_{m\beta}^{2}(\mathbf{x})\,d\Gamma(\mathbf{x}) \right.$$
$$\left. + \fint_{\Gamma} T_{\alpha\beta}^{*}(\mathbf{x}',\mathbf{x})\hat{u}_{m\beta}^{2}(\mathbf{x})\,d\Gamma(\mathbf{x})\right]\mathbf{F}^{-1}w_3 \qquad (1.113)$$

The domain integral $I_{7\alpha}^{D}$ can be evaluated from the particular solution $\hat{u}_{\alpha\gamma}^{m}$ for two-dimensional plane stress elasticity problems, that satisfy the differential equation

$$L_{\alpha\beta}^{m,adj}\hat{u}_{m\beta}^{\gamma} = f^m(r)\delta_{\alpha\gamma}$$

to give

$$I_{7\alpha}^{D} = \sum_{m=1}^{M}\left[\hat{u}_{m\alpha}^{\gamma}(\mathbf{X}') - \int_{\Gamma} U_{\alpha\beta}^{*}(\mathbf{X}',\mathbf{x})\hat{t}_{m\beta}^{\gamma}(\mathbf{x})d\Gamma(\mathbf{x}) \right.$$
$$\left. + \int_{\Gamma} T_{\alpha\beta}^{*}(\mathbf{X}',\mathbf{x})\hat{u}_{m\beta}^{\gamma}(\mathbf{x})d\Gamma(\mathbf{x})\right]\mathbf{F}^{-1}q_{\gamma}^{*} \qquad (1.114)$$

The domain integrals in Eq. (1.99) are:

$$I_{8\theta}^{D} = \int_{\Omega} W_{33,\theta}^{*}w_3\,d\Omega, \quad I_{9\theta}^{D} = \int_{\Omega} W_{33,\theta}^{*}\frac{\partial u_1}{\partial x_1}d\Omega,$$
$$I_{10\theta}^{D} = \int_{\Omega} W_{33,\theta}^{*}\frac{\partial u_2}{\partial x_2}d\Omega, \quad I_{11\theta}^{D} = \int_{\Omega} W_{33,\theta}^{*}q_3^{*}d\Omega \qquad (1.115)$$

and the domain integrals in Eqs. (1.79–1.81) and (1.100) are:

$$I_{12i\beta}^{D} = \int_{\Omega} W_{i\beta 3}^{*}w_3\,d\Omega, \quad I_{13i\beta}^{D} = \int_{\Omega} W_{i\beta 3}^{*}\frac{\partial u_1}{\partial x_1}d\Omega,$$
$$I_{14i\beta}^{D} = \int_{\Omega} W_{i\beta 3}^{*}\frac{\partial u_2}{\partial x_2}d\Omega, \quad I_{15i\beta}^{D} = \int_{\Omega} W_{i\beta 3}^{*}q_3^{*}d\Omega,$$
$$I_{16\alpha\beta}^{D} = \int_{\Omega} U_{\alpha\beta 1}^{*}\frac{\partial w_3}{\partial x_1}d\Omega, \quad I_{17\alpha\beta}^{D} = \int_{\Omega} U_{\alpha\beta 2}^{*}\frac{\partial w_3}{\partial x_2}d\Omega$$
$$I_{18\alpha\beta}^{D} = \int_{\Omega} U_{\alpha\beta\gamma}^{*}q_{\gamma}^{*}d\Omega \qquad (1.116)$$

The same procedure and also the same particular solutions from the displacement integral equations can be used to transfer these integrals to the boundary.[9,47] The domain integrals in Eqs. (1.115–1.116) are transferred to the boundary as follows,

$$I_{8\theta}^{D} = \sum_{m=1}^{M} \left[\hat{w}_{m3,\theta}^{3}(\mathbf{X}') - \int_{\Gamma} W_{3j,\theta}^{*}(\mathbf{X}', \mathbf{x}) \hat{p}_{mj}^{3}(\mathbf{x})\, d\Gamma(\mathbf{x}) \right.$$
$$\left. + \int_{\Gamma} P_{3j,\theta}^{*}(\mathbf{X}', \mathbf{x}) \hat{w}_{mj}^{3}(\mathbf{x})\, d\Gamma(\mathbf{x}) \right] \mathbf{F}^{-1} w_{3} \tag{1.117}$$

$$I_{9\theta}^{D} = \sum_{m=1}^{M} \left[\hat{w}_{m3,\theta}^{1}(\mathbf{X}') - \int_{\Gamma} W_{3j,\theta}^{*}(\mathbf{X}', \mathbf{x}) \hat{p}_{mj}^{1}(\mathbf{x})\, d\Gamma(\mathbf{x}) \right.$$
$$\left. + \int_{\Gamma} P_{3j,\theta}^{*}(\mathbf{X}', \mathbf{x}) \hat{w}_{mj}^{1}(\mathbf{x})\, d\Gamma(\mathbf{x}) \right] \mathbf{F}^{-1} u_{1} \tag{1.118}$$

$$I_{10\theta}^{D} = \sum_{m=1}^{M} \left[\hat{w}_{m3,\theta}^{2}(\mathbf{X}') - \int_{\Gamma} W_{3j,\theta}^{*}(\mathbf{X}', \mathbf{x}) \hat{p}_{mj}^{2}(\mathbf{x})\, d\Gamma(\mathbf{x}) \right.$$
$$\left. + \int_{\Gamma} P_{3j,\theta}^{*}(\mathbf{X}', \mathbf{x}) \hat{w}_{mj}^{2}(\mathbf{x})\, d\Gamma(\mathbf{x}) \right] \mathbf{F}^{-1} u_{2} \tag{1.119}$$

and,

$$I_{11\theta}^{D} = \sum_{m=1}^{M} \left[\hat{w}_{m3,\theta}^{3}(\mathbf{X}') - \int_{\Gamma} W_{3j,\theta}^{*}(\mathbf{X}', \mathbf{x}) \hat{p}_{mj}^{3}(\mathbf{x})\, d\Gamma(\mathbf{x}) \right.$$
$$\left. + \int_{\Gamma} P_{3j,\theta}^{*}(\mathbf{X}', \mathbf{x}) \hat{w}_{mj}^{3}(\mathbf{x})\, d\Gamma(\mathbf{x}) \right] \mathbf{F}^{-1} q_{3}^{*} \tag{1.120}$$

$$I_{12\alpha\beta}^{D} = \sum_{m=1}^{M} \left[\hat{M}_{m\alpha\beta}^{3}(\mathbf{X}') + \int_{\Gamma} P_{\alpha\beta j}^{*}(\mathbf{X}', \mathbf{x}) \hat{w}_{mj}^{3}(\mathbf{x}) d\Gamma(\mathbf{x}) \right.$$
$$\left. - \int_{\Gamma} W_{\alpha\beta j}^{*}(\mathbf{X}', \mathbf{x}) \hat{p}_{mj}^{3}(\mathbf{x}) d\Gamma(\mathbf{x}) \right] \mathbf{F}^{-1} w_{3} \tag{1.121}$$

$$I_{123\beta}^{D} = \sum_{m=1}^{M} \left[\hat{Q}_{m\beta}^{3}(\mathbf{X}') + \int_{\Gamma} P_{3\beta j}^{*}(\mathbf{X}', \mathbf{x}) \hat{w}_{mj}^{3}(\mathbf{x}) d\Gamma(\mathbf{x}) \right.$$
$$\left. - \int_{\Gamma} W_{3\beta j}^{*}(\mathbf{X}', \mathbf{x}) \hat{p}_{mj}^{3}(\mathbf{x}) d\Gamma(\mathbf{x}) \right] \mathbf{F}^{-1} w_{3} \tag{1.122}$$

The particular solutions $\hat{w}^{\alpha}_{mk}$ and $\hat{p}^{\alpha}_{mk}$, can be used for the domain integrals I^D_{13} and I^D_{14},

$$I^D_{13\alpha\beta} = \sum_{m=1}^{M}\left[\hat{M}^1_{m\alpha\beta}(\mathbf{X}') + \int_\Gamma P^*_{\alpha\beta j}(\mathbf{X}',\mathbf{x})\hat{w}^1_{mj}(\mathbf{x})d\Gamma(\mathbf{x}) - \int_\Gamma W^*_{\alpha\beta j}(\mathbf{X}',\mathbf{x})\hat{p}^1_{mj}(\mathbf{x})d\Gamma(\mathbf{x})\right]\mathbf{F}^{-1}u_1 \tag{1.123}$$

$$I^D_{133\beta} = \sum_{m=1}^{M}\left[\hat{Q}^1_{m\beta}(\mathbf{X}') + \int_\Gamma P^*_{3\beta j}(\mathbf{X}',\mathbf{x})\hat{w}^1_{m\gamma}(\mathbf{x})d\Gamma(\mathbf{x}) - \int_\Gamma W^*_{3\beta j}(\mathbf{X}',\mathbf{x})\hat{p}^1_{mj}(\mathbf{x})d\Gamma(\mathbf{x})\right]\mathbf{F}^{-1}u_1 \tag{1.124}$$

and

$$I^D_{14\alpha\beta} = \sum_{m=1}^{M}\left[\hat{M}^2_{m\alpha\beta}(\mathbf{X}') + \int_\Gamma P^*_{\alpha\beta\gamma}(\mathbf{X}',\mathbf{x})\hat{w}^2_{m\gamma}(\mathbf{x})d\Gamma(\mathbf{x}) - \int_\Gamma W^*_{\alpha\beta j}(\mathbf{X}',\mathbf{x})\hat{p}^2_{mj}(\mathbf{x})d\Gamma(\mathbf{x})\right]\mathbf{F}^{-1}u_2 \tag{1.125}$$

$$I^D_{143\beta} = \sum_{m=1}^{M}\left[\hat{Q}^2_{m\beta}(\mathbf{X}') + \int_\Gamma P^*_{3\beta j}(\mathbf{X}',\mathbf{x})\hat{w}^2_{mj}(\mathbf{x})d\Gamma(\mathbf{x}) - \int_\Gamma W^*_{3\beta j}(\mathbf{X}',\mathbf{x})\hat{p}^2_{mj}(\mathbf{x})d\Gamma(\mathbf{x})\right]\mathbf{F}^{-1}u_2 \tag{1.126}$$

Domain integral I^D_{15} can be obtained from Eqs. (1.121–1.122) by replacing w_3 with q_3 to give

$$I^D_{15\alpha\beta} = \sum_{m=1}^{M}\left[\hat{M}^3_{m\alpha\beta}(\mathbf{X}') + \int_\Gamma P^*_{\alpha\beta j}(\mathbf{X}',\mathbf{x})\hat{w}^3_{mj}(\mathbf{x})d\Gamma(\mathbf{x}) - \int_\Gamma W^*_{\alpha\beta j}(\mathbf{X}',\mathbf{x})\hat{p}^3_{mj}(\mathbf{x})d\Gamma(\mathbf{x})\right]\mathbf{F}^{-1}q^*_3 \tag{1.127}$$

and

$$I^D_{153\beta} = \sum_{m=1}^{M}\left[\hat{Q}^3_{m\beta}(\mathbf{X}') + \int_\Gamma P^*_{3\beta j}(\mathbf{X}',\mathbf{x})\hat{w}^3_{mj}(\mathbf{x})d\Gamma(\mathbf{x}) - \int_\Gamma W^*_{3\beta j}(\mathbf{X}',\mathbf{x})\hat{p}^3_{mj}(\mathbf{x})d\Gamma(\mathbf{x})\right]\mathbf{F}^{-1}q^*_3 \tag{1.128}$$

I^D_{16} can be evaluated from the particular solutions $\hat{u}^1_{m\alpha}$ and $\hat{t}^1_{m\alpha}$ for two-dimensional plane stress elasticity problems, to give

$$I^D_{16\alpha\beta} = \sum_{m=1}^{M} \left[\hat{N}^1_{m\alpha\beta}(\mathbf{X}') - \int_\Gamma U^*_{\alpha\beta\gamma}(\mathbf{X}',\mathbf{x})\hat{t}^1_{m\gamma}(\mathbf{x})\, d\Gamma(\mathbf{x}) + \int_\Gamma T^{(i)*}_{\alpha\beta\gamma}(\mathbf{X}',\mathbf{x})\hat{u}^1_{m\gamma}(\mathbf{x})\, d\Gamma(\mathbf{x}) \right] \mathbf{F}^{-1} w_3 \quad (1.129)$$

The domain integral I^D_{17} can be obtained from the particular solutions $\hat{u}^2_{m\alpha}$ and $\hat{t}^2_{m\alpha}$, to give

$$I^D_{17\alpha\beta} = \sum_{m=1}^{M} \left[\hat{N}^2_{m\alpha\beta}(\mathbf{X}') - \int_\Gamma U^*_{\alpha\beta\gamma}(\mathbf{X}',\mathbf{x})\hat{t}^2_{m\gamma}(\mathbf{x})\, d\Gamma(\mathbf{x}) + \int_\Gamma T^{(i)*}_{\alpha\beta\gamma}(\mathbf{X}',\mathbf{x})\hat{u}^2_{m\gamma}(\mathbf{x})\, d\Gamma(\mathbf{x}) \right] \mathbf{F}^{-1} w_3 \quad (1.130)$$

And finally,

$$I^D_{18\alpha\beta} = \sum_{m=1}^{M} \left[\hat{N}^\theta_{m\alpha\beta}(\mathbf{X}') - \int_\Gamma U^*_{\alpha\beta\gamma}(\mathbf{X}',\mathbf{x})\hat{t}^\theta_{m\gamma}(\mathbf{x})\, d\Gamma(\mathbf{x}) + \int_\Gamma T^{(i)*}_{\alpha\beta\gamma}(\mathbf{X}',\mathbf{x})\hat{u}^\theta_{m\gamma}(\mathbf{x})\, d\Gamma(\mathbf{x}) \right] \mathbf{F}^{-1} q^*_\theta \quad (1.131)$$

All the particular solutions for shear deformable plate bending ($\hat{w}^i_{mk}$ and $\hat{p}^i_{mk}$) and plane stress ($\hat{u}^\beta_{m\alpha}$ and $\hat{t}^\beta_{m\alpha}$) were derived for a radial basis function $F_m(r) = 1 + r$ by Wen *et al.*[46]

It is important to notice at this stage that the selected points M_T used in Eqs. (1.108–1.110) and (1.112–1.113) include domain and boundary points $M_T = N + L$, while the selected points for the application of the dual reciprocity approximation in all the other cases involve only the DRM domain points $M = L$.

The best approximation for the application of the DRM is typically based on the use of a series of particular solutions at selected boundary N and domain L nodes. In some applications, boundary nodes (N) are sufficient for a good approximation, but in most cases a series of selected boundary N and domain L nodes are necessary.

The main reason for the present implementation is the level of the singularities obtained when the collocation point ($\mathbf{X}'$) approaches the boundary ($\mathbf{x}'$). As will be shown by the numerical examples, there is no great loss in accuracy. Additionally, the use of DRM domain points means that less computational time is required for the evaluation of such expressions.

1.9. Numerical Implementation

In order to solve the integral equations presented in the previous sections, the boundary Γ and the domain Ω must be discretised. In the present work, quadratic isoparametric boundary elements are used to describe the boundary; for the domain, several uniformly distributed domain points are used for the implementation of the DRM. On the boundary, semi-discontinuous elements are used for corners to avoid difficulties with discontinuity of the tractions at corners.

From the implementation point of view, the use of semi-discontinuous elements requires the consideration of two different meshes: the geometrical mesh, defined by the geometrical nodes which always lie on the boundary of the element; and the functional mesh, defined by the functional nodes which can exist anywhere within the element boundaries.

1.9.1. *Discretization*

Equation (1.76) can be rewritten in a discretised form as follows:

$$\begin{aligned} c_{ij}(\mathbf{x}')w_j(\mathbf{x}') &+ \sum_{n=1}^{N_e}\sum_{l=1}^{3} w_j \int_{\xi=-1}^{\xi=+1} P^*_{ij}(\mathbf{x}',\mathbf{x}(\xi))\Phi^l(\xi)J_n(\xi)d\xi \\ &= \sum_{n=1}^{N_e}\sum_{l=1}^{3} p_j \int_{\xi=-1}^{\xi=+1} W^*_{ij}(\mathbf{x}',\mathbf{x}(\xi))\Phi^l(\xi)J_n(\xi)d\xi \\ &- \sum_{m=1}^{M_T} B(k_{11}+\nu k_{22})I^{Dm}_{2i} - \sum_{m=1}^{M_T} B(\nu k_{11}+k_{22})I^{Dm}_{3i} \\ &- \sum_{m=1}^{M_T} k_{\alpha\beta}B\left[(1-\nu)k_{\alpha\beta}+\nu\delta_{\alpha\beta}k_{\gamma\gamma}\right]I^{Dm}_{1i} + \sum_{m=1}^{M} I^{Dm}_{4i} \end{aligned} \quad (1.132)$$

where N_e is the number of boundary elements; Φ^l are the boundary shape functions; ξ is the local coordinate and J_n is the Jacobian of transformation. All other integral equations also have to be discretised in the same way as equation (1.132).

1.9.2. *Treatment of the integrals*

All the singular integrals appearing in the integral equations are dealt with by using well-established techniques and are treated separately, based on their order of singularity. A detailed explanation is given by Aliabadi.[1]

1.10. Solution of the Non-linear Integral Equations for Shallow Shells and Plates

1.10.1. *System of equations*

After discretization and point collocation of nodes on the boundary Γ and in the domain Ω, the system matrix can be written as

$$[H]\{u\} = [G]\{p\} + \{Q\} + [C]\{nl\} \tag{1.133}$$

where $[H]$ and $[G]$ are the well-known boundary element influence matrices,[1] $[C]$ is the influence matrix for non-linear terms, $\{u\}$ is the boundary and domain displacement vector, $\{p\}$ is the boundary traction vector, $\{nl\}$ is the non-linear term and $\{Q\}$ is the domain load vector. After imposing boundary conditions, equation (1.133) can be written as:

$$[A]\{x\} = \{b\} + [C]\{nl\} \tag{1.134}$$

where $[A]$ is the system matrix, $\{x\}$ is the unknown vector and $\{b\}$ is the vector of prescribed boundary values.

1.10.2. *Load increment technique*

Equations (1.76), (1.98) and (1.99–1.100) were first introduced by Dirgantara and Aliabadi.[9] A simple and efficient load incremental technique was developed in Wen *et al.*[47,48] and it can be summarised as follows:

- Step 1: Let $n = 1$, load $q_i^1 = \Delta q_i$ and non-linear terms in equations (1.76) and (1.98) be set to zero.
- Step 2: Solve linear system of equations (1.76) and (1.98).
- Step 3: Obtain values at domain points (derivatives of the out-of-plane displacement $w_{3,\beta}$ and linear membrane stress resultants $N_{\alpha\beta}^{(l)}$) from equations (1.99–1.100).
- Step 4: Calculate non-linear terms $N_{\alpha\beta}^{(n)}(\mathbf{X}')$ and $N_{\alpha\beta}(\mathbf{X})\, w_{3,\beta}(\mathbf{X})$, and derivatives of non-linear terms $N_{\gamma\theta,\theta}^{(n)}(\mathbf{X}')$ and $\left(N_{\alpha\beta}(\mathbf{X})\, w_{3,\beta}(\mathbf{X})\right)_{,\alpha}$ (derivatives of non-linear terms are obtained using the procedure

described in section 1.7.3). All these values are used in the next load increment.

- Step 5: If the final load is reached ($q_i^n = q_i$), go to Step 7, otherwise let $n = n + 1$.
- Step 6: Set $q_i^n = \Delta q_i + q_i^{n-1}$, go back to Step 2.
- Step 7: Finish.

More complex and expensive (iterative) solution procedures could be used, as shown by Zhang and Atluri[53] and more recently by Aliabadi and Baiz[2] for post-buckling analysis, but the present pure incremental procedure has proven to be an attractive alternative for a good range of large deflection problems.

1.11. Numerical Examples

Several examples were chosen to demonstrate the application of the boundary element method to the large deflection problem of plates and shallow shells. Comparisons are made with available numerical solutions.

1.11.1. *Clamped circular plate subjected to uniform load q_0*

A clamped circular plate of radii a as shown in Fig. 1.8 was analysed by Wen *et al.*[47] The plate is subjected to a uniform load $q_3 = q_0$, and the applied membrane body forces are $q_\alpha = 0$. The ratio of thickness and radius is

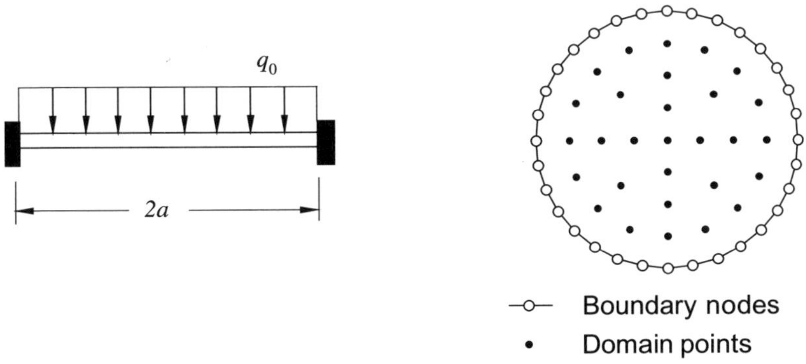

Fig. 1.8. Clamped circular plate and boundary element mesh with selected domain points.

$h/a = 0.02$. Material constants are chosen as $E = 10^7, \nu = 0.3$. Assuming that the edge of the plate is totally restrained from moving, the boundary conditions can be written as:

$$u_\alpha(a) = 0, \quad w_i(a) = 0.$$

A BEM mesh with 16 quadratic boundary elements and 25 DRM domain points are used (as shown in Fig. 1.8). The load is applied in small increments of $\delta q = 0.05q_0$. The maximum values of deflection (at the centre of the plate) are plotted in Fig. 1.9, along with the results obtained with the finite element method by Weil *et al.*[49] and the domain-boundary element method formulation by Lei *et al.*[18]

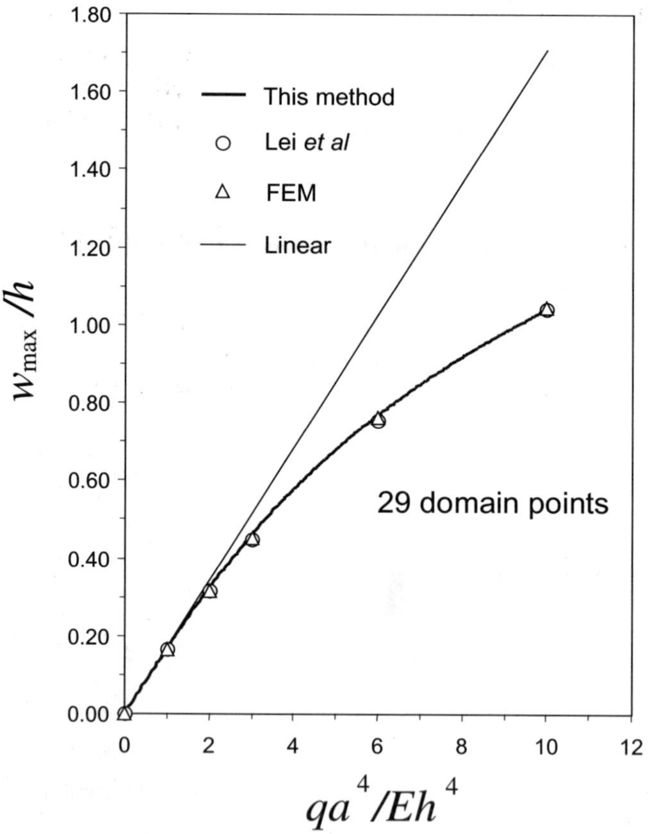

Fig. 1.9. Centre deflections for clamped circular plate.

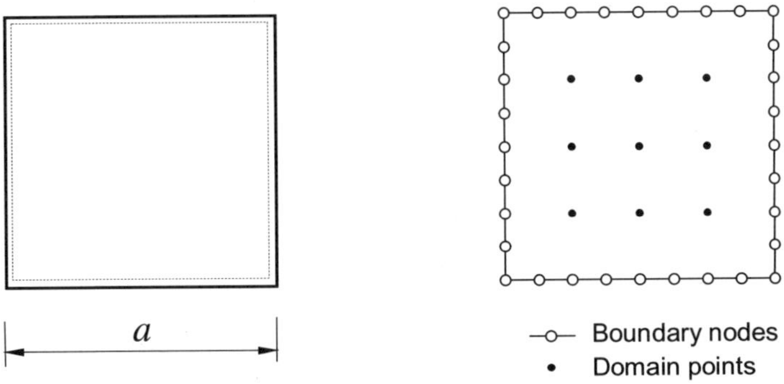

Fig. 1.10. Simply supported square plate and boundary element mesh with selected domain points.

1.11.2. *Simply supported square plate subjected to uniform load q_0*

In this second example, Wen *et al.*[47] consider a simply supported square plate ($a \times a$) subjected to uniform distribution load $q_3 = q_0$ (see Fig. 1.10). Two cases were considered in this example: thin plate (a) and moderately thick (b).

- case (a): $h/a = 0.01, E = 10^7, \nu = 0.316$
- case (b): $h/a = 0.05, E = 0.3 \times 10^7, \nu = 0.3$

In the case of simply supported edges, it is assumed that the displacement and deflection of the edge are zero but that it is free to rotate, i.e.

$$u_\alpha = 0, \quad w_2 = w_3 = 0 \quad \text{for} \quad x_1 = \pm a/2$$
$$u_\alpha = 0, \quad w_1 = w_3 = 0 \quad \text{for} \quad x_2 = \pm a/2$$

The boundary element mesh and distribution of DRM domain points are shown in Fig. 1.10. The load is applied in increments of $\Delta q = 0.05q_0$. The maximum values of deflection at the centre point of the plate are plotted in Figs. 1.11 and 1.12 for the different cases considered. Comparisons were made with the finite element method and good agreement is achieved.

1.11.3. *Clamped shallow cylindrical shells: uniformly loaded*

In this example, Dirgantara and Aliabadi[9] considered a clamped shallow cylindrical shell as shown in Fig. 1.13. The properties of the cap are as

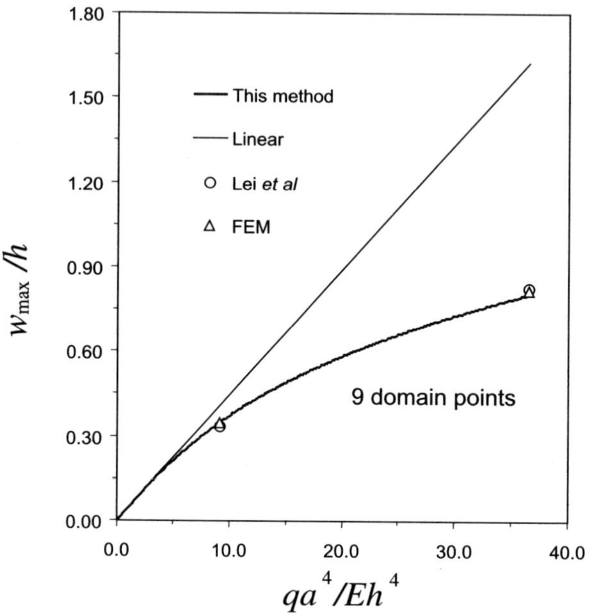

Fig. 1.11. Centre deflection for simply supported square plate (case a).

follows: shell thickness $h = 0.125$ in; in-plane projection dimensions of the shell $a = b = 20$ in; the curvature of the shell in the x_1^- direction $k_{11} = 0$, while $k_{22} = 1/R = 0.01$ in^{-1}; modulus of elasticity $E = 450{,}000\,\text{lbs/in}^2$ and Poisson's ratio $\nu = 0.3$. The cap is loaded with external uniform pressure of q_0. The BEM mesh has 40 boundary elements and 81 domain points.

All sides of the cylinder are clamped,

$$u_\alpha = 0, \quad w_i = 0$$

To study the convergence of the proposed method, several incremental load steps were chosen: $\delta q_0 = 0.005; 0.0025; 0.00125$ and 0.000625 lb.

The results are compared with non-linear finite element result reported by Sabir and Lock[34] and Brebbia and Connor,[6] and presented in Fig. 1.14. As can be seen from the result, the BEM model is able to simulate the progressive softening of the shell under a load well below the one given by linearised buckling. It can also be seen from the figure that for any further increase of the pressure, the stiffness of the shell starts to rise again.

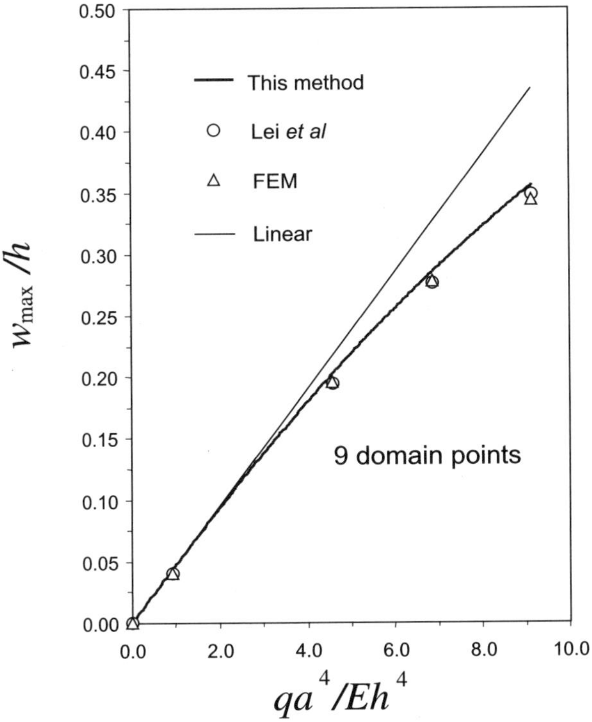

Fig. 1.12. Centre deflection for simply supported square plate (case b).

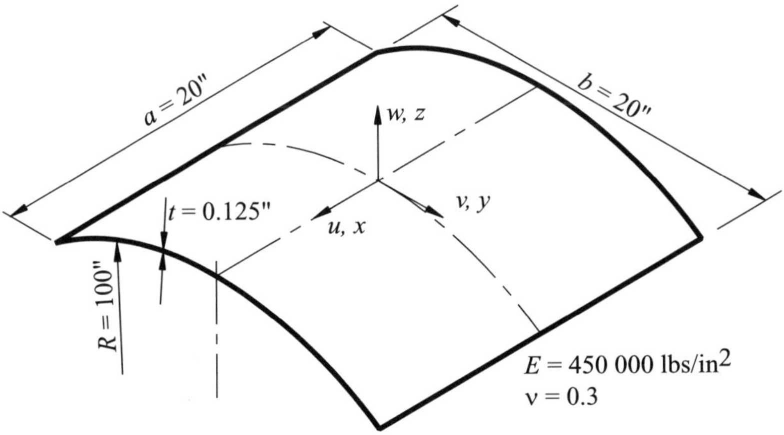

Fig. 1.13. Geometry of a clamped shallow cylindrical shell.

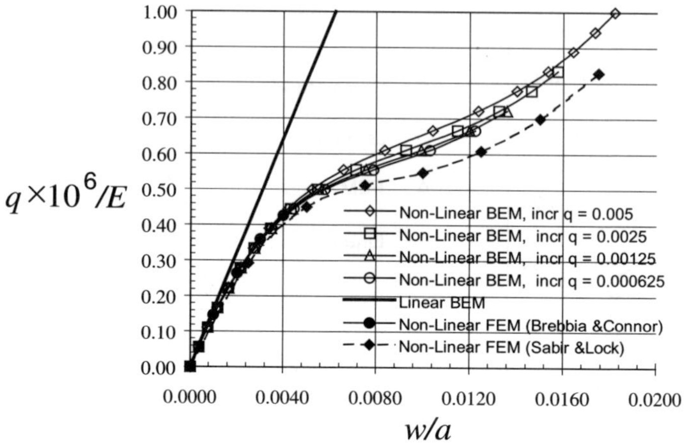

Fig. 1.14. Non linear central displacement of a clamped shallow cylindrical shell, uniformly loaded.

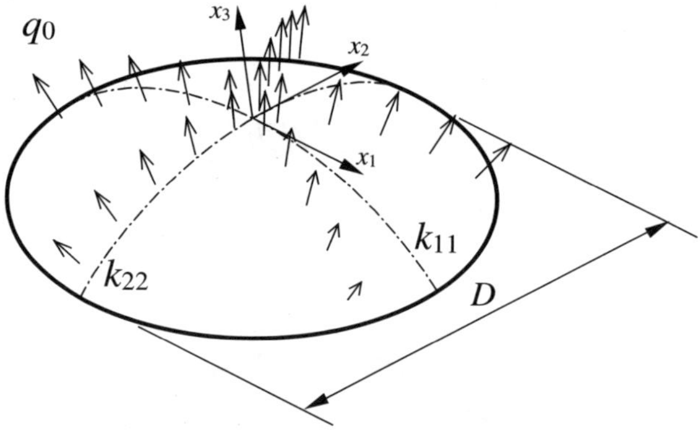

Fig. 1.15. A shallow spherical shell, uniformly loaded.

1.11.4. *Simply supported shallow spherical shells: uniformly loaded*

In this final example, a simply supported shallow spherical shell, as shown in Fig. 1.15, was analysed by Dirgantara and Aliabadi.[9] The properties of the cap are as follows: $h = 0.3\,\text{m}$; $k_{11} = k_{22} = 1/R = 1/15\,\text{m}^{-1}$; $E = 70{,}000\,\text{MPa}$ and $\nu = 0.33$. The diameter of the plane projection of the cap is $D = 10\,\text{m}$. The cap is loaded with an internal uniform pressure of q_0. In this example, an incremental load step $\delta q_0 = 0.125$ was used.

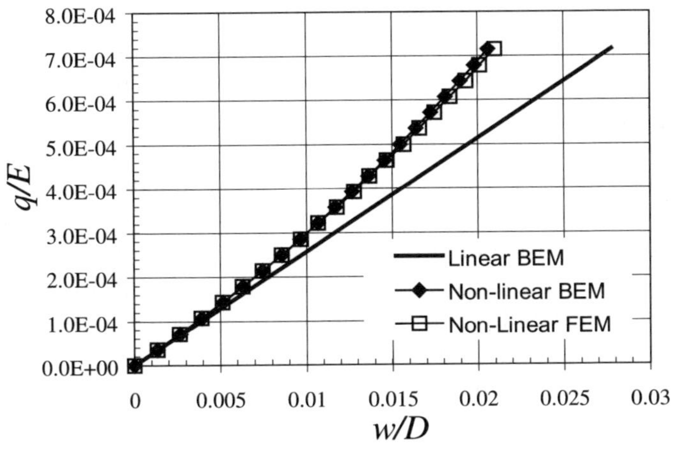

Fig. 1.16. Out-of-plane deflection at the centre of the cap.

All sides of the cap are simply supported, so that the translations $w_3 = 0$ and $u_\alpha = 0$, while the rotations are free. The BEM mesh has 24 quadratic boundary elements and 37 domain points.

The results are compared with non-linear finite element results obtained using commercial software, and are presented in Fig. 1.16. Contrary to the first example, because the load direction is opposite to the previous one, the shell becomes stiffer as the load is increased. As can be seen from the result, the BEM model is in agreement with the Finite Element Method (FEM) solution.

It was observed from this example that the difference between the BEM and FEM solutions, regarding the computational time, was negligible. However, it is important to keep in mind that the developed boundary element code has not been optimised, while the commercial FEM software can be regarded as optimised.

1.12. Conclusions

In this chapter, a boundary element formulation for the solution of large deflection problems of shear deformable thin-walled structures was reviewed. Domain integrals were transformed to equivalent boundary integrals by using the DRM. The BEM for shear deformable linear shallow shells analysis was reviewed. The derivation of the integral equations was given in such a way that closely follows the plate derivation by considering curvature and non-linear terms as pseudo body forces. This derivation allows an easy extension to boundary element formulations for large deflection of plates and shallow shells. The formulation was developed by coupling boundary element formulations of shear deformable plate bending with two-dimensional plane stress, thereby obtaining five independent integral equations.

Integral equations were discretised using quadratic isoparametric elements to describe the geometry along the boundary. For the domain, several uniformly distributed domain points are necessary for application of the DRM. Boundary and domain stress calculation procedures for non-linear shallow shell and plate structures were presented. Integral equations for stresses were obtained from the derivative of the displacement integral equations. An economic indirect approach was also presented for boundary stress resultants.

Non-linearity was dealt with by increasing the load incrementally. In every increment, non-linear terms from the previous increment were added to the right hand side of the equation. LU-decomposition was used in the first increment and, in the subsequent increments, only back substitution was performed. By applying this procedure, computational time can be reduced significantly.

To demonstrate the capability of the method, non-linear analyses of flat plates and cylindrical and spherical shells were presented and compared with FEM results. BEM models were able to simulate the progressive softening and stiffening of the shells, which could also be observed in FEM simulation results.

It is clear that the present boundary element formulation is highly accurate when analysing geometrically non-linear problems of plates and shallow shells. One of its main advantages is that no iterative procedure is required in order to achieve a good level of accuracy.

References

1. M.H. Aliabadi, *The Boundary Element Method, Vol II: Application to Solids and Structures* (Wiley, Chichester, 2001).
2. M.H. Aliabadi and P.M. Baiz, The boundary element method for buckling and postbuckling analysis of plates and shells, In: *Buckling and Postbuckling Structure*, Falzon B.G. and Aliabadi M.H. (eds.), pp. 375–412, (Imperial College Press, London, 2008).
3. S.N. Atluri and D.S. Pipkins, Large deformation analysis of plates and shells, In: *Boundary Element Analysis of Plate and Shells*, Beskos D.E. (ed.), pp. 141–166 (Spinger Verlag, Berlin, 1991).

4. C.A. Barcellos and L.H.M. Silva, A boundary element formulation for the Mindlin's plate model, In: *Boundary Element Technology*, Brebbia C.A. and Venturini W.S. (eds.), pp. 123–130, Computational Mechanics Publications, Southampton (1989).
5. D.E. Beskos (Ed), *Boundary Element Analysis of Plates and Shells* (Springer Verlag, Berlin, 1991).
6. C.A. Brebbia and J. Connor, Geometrically non-linear finite element analysis, *Proc. of the American Society of Civil Engineers* **95**, pp. 463–483, (1969).
7. T. Dirgantara and M.H. Aliabadi, A new boundary element formulation for shear deformable shells analysis, *International Journal for Numerical Methods in Engineering* **45**, pp. 1257–1275, (1999).
8. T. Dirgantara, *Boundary Element Analysis of Crack in Shear Deformable Plates and Shells (Topics in Engineering vol 43)*, WIT Press, Southampton, Computational Mechanics Publications (2002).
9. T. Dirgantara and M.H. Aliabadi, A boundary element formulation for geometrically nonlinear analysis of shear deformable shells, *Computer Methods in Applied Mechanics and Engineering* **195**, pp. 4635–4654, (2006).
10. A. Elzein and S. Syngellakis, High-order elements for the BEM stability analysis of imperfect plates, In: *Advances in Boundary Elements XI, vol 3*, Brebbia C.A. and Connor J.J. (eds.), pp. 269–284, Springer-Verlag, Berlin, (1989).
11. D.J. Forbes and A.R. Robinson, *Numerical Analysis of Elastic Plates and Shallow Shells by an Integral Equation Method*, Structural Research Series Report no. 345, University of Illinois, Urbana, (1969).
12. N. Kamiya and Y. Sawaki, An integral equation approach to finite deflection of elastic plates, *International Journal for Non-linear Mechanics*, **17**, pp. 187–194, (1982).
13. N. Kamiya and Y. Sawaki, Boundary element method applied to finite deflections of shallow elastic shells, *Methodes Numeriques de l'Ingenieur, Comptes Rendus du Troisieme Congres International*, pp. 63–76, (1983).
14. N. Kamiya, Y. Sawaki and Y. Nakamura, Postbuckling analysis by the boundary element method, *Engineering Analysis* **1**, pp. 40–44, (1984).
15. V.J. Karam and J.C.F. Telles, On boundary elements for Reissner's plate theory, *Engineering Analysis* **5**, pp. 21–27, (1988).
16. H. Kawabe, Plate buckling analysis by the boundary element method, In: *Theory and Application of Boundary Element Methods*, Tanaka M. and Du Q. (eds.), pp. 367–374, (Pergamon Press, Oxford, 1987).
17. G. Kirchhoff, Uber das gleichgewicht und die bewegung einer elastischen scheibe, *J. Reine Angew. Math.* **40**, pp. 51–88, (1850).
18. X.Y. Lei, M.K. Huang and X. Wang, Geometrically non-linear analysis of a Reissner type plate by the boundary element method, *Computers & Structures* **37**, pp. 911–916, (1990).
19. X.Y. Lei, M.K. Huang and X. Wang, Analysis of the explicit fundamental solution of a shallow spherical shell involving shear deformation, *Applied Mathematical Modelling* **19**, pp. 194–200, (1995).

20. J. Lin and S. Long, Geometrically non-linear analysis of the shallow shell by the displacement-based boundary element formulation, *Engineering Analysis with Boundary Elements* **18**, pp. 63–70, (1996).
21. A.E.H. Love, *A Treatise on the Mathematical Theory of Elasticity*, 4th edition, (Dover Publications, New York, 1944).
22. P. Lu and M. Huang, Computation of the fundamental solution for shallow shells involving shear deformation, *International Journal of Solids and Structures* **28**, pp. 631–645, (1991).
23. R.D. Mindlin, Influence of rotatory inertia and shear on flexural motions of isotropic, elastic plates, *Journal of Applied Mechanics* **18**, pp. 31–38, (1951).
24. P.M. Naghdi, On the theory of thin elastic shells, *Quarterly of Applied Mathematics* **14**, pp. 369–380, (1956).
25. D. Nardini and C.A. Brebbia, A new approach to free vibration analysis using a boundary element method, In: *Boundary Element Methods in Engineering*, Brebbia C.A. (ed.), pp. 312326, (Springer, New York, 1982).
26. D.A. Newton and H. Tottenham, Boundary value problems in thin shallow shells of arbitrary plan form, *Journal of Engineering Mathematics* **2**, pp. 211–224, (1968).
27. P.E. O'Donoghue and S.N. Atluri, Field/boundary element approach to the large deflection of thin flat plates, *Computers & Structures* **27**, pp. 427–435, (1987).
28. P.W. Partridge, C.A. Brebbia and L.C. Wrobel, *The Dual Reciprocity Boundary Element Method*, (Computational Mechanics Publications, Southampton, 1992).
29. Q. Qin and Y. Huang, BEM of post-buckling analysis of thin plates, *Applied Mathematical Modelling* **14**, pp. 544–548, (1990).
30. E. Reissner, The effect of transverse shear deformation on the bending of elastic plates, *Journal of Applied Mechanics* **12**, pp. 69–77, (1945).
31. E. Reissner, On bending of elastic plates, *Quarterly of Applied Mathematics* **5**, pp. 55–68, (1947).
32. E. Reissner, On a variational theorem in elasticity, *Journal of Mathematics and Physics* **29**, pp. 90–95, (1950).
33. E. Reissner, Stress strain relations in the theory of thin elastic shells, *Journal of Mathematics and Physics* **31**, pp. 109–119, (1952).
34. A.B. Sabir and A.C. Lock, The application of finite elements to the large deflection of geometrically non-linear behaviour of cylindrical shells, In: *Variational Methods in Engineering*, Brebbia C.A. and Tottenham H. (eds.), University of Southampton Press, Southampton, 1972).
35. Supriyono and M.H. Aliabadi, Boundary element method for shear deformable plates with combined geometric and material nonlinearities, *Engineering Analysis with Boundary Elements* **30**, pp. 31–42, (2006).
36. Y.B. Sun, X.Q. He and Q.H. Qin, A new procedure for the non-linear analysis of Reissner plate by Boundary Element, *Computers & Structures* **53**, pp. 649–652, (1994).
37. M. Tanaka, Large deflection analysis of thin elastic plates, In: *Developments in Boundary Element Methods*, Brebbia C.A. (ed.), Berlin, pp. 115–136, (Springer-Verlag, Berlin, 1984).

38. M. Tanaka, T. Matsumoto and Z. Zheng, Incremental analysis of finite deflection of elastic plates via boundary-domain-element method, *Engineering Analysis with Boundary Elements* **17**, pp. 123–131, (1996).
39. M. Tanaka, T. Matsumoto and Z. Zheng, Application of the boundary-domain element method to the pre/post-buckling problem of Von Karman plates, *Engineering Analysis with Boundary Elements* **23**, pp. 399–404, (1999).
40. N. Tosaka and S. Miyake, Non-linear analysis of elastic shallow shells by boundary element method, In: *Boundary Elements VII*, Brebbia C.A. (ed.), pp. 4–43/4–52, (Springer-Verlag, Berlin, 1985).
41. H. Tottenham, The boundary element method for plates and shells, In: *Developments in Boundary Element Methods - 1*, Banarjee P.K. and Butterfield R. (eds.), pp. 173–205, (Applied Science Publishers, London, 1979).
42. F. Van der Weeën, Application of the boundary integral equation method to Reissner's plate model, *International Journal for Numerical Methods in Engineering* **18**, pp. 1–10, (1982).
43. V.Z. Vlasov, *General Theory of Shells and its Applications in Engineering*, National Technical Information Services, US Department of Commerce, Washington, (1964).
44. W. Wang, X. Ji and M. Tanaka, A dual reciprocity boundary element approach for the problems of large deflection of thin elastic plates, *Computational Mechanics* **26**, pp. 58–65, (2000).
45. P.H. Wen, M.H. Aliabadi and A. Young, Plane stress and plate bending coupling in BEM analysis of shallow shells, *International Journal for Numerical Methods in Engineering* **48**, pp. 1107–1125, (2000).
46. P.H. Wen, M.H. Aliabadi and A. Young, Application of dual reciprocity method to plates and shells, *Engineering Analysis with Boundary Element* **24**, pp. 583–590, (2000).
47. P.H. Wen, M.H. Aliabadi and A. Young, Large deflection analysis of Reissner's plate by boundary element method, *Computers & Structures* **83**, pp. 870–879, (2005).
48. P.H. Wen, M.H. Aliabadi and A. Young, Post-buckling analysis of shear deformable plates by BEM, *Journal of Strain Analysis for Engineering Design* **41**, pp. 239–252, (2006).
49. N.A. Weil and N.M. Newmark, Large deflections of elliptical plates, *Journal of Applied Mechanics* **23**, pp. 21–26, (1956).
50. T.Q. Ye and Y.J. Liu, Finite deflection analysis of elastic plate by boundary element method, *Applied Mathematical Modelling* **9**, pp. 183–188, (1985).
51. J. Ye, Axisymmetric non-linear analysis of shallow spherical shells by a combined boundary element and finite difference method, *Engineering Analysis with Boundary Elements* **9**, pp. 283–287, (1992).
52. J.D. Zhang and S.N. Atluri, A boundary/interior element method for quasi static and transient response analysis of shallow shells, *Computers & Structures* **24**, pp. 213–223, (1986).
53. J.D. Zhang and S.N. Atluri, Post-buckling analysis of shallow shells by the field-boundary element method, *International Journal for Numerical Methods in Engineering* **26**, pp. 571–587, (1988).

Chapter 2

TIME-DOMAIN BEM TECHNIQUES

W.J. Mansur
Civil Engineering Programme
COPPE/Federal University of Rio de Janeiro
P.O. Box 68506, CEP 21945-970, Rio de Janeiro, Brazil
webe@coc.ufrj.br

J.A.M. Carrer
Post-Graduate Programme in Numerical Methods in Engineering
PPGMNE/Federal University of Paraná
P.O. Box 19011, CEP 81531-990, Curitiba, Paraná, Brazil
carrer@ufpr.br

P.E. Oyarzún
Civil Engineering Department, Austral University of Chile
CEP 5111187, Valdivia, Chile
pablooyarzun@uach.cl

This chapter discusses time-domain BEM (TD-BEM) techniques. Among the various formulations developed so far, special attention is given to the standard TD-BEM formulation, for which a modified scheme for time marching is presented, and to the so-called D-BEM formulation (D meaning domain), for which the time marching is carried out by employing not only the Houbolt scheme but also the Newmark scheme. Special attention is devoted to the development of a procedure that allows the computation of initial conditions contributions. Examples are included and discussed in order to verify the potentialities of the proposed developments; results are compared with analytical solutions.

2.1. Introduction

This chapter presents a discussion concerning different BEM formulations for time-dependent problems. The development of BEM formulations for solving time-dependent problems is a very attractive and challenging area of research. Many works concerned with this matter have been presented so far, increasing the range of applications of the BEM and the literature concerning numerical methods in engineering. For general purposes, when time appears explicitly in numerical analysis (in other words, when the response of a given problem is a time function), the BEM approaches can be classified in two main kinds: the first employs time-dependent fundamental solutions, and the second employs static fundamental solutions. Before proceeding, it is important to mention that although the use of static fundamental solutions may seem strange to some researchers, it demonstrates the versatility of the BEM in numerical analyses and, as expected when dealing with research work, also brings new challenges to researchers. After this comment, a discussion concerning the different BEM approaches shall be carried out.

The formulations that employ time-dependent fundamental solutions, designated by TD-BEM (TD meaning time-domain), are very elegant from the mathematical point of view, e.g. Mansur,[1] Dominguez,[2] Carrer and Mansur,[3] and Yu *et al.*[4] The fulfillment of the radiation condition makes the TD-BEM formulations suitable for infinite domain analyses. In addition, good representation of causality gives very accurate time response results. These advantageous characteristics, however, are counterbalanced by the high computational effort required to compute the time convolution integrals that appear in the TD-BEM integral equations. In order to overcome this difficulty, some compression algorithms concerning the reduction of computational cost in time integration were developed (Demirel and Wang,[5] Mansur and de Lima-Silva,[6] Soares and Mansur,[7] and Carrer and Mansur[8]). Focusing the attention on TD-BEM formulation, a modified approach for a 2-D scalar wave problem is presented in this chapter. The approach is quite simple: in the computation of the potential and its normal derivative at time t_n, the results related to the potential and its time derivative at time t_{n-1} play the role of 'initial conditions', thus generating a step-by-step procedure. As a consequence, due to the time translation property of the kernels,[1] the convolution integrals, which appear in the standard TD-BEM formulation to represent the whole time history, are no longer evaluated (as a matter of fact, time integration is restricted only to the first discrete interval $[0, \Delta t]$). As the potential time derivative plays a very important role in this formulation, it is computed through an appropriate integral equation, easily obtained by taking the time derivative of the potential integral equation. As the potential and its time derivative at time t_{n-1} play the role of 'initial conditions' for the potential and its normal and time derivatives at time t_n, the domain integrals (which in the standard TD-BEM formulation are restricted only to the part of the domain with non-homogeneous initial conditions) are now extended to the entire domain. One question can be formulated at this time: can this approach be used for infinite domain analysis? If results at late times are required, the answer is no: in this case the interruption of the cell mesh generates a fictitious boundary and, consequently, reflected waves whose influence tends to invalidate the results. An alternative in this case is to use the standard TD-BEM as a transmitting boundary, employing coupling algorithms similar to those presently employed in Finite Element Method-BEM coupling, e.g. Soares Jr. *et al.*[9] Two examples are presented at the end of the corresponding section, illustrating the potentialities of the proposed formulation and encouraging future research in this area.

The use of static fundamental solutions generates two lines of research: one that keeps the domain integrals in the equations of the method, designated D-BEM, e.g. Carrer and Mansur,[10] and Hatzigeorgiou and Beskos;[11] and one that transforms the domain integrals into boundary integrals by employing suitable interpolation functions, designated DR-BEM (DR meaning dual reciprocity), e.g. Kontoni and Beskos,[12] and Partridge *et al.*[13] As time does not appear explicitly in the integral equations, an approximation to the second order time derivative is required for the time marching. The Houbolt method[14] has become the most widely employed approximation. In spite of alternative time marching schemes recently proposed by Carrer and Mansur,[10] and Souza *et al.*,[15] the search for other approximations is a task that still deserves attention: among the various schemes presented in the finite element method literature, e.g. Bathe,[16] and Cook *et al.*,[17] the Newmark family is one of the methods that can be cited. As it is well known, the Newmark method[18] has been widely used in FEM formulations (more than the Houbolt method), and presents a better control of accuracy, according to the values of the parameters β and γ (see Bathe,[16] and Cook *et al.*[17]). For this reason, the Newmark scheme was implemented in the D-BEM formulation, the range of variation of the parameters β and γ being determined empirically. Note that for infinite domain analyses, if results at late times are required, the D-BEM formulation bears the same limitation as the FEM and any other domain discretization method; in spite of this disadvantage, it can provide reliable results, as demonstrated by the two examples included at the end of the corresponding section. For the solution of problems with non-homogeneous initial conditions, a general approach is presented.

Transformed-domain BEM formulations have also been the subject of intense research work and are mainly related either to the frequency-domain approach (e.g. Gaul and Wenzel,[19] and Mansur *et al.*[20]), or to the Laplace-domain approach (e.g. Gaul and Schanz[21]). When employing these formulations, the problem is initially solved in the transformed-domain for a suitable number of discrete values of the transform parameter. If required, the solution in the time-domain can be obtained by means of an inverse transformation procedure, e.g. Dubner and Abate,[22] and Durbin[23] for Laplace transform, and by means of the well known DFT (or FFT) algorithms for Fourier transforms. Another TD-BEM formulation, which has become popular in recent years, is the so-called *operational quadrature method* or *convolution quadrature method*, based on the works by Lubich,[24,25] e.g. Schanz and Antes,[26] Schanz,[27] Abreu *et al.*,[28,29] and Antes *et al.*[30] The main characteristics of this formulation, which can be applied to problems where time-domain fundamental solutions are not available, are: i) the use of fundamental solutions in the Laplace transformed domain and ii) the numerical approximation of the time convolution process, presented in the TD-BEM equations, by a quadrature formula based on a linear multistep method, which provides direct solution in the time-domain. For a complete discussion concerning dynamic analysis by the BEM until 1996, the reader may refer to Beskos.[31]

2.2. Standard TD-BEM Integral Equations

The TD-BEM equations can be written by employing either the kernel regularization procedure[1] or the concept of finite part of integrals (FPI), formulated by Hadamard.[32] If time integration is performed analytically, as is usual in TD-BEM formulations, the resulting time integrated kernels from both representations are the same,[33] that is, both representations are entirely equivalent. The use of Hadamard's concept, however, leads to more compact expressions and provides an elegant representation of the equations involved in the analysis. A brief summary of the TD-BEM equations is given below.

2.2.1. *Potential integral equation*

Time-domain integral representation of the 2-D scalar wave propagation: problem is written as[34]:

$$\begin{aligned}4\pi c(\xi)u(\xi,t) = &\int_\Gamma\int_0^{t^+} u^*(X,t,\xi,\tau)p(X,\tau)d\tau d\Gamma(X)\\
&-\int_\Gamma \frac{\partial r}{\partial n}\fint_0^{t^+} u_r^*(X,t,\xi,\tau)u(X,\tau)d\tau d\Gamma(X)\\
&-\frac{1}{c}\int_\Gamma \frac{\partial r}{\partial n}u_0^*(X,t,\xi)u_0(X)d\Gamma(X)\\
&+\frac{1}{c^2}\int_\Omega u_0^*(X,t,\xi)v_0(X)d\Omega(X)\\
&+\frac{1}{c}\int_\Omega u_0^*(X,t,\xi)\frac{\partial u_0(X)}{\partial r}d\Omega(X)\\
&+\frac{1}{c}\int_\Omega\left[\frac{u_0^*(X,t,\xi)}{r} - b_0^*(X,t,\xi)\right]u_0(X)d\Omega(X)\end{aligned} \tag{2.1}$$

In Eq. (2.1), $u(X,\tau)$ is the potential function and $p(X,\tau)$ is the flux, Γ is the boundary and Ω the domain (or the part of the domain that presents non-homogeneous initial conditions). The coefficient $c(\xi)$ adopts the same values used in the static case: $c(\xi) = 1$ for $\xi \in \Omega$ and $c(\xi) = \alpha/2\pi$ for $\xi \in \Gamma$ (α is the internal angle in Fig. 2.1); and the subscript 0 means $\tau = 0$.

The fundamental solution, $u^*(X,t,\xi,\tau)$, that corresponds to the effect at a field point X at time t of a source represented by a unit impulse applied at a source point ξ at time τ, reads:

$$u^*(X,t,\xi,\tau) = U^*(X,t,\xi,\tau)H[c(t-\tau)-r] \tag{2.2}$$

where:

$$U^*(X,t,\xi,\tau) = \frac{2c}{\sqrt{c^2(t-\tau)^2 - r^2}} \tag{2.3}$$

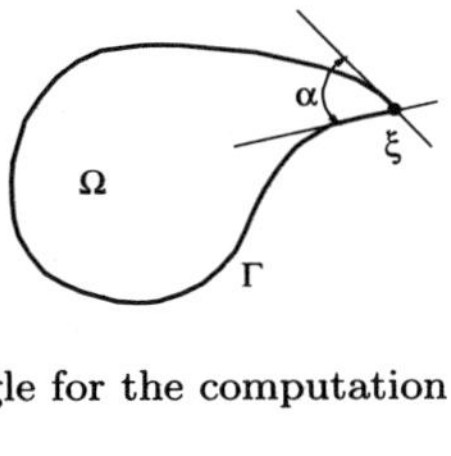

Fig. 2.1. Internal angle for the computation of the coefficient $c(\xi)$.

In Eq. (2.2), c is the wave propagation velocity, $H[c(t-\tau)-r]$ stands for the Heaviside function and $r = r(X,\xi)$ is the distance between the field and source points.

The functions $b^*(X,t,\xi,\tau)$ and $u_r^*(X,t,\xi,\tau)$, in Eq. (2.1), are given by:

$$b^*(X,t,\xi,\tau) = B^*(X,t,\xi,\tau)H[c(t-\tau)-r] \tag{2.4}$$

where:

$$B^*(X,t,\xi,\tau) = \frac{2c[c(t-\tau)-r]}{[c^*(t-\tau)^2 - r^2]^{3/2}} \tag{2.5}$$

and

$$u_r^*(X,t,\xi,\tau) = U_r^*(X,t,\xi,\tau)H[c(t-\tau)-r] \tag{2.6}$$

by considering:

$$U_r^*(X,t,\xi,\tau) = \frac{2cr}{[c^2(t-\tau)^2 - r^2]^{3/2}} \tag{2.7}$$

The FPI[32] on the second term on the right-hand side of Eq. (2.1) is given by:

$$\begin{aligned}
&\fint_0^{t^+} u_r^*(X,t,\xi,\tau)u(X,\tau)d\tau \\
&\quad = \lim_{\tau\to t-r/c}\left\{\int_0^{\tau} U_r^*(X,t,\xi,\tau)u(X,\tau)d\tau - \frac{1}{c}U^*(X,t,\xi,\tau)u(X,\tau)\right\}
\end{aligned} \tag{2.8}$$

2.2.2. *Numerical solution*

Referring to the solution of the described problem, linear and constant time variations can be assumed, respectively, for the potential and its normal derivative. Hence, time integrals are usually computed in analytic form[1,34] and linear elements and linear triangular cells are employed for the boundary and the domain discretizations[1,10] respectively.

In the standard TD-BEM formulation, Eq. (2.1) is applied to all boundary nodes and, after solving the resulting system of equations, it can be applied to internal points. In the modified version of the standard TD-BEM formulation (previously mentioned at the Introduction to this chapter and to be discussed in Section 2.3), denoted here as SSTD-BEM (SS meaning step-by-step), Eq. (2.1) is applied simultaneously to all boundary nodes and internal points, generating the enlarged system of equations written below:

$$\begin{bmatrix} \mathbf{C} & \mathbf{0} \\ \mathbf{0} & \mathbf{I} \end{bmatrix} \begin{Bmatrix} \mathbf{u}_n^b \\ \mathbf{u}_n^d \end{Bmatrix} + \sum_{m=0}^{n} \begin{bmatrix} \mathbf{H}_{nm}^{bb} & \mathbf{0} \\ \mathbf{H}_{nm}^{db} & \mathbf{0} \end{bmatrix} \begin{Bmatrix} \mathbf{u}_m^b \\ \mathbf{u}_m^d \end{Bmatrix}$$
$$= \sum_{m=0}^{n} \begin{bmatrix} \mathbf{G}_{nm}^{bb} \\ \mathbf{G}_{nm}^{db} \end{bmatrix} \{\mathbf{p}_m^b\} + \begin{bmatrix} \mathbf{U}_n^{bb} & \mathbf{U}_n^{bd} \\ \mathbf{U}_n^{db} & \mathbf{U}_n^{dd} \end{bmatrix} \begin{Bmatrix} \mathbf{u}_0^b \\ \mathbf{u}_0^d \end{Bmatrix} + \begin{bmatrix} \mathbf{V}_n^{bb} & \mathbf{V}_n^{bd} \\ \mathbf{V}_n^{db} & \mathbf{V}_n^{dd} \end{bmatrix} \begin{Bmatrix} \mathbf{v}_0^b \\ \mathbf{v}_0^d \end{Bmatrix} \tag{2.9}$$

In Eq. (2.9), the superscripts b and d correspond to the boundary and to the domain (internal points), respectively. In the sub-matrices, the first and second superscripts correspond, respectively, to the positions of the source and of the field points. In the first matrix on the left-hand side of Eq. (2.9), the $\mathbf{C}$ matrix is related to the coefficients $c(\xi)$ of the boundary nodes whereas the identity matrix, $\mathbf{I} = \mathbf{H}^{dd}$, is related to the coefficients $c(\xi) = 1$ of the internal points. Equation (2.9) can be written in a more compact way as:

$$\mathbf{C}\mathbf{u}_n + \sum_{m=0}^{n} \mathbf{H}_{nm}\mathbf{u}_m = \sum_{m=0}^{n} \mathbf{G}_{nm}\mathbf{p}_m + \mathbf{U}_n\mathbf{u}_0 + \mathbf{V}_n\mathbf{v}_0 \tag{2.10}$$

2.2.3. *Velocity integral equation*

The integral equation associated with the time derivative of the potential, to be used in the step-by-step procedure, is obtained by taking the time derivative of Eq. (2.1) and can be written as[34]:

$$\begin{aligned} 4\pi c(\xi)\frac{\partial u(\xi,t)}{\partial t} &= \int_\Gamma \fint_0^{t^+} u_t^*(X,t,\xi,\tau)p(X,\tau)d\tau d\Gamma(X) \\ &\quad - \int_\Gamma \frac{\partial}{\partial t}\left(\fint_0^{t^+} u_r^*(X,t,\xi,\tau)u(X,\tau)d\tau\right)\frac{\partial r}{\partial n}d\Gamma(X) \\ &\quad - \frac{1}{c}\int_\Gamma \frac{\partial r}{\partial n}\frac{\partial u_0^*(X,t,\xi)}{\partial t}u_0(X)d\Gamma(X) \\ &\quad + \frac{\partial}{\partial t}\left(I_{\Omega v_0} + I_{\Omega u_0}\right) \end{aligned} \tag{2.11}$$

where the time derivative of the domain integrals related to the non-homogeneous initial conditions is only indicated by the last term on the right-hand side of Eq. (2.11), because of the fact that the domain integrals do not require any special treatment.

The FPI in the first term on the right-hand side of Eq. (2.11) is given by the following expression:

$$\begin{aligned}&\;⨎_0^{t^+} u_t^*(X,t,\xi,\tau)p(X,\tau)d\tau\\&= \lim_{\tau\to t-r/c}\left\{\int_0^{\tau} U_t^*(X,t,\xi,\tau)p(X,\tau)d\tau + U^*(X,t,\xi,\tau)p(X,\tau)\right\}\end{aligned} \tag{2.12}$$

The function $u_t^*(X,t,\xi,\tau)$ in Eq. (2.12) is given by:

$$u_t^*(X,t,\xi,\tau) = U_t^*(X,t,\xi,\tau)H[c(t-\tau)-r] \tag{2.13}$$

with:

$$U_t^*(X,t,\xi,\tau) = -\frac{2c^3(t-\tau)}{[c^2(t-\tau)^2 - r^2]^{3/2}} \tag{2.14}$$

The time derivative of the FPI on the second term on the right-hand side of Eq. (2.11) is defined as:

$$\begin{aligned}&\frac{\partial}{\partial t}\left(⨎_0^{t^+} u_r^*(X,t,\xi,\tau)u(X,\tau)d\tau\right)\\&= \lim_{\tau\to t-r/c}\left\{\int_0^{\tau}\frac{\partial U_r^*(X,t,\xi,\tau)}{\partial t}u(X,\tau)d\tau + U_r^*(X,t,\xi,\tau)u(X,\tau)\right.\\&\quad\left.-\frac{1}{c}U^*(X,t,\xi,\tau)\frac{\partial u(X,\tau)}{\partial\tau}\right\}\end{aligned} \tag{2.15}$$

The system of equation that corresponds to the application of Eq. (2.11) to all boundary nodes and internal points can be represented, in a compact form, as:

$$\mathbf{C}\mathbf{v}_n = \sum_{m=0}^{n}\overline{\mathbf{G}}_{nm}\mathbf{p}_m - \sum_{m=0}^{n}\overline{\mathbf{H}}_{nm}\mathbf{u}_m + \overline{\mathbf{U}}_n\mathbf{u}_0 + \overline{\mathbf{V}}_n\mathbf{v}_0 \tag{2.16}$$

After the solution of Eq. (2.10), Eq. (2.16) can easily be solved for $\mathbf{v}_n$, since matrix $\mathbf{C}$ is a diagonal matrix.

2.3. Step-by-Step Procedure: SSTD-BEM Formulation

The step-by-step procedure was developed as an alternative approach to the standard TD-BEM. The procedure is based on the time translation property,[1] which can be written as:

$$u^*(X, t, \xi, \tau) = u^*(X, t + \Delta t, \xi, \tau + \Delta t) \tag{2.17}$$

This property, illustrated by Eq. (2.17), allows the time-dependent problem to be solved by considering the results related to $\mathbf{u}_{n-1}$ and $\mathbf{v}_{n-1}$ as 'initial conditions' in the computation of the results at time t_n. That is, in the computation of $\mathbf{u}_n$, $\mathbf{p}_n$ and $\mathbf{v}_n$, the convolution integrals in Eq. (2.1) and Eq. (2.11), represented by the summation symbol in Eq. (2.10) and Eq. (2.16), are no longer evaluated; the following versions of Eq. (2.10) and Eq. (2.16) can be written:

$$\mathbf{C}\mathbf{u}_n + \mathbf{H}_{nn}\mathbf{u}_n = \mathbf{G}_{nn}\mathbf{p}_n + \mathbf{U}_n\mathbf{u}_{n-1} + \mathbf{V}_n\mathbf{v}_{n-1} \tag{2.18}$$

$$\mathbf{C}\mathbf{v}_n = \overline{\mathbf{G}}_{nn}\mathbf{p}_n - \overline{\mathbf{H}}_{nn}\mathbf{u}_n + \overline{\mathbf{U}}_n\mathbf{u}_{n-1} + \overline{\mathbf{V}}_n\mathbf{v}_{n-1} \tag{2.19}$$

The known values of the 'initial conditions' are stored into the vectors $\mathbf{u}_{n-1}$ and $\mathbf{v}_{n-1}$. Therefore, the entire domain needs to be discretized. This feature generates the main limitation of the proposed approach, so it is suitable for finite domain applications and for infinite domains analyses only if early time responses are requested or if a transmitting boundary is employed.

2.3.1. *Examples*

2.3.1.1. *One-dimensional rod under a Heaviside-type forcing function*

This is the classical example of a one-dimensional rod under a Heaviside-type forcing function, applied instantaneously at $t = 0$ and kept constant in time. Figure 2.2 depicts the boundary conditions and geometry.

The analytical solution for this problem, by considering the boundary conditions depicted in Fig. 2.2, is given by[35]:

$$u(x,t) = \frac{Pa}{E}\left[\frac{x}{a} + \frac{8}{\pi^2}\sum_{n=1}^{\infty}\frac{(-1)^n}{(2n-1)^2} \times \sin\left(\frac{(2n-1)\pi x}{2a}\right)\cos\left(\frac{(2n-1)\pi ct}{2a}\right)\right] \tag{2.20}$$

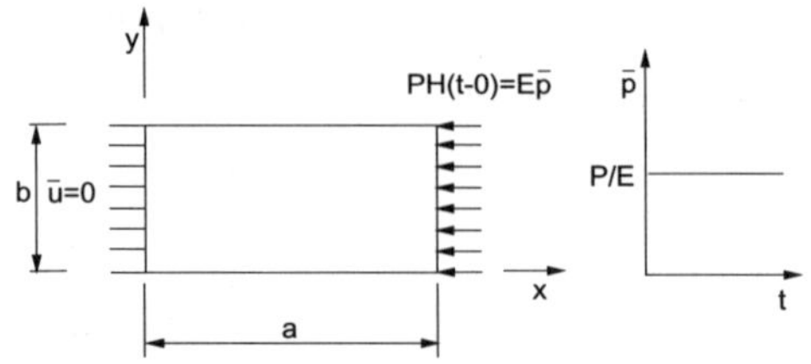

Fig. 2.2. One-dimensional rod under a Heaviside-type forcing function: geometry and boundary conditions.

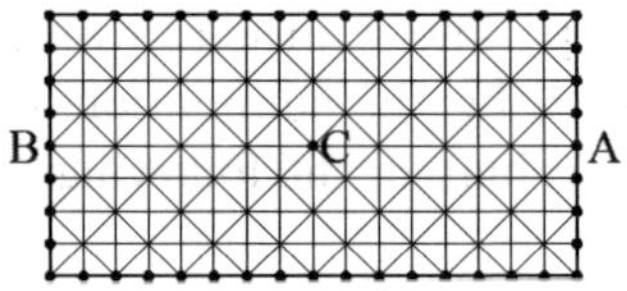

Fig. 2.3. Boundary and domain discretizations: basic mesh with selected nodes.

With the purpose of verifying the accuracy and convergence of the numerical results, according to the available analytical solutions, three analyses are presented with different mesh refinements. The poorest of the three meshes employs 48 boundary elements and 256 internal cells in accordance with the Fig. 2.3. Subsequent meshes were obtained by duplicating the number of boundary elements and by quadrupling the number of internal cells. In this way, the second analysis was carried out with 96 boundary elements and 1,024 cells and the third analysis, with 192 elements and 4,096 cells. As is usual in time-domain BEM analyses, the dimensionless parameter[1] $\beta_{\Delta t} = c\Delta t/l$ (which relates the wave velocity c with the smallest element length (l) and the time interval Δt assumed constant) is used to estimate the time-step length. It is important to mention that, for the proposed approach, the optimum time-step length was obtained using $\beta_{\Delta t} = 0.5$. One must observe that this is not the value recommended by Mansur[1] for the standard TD-BEM formulation ($\beta_{\Delta t} = 0.6$).

Results concerning the potential at the boundary node A(a,b/2) are compared with the analytical solution in Fig. 2.4. In Fig. 2.5 the results related to the velocity $v = \partial u/\partial t$, at the boundary node A(a,b/2), are compared with its analytical solution obtained from the time derivative of Eq. (2.20). In Fig. 2.6 the results related to the flux ($p = \partial u/\partial n$) at the boundary node B(0,b/2) are compared with the analytical solution,

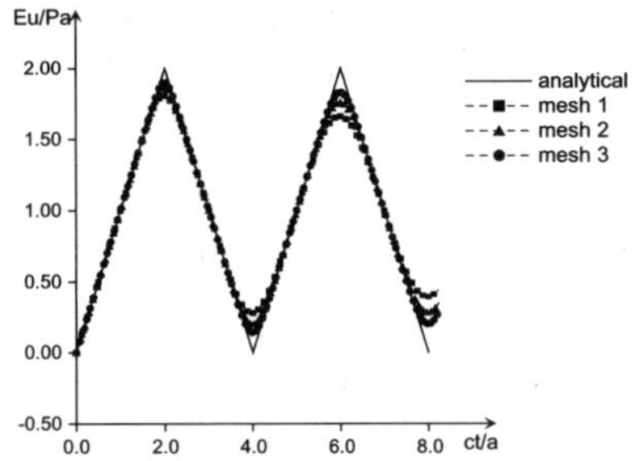

Fig. 2.4. Rod under a Heaviside-type forcing function: potential at A(a,b/2).

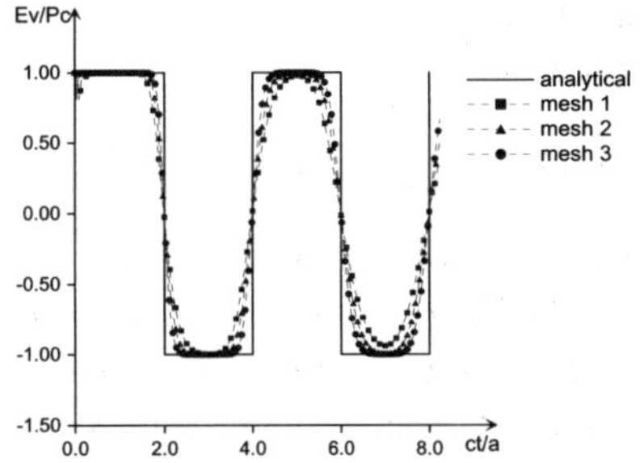

Fig. 2.5. Rod under a Heaviside-type forcing function: velocity at A(a,b/2).

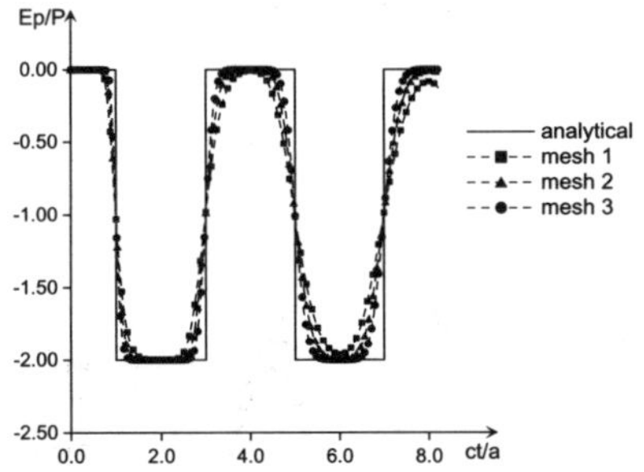

Fig. 2.6. Rod under a Heaviside-type forcing function: flux at node B(0,b/2).

obtained from the normal derivative of Eq. (2.20). Numerical damping is observed in the results of the three analyses, which diminishes with the refinement of the mesh. Although numerical damping also appears in the standard TD-BEM analyses,[1] its effect is not so pronounced there.

For the proposed approach, more refined meshes become necessary in order to achieve good results (in the sense of those furnished by the standard TD-BEM).

2.3.1.2. *One-dimensional rod under sinusoidal initial conditions*

In this example, the same rod depicted previously, now fixed at both ends, is analysed when subjected to initial conditions described below:

In Analysis (a) one has: $u_0 = b_0 \sin(\pi x/a)$ and $v_0 = 0$. The analytical solution[35] to this problem is:

$$u(x,t) = b_0 \cos(\pi ct/a) \sin(\pi x/a) \tag{2.21}$$

In Analysis (b) one has: $v_0 = d_0 \sin(\pi x/a)$ and $u_0 = 0$. The analytical solution[35] to this problem is:

$$u(x,t) = d_0 \frac{a}{\pi c} \sin(\pi ct/a) \sin(\pi x/a) \tag{2.22}$$

The three meshes adopted to perform analyses (a) and (b) are the same as in the previous example. By considering Analysis (a), the results related to the potential and the velocity at point C of coordinates (a/2,b/2) are depicted in Figs. 2.7 and 2.8, respectively. The results related to the flux at point A(a,b/2) are depicted in Fig. 2.9. The results corresponding to the Analysis (b) are presented in Figs. 2.10–2.12, in the same sequence that they were presented for Analysis (a). As in the first example, one can observe that the mesh refinement is followed by a reduction in the numerical damping.

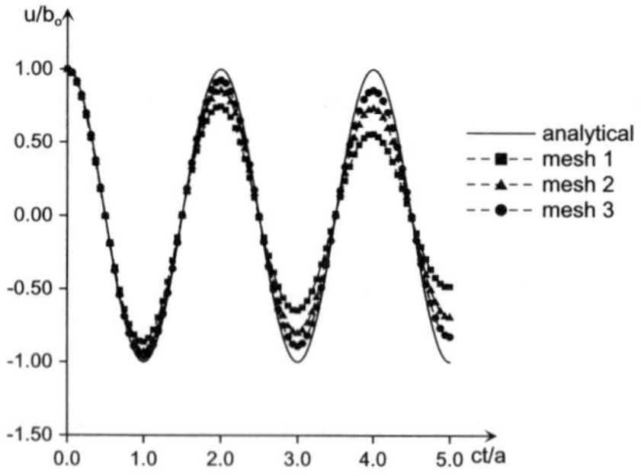

Fig. 2.7. Rod under sinusoidal initial conditions: potential at C(a/2,b/2), Analysis (a).

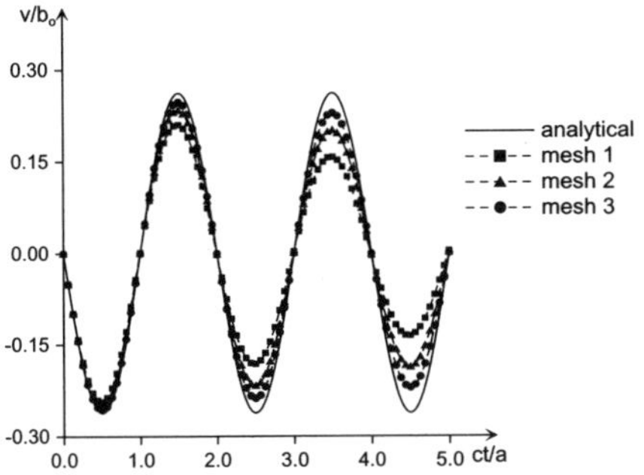

Fig. 2.8. Rod under sinusoidal initial conditions: velocity at C(a/2,b/2), Analysis (a).

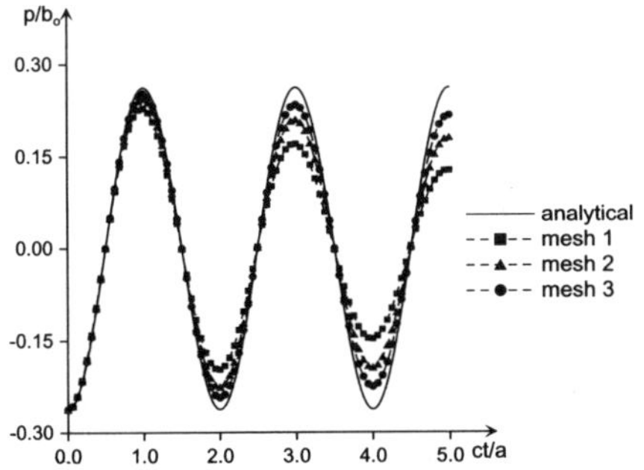

Fig. 2.9. Rod under sinusoidal initial conditions: flux at A(a,b/2), Analysis (a).

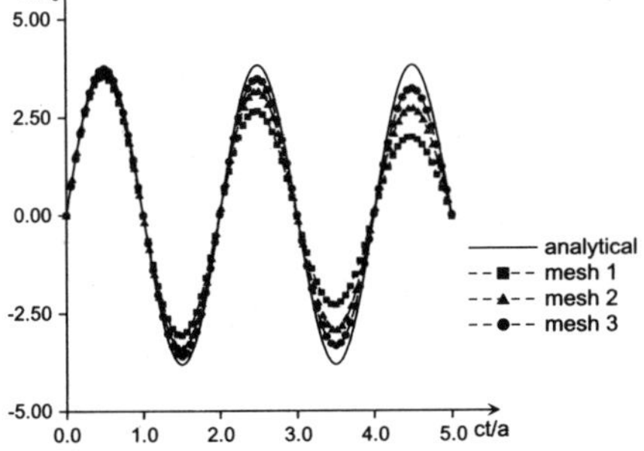

Fig. 2.10. Rod under sinusoidal initial conditions: potential at C(a/2,b/2), Analysis (b).

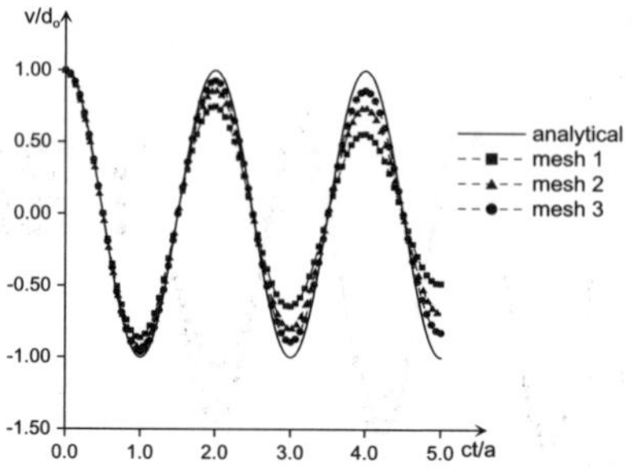

Fig. 2.11. Rod under sinusoidal initial conditions: velocity at C(a/2,b/2), Analysis (b).

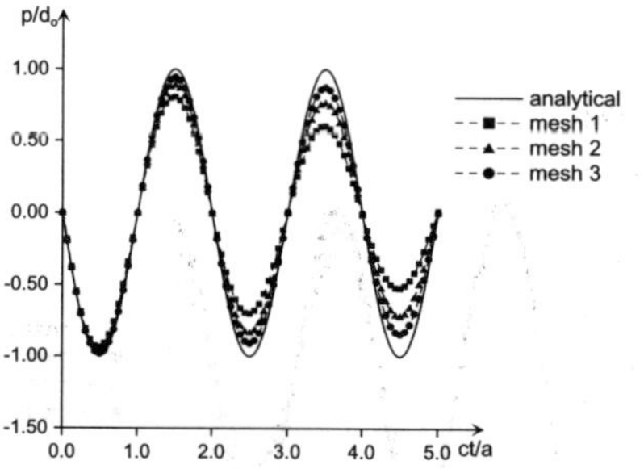

Fig. 2.12. Rod under sinusoidal initial conditions: flux at A(a,b/2), Analysis (b).

2.4. D-BEM Formulation

The starting D-BEM equation is written as follows:

$$\begin{aligned} c(\xi)u(\xi,t) &= \int_\Gamma u^*(\xi,X)p(X,t)d\Gamma(X) \\ &\quad - \int_\Gamma p^*(\xi,X)u(X,t)d\Gamma(X) - \frac{1}{c^2}\int_\Omega u^*(\xi,X)\ddot{u}(X,t)d\Omega(X) \end{aligned} \tag{2.23}$$

The symbols used in Eq. (2.23) are the same as those used in Eq. (2.1). Now, the two-dimensional fundamental solution, $u^*(\xi,X)$, is given by:

$$u^*(\xi,X) = \frac{1}{2\pi}\ln\left(\frac{1}{r}\right) \tag{2.24}$$

The domain integral that involves the second order time derivative $\ddot{u}$ is maintained in the subsequent developments, giving rise to the D-BEM formulation. The search for an adequate approximation of the acceleration plays an essential role in the success of the method.

2.4.1. *Numerical procedure*

As mentioned in Section 2.1, after the spatial discretization is accomplished, Eq. (2.23) is applied to all nodes generating an enlarged system of equations, written below, for which most of the notation employed before in Section 2.1 remains valid:

$$\begin{bmatrix} \mathbf{H}^{bb} & \mathbf{0} \\ \mathbf{H}^{db} & \mathbf{I} \end{bmatrix} \begin{Bmatrix} \mathbf{u}^{b}_{n+1} \\ \mathbf{u}^{d}_{n+1} \end{Bmatrix} = \begin{bmatrix} \mathbf{G}^{bb} \\ \mathbf{G}^{db} \end{bmatrix} \{\mathbf{p}^{b}_{n+1}\} - \frac{1}{c^2} \begin{bmatrix} \mathbf{M}^{bb} & \mathbf{M}^{bd} \\ \mathbf{M}^{db} & \mathbf{M}^{dd} \end{bmatrix} \begin{Bmatrix} \ddot{\mathbf{u}}^{b}_{n+1} \\ \ddot{\mathbf{u}}^{d}_{n+1} \end{Bmatrix} \tag{2.25}$$

In Eq. (2.25), in order to simplify the notation, the subscript $(n+1)$ represents the time $t_{n+1} = (n+1)\Delta t$, where Δt is the selected time interval. It is important to note that the assembly of such an enlarged system of equations is necessary because the domain integral relates boundary values to domain values. This remark is confirmed by matrix $\mathbf{M}^{bd}$ in Eq. (2.25).

2.4.1.1. *Time marching with the Houbolt method*

The Houbolt method[14] is obtained by performing cubic Lagrange interpolation of $\mathbf{u} = \mathbf{u}(t)$ from time $t_{n-2} = (n-2)\Delta t$ to time $t_{n+1} = (n+1)\Delta t$. Exact differentiation with respect to time gives the approximations for the velocity and acceleration described below:

$$\dot{\mathbf{u}}_{n+1} = \frac{1}{6\Delta t}\left[11\mathbf{u}_{n+1} - 18\mathbf{u}_n + 9\mathbf{u}_{n-1} - 2\mathbf{u}_{n-2}\right] \tag{2.26}$$

$$\ddot{\mathbf{u}}_{n+1} = \frac{1}{\Delta t^2}\left[2\mathbf{u}_{n+1} - 5\mathbf{u}_n + 4\mathbf{u}_{n-1} - \mathbf{u}_{n-2}\right] \tag{2.27}$$

After substituting Eq. (2.27) in Eq. (2.25), one can write:

$$\begin{bmatrix} ((c\Delta t)^2\mathbf{H}^{bb} + 2\mathbf{M}^{bb}) & 2\mathbf{M}^{bd} \\ ((c\Delta t)^2\mathbf{H}^{db} + 2\mathbf{M}^{db}) & ((c\Delta t)^2\mathbf{I} + 2\mathbf{M}^{dd}) \end{bmatrix} \begin{Bmatrix} \mathbf{u}^{b}_{n+1} \\ \mathbf{u}^{d}_{n+1} \end{Bmatrix}$$
$$= \begin{bmatrix} (c\Delta t)^2\mathbf{G}^{bb} \\ (c\Delta t)^2\mathbf{G}^{db} \end{bmatrix} \{\mathbf{p}^{b}_{n+1}\} - \begin{bmatrix} \mathbf{M}^{bb} & \mathbf{M}^{bd} \\ \mathbf{M}^{db} & \mathbf{M}^{dd} \end{bmatrix} \begin{Bmatrix} -5\mathbf{u}^{b}_{n} + 4\mathbf{u}^{b}_{n-1} - \mathbf{u}^{b}_{n-2} \\ -5\mathbf{u}^{d}_{n} + 4\mathbf{u}^{d}_{n-1} - \mathbf{u}^{d}_{n-2} \end{Bmatrix} \tag{2.28}$$

Equation (2.28) can be represented in a concise manner as:

$$\overline{\mathbf{H}}\mathbf{u}_{n+1} = \overline{\mathbf{G}}\mathbf{p}_{n+1} + \mathbf{g}_n \tag{2.29}$$

The contributions of the previous times are stored in the vector $\mathbf{g}_n$. After the boundary conditions are imposed, the final system of equations that arise from Eq. (2.28) can be solved.

The contribution of the Initial Conditions with Houbolt Method
In the Houbolt method,[14] the computation of the velocity and acceleration at time $t_{n+1} = (n+1)\Delta t$ requires the knowledge of the values of $\mathbf{u}$ between the times $t_{n-2} = (n-2)\Delta t$ and $t_{n+1} = (n+1)\Delta t$. At the beginning of the time marching process, $n = 0$ and consequently the values $\mathbf{u}_{-2}$ and $\mathbf{u}_{-1}$ must be computed appropriately in order to provide enough accuracy in the analysis.

For the determination of $\mathbf{u}_{-1}$, $\dot{\mathbf{u}}_0$ is computed by employing the forward and the backward finite difference formulae at $t = 0$, respectively. Assuming that the resulting finite difference expressions are equal:

$$\dot{\mathbf{u}}_0 = \frac{\mathbf{u}_1 - \mathbf{u}_0}{\Delta t} = \frac{\mathbf{u}_0 - \mathbf{u}_{-1}}{\Delta t} \tag{2.30}$$

Solving Eq. (2.30) for $\mathbf{u}_1$, gives:

$$\mathbf{u}_1 = 2\mathbf{u}_0 - \mathbf{u}_{-1} \tag{2.31}$$

One can also assume that $\dot{\mathbf{u}}_0$ can be computed by employing a central finite difference formula, which gives:

$$\dot{\mathbf{u}}_0 = \frac{\mathbf{u}_1 - \mathbf{u}_{-1}}{2\Delta t} \tag{2.32}$$

Solving Eq. (2.32) for $\mathbf{u}_1$:

$$\mathbf{u}_1 = 2\Delta t\dot{\mathbf{u}}_0 + \mathbf{u}_{-1} \tag{2.33}$$

From (2.31) and (2.33) one has $\mathbf{u}_{-1}$ as a function of $\dot{\mathbf{u}}_0$ and $\mathbf{u}_0$:

$$\mathbf{u}_{-1} = \mathbf{u}_0 - \Delta t\dot{\mathbf{u}}_0 \tag{2.34}$$

Now, for the determination of $\mathbf{u}_{-2}$, initially $\dot{\mathbf{u}}_{-1}$ is computed by employing the forward and the backward finite difference formulae at $t = -\Delta t$, respectively. Assuming that the resulting finite difference expressions are equal:

$$\dot{\mathbf{u}}_{-1} = \frac{\mathbf{u}_0 - \mathbf{u}_{-1}}{\Delta t} = \frac{\mathbf{u}_{-1} - \mathbf{u}_{-2}}{\Delta t} \tag{2.35}$$

Solving Eq. (2.35) for $\mathbf{u}_{-2}$:

$$\mathbf{u}_{-2} = 2\mathbf{u}_{-1} - \mathbf{u}_0 \tag{2.36}$$

Substituting (2.34) in (2.36) one obtains the value of $\mathbf{u}_{-2}$ as a function of $\dot{\mathbf{u}}_0$ and $\mathbf{u}_0$:

$$\mathbf{u}_{-2} = \mathbf{u}_0 - 2\Delta t\dot{\mathbf{u}}_0 \tag{2.37}$$

2.4.1.2. *Time marching with the Newmark method*

For the Newmark family methods,[18] the approximations are:

$$\dot{\mathbf{u}}_{n+1} = \frac{\gamma}{\beta\Delta t}[\mathbf{u}_{n+1} - \mathbf{u}_n] + \frac{(\beta-\gamma)}{\beta}\dot{\mathbf{u}}_n - \frac{(\gamma-2\beta)}{2\beta}\Delta t\ddot{\mathbf{u}}_n \tag{2.38}$$

$$\ddot{\mathbf{u}}_{n+1} = \frac{1}{\beta\Delta t^2}[\mathbf{u}_{n+1} - \mathbf{u}_n] - \frac{1}{\beta\Delta t}\dot{\mathbf{u}}_n - \frac{(1-2\beta)}{2\beta}\ddot{\mathbf{u}}_n \tag{2.39}$$

The stability and accuracy of the Newmark method depend on the correct choice of the parameters β and γ. Some remarks concerning the Newmark method are useful: i) the method is unstable for $\gamma < 1/2$; ii) unconditional stability occurs when $2\beta \geq \gamma \geq 1/2$; iii) taking $\gamma > 1/2$ introduces artificial damping but reduces the accuracy of the method to first order. If one takes $\beta = 1/6$ and $\gamma = 1/2$, Eq. (2.38) and Eq. (2.39) correspond to the linear acceleration method. For $\beta = 1/4$ and $\gamma = 1/2$, Eq. (2.38) and Eq. (2.39) correspond to the average acceleration method. These observations can be found in FEM textbooks, e.g. Bathe,[16] and Cook *et al.*[17] However, for the D-BEM formulation developed here, reliable results were obtained only with the adoption of experimentally determined values for β and γ.

After substituting Eq. (2.39) into Eq. (2.25), one has:

$$\begin{aligned}
&\begin{bmatrix} (\beta(c\Delta t)^2\mathbf{H}^{bb} + \mathbf{M}^{bb}) & \mathbf{M}^{bd} \\ (\beta(c\Delta t)^2\mathbf{H}^{db} + \mathbf{M}^{db}) & (\beta(c\Delta t)^2\mathbf{I} + \mathbf{M}^{dd}) \end{bmatrix} \begin{Bmatrix} \mathbf{u}^b_{n+1} \\ \mathbf{u}^d_{n+1} \end{Bmatrix} \\
&\quad = \begin{bmatrix} \beta(c\Delta t)^2\mathbf{G}^{bb} \\ \beta(c\Delta t)^2\mathbf{G}^{db} \end{bmatrix} \{\mathbf{p}^b_{n+1}\} + \begin{bmatrix} \mathbf{M}^{bb} & \mathbf{M}^{bd} \\ \mathbf{M}^{db} & \mathbf{M}^{dd} \end{bmatrix} \\
&\qquad \times \begin{Bmatrix} \mathbf{u}^b_n + \Delta t\dot{\mathbf{u}}^b_n + \frac{(1-2\beta)}{2}\Delta t^2\ddot{\mathbf{u}}^b_n \\ \mathbf{u}^d_n + \Delta t\dot{\mathbf{u}}^d_n + \frac{(1-2\beta)}{2}\Delta t^2\ddot{\mathbf{u}}^d_n \end{Bmatrix}
\end{aligned} \tag{2.40}$$

Equation (2.40) can be represented in a concise manner as:

$$\overline{\mathbf{H}}\mathbf{u}_{n+1} = \overline{\mathbf{G}}\mathbf{p}_{n+1} + \mathbf{g}_n \tag{2.41}$$

The contributions from previous times are stored in vector $\mathbf{g}_n$.

The contribution of the Initial Conditions with Newmark Method
At the beginning of the time marching process, by taking $n = 0$ in Eq. (2.38) and in Eq. (2.39) one can write:

$$\dot{\mathbf{u}}_1 = \frac{\gamma}{\beta\Delta t}\left[\mathbf{u}_1 - \mathbf{u}_0\right] + \frac{(\beta - \gamma)}{\beta}\dot{\mathbf{u}}_0 - \frac{\gamma - 2\beta}{2\beta}\Delta t\ddot{\mathbf{u}}_0 \tag{2.42}$$

and

$$\ddot{\mathbf{u}}_1 = \frac{1}{\beta\Delta t^2}\left[\mathbf{u}_1 - \mathbf{u}_0\right] - \frac{1}{\beta\Delta t}\dot{\mathbf{u}}_0 - \frac{(1 - 2\beta)}{2\beta}\ddot{\mathbf{u}}_0 \tag{2.43}$$

In Eqs. (2.42) and (2.43), $\ddot{\mathbf{u}}_0$ is the only unknown; in order to determine this value, one can assume that:

$$\ddot{\mathbf{u}}_0 = \frac{\dot{\mathbf{u}}_1 - \dot{\mathbf{u}}_0}{\Delta t} \tag{2.44}$$

After solving Eq. (2.44) for $\dot{\mathbf{u}}_1$ and substituting the resulting expression in (2.42), one has:

$$\ddot{\mathbf{u}}_0 = \frac{2}{\Delta t^2}\left[\mathbf{u}_1 - \mathbf{u}_0\right] - \frac{2}{\Delta t}\dot{\mathbf{u}}_0 \tag{2.45}$$

The substitution of Eq. (2.45) into Eq. (2.43) produces the following expression:

$$\ddot{\mathbf{u}}_1 = \frac{2}{\Delta t^2}\left[\mathbf{u}_1 - \mathbf{u}_0\right] - \frac{2}{\Delta t}\dot{\mathbf{u}}_0 \tag{2.46}$$

Bearing Eq. (2.46) in mind, Eq. (2.40) at the first time step ($n = 0$) can be particularized and written accordingly:

$$\begin{bmatrix} ((c\Delta t)^2\mathbf{H}^{bb} + 2\mathbf{M}^{bb}) & 2\mathbf{M}^{bd} \\ ((c\Delta t)^2\mathbf{H}^{db} + 2\mathbf{M}^{db}) & ((c\Delta t)^2\mathbf{I} + 2\mathbf{M}^{dd}) \end{bmatrix} \begin{Bmatrix} \mathbf{u}_1^b \\ \mathbf{u}_1^d \end{Bmatrix}$$
$$= \begin{bmatrix} (c\Delta t)^2\mathbf{G}^{bb} \\ (c\Delta t)^2\mathbf{G}^{db} \end{bmatrix} \left\{\mathbf{p}_1^b\right\} + \begin{bmatrix} 2\mathbf{M}^{bb} & 2\mathbf{M}^{bd} \\ 2\mathbf{M}^{db} & 2\mathbf{M}^{dd} \end{bmatrix} \begin{Bmatrix} \mathbf{u}_0^b + \Delta t\dot{\mathbf{u}}_0^b \\ \mathbf{u}_0^d + \Delta t\dot{\mathbf{u}}_0^d \end{Bmatrix} \tag{2.47}$$

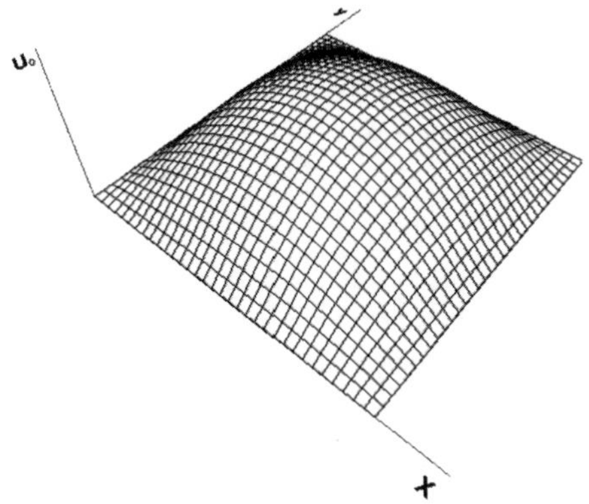

Fig. 2.13. Square membrane under initial displacement field over the entire domain.

2.4.2. *Examples*

2.4.2.1. *Square Membrane under prescribed initial displacement over the entire domain*

This example deals with a square membrane subjected to the following initial conditions distribution applied over the domain $0 \le x \le a$, $0 \le y \le a$ (Fig. 2.13):

$$u_0(x, y) = Ux(x - a)y(y - a); v_0(x, y) = 0 \tag{2.48}$$

The general analytical solution[36] for this problem, by considering a rectangular membrane with dimensions a and b, is:

$$u(x, y, t) = \frac{64Ua^2b^2}{\pi^6} \sum_{m=1}^{\infty} \sum_{n=1}^{\infty} \frac{1}{m^3n^3} \sin\left(\frac{m\pi x}{a}\right) \sin\left(\frac{n\pi y}{b}\right) \cos(\lambda_{mn}\pi ct) \tag{2.49}$$

with m and n odd and where:

$$\lambda_{mn} = \sqrt{\frac{m^2}{a^2} + \frac{n^2}{b^2}}$$

In this analysis, 80 boundary elements and 800 cells were employed in the discretization (see Fig. 2.14).

Results corresponding to the displacement at point A(a/2,a/2) and to the support reaction at node B(a,a/2), from Houbolt analysis, are shown in Figs. 2.15 and 2.16, respectively.

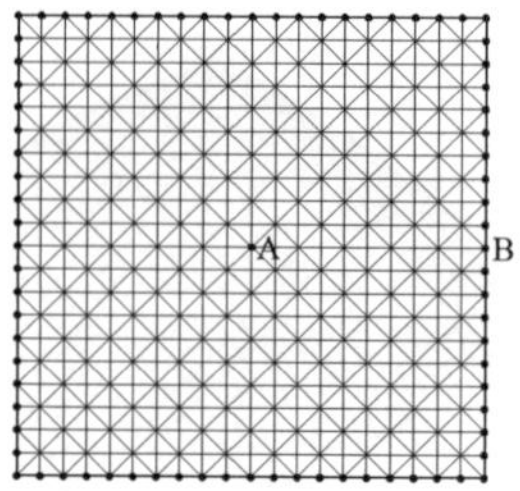

Fig. 2.14. Square membrane: boundary and domain discretizations with selected nodes.

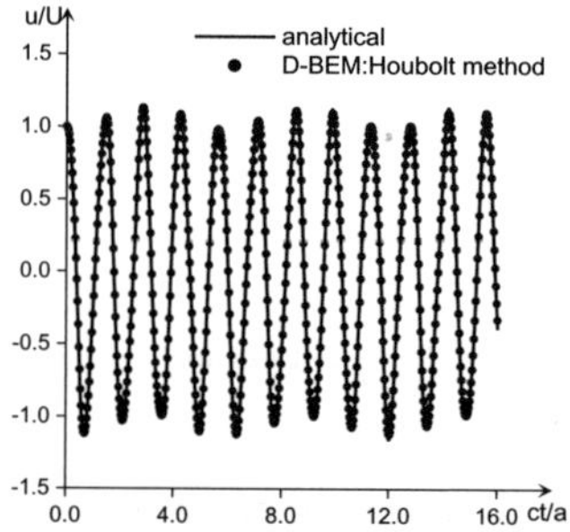

Fig. 2.15. Square membrane under initial displacement field over the entire domain: displacement at A(a/2,a/2) for Houbolt analysis.

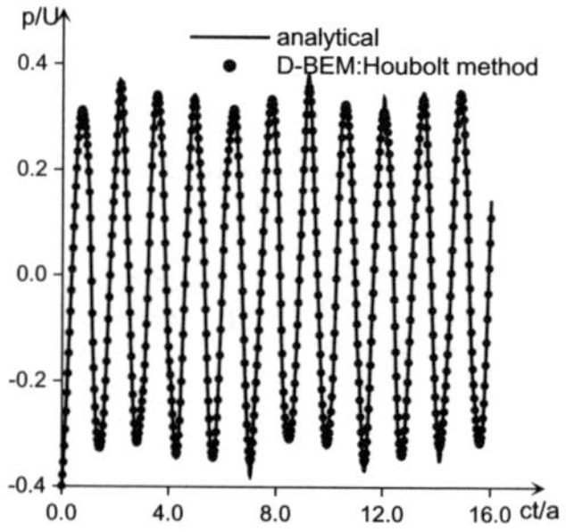

Fig. 2.16. Square membrane under initial displacement field over the entire domain: support reaction at B(a/,a/2) for Houbolt analysis.

Newmark analysis was carried out by adopting $\beta = 0.30$ and $\gamma = 0.52$. The results are shown in Figs. 2.17 and 2.18, respectively. Note that the values for β and γ were determined empirically.

Houbolt and Newmark analyses were carried out with the same time interval, computed by taking $\beta_{\Delta t} = 3/10$.

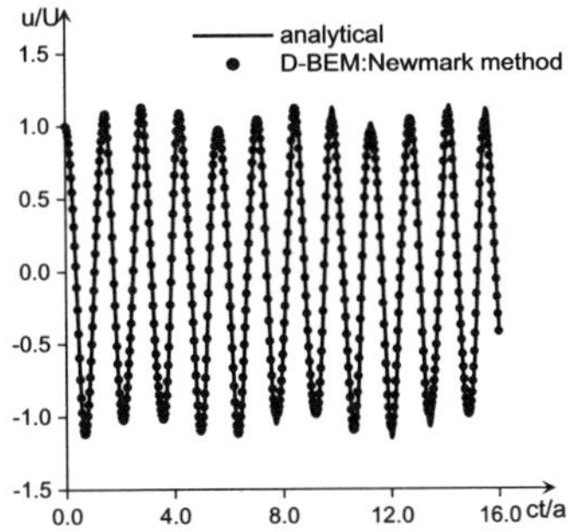

Fig. 2.17. Square membrane under initial displacement field over the entire domain: displacement at A(a/2,a/2) for Newmark analysis.

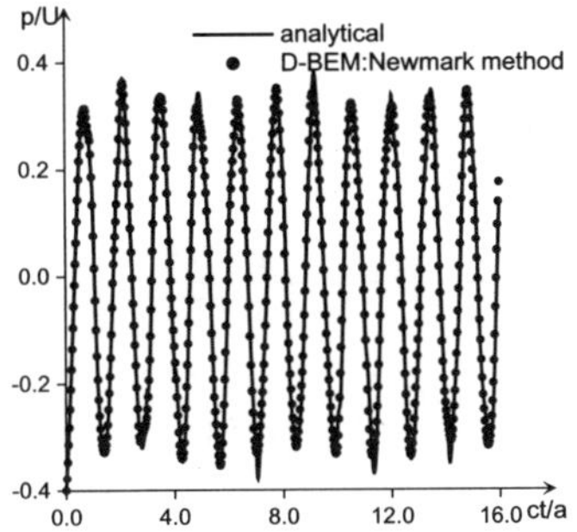

Fig. 2.18. Square membrane under initial displacement field over the entire domain: support reaction at B(a/,a/2) for Newmark analysis.

Good agreement between D-BEM and analytical results is observed in the previous figures.

2.4.2.2. *Square membrane under prescribed initial velocity over part of the domain*

The square membrane depicted in Fig. 2.19, with initial velocity field $v_0 = V$ prescribed over the sub-domain Ω_0 and null displacements prescribed all over the boundary, is analysed in this example. The analytical solution to this problem is given by[36]:

$$u(x, y, t) = \frac{4V}{c\pi^3} \sum_{m=1}^{\infty} \sum_{n=1}^{\infty} \frac{1}{mn\lambda_{mn}} \sin\left(\frac{m\pi x}{a}\right) \sin\left(\frac{n\pi y}{a}\right) G_{mn} \tag{2.50}$$

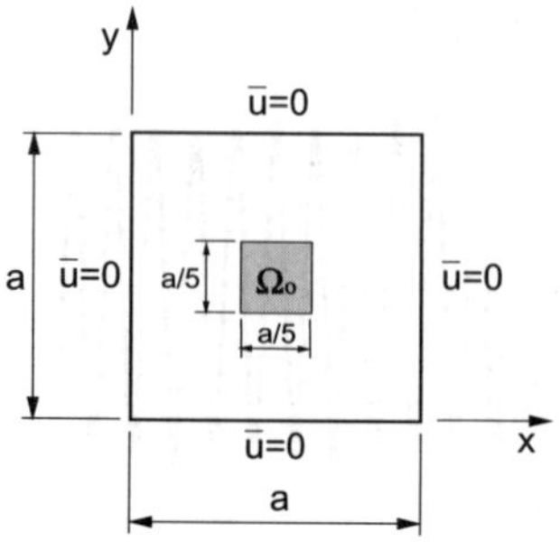

Fig. 2.19. Square membrane under prescribed initial velocity over part of the domain.

where:

$$\lambda_{mn} = \frac{\sqrt{m^2 + n^2}}{a}$$

and

$$\begin{aligned} G_{mn} &= \left(\cos\left(\frac{3n\pi}{5}\right) - \cos\left(\frac{2n\pi}{5}\right)\right) \\ &\quad \times \left(\cos\left(\frac{3m\pi}{5}\right) - \cos\left(\frac{2m\pi}{5}\right)\right) \sin\left(\lambda_{mn}\pi c t\right) \end{aligned}$$

This example was analysed previously with the use of the standard TD-BEM formulation[1] and, more recently, with the use of a BEM formulation based on the convolution quadrature method.[29]

Note that, as the initial conditions were assumed to have a linear variation in the internal cells, the use of double-nodes in the boundary of the sub-domain Ω_0 became necessary in order to avoid the spreading of the initial conditions to the cells in the neighbourhood of Ω_0. Aiming at providing a good representation of the jumps in the support reaction response, a more refined mesh constituted of 160 elements and 3,200 cells (not shown here) was employed.

The results of the Houbolt analysis, corresponding to the displacement at point A(a/2,a/2) and to the support reaction at node B(a,a/2), are depicted in Figs. 2.20 and 2.21, respectively. These results were obtained by adopting $\beta_{\Delta t} = 3/10$.

The Newmark analysis was carried out by adopting $\beta = 0.30$, $\gamma = 0.52$ and $\beta_{\Delta t} = 3/10$; the corresponding results are depicted in Figs. 2.22 and 2.23.

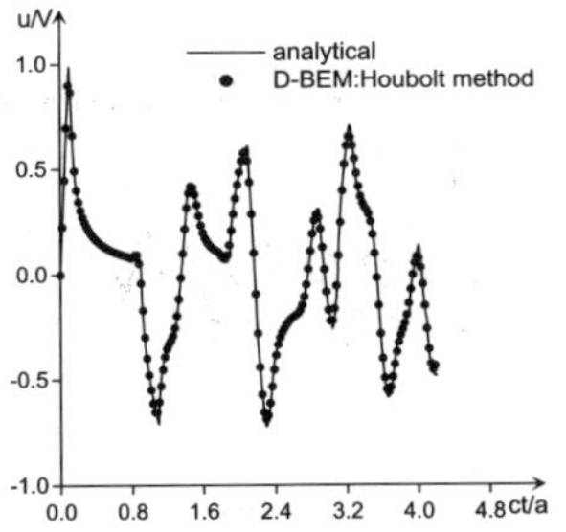

Fig. 2.20. Square membrane under initial displacement field over part of the domain: displacement at A(a/2,a/2) for Houbolt analysis.

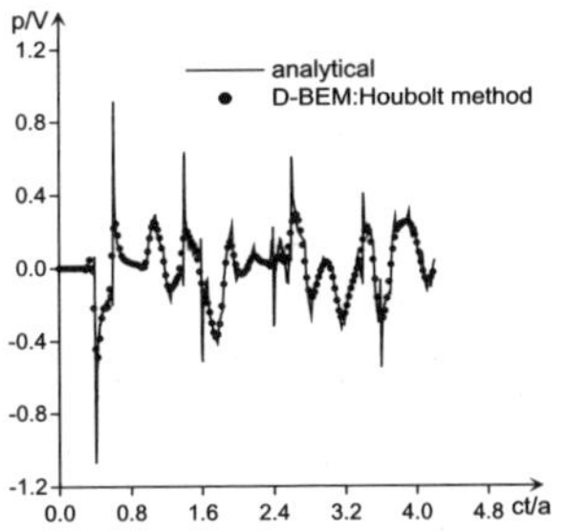

Fig. 2.21. SSquare membrane under initial displacement field over part of the domain: support reaction at B(a,a/2) for Houbolt analysis.

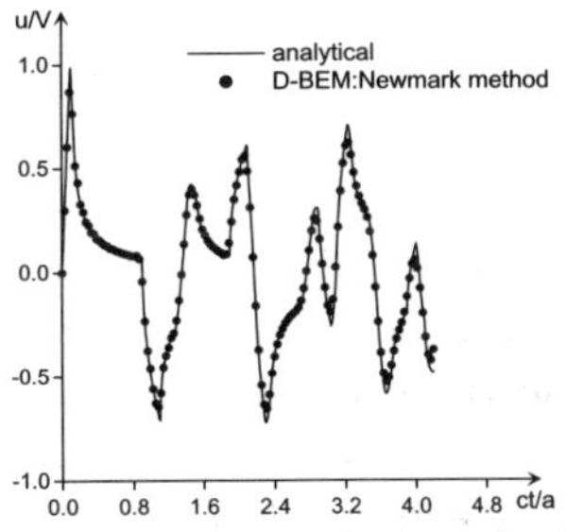

Fig. 2.22. Square membrane under initial displacement field over part of the domain: displacement at A(a/2,a/2) for Newmark analysis.

The D-BEM responses can be considered in good agreement with the analytical ones, although it must be pointed out that a better representation of the jumps in the support reaction response is provided by the standard TD-BEM formulation.[1] Nevertheless, the responses furnished

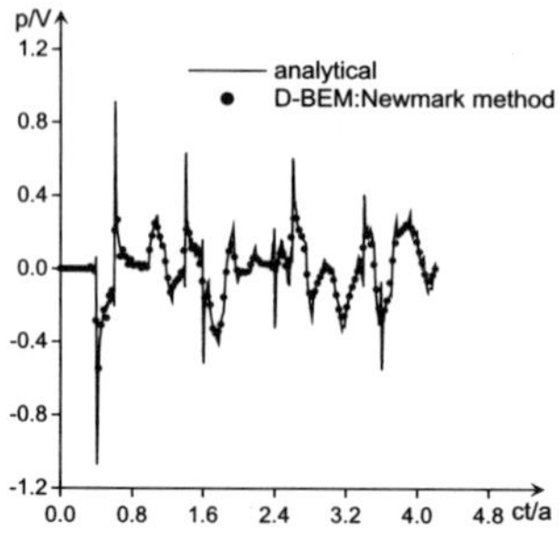

Fig. 2.23. Square membrane under initial displacement field over part of the domain: support reaction at B(a,a/2) for Newmark analysis.

by the D-BEM formulation demonstrate its applicability to the solution of these kinds of problems.

2.5. Conclusions and Outlook

In this chapter, a discussion concerning time-domain techniques for the solution of 2-D scalar wave problems is presented. The authors presented a general overview involving some research lines and focused their attention in two alternative approaches recently developed by them, in order to demonstrate how this research area is attractive and fecund. The first one, called SSTD-BEM is a step-by-step procedure based on the standard TD-BEM formulation. Its main advantage is that it avoids the computation of the time history contribution by evaluating a convolution integral through the whole analysis time. Its main drawbacks are the presence of a large amount of numerical damping (a matter that deserves more research in the future) and the need to employ transmitting boundaries for unbounded domain analyses. The second approach consists of a D-BEM formulation, for which the time marching employs both the Houbolt and the Newmark methods. A suggestion for the range of variation of the parameters of the Newmark method, determined empirically, is presented. General expressions for computing the initial conditions contributions are also included. Each development is followed by examples, which can provide an assessment of their potentialities and stimulate future research. The extension of the developments presented in this chapter to elastodynamics also looks very attractive.

References

1. W. J. Mansur, *A time-stepping technique to solve wave propagation problems using the boundary element method.* PhD thesis, University of Southampton, (1983).
2. J. Dominguez, *Boundary elements in dynamics.* (Computational Mechanics Publications, Southampton, 1993).
3. J. A. M. Carrer and W. J. Mansur, Time-dependent fundamental solution generated by a not impulsive source in the boundary element method analysis of the 2D scalar wave equation, *Communications in Numerical Methods in Engineering* **18**, 277–285, (2002).
4. G. Yu, W. J. Mansur, J. A. M. Carrer, and L. Gong, A linear θ method applied to 2D time-domain BEM, *Communications in Numerical Methods in Engineering* **14**, 1171–1179, (1998).
5. V. Demirel and S. Wang, An efficient boundary element method for two-dimensional transient wave propagation problems, *Applied Mathematical Modelling* **11**, 411–416, (1987).
6. W. J. Mansur and W. deLima Silva, Efficient time truncation in two-dimensional BEM analysis of transient wave propagation problems, *Earthquake Engineering and Structural Dynamics* **21**, 51–63, (1992).
7. D. Soares Jr and W. J. Mansur, Compression of time-generated matrices in two-dimensional time-domain elastodynamic BEM analysis, *International Journal for Numerical Methods in Engineering* **61**, 1209–1218, (2004).
8. J. A. M. Carrer and W. J. Mansur, Solution of the two-dimensional scalar wave equation by the time-domain boundary element method: Lagrange truncation strategy in time integration, *Structural Engineering and Mechanics* **23**, 263–278, (2006).
9. D. Soares Jr, W. J. Mansur, and O. Von Estorff, An efficient time-domain FEM/BEM coupling approach based on FEM implicit Green's functions and truncation of BEM time convolution process, *Computer Methods in Applied Mechanics and Engineering* **196**, 1816–1826, (2007).
10. J. A. M. Carrer and W. J. Mansur, Alternative time-marching schemes for elastodynamic analysis with the domain boundary element method formulation, *Computational Mechanics* **34**, 387–399, (2004).
11. G. D. Hatzigeorgiou and D. E. Beskos, Dynamic elastoplastic analysis of 3-D structures by the domain/boundary element method, *Computers & Structures* **80**, 339–347, (2002).
12. D. P. N. Kontoni and D. E. Beskos, Transient dynamic elastoplastic analysis by the dual reciprocity BEM, *Engineering Analysis with Boundary Elements* **12**, 1–16, (1993).
13. P. W. Partridge, C. A. Brebbia, and L. C. Wrobel, *The dual reciprocity boundary element method.* (Computational Mechanics Publications, Southampton, 1992).
14. J. C. Houbolt, A recurrence matrix solution for the dynamic response of elastic aircraft, *Journal of the Aeronautical Sciences* **17**, 540–550, (1950).

15. L. A. Souza, J. A. M. Carrer, and C. J. Martins, A fourth order finite difference method applied to elastodynamics: Finite element and boundary element formulations, *Structural Engineering and Mechanics* **17**, 735–749, (2004).
16. K. J. Bathe, *Finite element procedures* (Prentice Hall Inc., New Jersey, 1996).
17. R. D. Cook, D. S. Malkus, and M. E. Plesha, *Concepts and applications of finite element analysis* (John Wiley and Sons, New York, 1989).
18. N. M. Newmark, A method of computation for structural dynamics, ASCE, *Journal of Engineering Mechanics Division* **85**, 67–94, (1959).
19. L. Gaul and W. Wenzel, A coupled symmetric BE-FE method for acoustic fluid-structure interaction, *Engineering Analysis with Boundary Elements* **26**, 629–636, (2002).
20. W. J. Mansur, A. I. Abreu, and J. A. M. Carrer, Initial conditions contribution in frequency-domain BEM analysis, *Computer Modeling in Engineering & Sciences* **6**, 31–42, (2004).
21. L. Gaul and M. Schanz, A comparative study of three boundary element approaches to calculate the transient response of viscoelastic solids with unbounded domains, *Computer Methods in Applied Mechanics and Engineering* **179**, 111–123, (1999).
22. H. Dubner and J. Abate, Numerical inversion of Laplace transforms by relating them to the finite Fourier cosine transform, *Journal of the Association for Computing Machinery* **15**, 115–123, (1968).
23. F. Durbin, Numerical inversion of Laplace transforms: An efficient improvement to Dubner and Abate's method, *Computer Journal* **17**, 371–376, (1974).
24. C. Lubich, Convolution quadrature and discretized operational calculus I, *Numerische Mathematik* **52**, 129–145, (1988a).
25. C. Lubich, Convolution quadrature and discretized operational calculus II, *Numerische Mathematik* **52**, 413–425, (1988b).
26. M. Schanz and H. Antes, Application of 'operational quadrature methods' in time domain boundary element methods, *Meccanica* **32**, 179–186, (1997).
27. M. Schanz, Application of 3D time domain boundary element formulation to wave propagation in poroelastic solids, *Engineering Analysis with Boundary Elements* **25**, 363–376, (2001).
28. A. I. Abreu, J. A. M. Carrer, and W. J. Mansur, Scalar wave propagation in 2D: A BEM formulation based on the operational quadrature method, *Engineering Analysis with Boundary Elements* **27**, 101–105, (2003).
29. A. I. Abreu, W. J. Mansur, and J. A. M. Carrer, Initial conditions contribution in a BEM formulation based on the convolution quadrature method, *International Journal for Numerical Methods in Engineering* **67**, 417–434, (2006).
30. H. Antes, M. Schanz, and S. Alvermann, Dynamic analyses of plane frames by integral equations for bars and Timoshenko beams, *Journal of Sound and Vibration* **276**, 807–836, (2004).
31. D. E. Beskos, Boundary element method in dynamic analysis: Part II (1986–1996), *Applied Mechanics Reviews* **50**, 149–197, (1997).

32. J. Hadamard, *Lectures on Cauchy's Problem in Linear Partial Differential Equations* (Dover Publications, New York, 1952).
33. W. J. Mansur and J. A. M. Carrer, Two-dimensional transient BEM analysis for the scalar wave equation: kernels, *Engineering Analysis with Boundary Elements* **12**, 283–288, (1993).
34. W. J. Mansur, J. A. M. Carrer, and E. F. N. Siqueira, Time discontinuous linear traction approximation in time-domain BEM scalar wave propagation analysis, *International Journal for Numerical Methods in Engineering* **42**, 667–683, (1998).
35. G. Stephenson, *Partial Differential Equations for Scientists and Engineers* (Imperial College Press, London, 2007).
36. E. Kreyszig, *Advanced engineering mathematics*, 8th edition (John Wiley & Sons, New York, 1999).

Chapter 3

THE BOUNDARY ELEMENT METHOD FOR THE FRACTURE ANALYSIS OF THE GENERAL PIEZOELECTRIC SOLIDS

M. Denda
Rutgers University
Mechanical and Aerospace Engineering Department
98 Brett Road, Piscataway, New Jersey 08854-8058, U.S.A.
denda@rutgers.edu

The boundary element method (BEM), developed by the author, for the fracture analysis of the general piezoelectric solids in two-dimensions is presented. Several innovative approaches not used in the traditional BEM are introduced in the formulation of the method. These are (1) physical interpretation of the extended Somigliana's identity as the basis of the direct BEM formulation, (2) use of the extended dislocation dipoles for crack modeling, (3) analytical representation of the crack tip singularity, and (4) the non-linear solution algorithm for the multiple semipermeable cracks based on the linear impermeable and permeable crack solvers. The complex variable formalism used to derive the fundamental results necessary for the formulation of the BEM is also presented in full.

3.1. Introduction

The behavior of the piezoelectric solids can be described by the extended displacement and force vectors, both of which have four components: the first three are the mechanical and the last electrical. Thus, the extended displacement vector has three displacement and one electric potential components and the extended force vector has three force and one electric charge components. Similarly, the stress, strain and material constants can be extended from the anisotropic elastic to piezoelectric quantities by appending the electric quantities to the elastic ones. The situation is the same for the boundary element method (BEM). The BEM for the general piezoelectric solids can readily be obtained following the simple extension described above. As a matter of fact, the computer codes for the two solids are almost identical except for the range of the index: 3 for the anisotropic and 4 for the piezoelectric solids. Among many contributions to the anisotropic BEM (REF), most of them are dealing with specific anisotropic solids (such as the transverse isotropic and orthotropic) and those dealing with the general anisotropic solids are scarce. The state of the art for the the piezoelectric BEM is even more focused on specific materials rather than on the general piezoelectric solids. One of the goals of this article is to provide the theoretical foundation for the piezoelectricity in two-dimensions for the most general piezoelectric solids. The extended Lekhnitskii formalism, which is a complex variable formalism developed by Lekhnitskii[1] for the two-dimensional anisotropic solids and extended by the author to the general piezoelectric solids, is introduced to provide the powerful foundation for the analytical solution of the two-dimensional boundary value problems for the general piezoelectric solids. This formalism is used in the derivation of the fundamental solutions for our BEM.

The BEM is superior to the Finite Element Method (FEM) in the linear fracture analysis due to its capability to model the cracks by lines in two-dimensions. One of the approaches used is the multi-domain method,[2,3] which artificially breaks the original cracked region into multiple dissected domains. Another technique, using the single domain approach, is the dual reciprocal BEM.[4,5] This article introduces another single domain approach using the dislocation dipole distribution to model the crack. The basis of the dislocation dipole approach in crack modeling is the physical interpretation of the extended Somigliana's identity, which also serves as the guiding principle in the formulation of the BEM.

The challenge in the fracture analysis is modeling of the crack tip singularity. The standard technique used for the anisotropic solids is the use of the regular elements, with a progressive refinement toward the crack tip, for which the stress intensity factors (SIFs) are determined by the conservation integrals[6–8] in the post-processing. Others are techniques that embed the singularity at the crack tip and determine the SIFs directly in the main-processing: the quarter-point traction and displacement crack tip elements from which the analytical expressions for the SIFs are derived[9] and the Green's function approach.[10] The Green's function approach, although effective, depends on the availability of the analytical solution and is limited to the single crack. This limitation was overcome by the numerical Green's function approach[11] for multiple straight cracks in the general anisotropic solids. The dislocation dipoles are used to model the crack opening displacement (COD) with the embedded $\sqrt{r}$ COD behavior, which is integrated analytically to give the whole crack singular element (WCSE) with the $\sqrt{r}$ COD and the $1/\sqrt{r}$ crack tip stress singularity. Since the magnitude of the COD, which defines the Green's function, is determined numerically by applying the crack surface extended traction boundary condition, this approach is called the numerical Green's function method. Just like the Green's function method, this approach does not require the post-processing for determination of the SIFs. In this article, the extension of the numerical Green's function method to the general piezoelectric solids[12] is introduced.

As mentioned earlier, the BEM crack modeling scheme for the piezoelectric solids can be obtained as a direct extension of the technique developed for the anisotropic solids. However, one exception is the crack surface electric boundary condition (BC). The straightforward extension of the stress-free crack surface boundary condition to the piezoelectric solids is to specify the fourth extended traction component (electric induction), to zero. This gives rise to the impermeable BC, which does not correctly represent the electric behavior on the opened crack surface. The consistent condition for the opened crack, located along the x_1-axis, is given by the semipermeable BC,

$$D_2^+ = D_2^-; \quad D_2^+(u_2^+ - u_2^-) = -\kappa^C(\Phi^+ - \Phi^-), \tag{3.1}$$

proposed by Hao and Shen,[13] where u_2, D_2 and Φ are the normal displacement, electric induction and the electric potential, respectively, and κ^C is the electric permittivity. The symbol $\pm$ indicates the upper and lower crack surfaces. In general, the electric BC on the piezoelectric crack surface comes in different degrees of shielding the electric induction defined by the electric permittivity. The impermeable crack with the BC,

$$D_2^+ = D_2^- = 0, \tag{3.2}$$

shields the electric induction completely. The permeable crack given by the BC,

$$D_2^+ = D_2^-; \quad \Phi^+ - \Phi^- = 0, \tag{3.3}$$

does not shield the electric induction. If the crack is closed, then the permeable BC is correct, while the impermeable BC is correct for the opened crack if the electric permittivity κ^C of the crack medium is zero.

Since no medium has zero permittivity, the two boundary conditions are not correct for the opened crack. The majority of the earlier works adopted the impermeable BC[14–34] and the permeable BC.[35–40] Following the argument introduced above, the semipermeable BC is reduced to the impermeable BC when $\kappa^C = 0$ and to the permeable BC when $u_2^+ - u_2^- = 0$ and the impermeable and permeable cracks set the upper and lower bounds of the electric boundary condition for the semipermeable crack.[12] The progress in the use of the consistent semipermeable crack is slow and limited to the single crack problems in specific piezoelectric solids.[13,41–51] In this article, the crack with the semipermeable electric BC is often called the semipermeable crack and those with the permeable and impermeable electric BCs are referred to as permeable and impermeable cracks, respectively. Since the variation of the electric induction on the semipermeable crack surface is unknown, we need an iteration process to determine all unknowns including the extended COD and the electric induction. This is inherently a non-linear problem. A stable iteration scheme is proposed, for multiple straight semipermeable cracks, to obtain the solution using the linear BEM solvers for permeable and impermeable cracks. Numerical results of the impermeable, permeable and semipermeable cracks are presented for several multiple crack configurations.

3.2. Basic Equations in Piezoelectricity

3.2.1. *Piezoelectric equations in 3-D*

For the linear piezoelectric material, the stress σ_{ij} and the electric induction D_i are given, under the isothermal condition, by the strain ε_{ij} and the electric field E_i as

$$\sigma_{ij} = c_{ijkl}^E \varepsilon_{kl} - e_{kij} E_k, \quad D_i = e_{ikl}\varepsilon_{kl} + \kappa_{ik}^{\varepsilon} E_k, \tag{3.4}$$

where c_{ijkl}^E, e_{ikl}, and $\kappa_{ik}^{\varepsilon}$ are the elastic stiffness constants at constant electric field, piezoelectric stress constants, and dielectric permittivity constants at constant strain, respectively. The lower case Roman indices range from 1 to 3 and the summation convention for the repeated indices is implied. Inversion of the relations in Eq. (3.4) will give the strain and the electric field in terms of the stress and the electric induction by

$$\varepsilon_{ij} = s_{ijkl}^D \sigma_{kl} + g_{kij} D_k, \quad E_i = -g_{ikl}\sigma_{kl} + \beta_{ik}^{\sigma} D_k, \tag{3.5}$$

where s^D_{ijkl}, g_{ikl}, β^σ_{ik} are the elastic compliance constants at constant electric induction, piezoelectric constants, and dielectric impermeability constants at constant stress, respectively. The stress and the electric induction satisfy the equilibrium equations and Gauss's law,

$$\sigma_{ji,j} + f_i = 0, \quad D_{j,j} - \rho = 0, \tag{3.6}$$

where f_i and ρ are the body force and extrinsic bulk charge densities, respectively. The boundary conditions for the traction and the electric induction are given by

$$n_j \sigma_{ji} = t_i, \quad n_j D_j = -\omega, \tag{3.7}$$

where t_i and ω are the force and the charge per unit area of the boundary and n_j is the unit normal to the boundary. The strain and the electric field are given by the displacement u_i and the electric potential ϕ, respectively, by

$$\varepsilon_{ij} = \frac{1}{2}(u_{i,j} + u_{j,i}), \quad E_i = -\phi_{,i}, \tag{3.8}$$

where a comma followed by an index i indicates differentiation by the coordinate x_i.

Define the extended displacement U_I, body force F_I, traction T_I, stress Σ_{Ij}, and strain $\mathcal{E}_{Ij}$ according to

$$\begin{aligned} U_I &= \begin{cases} u_i \\ \phi \end{cases}, \quad F_I = \begin{cases} f_i \\ -\rho \end{cases}, \quad T_I = \begin{cases} t_i & (I = i = 1,2,3) \\ -\omega & (I = 4) \end{cases}, \\ \Sigma_{Ij} &= \begin{cases} \sigma_{ij} \\ D_j \end{cases}, \quad \mathcal{E}_{Ij} = \begin{cases} \epsilon_{ij} & (I = i = 1,2,3; j = 1,2,3) \\ E_j & (I = 4; j = 1,2,3) \end{cases}, \end{aligned} \tag{3.9}$$

where the upper case Roman indices range from 1 to 4. Introduction of these extended variables enables the direct extension of the governing equations from the anisotropic to piezoelectric solids.

3.2.2. *Piezoelectric equations in 2-D*

In dealing with the the stress, strain and compliance in two-dimensions (2-D), adopt the convention of replacing the pair of suffixes ij by a single suffix $\mathcal{M}$ (upper case Roman script) following the convention: ($11 \to 1$), ($22 \to 2$), ($33 \to 3$), ($23 \to 4$), ($31 \to 5$,) ($12 \to 6$). In 2-D, the extended displacement and other variables depend on coordinates x_1 and x_2 only.

The non zero components of the engineering strain are given by

$$e_1 = u_{1,1}, \quad e_2 = u_{2,2}, \quad e_4 = u_{3,2}, \quad e_5 = u_{3,1}, \quad e_6 = u_{2,1} + u_{1,2}, \tag{3.10}$$

and those for the electric field by

$$E_1 = -\phi_{,1} \quad E_2 = -\phi_{,2}. \tag{3.11}$$

The compatibility equations that follow from (3.10) and (3.11) are given by

$$e_{2,11} + e_{1,22} - e_{6,12} = 0, \quad e_{4,1} - e_{5,2} = 0, \quad E_{2,1} - E_{1,2} = 0. \tag{3.12}$$

Since $e_3 = E_3 \equiv 0$, we can eliminate σ_3 and D_3 in (3.5) to get the reduced extended strain and stress relations,

$$\begin{aligned} e_{\mathcal{M}} &= S_{\mathcal{M}\mathcal{N}}\, \sigma_{\mathcal{N}} + G_{\alpha\mathcal{M}}\, D_\alpha, \\ E_\alpha &= -G_{\alpha\mathcal{N}}\, \sigma_{\mathcal{N}} + B_{\alpha\gamma}\, D_\gamma, \end{aligned} \tag{3.13}$$

where $S_{\mathcal{M}\mathcal{N}}$, $G_{\alpha\mathcal{N}}$ and $B_{\alpha\beta}$ are the reduced elastic compliance, piezoelectric constants and dielectric impermeability constants, respectively, given by

$$\begin{aligned} S_{\mathcal{M}\mathcal{N}} &= s^D_{\mathcal{M}\mathcal{N}} - \frac{s^D_{\mathcal{M}3}s^D_{\mathcal{N}3}\beta^\sigma_{33} + (s^D_{\mathcal{M}3}g_{3\mathcal{N}} + s^D_{\mathcal{N}3}g_{3\mathcal{M}})g_{33} - g_{3\mathcal{M}}g_{3\mathcal{N}}s^D_{33}}{s^D_{33}\beta^\sigma_{33} + g^2_{33}}, \\ G_{\alpha\mathcal{M}} &= g_{\alpha\mathcal{M}} - \frac{s^D_{\mathcal{M}3}g_{\alpha 3}\beta^\sigma_{33} + (-s^D_{\mathcal{M}3}\beta^\sigma_{\alpha 3} + g_{3\mathcal{M}}g_{\alpha 3})g_{33} + g_{3\mathcal{M}}\beta^\sigma_{\alpha 3}s^D_{33}}{s^D_{33}\beta^\sigma_{33} + g^2_{33}}, \\ B_{\alpha\gamma} &= \beta^\sigma_{\alpha\gamma} + \frac{g_{\alpha 3}g_{\gamma 3}\beta^\sigma_{33} - (g_{\alpha 3}\beta^\sigma_{\gamma 3} + g_{\gamma 3}\beta^\sigma_{\alpha 3})g_{33} - \beta^\sigma_{\alpha 3}\beta^\sigma_{\gamma 3}s^D_{33}}{s^D_{33}\beta^\sigma_{33} + g^2_{33}}. \end{aligned} \tag{3.14}$$

$$(\mathcal{M}, \mathcal{N} = 1, 2, 4, 5, 6; \alpha, \gamma = 1, 2)$$

Notice that the upper case Roman script indices take up the values 1, 2, 4, 5 and 6. Greek indices range from 1 to 2 and the comma followed by a subscript α indicates differentiation by x_α.

The equilibrium equations in (3.6), assuming zero extended body force, can be reduced and given in terms of the extended stress components by

$$\frac{\partial \Sigma_{I1}}{\partial x_1} + \frac{\partial \Sigma_{I2}}{\partial x_2} = 0, \tag{3.15}$$

which is automatically satisfied by the extended stress function vector Ψ_I, defined by

$$\Sigma_{I1} = -\frac{\partial \Psi_I}{\partial x_2}, \quad \Sigma_{I2} = \frac{\partial \Psi_I}{\partial x_1}. \tag{3.16}$$

The extended traction on the boundary is given, from (3.7), by

$$T_I = \Sigma_{I\alpha} n_\alpha, \tag{3.17}$$

where n_α is the unit normal to the boundary.

3.2.3. *Lekhnitskii formalism for 2-D piezoelectricity*

The extended Lekhnitskii formalism uses the compatibility equations (3.12) by assuming the extended stress function vector of the form,

$$\Psi_I = L_I f(x_1 + p x_2). \tag{3.18}$$

An eighth order characteristic equation in p,

$$\begin{aligned} d^{(4)}(p)\, d^{(2)}(p)\, f^{(2)}(p) + d^{(4)}(p)\, e^{(2)}(p)\, e^{(2)}(p) - d^{(3)}(p)\, d^{(3)}(p)\, f^{(2)}(p) \\ - 2d^{(3)}(p)\, e^{(3)}(p)\, e^{(2)}(p) + d^{(2)}(p)\, e^{(3)}(p)\, e^{(3)}(p) = 0, \end{aligned} \tag{3.19}$$

is obtained from the compatibility equations in (3.12), using (3.18), (3.16) and (3.13), where

$$\begin{aligned} d^{(4)}(p) &= S_{11}p^4 - 2S_{16}p^3 + (2S_{12} + S_{66})p^2 - 2S_{26}p + S_{22}, \\ d^{(3)}(p) &= S_{15}p^3 - (S_{14} + S_{56})p^2 + (S_{25} + S_{46})p - S_{24}, \\ e^{(3)}(p) &= G_{11}p^3 - (G_{21} + G_{16})p^2 + (G_{12} + G_{26})p - G_{22}, \\ d^{(2)}(p) &= S_{55}p^2 - 2S_{45}p + S_{44}, \\ e^{(2)}(p) &= G_{15}p^2 - (G_{14} + G_{25})p + G_{24}, \\ f^{(2)}(p) &= B_{11}p^2 - 2B_{12}p + B_{22}. \end{aligned} \tag{3.20}$$

It can be shown that the characteristic equation (3.19) with real coefficients has no real roots and has four pairs of conjugate complex roots p_J, $\overline{p}_J$ (J=1,2,3,4). It is assumed that four roots p_J are distinct; coincident roots can be made different by slightly perturbing the material constants. Now the full form of the extended stress function vector (3.18) is given, in terms of four generalized complex variables and complex potential functions $f_J(z_J)$ $(J = 1, 2, 3, 4)$ by

$$\Psi_I = \sum_{J=1}^{4} [L_{IJ} f_J(z_J) + \overline{L}_{IJ} \overline{f_J(z_J)}] = 2\Re\left\{\sum_{J=1}^{4} L_{IJ} f_J(z_J)\right\}, \tag{3.21}$$

where $\Re$ indicates the real part of the complex number and an over bar the complex conjugate. The L_{IJ} matrix is given by

$$\mathbf{L} = [L_{IJ}] = [\mathbf{L}_1, \mathbf{L}_2, \mathbf{L}_3, \mathbf{L}_4]$$
$$= \begin{bmatrix} -p_1 L_{21} & -p_2 L_{22} & -p_3 l_3 L_{33} & -p_4 l_7 L_{44} \\ L_{21} & L_{22} & l_3 L_{33} & l_7 L_{44} \\ l_1 L_{21} & l_2 L_{22} & L_{33} & l_8 L_{44} \\ l_4 L_{21} & l_5 L_{22} & l_6 L_{33} & L_{44} \end{bmatrix}, \tag{3.22}$$

where

$$\begin{aligned} l_1 &= \frac{f_1^{(2)} d_1^{(3)} + e_1^{(2)} e_1^{(3)}}{d_1^{(2)} f_1^{(2)} + e_1^{(2)} e_1^{(2)}}, & l_2 &= \frac{f_2^{(2)} d_2^{(3)} + e_2^{(2)} e_2^{(3)}}{d_2^{(2)} f_2^{(2)} + e_2^{(2)} e_2^{(2)}}, \\ l_3 &= \frac{f_3^{(2)} d_3^{(3)} + e_3^{(3)} e_3^{(2)}}{d_3^{(4)} f_3^{(2)} + e_3^{(3)} e_3^{(3)}}, & l_4 &= \frac{e_1^{(2)} d_1^{(3)} - d_1^{(2)} e_1^{(3)}}{d_1^{(2)} f_1^{(2)} + e_1^{(2)} e_1^{(2)}}, \\ l_5 &= \frac{e_2^{(2)} d_2^{(3)} - d_2^{(2)} e_2^{(3)}}{d_2^{(2)} f_2^{(2)} + e_2^{(2)} e_2^{(2)}}, & l_6 &= \frac{d_3^{(4)} e_3^{(2)} - d_3^{(3)} e_3^{(3)}}{d_3^{(4)} f_3^{(2)} + e_3^{(3)} e_3^{(3)}}, \\ l_7 &= \frac{d_4^{(2)} e_4^{(3)} - d_4^{(3)} e_4^{(2)}}{d_4^{(4)} d_4^{(2)} - d_4^{(3)} d_4^{(3)}}, & l_8 &= \frac{d_4^{(3)} e_4^{(3)} - d_4^{(4)} e_4^{(2)}}{d_4^{(4)} d_4^{(2)} - d_4^{(3)} d_4^{(3)}}. \end{aligned} \tag{3.23}$$

The extended stress vectors are calculated by substituting (3.21) to(3.16) with the result,

$$\Sigma_{I2} = 2\Re\left\{\sum_{J=1}^{4} L_{IJ} f_J'(z_J)\right\}, \quad \Sigma_{I1} = -2\Re\left\{\sum_{J=1}^{4} p_J L_{IJ} f_J'(z_J)\right\}, \tag{3.24}$$

where $f_J'(z_J) = df_J(z_J)/dz_J$. To obtain the extended displacement, we first calculate the strain and electric field, using the stress (3.24) in (3.13), and then integrate the extended strain-displacement relations (3.10) and (3.11) with the result,

$$U_I = 2\Re\left\{\sum_{J=1}^{4} A_{IJ} f_J(z_J)\right\}, \tag{3.25}$$

where

$$\mathbf{A} = [A_{IJ}] = [\mathcal{A}_1 \mathbf{L}_1, \mathcal{A}_2 \mathbf{L}_2, \mathcal{A}_3 \mathbf{L}_3, \mathcal{A}_4 \mathbf{L}_4], \tag{3.26}$$

and

$$\mathcal{A}_J = \begin{bmatrix} S_{16} - S_{11}p_J, & S_{12}, & S_{14} - S_{15}p_J, & G_{21} - G_{11}p_J \\ \frac{S_{26}-S_{21}p_J}{p_J}, & \frac{S_{22}}{p_J}, & \frac{S_{24}-S_{25}p_J}{p_J}, & \frac{G_{22}-G_{12}p_J}{p_J} \\ S_{56} - S_{51}p_J, & S_{52}, & S_{54} - S_{55}p_J, & G_{25} - G_{15}p_J \\ G_{16} - G_{11}p_J, & G_{12}, & G_{14} - G_{15}p_J, & -B_{12} + B_{11}p_J \end{bmatrix}. \tag{3.27}$$

Note that the extended rigid body displacement is neglected in (3.25).

Since each column of L_{IJ} matrix is determined within an arbitrary multiplication constant, it is necessary to normalize the components of the L_{IJ} and A_{IJ} matrices. For this purpose we use the relations,

$$\begin{aligned} &\mathbf{L}^T\mathbf{A} + \mathbf{A}^T\mathbf{L} \\ &= \begin{bmatrix} 2\sum_{I=1}^4 L_{I1}A_{I1} & 0 & 0 & 0 \\ 0 & 2\sum_{I=1}^4 L_{I2}A_{I2} & 0 & 0 \\ 0 & 0 & 2\sum_{I=1}^4 L_{I3}A_{I3} & 0 \\ 0 & 0 & 0 & 2\sum_{I=1}^4 L_{I4}A_{I4} \end{bmatrix}, \\ &\mathbf{L}^T\bar{\mathbf{A}} + \mathbf{A}^T\bar{\mathbf{L}} = \mathbf{0}, \end{aligned} \tag{3.28}$$

where $\mathbf{0}$ is the 4×4 zero matrix and a bar and the superscript T indicate the complex conjugate and the transpose, respectively. These relations are obtained by extending the corresponding relations[52–56] for the anisotropic solids. The normalization is introduced by setting

$$2\sum_{I=1}^{4} L_{IJ}A_{IJ} = 1 \quad (\text{no sum on } J = 1, 2, 3, 4). \tag{3.29}$$

The resultant of the extended traction vector (per unit thickness) acting on a contour $L = \widehat{AB}$ from the material to the right toward the material to the left is given, through (3.16) and (3.21), by

$$R_I = \int_L T_I ds = -\int_L d\Psi = -[\Psi]_A^B = -\left[2\Re\left\{\sum_{J=1}^{4} L_{IJ}f_J(z_J)\right\}\right]_A^B. \tag{3.30}$$

3.3. Physical Interpretation of Extended Somigliana's Identity in 2-D Piezoelectricity

Consider a finite domain A in 2-D, surrounded by the boundary s subject to the extended boundary displacement U_I and traction T_I. The extended

Somigliana's identity in piezoelectricity, assuming zero extended body force, gives the extended displacement according to

$$U_M(\mathbf{x}) = \int_s [T_I(\boldsymbol{\xi})U_{IM}(\boldsymbol{\xi},\mathbf{x})]\, ds(\boldsymbol{\xi}) - \int_s [U_I(\boldsymbol{\xi})T_{IM}(\boldsymbol{\xi},\mathbf{x})]\, ds(\boldsymbol{\xi}), \quad (3.31)$$

where $U_{IM}(\boldsymbol{\xi},\mathbf{x})$ $(I, M = 1, 2, 3, 4)$ and $T_{IM}(\boldsymbol{\xi},\mathbf{x})$ are the I-th components of the extended displacement and traction, respectively, at $\boldsymbol{\xi} = (\eta_1, \eta_2)$ when the M-th component of the unit extended point force is applied at $\mathbf{x} = (x_1, x_2)$. The latter is defined on a line segment with the unit normal $n_\alpha(\boldsymbol{\xi})$.

An extended dislocation dipole is defined by an infinitesimal line segment ds with the unit normal $\mathbf{n} = (n_1, n_2)$ over which the extended displacement discontinuity is specified. This is the projection of the 3-D dislocation loop[57] defined by a narrow rectangular strip of infinite extent, in the out-of-plane direction, into the 2-D plane. Let $U_{MI}(\mathbf{x},\boldsymbol{\xi})^*$ be the M-th component of the extended displacement solution at $\mathbf{x}$ due to the extended dislocation dipole with the unit discontinuity in the I-th direction over ds at $\boldsymbol{\xi}$. Using the symmetry property of the displacement fundamental solution,

$$U_{IM}(\boldsymbol{\xi},\mathbf{x}) = U_{MI}(\mathbf{x},\boldsymbol{\xi}), \quad (3.32)$$

the following relation,

$$-T_{IM}(\boldsymbol{\xi},\mathbf{x}) = U_{MI}(\mathbf{x},\boldsymbol{\xi})^*, \quad (3.33)$$

can be derived.[58] By substituting these relations into (3.31) we get an alternative form of the extended Somigliana's identity in 2-D piezoelectricity,

$$U_M(\mathbf{x}) = \int_s [U_{MI}(\mathbf{x},\boldsymbol{\xi})T_I(\boldsymbol{\xi})]\, ds(\boldsymbol{\xi}) + \int_s [U_{MI}(\mathbf{x},\boldsymbol{\xi})^* U_I(\boldsymbol{\xi})]\, ds(\boldsymbol{\xi}). \quad (3.34)$$

According to this form of the identity, the extended displacement field in the body A is given by the extended point force T_I and extended dislocation dipole U_I distributions over the contour s embedded in the infinite domain. This is a physical interpretation of the Somigliana's identity in 2-D piezoelectricity.

3.4. Direct Formulation of the BEM in 2-D

3.4.1. *Fundamental solutions for 2-D piezoelectricity*

Consider the unit I-th component of the extended line force $r_I = 1$ located at $\boldsymbol{\xi} = (\eta_1, \eta_2)$. Introduce an arbitrary circuit Γ around $\boldsymbol{\xi}$ and determine the analytic function vector $f_I(z_I)$ such that the extended force resultant around this circuit is $-r_I$. Use of (3.30) along with the orthogonality relations (3.28) and (3.29) will give the fundamental solution,

$$U_{MI}(\mathbf{x}, \boldsymbol{\xi}) = \Im \frac{1}{\pi} \sum_{J=1}^{4} A_{MJ} A_{IJ} \ln(z_J - \xi_J), \tag{3.35}$$

which is the M-th component of the extended displacement at $\mathbf{x} = (x_1, x_2)$, where $z_J = x_1 + p_J x_2$ and $\xi_J = \eta_1 + p_J \eta_2$. The symbol $\Im$ indicates the imaginary part of a complex variable. Next, consider an extended line dislocation at $\boldsymbol{\xi} = (\eta_1, \eta_2)$ with the unit jump in the I-th component of the extended displacement. The solution is determined such that the extended displacement jump around the circuit Γ is $b_I = 1$. Using (3.25) and the orthogonality relations (3.28) and (3.29) we get the resulting M-th component of the extended displacement at $\mathbf{x} = (x_1, x_2)$,

$$P_{MI}(\mathbf{x}, \boldsymbol{\xi}) = \Im \frac{1}{\pi} \sum_{J=1}^{4} A_{MJ} L_{IJ} \ln(z_J - \xi_J). \tag{3.36}$$

The extended dislocation dipole is defined by an infinitesimal line segment $d\boldsymbol{\xi} = (d\eta_1, d\eta_2)$ at $\boldsymbol{\xi}$ along which the extended displacement is discontinuous. This configuration is produced by placing a pair of extended dislocations, with the same magnitude but opposite signs, at $\boldsymbol{\xi}$ and $\boldsymbol{\xi} + d\boldsymbol{\xi}$ and obtained by taking the total derivative of the fundamental solution of the extended dislocation with respect to the generalized complex variables $\xi_J = \eta_1 + p_J \eta_2$. For a dislocation dipole at $\boldsymbol{\xi} = (\eta_1, \eta_2)$ with the unit jump in the I-th extended displacement component, the resulting M-th extended displacement component at $\mathbf{x} = (x_1, x_2)$ is given by

$$U_{MI}(\mathbf{x}, \boldsymbol{\xi})^* \, ds = -\Im \frac{1}{\pi} \sum_{J=1}^{4} A_{MJ} L_{IJ} \frac{d\xi_J}{z_J - \xi_J}, \tag{3.37}$$

where $d\xi_J = d\eta_1 + p_j d\eta_2$ and $ds = \sqrt{d\eta_1^2 + d\eta_2^2}$ is the magnitude of $d\boldsymbol{\xi}$.

3.4.2. *Implementation*

In the BEM discretization, we approximate the whole boundary by a collection of straight elements. For a boundary element Γ of length L with the end nodes 1, 2 and the middle node 3, approximate the extended displacement and traction on the boundary by quadratic interpolation functions,

$$U_I(\xi_J) = \sum_{a=1}^{3} \varphi_a(\xi_J) U_I^a, \quad T_I(\xi_J) = \sum_{a=1}^{3} \varphi_a(\xi_J) T_I^a, \tag{3.38}$$

where $\varphi_a(\xi_J)$ $(a = 1, 2, 3)$ are the quadratic shape functions and U_I^a and T_I^a are the nodal values of U_I and T_I. The contribution of the boundary element Γ to the first integral of (3.34) is given, through the fundamental solution (3.35) and the interpolation (3.38), by

$$U_M^T(\mathbf{x}) = \sum_{a=1}^{3} \sum_{I=1}^{4} \left\{ \int_\Gamma \Im \frac{1}{\pi} \sum_{J=1}^{4} A_{MJ} A_{IJ} \, \ln(z_J - \xi_J) \varphi_a(\xi_J) ds \right\} T_I^a. \tag{3.39}$$

Similarly, the second integral in (3.34) is given, from (3.37) and (3.38), by

$$U_M^U(\mathbf{x}) = \sum_{a=1}^{3} \sum_{I=1}^{4} \left\{ \int_\Gamma -\Im \frac{1}{\pi} \sum_{J=1}^{4} A_{MJ} L_{IJ} \, \frac{1}{z_J - \xi_J} \varphi_a(\xi_J) d\xi_J \right\} U_I^a. \tag{3.40}$$

For the straight element both integrals can be evaluated analytically with the result,

$$U_M^T(\mathbf{x}) = \sum_{a=1}^{3} \sum_{I=1}^{4} U_{MI}^{Ta}(\mathbf{x}) \, T_I^a, \quad U_M^U(\mathbf{x}) = \sum_{a=1}^{3} \sum_{I=1}^{4} U_{MI}^{Ua}(\mathbf{x}) \, U_I^a, \tag{3.41}$$

with

$$\begin{aligned} U_{MI}^{Ta}(\mathbf{x}) &= \Im \left\{ \frac{1}{\pi} \sum_{J=1}^{4} A_{MJ} A_{IJ} \mathcal{U}_J^{Ta}(z_J) \right\}, \\ U_{MI}^{Ua}(\mathbf{x}) &= \Im \left\{ \frac{1}{\pi} \sum_{J=1}^{4} A_{MJ} L_{IJ} \mathcal{U}_J^{Ua}(z_J) \right\}, \end{aligned} \tag{3.42}$$

and

$$\begin{aligned} \mathcal{U}_J^{Ta}(z_J) &= \frac{1}{\cos\phi + p_J \sin\phi} \sum_{n=1}^{3} (-1)^{n-1} \left[\varphi_a^{(n-1)}(\xi_J) \ln^{[n]}(z_J - \xi_J) \right]_{\xi_J^1}^{\xi_J^2}, \\ \mathcal{U}_J^{Ua}(z_J) &= \sum_{n=0}^{2} (-1)^n \left[\varphi_a^{(n)}(\xi_J) \ln^{[n]}(z_J - \xi_J) \right]_{\xi_J^1}^{\xi_J^2}, \end{aligned} \tag{3.43}$$

where ξ_J^1 and ξ_J^2 are the values of ξ_J at end nodes 1 and 2, respectively, of Γ and $\varphi_a^{(n)}(\xi_J)$ is the nth derivative of the shape function with respect to ξ_J. Further, $\ln^{[n]}(z_J - \xi_J)$ is the n-th integral of $\ln(z_J - \xi_J)$ with respect to the argument ξ_J and is given by

$$\ln^{[n]}(z_J - \xi_J) = (-1)^n \frac{1}{n!}(z_J - \xi_J)^n \left\{ \ln(z_J - \xi_J) - \sum_{k=1}^{n} \frac{1}{k} \right\} \quad (n \geq 0). \tag{3.44}$$

For the negative integer n we interpret $\ln^{[n]}(z_J - \xi_J)$ as the derivative so that

$$\ln^{[-1]}(z_J - \xi_J) = -\frac{1}{z_J - \xi_J}, \quad \ln^{[-2]}(z_J - \xi_J) = -\frac{1}{(z_J - \xi_J)^2}. \tag{3.45}$$

As shown in Fig. 3.1, in each plane of the generalized complex variable $z_J = x_1 + p_J x_2$ $(J = 1, 2, 3, 4)$ the image Γ_J of the original boundary element Γ is defined. The branch cut of the logarithmic function $\ln(z_J - \xi_J)$, where ξ_J is a point on Γ_J is defined as follows. First, get the image ξ_J^1 of the node point ξ^1 (Fig. 3.1). Second, select the branch cut of the logarithmic function $\ln(z_J - \xi_J)$ to be a straight line emanating from ξ_J and extending indefinitely toward ξ_J^1. Next, calculate the principal value argument of $\xi_J^1 - \xi_J$ and set this value as the maximum $\arg_J^{max}$ of the angular range so that the argument of the logarithmic function $\ln(z_J - \xi_J)$ is defined by

$$\arg_J^{max} - \pi < \arg(z_J - \xi_J) \leq \arg_J^{max}. \tag{3.46}$$

Notice that the branch cut defined this way differs from element to element.

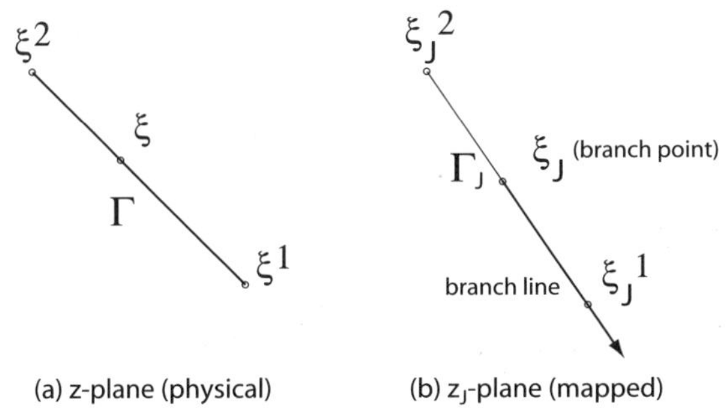

Fig. 3.1. Physical (z) and mapped ($z_J; J = 1, 2, 3, 4$) planes. The source point ξ is mapped to ξ_J. Images $\xi_J^1 (J = 1, 2, 3, 4)$ of the node point ξ^1 are used to define the branch cuts of $\log(z_J - \xi_J)$ in the mapped planes.

3.5. Numerical Green's Functions for Crack Modeling

Let the upper and lower faces of the 2-D crack be L^+ and L^-, respectively. Since the two faces of the crack coincide, the BEM formulation based on the mathematical interpretation of Somigliana's identity cannot determine the extended displacement uniquely along the crack surface. If we consider the crack as the limiting case of an ellipsoidal hole, then the physical interpretation of the extended Somigliana's identity leads naturally to a continuous distribution of extended dislocation dipoles along L^+ and L^- with zero extended traction. This distribution, when combined, will give the continuous distribution of the extended dislocation dipoles along $L = L^+$ with the magnitude $\delta_K = U_K^+ - U_K^-$, where U_K^+ and U_K^- are the extended displacement on the upper and lower surfaces of the crack and δ_K is the extended COD. The displacement due to the crack is given, by integrating (3.37) along the crack, by

$$U_J^{(d)}(\mathbf{x}) = -\Im\frac{1}{\pi}\int_L \sum_{N=1}^{4} A_{JN} \sum_{K=1}^{4} L_{KN}\delta_K \frac{d\xi_N}{z_N - \xi_N}. \tag{3.47}$$

Consider a straight crack in the interval $(-a, +a)$ on the x_1-axis where $\xi_N = \eta_1 + p_N\eta_2 = \eta_1$. Introduce non-dimensional variables $\eta = \eta_1/a$ and $Z_N = z_N/a$ and rewrite (3.47) as

$$U_J^{(d)}(\mathbf{x}) = \Im\frac{1}{\pi}\int_{-1}^{+1} \sum_{N=1}^{4} A_{JN} \sum_{K=1}^{4} L_{KN}\delta_K(\eta) \frac{d\eta}{\eta - Z_N}. \tag{3.48}$$

Interpolate the extended crack opening displacement by

$$\delta_K(\eta) = \sqrt{1-\eta^2}\sum_{m=1}^{M} \delta_K^{(m)} U_{m-1}(\eta), \tag{3.49}$$

where $U_{m-1}(\eta)$ is the Chebyshev polynomial of the second kind, to embed the crack tip singular behavior. Substitute (3.49) into (3.48) and evaluate the integral analytically to get,

$$U_J^{(d)}(\mathbf{x}) = -\Im\sum_{m=1}^{M}\sum_{N=1}^{4} A_{JN} \sum_{K=1}^{4} L_{KN}\delta_K^{(m)}\, R_m(Z_N), \tag{3.50}$$

where

$$R_m(Z_N) = \left(Z_N - \sqrt{(Z_N)^2 - 1}\right)^m \quad (m \geq 1). \tag{3.51}$$

Similarly, the extended stress function is given by

$$\Phi_J^{(d)}(\mathbf{x}) = -\Im \sum_{m=1}^{M} \sum_{N=1}^{4} L_{JN} \sum_{K=1}^{4} L_{KN}\delta_K^{(m)} \, R_m(Z_N), \tag{3.52}$$

from which the extended stress components are obtained, through (3.16), with the result,

$$\begin{aligned} \Sigma_{J2}^{(d)}(\mathbf{x}) &= -\frac{1}{a}\Im \sum_{m=1}^{M} \sum_{N=1}^{4} L_{JN} \sum_{K=1}^{4} L_{KN}\delta_K^{(m)} \, m\, G_{m-1}(Z_N), \\ \Sigma_{J1}^{(d)}(\mathbf{x}) &= \frac{1}{a}\Im \sum_{m=1}^{M} \sum_{N=1}^{4} p_N L_{JN} \sum_{K=1}^{4} L_{KN}\delta_K^{(m)} \, m\, G_{m-1}(Z_N), \end{aligned} \tag{3.53}$$

where

$$G^{(m-1)}(Z_N) = -\frac{\left(Z_N - \sqrt{(Z_N)^2 - 1}\right)^m}{\sqrt{(Z_N)^2 - 1}} \qquad (m \geq 1). \tag{3.54}$$

The traction on the top (+) and bottom (−) surfaces of the crack is given, through the limiting behavior of $\Sigma_{J2}^{(d)}(\mathbf{x})$ in (3.53), by

$$T_J^{\pm}(X) = \pm\frac{1}{a}\Im \sum_{m=1}^{M} \sum_{N=1}^{4} L_{JN} \sum_{K=1}^{4} L_{KN}\delta_K^{(m)} \, m\, U_{m-1}(X) \quad (|X| \leq 1), \tag{3.55}$$

where $X = x_1/a$. Observation of the limiting behavior in front of each crack tip at $X = \pm 1$ will yield

$$K_J(\pm 1) = \sqrt{\frac{\pi}{a}}\Im \sum_{m=1}^{M} (\pm)^{m+1} \sum_{N=1}^{4} L_{JN} \sum_{K=1}^{4} L_{KN}\delta_K^{(m)} \, m, \tag{3.56}$$

which give Mode I ($K_I = K_2$), Mode II ($K_{II} = K_1$), Mode III ($K_{III} = K_3$) and Mode IV ($K_{IV} = K_4$) extended SIFs. Inspired by this equation, we introduce an influence coefficient,

$$K_{JI} = \sqrt{\frac{\pi}{a}}\Im \sum_{N=1}^{4} L_{JN}L_{IN}, \tag{3.57}$$

which gives the extended stress intensity factor K_J due to the unit extended crack opening displacement component $\delta_I^{(1)}$.

Consider, for example, an impermeable crack with zero extended traction on the crack surface. Under a given set of the remote loading, we can use the superposition principle to determine the extended COD coefficients $\delta_K^{(m)}$ such that, along the crack contour L, the sum of the traction due to the remote loading and that by the extended dislocation dipole distribution, (3.55), satisfies the given extended crack surface traction BC. The value of the crack surface extended traction is zero for the impermeable crack, but non-zero for the permeable and semipermeable cracks. The extended COD coefficients $\delta_K^{(m)}$ ($m = 1, 2, \ldots, M$) are determined by evaluating the extended traction at M collocation points,

$$\begin{cases} \eta_1 = 0.0 & (M = 1) \\ \eta_1 = -1.0, \eta_M = 1.0 & (M \geq 2) \end{cases}, \tag{3.58}$$

obtained by dividing the crack interval $-1 \leq \eta \leq +1$ equally. Notice that the crack tips are included among the collocation points. This is possible since the extended traction is bounded at the crack tips as seen from (3.55).

Formulas developed above satisfy one requirement for the Green's function to have the built-in singularity at the crack tips. Another requirement to satisfy the crack surface boundary condition is not fulfilled automatically, but a procedure is set up to satisfy it numerically. This is a typical approach of the numerical Green's function and it makes sense for the fracture analysis of the piezoelectric solids, where multiple options for the crack surface BC exist. Equations (3.50) and (3.52) give the numerical Green's functions for the extended displacement and stress function vectors in terms of the unknown extended COD coefficients $\delta_K^{(m)}$. Once these coefficients are determined, through the procedure described above, the formula (3.56) gives the extended SIFs explicitly and there is no need for the post-processing. The crack element developed here is called the whole crack singular element (WCSE). For the inclined crack, we introduce a pair of rotated local coordinate axes parallel and perpendicular to the crack. In applying formulas (3.50)–(3.56), the compliance, characteristic roots and matrices $\mathbf{L}$ and $\mathbf{A}$ need to be redefined to reflect the coordinate rotation. The application of the WCSE can readily be extended to multiple cracks in the finite domain with the introduction of the boundary elements as described in the next section. Similar approach for the numerical Green's function by Telles and Guimarães[59] uses numerical integration of integrals similar to (3.47), which is more time consuming but flexible and not limited to the straight crack.

3.6. Crack Surface Electric BCs and Solution Algorithms

3.6.1. *Impermeable and permeable cracks*

Consider a finite body subject to the extended boundary displacement $\{\mathbf{U}\}$ and traction $\{\mathbf{T}\}$ vectors that contains N straight impermeable cracks. In addition to the global coordinate system, a local coordinate with its origin at the crack center and the horizontal axis along the crack is introduced for each crack. The extended COD of n-th crack $(n = 1, 2, \cdots, N)$ is interpolated by $M = M_n$ Chebyshev polynomials in (3.49) introducing a $4M_n$-dimensional extended COD coefficient vector $\{\boldsymbol{\delta}\}_n$, which is determined by setting the extended crack surface traction to zero,

$$\{\mathbf{T}^C\}_n = [\mathbf{H}]_n\{\mathbf{U}\} + [\mathbf{G}]_n\{\mathbf{T}\} + \sum_{j=1}^{N}[\mathbf{D}]_{nj}\{\boldsymbol{\delta}\}_j = \{\mathbf{0}\} \qquad (n = 1, 2, \cdots, N), \tag{3.59}$$

at M_n collocation points. Here $\{\mathbf{T}^C\}_n$ is the $4M_n$-dimensional global extended traction vector. The coefficient matrices $[\mathbf{H}]_n$, $[\mathbf{G}]_n$ and $[\mathbf{D}]_{nj}$ represent contributions from the boundary displacement, traction and j-th crack, respectively. An additional system of the extended boundary displacement equations is given by

$$\{\mathbf{U}\} = [\mathbf{H}]\{\mathbf{U}\} + [\mathbf{G}]\{\mathbf{T}\} + \sum_{j=1}^{N}[\mathbf{D}]_j\{\boldsymbol{\delta}\}_j, \tag{3.60}$$

where coefficient matrices $[\mathbf{H}]$, $[\mathbf{G}]$ and $[\mathbf{D}]_j$ represent contributions from the extended boundary displacement, traction and j-th crack, respectively.

For multiple permeable cracks, replace the last component, $\delta_4 = 0$, of the extended COD with the last component, T_4^C, of the extended crack surface traction to introduce a four-dimensional unknown vector $\boldsymbol{\delta}_n^* = (\delta_1, \delta_2, \delta_3, T_4^C)_n$ for n-th crack. Swap δ_4 and T_4^C in (3.59) to produce a modified system of equations for the new unknowns $\boldsymbol{\delta}_n^*$. In (3.60), the extended COD for each permeable crack has only three components since $\delta_4 = 0$. In (3.59) and (3.60) the local contributions from each crack, given in terms of the local components, must be converted to the global components before adding them for all cracks.

3.6.2. *Semipermeable cracks*

Consider a finite body with N straight semipermeable cracks. Interpolate, following (3.49), the extended COD for n-th crack ($n = 1, \ldots, N$) by

$$\delta_K^n(\eta) = \sqrt{1-\eta^2} \sum_{m=1}^{M} \delta_K^{n(m)} U_{m-1}(\eta) \tag{3.61}$$

and the extended traction component T_4^{Cn} on the upper surface of the crack by

$$T_4^{Cn}(\eta) = \sum_{m=1}^{M} T_4^{Cn(m)} U_{m-1}(\eta), \tag{3.62}$$

in terms of the normalized parameter $-1 \le \eta \le +1$ introduced along the crack. The number M of interpolation polynomials in (3.61) and (3.62) is assumed to be the same for all cracks. Notice that $\delta_4^{n(m)} \neq 0$ and $T_4^{Cn(m)} = 0$ for for the impermeable crack while $\delta_4^{n(m)} = 0$ and $T_4^{Cn(m)} \neq 0$ for the permeable crack. The proposed iteration procedure for the semipermeable multiple cracks consists of two stages: the preliminary and the refinement stages. First, apply the preliminary iteration to find the approximate location of the semipermeable solution in-between the impermeable and permeable solutions.

1. Use the impermeable BEM solver to obtain the impermeable solution $\delta_4^{n(m)[0]}$, where the superscript [0] indicates the initial value and, in the following, the superscript [k] attached to variables indicates k-th iteration values.
2. (a) With a control parameter $p = p^{[1]} < 1.0$, set a slightly reduced value of $\delta_4^{n(m)[1]} = p^{[1]} * \delta_4^{n(m)[0]}$.
   (b) Use the permeable BEM solver to calculate $\delta_2^{n(m)[1]}$ and $T_4^{Cn(m)[1]}$ corresponding to $\delta_4^{n(m)[1]}$.
   (c) Use (3.1) and the relation (3.17) to calculate

$$\kappa^{Cn[1]}(\zeta_i) = T_4^{Cn[1]}(\zeta_i) \frac{\delta_2^{n[1]}(\zeta_i)}{\delta_4^{n[1]}(\zeta_i)} \tag{3.63}$$

at M sample points ζ_i $(i = 1, \ldots, M)$ on each crack. The sample points ζ_i are given by the Gaussian quadrature points, which are different from (3.58) used for the extended traction calculation. Here $\delta_2^{n[1]}(\zeta_i)$, $\delta_4^{n[1]}(\zeta_i)$ and $T_4^{Cn[1]}(\zeta_i)$ are given by (3.61) and (3.62).

(d) Calculate the average

$$\bar{\kappa}^{Cn[1]} = \frac{\sum_{i=1}^{M} \kappa^{Cn[1]}(\zeta_i)}{M} \tag{3.64}$$

for each crack.

3. Repeat STEP 2, R times, for gradually reduced values of $\delta_4^{n(m)[k]} = p^{[k]} * \delta_4^{n(m)[0]}$ for the control parameter $p = p^{[k]} (k = 2, 3, 4, \ldots,$ R) and plot the $p^{[k]} - \bar{\kappa}^{Cn[k]}$ curve for each crack.
4. Get the intersection of the $p^{[k]} - \bar{\kappa}^{Cn[k]}$ curve with the horizontal line $\kappa^C = \kappa^{air}$ that gives the best fit parameter value $p^{n[best]}$ for each crack. Use the permeable BEM solver with

$$\delta_4^{n(m)[best]} = p^{n[best]} * \delta_4^{n(m)[0]} \quad \text{(no sum on n)} \tag{3.65}$$

to calculate $\delta_2^{n(m)[best]}$ and $T_4^{Cn(m)[best]}$.

In the preliminary iteration stage we scan R multiple solutions between the impermeable $(p^{[0]} = 1)$ and permeable $(p^{[R]} = 0)$ cracks keeping the modal shape of $\delta_4^{n(m)[k]} = p^{[k]} * \delta_4^{n(m)[0]}$ always similar. With the exception of the single crack, however, this strategy does not get the accurate semipermeable solution for multiple cracks. The refinement stage of the iteration starts with the solution, given by (3.65), and an effort is made to modify this modal shape.

5. Select the starting values of the refined iteration by adopting the best values obtained in the preliminary stage: $\delta_4^{n(m)[0]} = \delta_4^{n(m)[best]}$, $\delta_2^{n(m)[0]} = \delta_2^{n(m)[best]}$ and $T_4^{Cn(m)[0]} = T_4^{Cn(m)[best]}$. Notice that the notation $\delta_4^{n(m)[0]}$ is reused here with a different definition from STEP 1.
6. Enforce the semipermeable relation (3.1) to calculate

$$\delta_4^{n[prj]}(\eta_i) = \frac{T_4^{Cn[0]}(\eta_i) * \delta_2^{n[0]}(\eta_i)}{\kappa^C} \tag{3.66}$$

at M collocation points η_i defined by (3.58), where $\delta_2^{n[0]}(\eta_i)$ and $T_4^{Cn[0]}(\eta_i)$ are given by (3.61) and (3.62). Solve the system of equations,

$$\sum_{m=1}^{M} U_{m-1}(\eta_i)\delta_4^{n(m)[prj]} = \delta_4^{n[prj]}(\eta_i),$$

to calculate M projected coefficients $\delta_4^{n(m)[prj]}$ for each crack n. Although the projected modal shape, $\delta_4^{n(m)[prj]}$, pushes the solution in the right direction, it is not close enough from the semipermeable solution requiring to sweep another set of multiple solutions between $\delta_4^{n(m)[prj]}$ and $\delta_4^{n(m)[0]}$, where $\delta_4^{n(m)[0]} = \delta_4^{n(m)[best]}$ defined in STEP 5 is the best solution after the preliminary iteration.

7. Divide the interval $\|\delta_4^{n(m)}\| = \delta_4^{n(m)[prj]} - \delta_4^{n(m)[0]}$ by L equal subintervals of magnitude $\Delta\delta_4^{n(m)} = \|\delta_4^{n(m)}\|/L$. Run the permeable BEM solver for L values of $\delta_4^{n(m)[l]} = \delta_4^{n(m)[0]} + l * \Delta\delta_4^{n(m)}$ $(l = 1, \ldots, L)$ to calculate $\delta_2^{n(m)[l]}$ and $T_4^{Cn(m)[l]}$.
8. Follow STEPs 2 (c) to (d) to calculate the average $\bar{\kappa}^{Cn[l]}$ for each crack according to (3.64).
9. Following the step similar to STEP 4, obtain the best fit $\delta_4^{n(m)[best]}$ for each crack and calculate $\delta_2^{n(m)[best]}$ and $T_4^{Cn(m)[best]}$.
10. Repeat STEPs 5 to 9 a total of S times if necessary.

Since the extended COD for the single crack is given by one term (i.e. $M = 1$) of the interpolation in (3.49), the ratio of δ_2/δ_4 in (3.63) is constant $\delta_2^{(1)}/\delta_4^{(1)}$ over the entire crack surface along with the electric induction. Therefore, the calculation of κ^C by (3.63) requires only one sample point on the crack, typically at the center. On the other hand, the interpolation of the extended COD for multiple cracks requires multiple polynomial terms $U_{m-1}(\eta)$ in (3.49), and the ratio δ_2/δ_4 is not constant over the crack surface.

3.7. Material Constants

In addition to the constitutive equations, (3.4) and (3.5), there exist another set of constitutive equations that gives the strain and electric induction components in terms of the stress and electric field components,

$$\epsilon_{ij} = s_{ijkl}^{E}\sigma_{kl} + d_{kij}E_k, \quad D_i = d_{ikl}\sigma_{kl} + \kappa_{ik}^{\sigma}E_k, \qquad (3.67)$$

where s_{ijkl}^{E}, d_{kij}, and κ_{ik}^{σ} are the elastic compliance constants at constant electric field, piezoelectric strain constants, and dielectric permittivity constants at constant stress, respectively. The constants required in our analysis, $(s^D_{ijkl}, g_{kij}, \beta^\sigma_{ik})$, need to be obtained in terms of the other sets of constants, $(s^E_{ijkl}, d_{kij}, \kappa^\sigma_{ik})$ and $(c^E_{ijkl}, e_{ikl}, \kappa^\varepsilon_{ik})$, which are available in the literature.[60] For example, given the material constants $(c^E_{ijkl}, e_{ikl}, \kappa^\varepsilon_{ik})$ as the input, we can calculate other constants as follows:

$$\begin{aligned} s^E_{ijkl} &= (c^E_{ijkl})^{-1} \rightarrow d_{ijk} = e_{ilm}s^E_{lmjk} \rightarrow \kappa^\sigma_{ij} = \kappa^\varepsilon_{ij} + d_{ikl}e_{jkl} \\ &\rightarrow \beta^\sigma_{ij} = (\kappa^\sigma_{ij})^{-1} \rightarrow g_{ijk} = \beta^\sigma_{il}d_{ljk} \rightarrow s^D_{ijkl} = s^E_{ijkl} - d_{mij}g_{mkl}. \end{aligned} \tag{3.68}$$

The stress, strain, electric induction and electric field are of the order of $\sigma_0 = 10^8$ (N/m^2), $\varepsilon_0 = 10^{-3}$, $D_0 = 10^{-2}$ (C/m^2), and $E_0 = 10^7$ (V/m), respectively, and the elastic stiffness, piezoelectric stress and dielectric permittivity constants have the order of $c_0 = 10^{11}$ (N/m^2), $e_0 = 10^1$ (C/m^2) and $\kappa_0 = 10^{-9}$ $(C/(mV))$, respectively. The truncation error, caused by the wide variation in the order of magnitudes for the piezoelectric variables and coefficients, can be avoided by their normalization. For example, $\bar{q} = q/q_0$ gives the normalization of a dimensional quantity q by its reference quantity q_0, examples of which are given above. Given the reference values of the stress, strain, electric induction and electric field, those for other quantities are determined in terms of these four reference variables such that the normalized governing equations remain exactly the same as the original equations. See Table 3.1 for a complete list of the

Table 3.1. The reference values for the material constants and the field variables in piezoelectricity.

Displacement	$u_0 = x_0\varepsilon_0 = 10^{-3}x_0$ (m)	Elec. Potential	$\phi_0 = x_0E_0 = 10^7x_0$ (V)
Stiffness	$c_0 = \dfrac{\sigma_0}{\varepsilon_0} = 10^{11}$ (N/m^2)	Permittivity	$\kappa_0 = \dfrac{D_0}{E_0} = 10^{-9}$ (C/mV)
Compliance	$s_0 = \dfrac{\varepsilon_0}{\sigma_0} = 10^{-11}$ (m^2/N)	Impermeability	$\beta_0 = \dfrac{E_0}{D_0} = 10^9$ (mV/C)
Piezoelectric Stress	$e_0 = \dfrac{\sigma_0}{E_0} = 10^1$ (N/mV) $= \dfrac{D_0}{\varepsilon_0} = 10^1$ (C/m^2)	Piezoelectric Strain	$g_0 = \dfrac{E_0}{\sigma_0} = 10^{-1}$ (mV/N) $= \dfrac{\varepsilon_0}{D_0} = 10^{-1}$ (m^2/C)
$K_{I,II,III}$	$K_{I0,II0,III0} = 10^{13/2}x_0^{1/2}(N/m^{3/2})$	K_{IV}	$K_{IV0} = 10^{-7/2}x_0^{1/2}(C/m^{3/2})$

reference values, where $x_0(m)$ is the characteristic length of the problem. In this paper $x_0 = 1(m)$ is used.

The piezoelectric solids considered here is Barium Sodium Niobate ($Ba_2NaNb_5O_{15}$, Orthogonal $2mm$) -BSN- with the material constants

$$[c^E_{\mathcal{IJ}}] = \begin{bmatrix} 23.9, & 10.4, & 5.0, & 0.0, & 0.0, & 0.0 \\ & 24.7, & 5.2, & 0.0, & 0.0, & 0.0 \\ & & 13.5, & 0.0, & 0.0, & 0.0 \\ & & & 6.5, & 0.0, & 0.0 \\ & & & & 6.6, & 0.0 \\ & & & & & 7.6 \end{bmatrix} \times 10^{10}\,(N/m^2),$$

$$[e_{i\mathcal{J}}] = \begin{bmatrix} 0.0, & 0.0, & 0.0, & 0.0, & 2.8, & 0.0 \\ 0.0, & 0.0, & 0.0, & 3.4, & 0.0, & 0.0 \\ -0.4, & -0.3, & 4.3, & 0.0, & 0.0, & 0.0 \end{bmatrix} (C/m^2),$$

$$[\kappa^\varepsilon_{ij}] = \begin{bmatrix} 222.0, & 0.0, & 0.0 \\ & 227.0, & 0.0 \\ & & 32.0 \end{bmatrix} \times 8.854 \times 10^{-12}\,(farads/m). \quad (3.69)$$

Only the upper triangular half of the symmetric matrices are shown. BSN has the electrical/out-of-plane mechanical coupling with no in-plane/out-of-plane mechanical and electrical/in-plane mechanical couplings as can be seen from its influence coefficients, defined by (3.57),

$$[\bar{K}_{JI}] = \begin{bmatrix} 0.900143, & 0.0, & 0, & 0.0 \\ 0.0, & 0.915084, & 0.0, & 0.0 \\ 0.0, & 0.0, & 0.580679, & 0.274588 \\ 0.0, & 0.0, & 0.274588, & -1.7619 \end{bmatrix}, \quad (3.70)$$

in which $\bar{K}_{34}$ and $\bar{K}_{43}$ are nonzero.

3.8. Numerical Results

3.8.1. *Upper and lower bound analysis*

For all piezoelectric multiple crack configurations, the permeable and impermeable crack solutions set the upper and lower bounds for the crack surface electric induction, electric potential jump and the electric induction intensity factor $\bar{K}_{IV}$. The Mode I, II, and III SIFs and the crack opening displacement components $\bar{\delta}_1$, $\bar{\delta}_2$, and $\bar{\delta}_3$ are identical for permeable and impermeable cracks. The semipermeable crack solution should be located somewhere in-between these two solutions. Even though they are not correct, their upper and lower bounds property can be exploited in the fracture analysis if the semipermeable crack solution is not available. Detailed upper and lower bounds analysis can be found in Denda and Mansukh.[12]

3.8.2. *Semipermeable cracks*

3.8.2.1. *Single crack in an infinite body*

Consider the single crack in an infinite BSN subjected to the combined remote loadings $\bar{\sigma}_{22}^{\infty} = 1$ (tension) and $\bar{D}_2^{\infty} = 1$ (electric induction), resulting in $\bar{\delta}_2$ (opening) and $\bar{\delta}_4$ (potential jump) at the same time. Due to the absence of the electrical/in-plane coupling the single application of either $\bar{\sigma}_{22}^{\infty}$ or $\bar{D}_2^{\infty}$ for the impermeable crack produces $\bar{\delta}_2$ or $\bar{\delta}_4$ only, which creates no inconsistency. For the permeable crack, since the electric potential jump is always zero for both loadings, each crack solution for individual loading is always consistent. Therefore, although the impermeable and permeable BCs produce different electric induction on the crack faces, both solutions are consistent and there is no need to come up with the semipermeable BC if only one of the loadings, $\bar{\sigma}_{22}^{\infty}$ or $\bar{D}_2^{\infty}$, is separately applied. The consideration of the semipermeable boundary condition becomes necessary only when the both loadings are applied.

For the single crack in the infinite body, only the preliminary iteration, consisting of STEPs 1 to 4, is needed to get the accurate semipermeable solution. This is because only one term in the interpolation (3.49) is sufficient to describe the extended crack opening displacement, which remains similar among the impermeable, permeable and semipermeable cracks. The iteration was performed $R = 1000$ times decreasing the parameter p evenly from 1 to 0. The variation of $\bar{\kappa}^C$ as the function of the control parameter p is shown in Fig. 3.2(a), in which the intersection of this curve with the horizontal line $\bar{\kappa}^{air} = 8.854$ occurs at $p^{(smp)} = 0.189292$. The variation of $\bar{K}_{IV}$ and $\bar{D}_2^C$ is shown in Figs. 3.2(b) and (c). Since they vary linearly between the impermeable and permeable conditions, the semipermeable solutions are given by $\bar{K}_{IV}^{(smp)} = p^{(smp)}\bar{K}_{IV}^{(imp)}$ and $\bar{D}_2^{C(smp)} = (1 - p^{(smp)})\bar{D}_2^{C(per)}$, where $\bar{K}_{IV}^{(imp)} = \sqrt{\pi}$ and $\bar{D}_2^{C(per)} = 1$. The value of $\bar{D}_2^{C(smp)} = 0.810708$ agrees perfectly with the theoretical value $\bar{D}_4^{C(smp)} = 0.810709$ given by Denda and Mansukh.[12] Note that $\bar{K}_{III}$ is very small, but not zero as shown in Fig. 3.2(d); this comes from the electrical/out-of-plane coupling. Although not shown, $\bar{K}_I$ and $\bar{\delta}_2$ remain

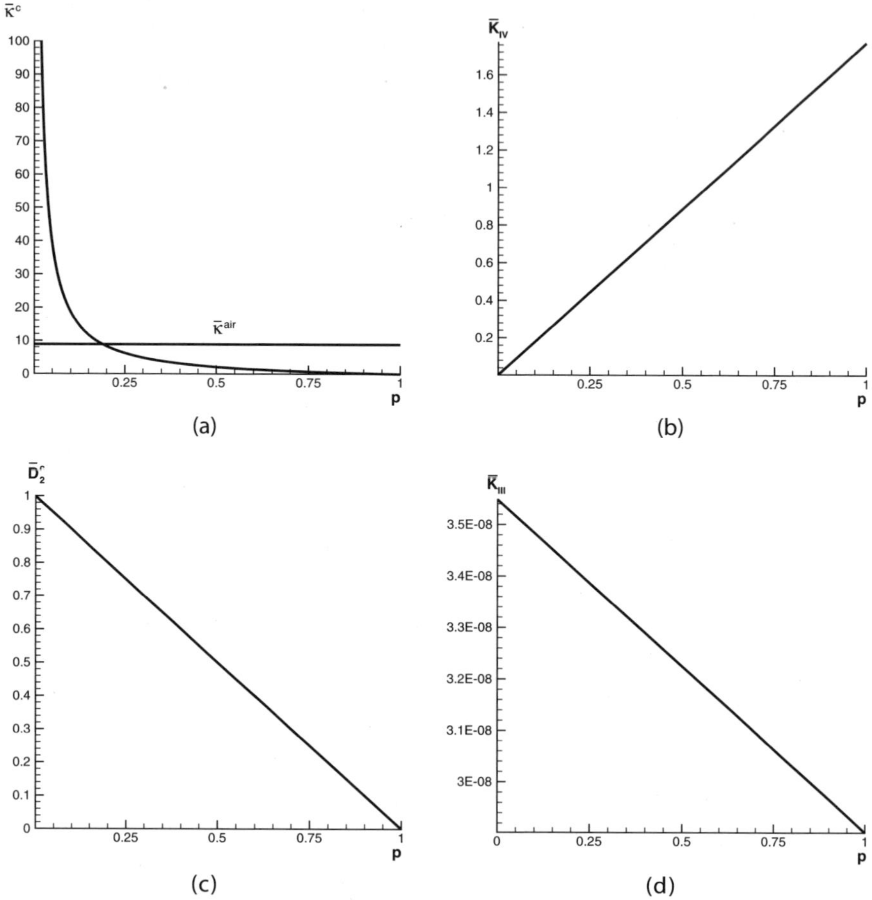

Fig. 3.2. Variation of (a) $\bar{\kappa}^C$, (b) $\bar{K}_{IV}$, (c) $\bar{D}_2^C$ and (d) $\bar{K}_{III}$ in the iteration when p is varied from 1 to 0 for BSN under $\bar{\sigma}_{22}^{\infty} = 1$ (tension) and $\bar{D}_2^{\infty} = 1$ (electric induction).

constant throughout the iteration and $\bar{K}_{II}$ is zero. The variation of (a) $\bar{\delta}_4$ and (b) $\bar{D}_2^C$ along the crack for impermeable, semipermeable and permeable cracks, respectively, is shown in Figs. 3.3(a) and (b).

3.8.2.2. *Single crack in a finite body*

Consider a crack ($\bar{a} = 1$) in a finite BSN body ($\bar{H} = \bar{W} = 4$) under the unit normalized tension $\bar{\sigma}_{22} = 1$ and electric induction $\bar{D}_2 = 1$, as shown in Fig. 3.4(a). We have used $M = 7$ terms in the interpolation (3.49) and $\bar{\kappa}^C$, calculated by (3.63), is not constant along the crack. As

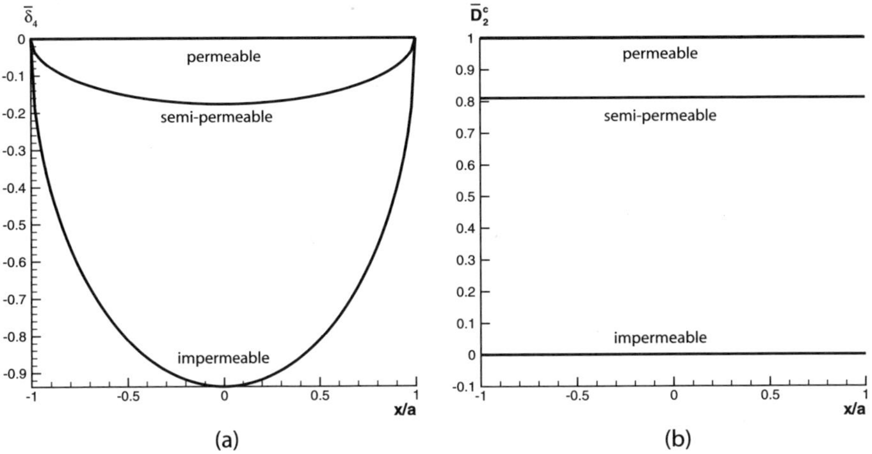

Fig. 3.3. Variation of (a) $\bar{\delta}_4$ and (b) $\bar{D}_2^C$ along the crack for impermeable, semipermeable and permeable cracks for BSN under $\bar{\sigma}_{22}^{\infty} = 1$ (tension) and $\bar{D}_2^{\infty} = 1$ (electric induction).

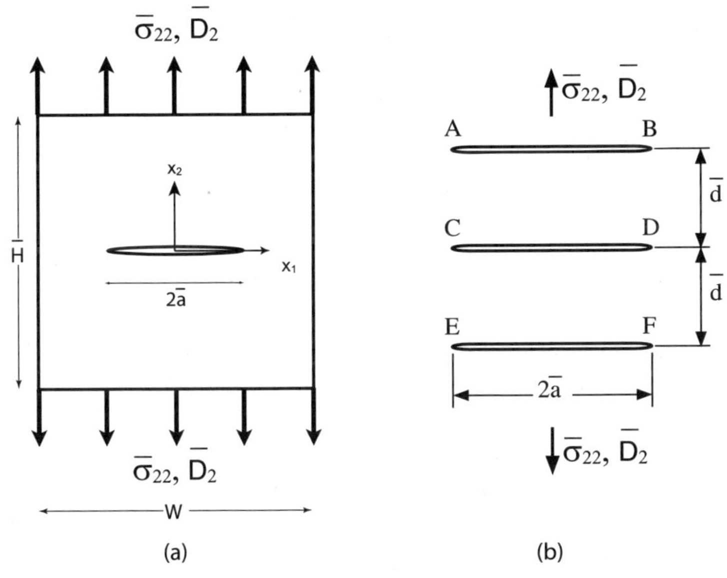

Fig. 3.4. (a) A center crack in a finite body ($\bar{H} = \bar{W} = 4$) under uniaxial tension $\bar{\sigma}_{22}$ and electric induction $\bar{D}_2$. (b) Three aligned parallel cracks ($2\bar{a}/\bar{d} = 4$) in an infinite body under uniaxial tension $\bar{\sigma}_{22}$ and electric induction $\bar{D}_2$.

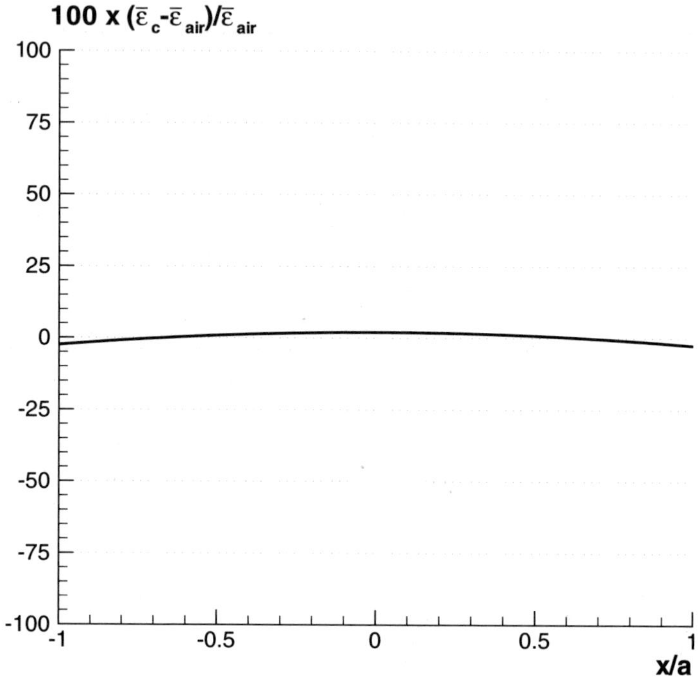

Fig. 3.5. Relative error in $\bar{\epsilon}_C$ over the semipermeable crack surface for a crack in a finite body (Fig. 3.4(a)) under uniaxial tension and electric induction. Only the preliminary iteration is applied.

in (3.64) the average value was used to calculate $\bar{\kappa}^C$. The preliminary stage of the iteration using STEPs 1 to 4 with $R = 1,000$ iterations gave the best approximate solution for $\bar{\kappa}^C$, which has more than 2% deviation from the required constant value $\bar{\kappa}^{air}$ as shown in Fig. 3.5. The refinement stage following the preliminary (with $R = 10$ iterations)was applied, using $S = 10$ iteration in STEPs 5 to 10 (with $L = 20$ in STEP 7). The value of $\bar{\kappa}^C$ has agreed with that for $\bar{\kappa}^{air}$ up to four significant digits for the entire crack length. The extended stress intensity factor is given by $\bar{K}_I^{(smp)} = 2.36716$, $\bar{K}_{II}^{(smp)} = 0.0$, $\bar{K}_{III}^{(smp)} = -0.802325 \times 10^{-4}$ and $\bar{K}_{IV}^{(smp)} = 0.437143$. The electric intensity factors of the impermeable and permeable cracks are given by $\bar{K}_{IV}^{(imp)} = 2.00409$ and $\bar{K}_{IV}^{(per)} = 0.0$, respectively. Other stress intensity factors ($\bar{K}_I$, $\bar{K}_{II}$ and $\bar{K}_{III}$) remain the same for three crack BCs. The electric potential jump and electric induction field over the crack are shown

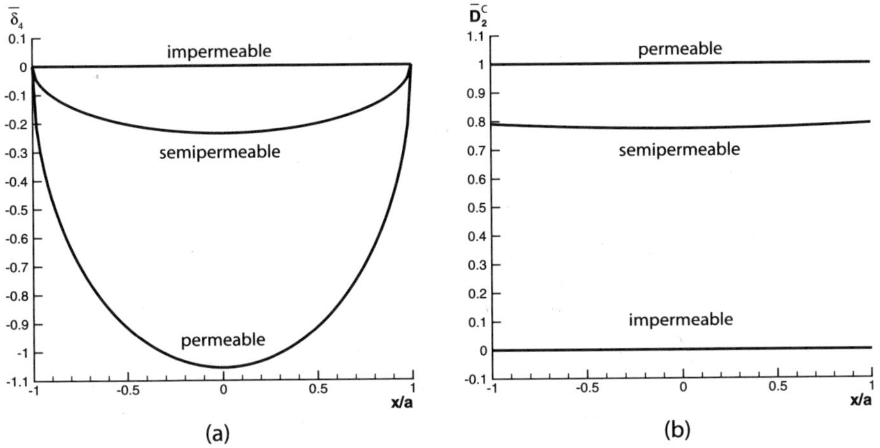

Fig. 3.6. Variation of (a) $\bar{\delta}_4$ and (b) $\bar{D}_2^C$ over impermeable, semipermeable and permeable conditions for a crack in a finite body (Fig. 3.4(a)) under uniaxial tension and electric induction.

in Figs. 3.6(a) and (b). Notice that the $\bar{D}_2$ for the semipermeable crack is not constant along the crack in comparison to the constant values for the impermeable and permeable cracks. The crack opening $\bar{\delta}_2$ is the same for all three cracks.

3.8.2.3. *Three parallel cracks*

The problem of three parallel cracks of Fig. 3.4(b) was considered since the solutions for the upper (or lower) and the middle cracks are different. The number of polynomials used in the interpolation (3.49) is $M = 7$ for each crack. The preliminary and refined iteration schemes were applied using $R = 10$ iterations in STEPs 1 to 4, $S = 10$ iterations in STEPs 5 to 10 and $L = 20$ in STEP 7. The value of $\bar{\kappa}^C$ calculated by (3.63) along three cracks agrees with $\bar{\kappa}^{air}$ up to three significant digits. Remote loading consisits of $\bar{\sigma}_{22}^{\infty} = 1$ (tension) and $\bar{D}_2^{\infty} = 1$ (electric induction). As seen in Table 3.2, the upper and lower bounds properties of the electric intensity factor for the impermeable (IP) and permeable (P) cracks are clearly demonstrated. The variation of (a) $\bar{\delta}_4$ and (b) $\bar{D}_2^c$ along the top crack AB (or bottom crack EF) in Fig. 3.4(b) for impermeable, semipermeable and permeable cracks, respectively, is shown in Fig. 3.7(a) and (b). Figure 3.8 shows similar figures for the middle crack CD. Although not shown, $\bar{K}_I$ and $\bar{\delta}_2$ remain constant throughout the iteration and $\bar{K}_{III}$ is very small.

Table 3.2. Extended SIFs for three parallel cracks (Fig. 3.4(b)) in infinite BSN under $\bar{\sigma}_{22}^{\infty} = 1$ and $\bar{D}_2^{\infty} = 1$.

SIF	BC	Tip A	Tip C
$\bar{K}_I/\sqrt{\pi}$	impermeable	0.6955	0.4516
	semipermeable	0.6955	0.4516
	permeable	0.6955	0.4516
$\bar{K}_{II}/\sqrt{\pi}$	impermeable	−0.1465	0.0000
	semipermeable	−0.1465	0.0000
	permeable	−0.1465	0.0000
$\bar{K}_{IV}/\sqrt{\pi}$	impermeable	0.7490	0.6160
	semipermeable	0.1331	0.08525
	permeable	0.0000	0.0000

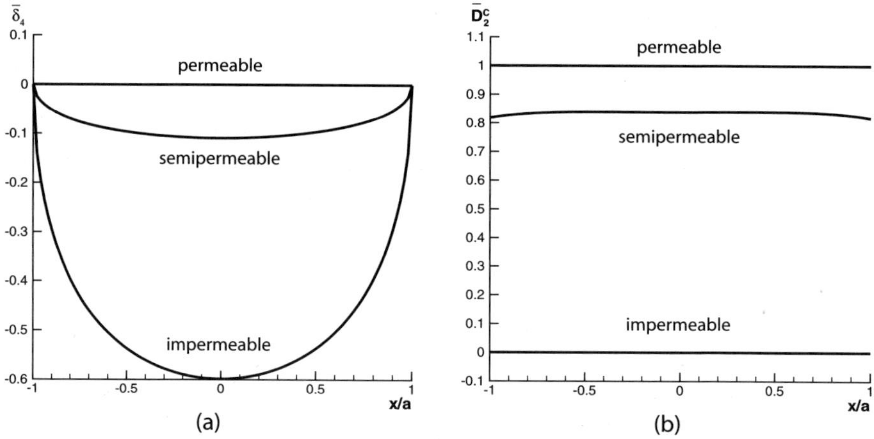

Fig. 3.7. Variation of (a) $\bar{\delta}_4$ and (b) $\bar{D}_2^C$ along the top crack AB (or bottom crack EF) in Fig. 3.4(b) for impermeable, semipermeable and permeable cracks for BSN under $\bar{\sigma}_{22}^{\infty} = 1$ (tension) and $\bar{D}_2^{\infty} = 1$ (electric induction).

3.9. Concluding Remarks

In this article the boundary element method for the fracture analysis of the general piezoelectric solids in two-dimensions has been presented along with the solution algorithms for impermeable, permeable and semipermeable crack face electric boundary conditions. Several innovative approaches not used in the traditional boundary element methods have been introduced: physical interpretation of the extended Somigliana's identity,

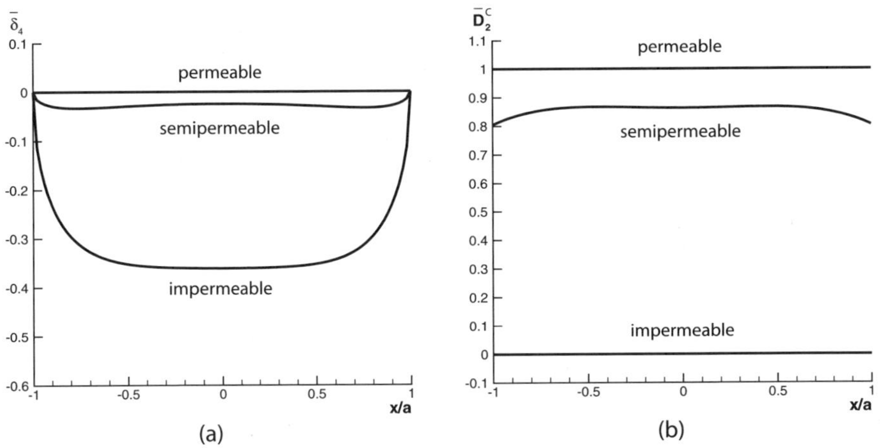

Fig. 3.8. Variation of (a) $\bar{\delta}_4$ and (b) $\bar{D}_2^C$ along the middle crack CD in Fig. 3.4(b) for impermeable, semipermeable and permeable cracks for BSN under $\bar{\sigma}_{22}^{\infty} = 1$ (tension) and $\bar{D}_2^{\infty} = 1$ (electric induction).

use of the extended dislocation for crack modeling, analytical representation of the crack tip singularity, and the solution algorithm for the multiple semipermeable cracks based on the impermeable and permeable crack solvers.

The numerical Green's function based on the WCSE developed here can deal with straight center cracks only. The extension of the numerical Green's function for edge cracks and non-straight cracks, developed for the general anisotropic solids (Crack-Tip Singular Element, CTSE, by Denda and Marante[61]), is in the list of future work. Note that the general solution algorithms developed for the impermeable, permeable and semipermeable cracks here are applicable to the edge and/or curvilinear crack modeled by the CTSE.

References

1. S. Lekhnitskii, *Theory of Elasticity of an Anisotropic Elastic Body* (Hoden-Day, San Francisco, 1963).
2. G. Blandford, A. Ingraffea, and J. Liggett, Two-dimensional stress intensity factor computations using the boundary element method, *Int. J. Numer. Methods Eng.* **17**, 387–404, (1981).
3. G. Kuhn, Numerische behandlung von mehrfachrissen in ebenen scheiben, *ZAMM.* **61**, 105–106, (1981).

4. H. Hong and J. Chen, Derivatives of integral equations of elasticity, *J. Engng. Mech.* **114**(6), 1028–1044, (1988).
5. A. Portela and M. Aliabadi, The dual boundary element method: Effective implementation for crack problems, *Int. J. Numer. Methods Eng.* **33**, 1269–1287, (1992).
6. P. Sollero and M. Aliabadi, Fracture mechanics analysis of anisotropic plates by the boundary element method, *Int. J. Fract.* **64**(4), 269–284, (1993).
7. M. Denda, A dislocation and point force approach to the boundary element method for mixed mode crack analysis of plane anisotropic solids, *J. Chinese Institute of Engineers* **22**(6), 677–693, (1999).
8. M. Denda, Mixed mode I, II and III analysis of multiple cracks in plane anisotropic solids by the BEM: A dislocation and point force approach, *Engng. Anal. with Boundary Elements* **25**(4–5), 267–278, (2001).
9. C. Tan and Y. Gao, Boundary element analysis of plane anisotropic bodies with stress concentrations and cracks, *Composite Structures* **20**, 17–28, (1992).
10. M. Snyder and T. Cruse, Boundary integral equation analysis of cracked anisotropic plates, *Int. J. Frac.* **11**, 315–328, (1975).
11. M. Denda and E. Mattingly, The whole crack singular element for 2-D boundary element analysis of multiple straight cracks in the general anisotropic solids, *Electronic J. Boundary Elements* **1**(3), 404–417, (2003).
12. M. Denda and M. Mansukh, Upper and lower bounds analysis of electric induction intensity factors for multiple piezoelectric cracks by the BEM, *Engng. Anal. with Boundary Elements* **29**, 533–550, (2005).
13. T. Hao and Z. Shen, A new electric boundary condition of electric fracture mechanics and its applications, *Eng. Frac. Mech.* **47**, 793–802, (1994).
14. W. Deeg, *The analysis of dislocation, crack and inclusion in piezoelectric solids.* PhD thesis, Stanford University, (1980).
15. H. Sosa and Y. Pak, Three-dimensional eigenfunction analysis of a crack in a piezoelectric material, *Int. J. Solids Struct.* **26**, 1–15, (1990).
16. H. Sosa, Plane problems in piezoelectric media with defects, *Int. J. Solids Struct.* **28**, 491–505, (1991).
17. H. Sosa, On the fracture mechanics of piezoelectric solids, *Int. J. Solids Struct.* **29**, 2613–2622, (1992).
18. Y. Pak, Linear electro-elastic fracture mechanics of piezoelectric materials, *Int. J. Fract.* **54**, 79–100, (1992).
19. P. Park and C. Sun, Effect of electric field on fracture of piezoelectric ceramics, *Int. J. Fract.* **70**, 203–216, (1995).
20. P. Park and C. Sun, Fracture criteria for piezoelectric ceramics, *J. Am. Ceramic Soc.* **78**, 1475–1480, (1995).
21. S. Kumar and R. Singh, Energy release rate and crack propagation in piezoelectric materials. Part I: Mechanical/electrical load, *Acta Mater.* **45**, 849–857, (1997).
22. S. Kumar and R. Singh, Energy release rate and crack propagation in piezoelectric materials. Part II: Combined mechanical and electrical loads, *Acta Mater.* **45**, 859–868, (1997).

23. S. Kumar and R. Singh, Influence of applied electric field and mechanical boundary condition on the stress distribution at the crack tip in piezoelectric material, *Mater. Sci. Eng. A.* **231**, 1–9, (1997).
24. S. Kumar and R. Singh, Effect of mechanical boundary condition at the crack surfaces on the stress distribution at the crack tip in piezoelectric material, *Mater. Sci. Eng. A.* **231**, 64–77, (1998).
25. R. Singh and S. Kumar, Electromechanical response of cracks in piezoelectric materials under combined mechanical and electrical loadings, *Key Eng. Mater.* **145–149**, 1005–1010, (1998).
26. Q. Qin and Y. Mai, A closed crack model for interface cracks in thermopiezoelectric materials, *Int. J. Solids Struct.* **36**, 2463–2479, (1999).
27. Q. Qin, General solutions for thermopiezoelectrics with various holes under thermal loading, *Int. J. Solids Struct.* **37**, 5561–5578, (2000).
28. B. Wang, J. Han, and S. Du, Electroelastic fracture dynamics for multilayered piezoelectric materials under dynamic anti-plane shearing, *Int. J. Solids Struct.* **37**, 4969–4986, (2000).
29. B. Wang and N. Noda, Mixed mode crack initiation in piezoelectric ceramic strip, *Theoret. Appl. Fract. Mech.* **34**, 35–47, (2000).
30. B. Wang and N. Noda, Axisymmetric deformation of piezoelectric multilayers, *Phil. Mag. A.* **81**(4), 1009–1019, (2001).
31. H. Gao and D. Barnett, An invariance property of local energy release rate in a strip saturation model of piezoelectric fracture, *Int. J. Fract.* **79**, R25–R29, (1996).
32. C. Fulton and H. Gao, Electrical nonlinearity in fracture of piezoelectric ceramics, *Appl. Mech. Rev.* **50**(11), 556–563, (1997).
33. H. Gao, T. Zhang, and P. Tong, Local and global energy release rate for an electrically yielded crack in a piezoelectric ceramic, *J. Mech. Phys. Solids.* **45**, 491–510, (1997).
34. K. Herrmann, V. Loboda, and V. Govorukha, On contact zone model for an electrically impermeable interface crack in a piezoelectric bimaterial, *Int. J. Fract.* **111**, 203–227, (2001).
35. R. McMeeking, Electrostrictive stresses near crack-like flaws, *J. Appl. Math. Phys.* **40**, 615–627, (1989).
36. G. Mikahailov and V. Parton, *Electromagnetoelasticity.* (Hemisphere, New York, 1990).
37. S. Kwon and K. Lee, Analysis of stress and electric fields in a rectangular piezoelectric body with a center crack under anti-plane shear loading, *Int. J. Solids Struct.* **37**, 4859–4869, (2000).
38. Y. Shindo, E. Ozawa, and J. Nowacki, Singular stress and electric fields of a cracked piezoelectric strip, *Int. J. Appl. Electromagn. Mater.* **1**, 77–87, (1990).
39. Y. Shindo, K. Watanabe, and F. Narita, Electroelastic analysis of a piezoelectric ceramic strip with a central crack, *Int. J. Eng. Sci.* **38**, 1–19, (2000).
40. F. Narita, Y. Shindo, and K. Horiguchi, Electroelastic fracture mechanics of piezoelectric ceramics. In Y. Shindo (ed.), *Mechanics of Electromagnetic*

Material Systems and Structures, pp. 89–101, WIT Press, Southampton, (2002).
41. M. Dunn, The effects of crack face boundary conditions on the fracture of piezoelectric solids, *Eng. Fract. Mech.* **48**, 25–39, (1994).
42. T. Zhang and P. Tong, Fracture mechanics for a mode-III crack in a piezoelectric material, *Int. J. Solids Struct.* **33**, 343–359, (1996).
43. T. Zhang, C. Qian, and P. Tong, Linear electro-elastic analysis of a cavity or a crack in a piezoelectric material, *Int. J. Solids Struct.* **35**, 2122–2149, (1998).
44. C. Gao and W. Fan, Exact solutions for the plane problem in piezoelectric materials with an elliptic hole or a crack, *Int. J. Solids Struct.* **36**, 2527–2540, (1999).
45. R. McMeeking, Crack tip energy release rate for a piezoelectric compact tension specimen, *Eng. Fract. Mech.* **64**, 217–244, (1999).
46. B. Wang, J. Han, and S. Du, New considerations for the fracture of piezoelectric materials under electromechanical loading, *Mech. Res. Commun.* **27**, 435–444, (2000).
47. C. Fulton and H. Gao, Effect of local polarization on piezoelectric fracture, *J. Mech. Phys. Solids.* **49**, 927–952, (2001).
48. R. McMeeking, Towards a fracture mechanics for brittle piezoelectric and dielectric materials, *Int. J. Fract.* **108**, 25–41, (2001).
49. T. Zhang, M. Zhao, and P. Tong, Fracture of piezoelectric ceramics, *Advance in Applied Mechanics.* **38**, 148–289, (2001).
50. B. Wang and Y. Mai, On the electrical boundary conditions on the crack surfaces in piezoelectric ceramics, *Int. J. Eng. Sci.* **41**, 633–652, (2003).
51. O. Gruebner, M. Kamlah, and D. Munz, Finite element analysis of cracks in piezoelectric materials taking into account the permittivity of the crack medium, *Engng. Frac. Mech.* **70**, 1399–1413, (2003).
52. A. Stroh, Dislocations and cracks in anisotropic elasticity, *Phil. Mag.* **7**, 625–646, (1958).
53. A. Stroh, Steady state problems in anisotropic elasticity, *J. Math. Phys.* **41**, 77–103, (1962).
54. T. Ting, *Anisotropic Elasticity: Theory and Applications* (Oxford University Press, New York, 1996).
55. Z. Suo, Singularities, interfaces and cracks in dissimilar anisotropic media, *Proc. R. Soc. Lond.* **A427**, 331–358, (1990).
56. L. Ni and S. Nemat-Nasser, General duality principle in elasticity, *Mech. Materials* **24**(2), 87–123, (1996).
57. T. Mura, *Micromechanics of Defects in Solids* (Martinus Nijhoff, Dordrecht, 1987).
58. M. Denda and I. Kosaka, Dislocation and point-force-based approach to the special Green's function BEM for elliptic hole and crack problems in two dimensions, *Int. J. Numer. Methods Eng.* **40**, 2857–2889, (1997).
59. J. Telles and S. Guimares, Green's function: A numerical generation for fracture mechanics problems via boundary elements, *Compt. Methods Appl. Mech. Engng.* **141**, 847–858, (1997).

60. B. Auld, *Acoustic Fields and Waves in Solids: Vol. I* (John Wiley, New York, 1973).
61. M. Denda and M. Marante, Mixed mode BEM analysis of multiple curvilinear cracks in the general anisotropic solids by the crack tip singular element, *Int. J. Solids and Struct.* **41**(5–6), 1473–1489, (2004).

Chapter 4

BOUNDARY INTEGRAL ANALYSIS FOR THREE-DIMENSIONAL EXPONENTIALLY GRADED ELASTICITY

J.E. Ortiz and V. Mantič
Group of Elasticity and Strength of Materials
University of Seville, Camino de los Descubrimientos s/n
41092 Sevilla, Spain

L.J. Gray, S. Nintcheu Fata and O. Sallah
Computer Science and Mathematics Division
Oak Ridge National Laboratory
Oak Ridge, TN 37831-6367, USA

Recent work on the development of boundary integral equation methods for 3-D exponentially graded elasticity is summarized. A synopsis of the derivation of the fundamental displacement and traction kernels is presented, and the difficulties inherent in dealing with these functions are highlighted. Results of new calculations utilizing a parallel domain decomposition boundary integral algorithm are reported. A possible new approach to representing the graded Green's function and its derivatives is discussed.

4.1. Introduction

The unique and powerful features of numerical methods based upon boundary integral equations are well known: a boundary-only formulation (therefore requiring only a boundary mesh), fast solution methods for highly complex geometries,[1] and a solution that, while approximately satisfying the boundary conditions, is nevertheless an exact solution of the governing differential equation. However, the price that is paid for these advantages is generality — the foundation of the numerical method is having a manageable Green's function,[2] and these fundamental solutions are simply not widely available.

Boundary integral methods have therefore been limited almost exclusively to the simplest partial differential equations posed in homogeneous domains, and an important research goal is to try to expand the range of applications of the method. For *non-homogeneous* media, wherein the material properties vary in a smooth but otherwise arbitrary manner, analytic forms for Green's functions are virtually non-existent (see Chapter 2 in Ref. 3). Work in this area, with at best partial success, has been carried out by Shaw and co-workers,[4–6] Clements *et al.*,[7,8] and Kassab and Divo[9,10] (regarding this last work concerning an arbitrary spatially varying thermal conductivity, see also Refs. 11 and 12).

For general inhomogeneity, the best that appears possible is the numerical construction of an 'approximate Green's function' (called a Levi function).[3] Instead of solving the equation $A\mathcal{G} = \boldsymbol{\delta}$, where A is the differential operator and $\mathcal{G}$ the Green's function, one solves $A\mathcal{L} = \boldsymbol{\delta} + \mathbf{R}$, where $\mathbf{R}$ is a 'small remainder'. In this case, the integral equation formulation will necessarily retain a volume integral; however, with the recent development of fast volume methods[13–15] this is not necessarily a major obstacle.

One path that has met with some success is to assume a specific exponential form for the variation in material properties. This chapter summarizes recent work on the integral equation formulation for the elastic analysis of an isotropic solid with Lamé moduli varying as

$$\lambda(\boldsymbol{x}) = \lambda_0 \exp(2\boldsymbol{\beta} \cdot \boldsymbol{x}) \quad \text{and} \quad \mu(\boldsymbol{x}) = \mu_0 \exp(2\boldsymbol{\beta} \cdot \boldsymbol{x}). \tag{4.1}$$

Here, λ_0 and μ_0 are constants, $\boldsymbol{\beta}$ is a given constant vector and the solid is called *exponentially graded* in the direction $\boldsymbol{\beta}$. For Laplace problems, steady state and transient thermal modelling in exponentially graded solids, Green's functions and boundary integral analysis were considered in Refs. 16–19.

Exponential grading is not only mathematically tractable, but it is the typical assumption in the engineering literature on *functionally graded materials* (FGMs). A partial list of applications of FGMs includes biomechanics and natural materials,[20] thermal barrier coatings,[21] graded piezoelectric and smart materials,[22] and aerospace[23] (see Ref. 24 for a more complete discussion and references). The papers by Hirai[25] and Markworth *et al.*,[26] and the books by Suresh and Mortensen[27] and Miyamoto *et al.*,[28] provide a good overview of the initial efforts on FGMs. A recent review article[24] provides an excellent review of work after 2000 and an extensive bibliography.

In addition to the above references, it should also be mentioned that Vrettos[29,30] considered wave propagation in an exponentially graded model of the earth, while V. and J. Sládek and co-workers have applied integral equation methods to study thermal and elasticity problems in graded materials (see Refs. 31 and 32 and references therein).

The work reviewed herein is aimed at developing a 'standard' boundary integral capability for modelling exponentially graded elastic materials. This must naturally start with the Green's function $\mathcal{U}$, the analogue of the well known Kelvin solution $\mathcal{U}^0$ for isotropic homogeneous materials, and an abbreviated version of the derivation of $\mathcal{U}$ is first presented. As shown by Martin *et al.*[33] this fundamental displacement tensor can be written as

$$\mathcal{U}(\boldsymbol{x}, \boldsymbol{x}') = \exp\{-\boldsymbol{\beta} \cdot (\boldsymbol{x} + \boldsymbol{x}')\} \left\{\mathcal{U}^0(\boldsymbol{x}, \boldsymbol{x}') + \mathcal{U}^g(\boldsymbol{x}, \boldsymbol{x}')\right\}, \tag{4.2}$$

where the additional *grading term* $\mathcal{U}^g$ is nonsingular at $\boldsymbol{x} = \boldsymbol{x}'$ and can be computed. The corresponding traction kernel $\mathcal{T}$ in a direct boundary integral equation (for surface displacement) is defined in terms of first order derivatives of $\mathcal{U}$ and takes the analogous form

$$\mathcal{T}(\boldsymbol{x}, \boldsymbol{x}') = \exp\{\boldsymbol{\beta} \cdot (\boldsymbol{x} - \boldsymbol{x}')\} \left\{\mathcal{T}^0(\boldsymbol{x}, \boldsymbol{x}') + \mathcal{T}^g(\boldsymbol{x}, \boldsymbol{x}')\right\}. \tag{4.3}$$

The homogeneous Kelvin term is $\mathcal{T}^0$, and formulas for the grading component $\mathcal{T}^g$ can be derived and computed.[34]

Nevertheless, the expressions for $\mathcal{U}^g$ and $\mathcal{T}^g$ are quite complicated, and this has two serious consequences. First, the numerical evaluation of the kernels, while possible, is sufficiently expensive that the practicality of a boundary integral FGM calculation based upon these formulas comes into question, especially for a Galerkin approximation. Second, an ultimate goal of this work is to develop an effective boundary integral capability for fracture analysis, and this requires the equation for surface traction. The traction equation, however, involves a kernel function comprised of derivatives of $\mathcal{T}$, and it is highly dubious that straightforward continuation of the techniques developed thus far would lead to a successful implementation of this equation. It therefore seems likely that an effective treatment of graded elasticity will require a new approach to representing the Green's function and its derivatives.

In the summary below, as much detail as possible will be omitted, retaining only enough to convince the reader of the inherent difficulties. This will also permit, in Sec. 4.6, a brief discussion of a possible new approach that can successfully bypass these roadblocks.

4.2. Graded Green's Function $\mathcal{U}$

The fundamental displacement tensor $\mathcal{U}(\boldsymbol{x},\boldsymbol{x}')$ for exponentially graded elasticity was derived in Martin *et al.*,[33] and one part of the expression for $\mathcal{U}$ was subsequently corrected in Criado *et al.*[35] This complete derivation will not be repeated herein; it is not necessary to repeat the lengthy analysis needed to invert the Fourier transform of $\mathcal{U}^g$. The version presented below will suffice to give an indication of the basic procedure and, as noted above, it will allow a discussion of the difficulties inherent in the current formulation, as well as possible remedies.

Neglecting body forces, the equations of equilibrium are, as usual,

$$\sigma_{ij,j} = 0, \tag{4.4}$$

and the linear elastic constitutive law relating stresses σ_{ij} to strains $\epsilon_{k\ell}$ is

$$\sigma_{ij}(\boldsymbol{x}) = c_{ijk\ell}(\boldsymbol{x})\epsilon_{k\ell}(\boldsymbol{x}). \tag{4.5}$$

The strains $\epsilon_{k\ell}$ are expressed through displacements u_k as

$$\epsilon_{k\ell} = \frac{1}{2}\left(u_{k,\ell} + u_{\ell,k}\right), \tag{4.6}$$

and for an exponentially graded material the elastic constants are assumed to satisfy

$$c_{ijk\ell}(\boldsymbol{x}) = C_{ijk\ell}\, e^{2\boldsymbol{\beta}\cdot\boldsymbol{x}}, \tag{4.7}$$

where $\boldsymbol{x} = (x_1, x_2, x_3)$ (or $\boldsymbol{x} = (x, y, z)$). Here, $\boldsymbol{\beta}$ is the vector defining the direction of the grading, and $\beta = \|\boldsymbol{\beta}\|$ governs the strength of the exponential variation. The 2 appears in the exponent for algebraic convenience; this will prevent having $\beta/2$ appear in the expressions for $\mathcal{U}(\boldsymbol{x},\boldsymbol{x}')$.

The quantity of interest is the Green's function or fundamental displacement solution $\mathcal{U}(\boldsymbol{x},\boldsymbol{x}')$, which satisfies

$$\frac{\partial}{\partial x_j}\left\{c_{ijk\ell}(\boldsymbol{x})\,\frac{\partial U_{\ell m}}{\partial x_k}\right\} = -\delta_{im}\,\delta(\boldsymbol{x}-\boldsymbol{x}'), \quad i = 1, 2, 3. \tag{4.8}$$

Here, δ_{ij} is the Kronecker delta, $\delta(\boldsymbol{x})$ is the three-dimensional Dirac delta function and $U_{ij}(\boldsymbol{x},\boldsymbol{x}')$ represents the i-th component of the displacement at $\boldsymbol{x}$ due to a point force acting in the j-th direction at $\boldsymbol{x}'$. Expanding the left-hand side of Eq. (4.8) gives for $i = 1, 2, 3$

$$c_{ijk\ell}(\boldsymbol{x})\,\frac{\partial^2 U_{\ell m}}{\partial x_j\,\partial x_k} + \left(\frac{\partial}{\partial x_j}c_{ijk\ell}(\boldsymbol{x})\right)\frac{\partial U_{\ell m}}{\partial x_k} = -\delta_{im}\,\delta(\boldsymbol{x}-\boldsymbol{x}'), \tag{4.9}$$

and from Eq. (4.7),

$$(\partial/\partial x_j)c_{ijk\ell}(\boldsymbol{x}) = 2C_{ijk\ell}\,\beta_j\,\exp(2\boldsymbol{\beta}\cdot\boldsymbol{x}) = 2\beta_j\,c_{ijk\ell}(\boldsymbol{x}). \tag{4.10}$$

Thus (4.9) becomes,

$$\begin{aligned} C_{ijk\ell}\,\frac{\partial^2 U_{\ell m}}{\partial x_j\,\partial x_k} + 2\beta_j\,C_{ijk\ell}\,\frac{\partial U_{\ell m}}{\partial x_k} &= -\delta_{im}\,\exp(-2\boldsymbol{\beta}\cdot\boldsymbol{x})\,\delta(\boldsymbol{x}-\boldsymbol{x}') \\ &= -\delta_{im}\,\exp\{-\boldsymbol{\beta}\cdot(\boldsymbol{x}+\boldsymbol{x}')\}\,\delta(\boldsymbol{x}-\boldsymbol{x}'). \end{aligned} \tag{4.11}$$

The last manipulation of the right-hand side has been executed to ensure that $\mathcal{U}$ retains the usual symmetry of the Green's function,

$$U_{ij}(\boldsymbol{x},\boldsymbol{x}') = U_{ji}(\boldsymbol{x}',\boldsymbol{x}). \tag{4.12}$$

Introduce the Green's function for a *homogeneous* solid with constant stiffnesses $C_{ijk\ell}$, $\mathcal{U}^0$, defined by

$$C_{ijk\ell}\,\frac{\partial^2}{\partial x_j\,\partial x_k}U^0_{\ell m} = -\delta_{im}\,\delta(\boldsymbol{x}-\boldsymbol{x}'), \quad i=1,2,3. \tag{4.13}$$

Comparing this equation with (4.11) suggests decomposing the Green's function as

$$\mathcal{U}(\boldsymbol{x},\boldsymbol{x}') = \exp\{-\boldsymbol{\beta}\cdot(\boldsymbol{x}+\boldsymbol{x}')\}\,\left\{\mathcal{U}^0(\boldsymbol{x},\boldsymbol{x}') + \mathcal{U}^g(\boldsymbol{x},\boldsymbol{x}')\right\}, \tag{4.14}$$

with the *grading term* $\mathcal{U}^g$ satisfying the equation

$$C_{ijk\ell}\,\frac{\partial^2}{\partial x_j\,\partial x_k}U^g_{\ell m}(\boldsymbol{x},\boldsymbol{x}') + L_{i\ell}\,U^g_{\ell m}(\boldsymbol{x},\boldsymbol{x}') = -L_{i\ell}\,U^0_{\ell m}(\boldsymbol{x},\boldsymbol{x}'), \tag{4.15}$$

where $i=1,2,3$, and the differential operator $L_{i\ell}$ is defined by

$$L_{i\ell} = (C_{ijk\ell} - C_{ikj\ell})\beta_j(\partial/\partial x_k) - C_{ijk\ell}\beta_j\beta_k. \tag{4.16}$$

In what follows, the objective is therefore to develop formulas for the grading term $\mathcal{U}^g$, and this is accomplished by employing the three-dimensional Fourier transform and its inverse, defined by

$$\begin{aligned} \mathcal{F}\{f\} &= \widehat{f}(\boldsymbol{\xi}) = \int f(\boldsymbol{x})\,\exp(\mathrm{i}\,\boldsymbol{\xi}\cdot\boldsymbol{x})\,\mathrm{d}\boldsymbol{x}, \\ \mathcal{F}^{-1}\{\widehat{f}\} &= f(\boldsymbol{x}) = (2\pi)^{-3}\int \widehat{f}(\boldsymbol{\xi})\,\exp(-\mathrm{i}\,\boldsymbol{\xi}\cdot\boldsymbol{x})\,\mathrm{d}\boldsymbol{\xi}. \end{aligned} \tag{4.17}$$

Applying the transform to Eq. (4.15),

$$\{Q_{i\ell}(\boldsymbol{\xi}) + B_{i\ell}(\boldsymbol{\beta},\boldsymbol{\xi})\}\,\widehat{U}^g_{\ell m}(\boldsymbol{\xi},\boldsymbol{x}') = -B_{i\ell}(\boldsymbol{\beta},\boldsymbol{\xi})\,\widehat{U}^0_{\ell m}(\boldsymbol{\xi},\boldsymbol{x}'), \tag{4.18}$$

where

$$Q_{i\ell}(\boldsymbol{\xi}) = C_{ijk\ell}\xi_j\xi_k, \qquad (4.19)$$
$$B_{i\ell}(\boldsymbol{\beta},\boldsymbol{\xi}) = \mathrm{i}(C_{ijk\ell} - C_{ikj\ell})\beta_j\xi_k + C_{ijk\ell}\beta_j\beta_k.$$

Note that if we define a complex vector $\boldsymbol{\gamma}$ by $\boldsymbol{\gamma} = \boldsymbol{\xi} + \mathrm{i}\boldsymbol{\beta}$, then

$$Q_{i\ell} + B_{i\ell} = C_{ijk\ell}\gamma_j\overline{\gamma_k}, \qquad (4.20)$$

where the overbar denotes complex conjugation: $\overline{\gamma} = \boldsymbol{\xi} - \mathrm{i}\boldsymbol{\beta}$.

From Eq. (4.13) we have

$$Q_{i\ell}(\boldsymbol{\xi})\,\widehat{U}^0_{\ell m}(\boldsymbol{\xi},\boldsymbol{x}') = \delta_{im}\,\exp(\mathrm{i}\,\boldsymbol{\xi}\cdot\boldsymbol{x}'), \qquad (4.21)$$

whence

$$\widehat{U}^g_{\ell m}(\boldsymbol{\xi},\boldsymbol{x}') = E_{\ell m}(\boldsymbol{\beta},\boldsymbol{\xi})\,\exp(\mathrm{i}\,\boldsymbol{\xi}\cdot\boldsymbol{x}'), \qquad (4.22)$$

where

$$\mathbf{E}(\boldsymbol{\beta},\boldsymbol{\xi}) = -\{\mathbf{Q}(\boldsymbol{\xi}) + \mathbf{B}(\boldsymbol{\beta},\boldsymbol{\xi})\}^{-1}\mathbf{B}(\boldsymbol{\beta},\boldsymbol{\xi})\,[\mathbf{Q}(\boldsymbol{\xi})]^{-1}. \qquad (4.23)$$

Inverting the Fourier transform, we obtain

$$\mathcal{U}^g(\boldsymbol{x},\boldsymbol{x}') = (2\pi)^{-3}\int \mathbf{E}(\boldsymbol{\beta},\boldsymbol{\xi})\,\exp(-\mathrm{i}\,\boldsymbol{r}\cdot\boldsymbol{\xi})\,\mathrm{d}\boldsymbol{\xi} \qquad (4.24)$$

where $\boldsymbol{r} = \boldsymbol{x} - \boldsymbol{x}'$. It remains to evaluate this three-dimensional integral over $\boldsymbol{\xi}$.

At this point, it is useful to further simplify by assuming the material is an isotropic solid. The elastic constants can then be written in terms of Lamé parameters

$$C_{ijk\ell} = \lambda_0\delta_{ij}\delta_{k\ell} + \mu_0\left(\delta_{ik}\delta_{j\ell} + \delta_{i\ell}\delta_{jk}\right), \qquad (4.25)$$

and the exponential grading can be expressed in terms of Lamé constants as

$$\lambda(\boldsymbol{x}) = \lambda_0 e^{2\boldsymbol{\beta}\cdot\boldsymbol{x}}, \quad \mu(\boldsymbol{x}) = \mu_0 e^{2\boldsymbol{\beta}\cdot\boldsymbol{x}},$$

where $\lambda/\mu = \lambda_0/\mu_0 = 2\nu/(1-2\nu)$, indicating that the Poisson ratio ν is constant. In this case, the function $\mathcal{U}^0$ is the well known Kelvin solution for a homogeneous isotropic material (with elastic constants μ_0 and ν),

$$U^0_{j\ell}(\boldsymbol{x}-\boldsymbol{x}') = \frac{1}{16\pi\mu_0(1-\nu)r}\left((3-4\nu)\,\delta_{j\ell} + r_{,j}\,r_{,\ell}\right). \qquad (4.26)$$

Here $r = \|\boldsymbol{x} - \boldsymbol{x}'\|$ and $r_{,j}$ denotes the derivative of r with respect to the jth coordinate of $\boldsymbol{x}$.

Even with the isotropic simplification, the algebra required to obtain the desired expressions for the Fourier transform $\mathbf{E}(\boldsymbol{\beta}, \boldsymbol{\xi})$ in Eq. (4.24) is not insignificant. Introducing the notation $\xi^2 = |\boldsymbol{\xi}|^2$, $\beta^2 = |\boldsymbol{\beta}|^2$, $\gamma^2 = \gamma_j \overline{\gamma_j}$, $C = \boldsymbol{\xi} \cdot \boldsymbol{\beta} = \xi\beta\cos\theta$, $\alpha = (1-2\nu)^{-1}$, $\kappa = (1-\nu)^{-1}$, $\sigma = 4\nu - 1$ and setting $q = 1 + 2\nu\kappa\sin^2\theta$ and

$$\Delta = \xi^4 + 2\xi^2\beta^2 q + \beta^4, \tag{4.27}$$

the end result[33] is

$$E_{j\ell}(\boldsymbol{\beta}, \boldsymbol{\xi}) = -\frac{\beta^2}{\mu_0\gamma^2\xi^2}\delta_{j\ell} - \frac{\kappa\Omega_{j\ell}}{2\mu_0\gamma^2\xi^4\Delta} \equiv E\,\delta_{j\ell} + H_{j\ell}, \tag{4.28}$$

where

$$\Omega_{j\ell}(\boldsymbol{\beta}, \boldsymbol{\xi}) = \mathcal{A}\xi_j\xi_\ell + \mathcal{B}\xi_j\beta_\ell + \overline{\mathcal{B}}\beta_j\xi_\ell + \mathcal{C}\beta_j\beta_\ell, \tag{4.29}$$

and

$$\begin{aligned} \mathcal{A} &= \xi^4(\gamma^2 + 8\nu\beta^2) - \gamma^2\Delta, \\ \mathcal{B} &= -\xi^4(8\nu C + \mathrm{i}\sigma\gamma^2), \\ \mathcal{C} &= \xi^4(\gamma^2 + 8\nu\xi^2). \end{aligned} \tag{4.30}$$

The inverse transform of the diagonal contribution E is easily obtained using spherical polar coordinates (ξ, φ, χ), with the polar axis $(\varphi = 0)$ in the direction of $\boldsymbol{r}$. The result is

$$\frac{1}{(2\pi)^3}\int E\,\exp(-\mathrm{i}\boldsymbol{r}\cdot\boldsymbol{\xi})\,\mathrm{d}\boldsymbol{\xi} = \frac{1 - \mathrm{e}^{-\beta r}}{4\pi\mu_0 r}. \tag{4.31}$$

The inversion of the $H_{j\ell}$ term in Eq. (4.28) is far more complicated,[33] and its lengthy derivation will not be repeated here. However, it is worth noting that the analysis of this term employs a spherical coordinate system $\{r, \Theta, \Phi\}$ with $\boldsymbol{\beta}$, and not $\boldsymbol{r}$, as the polar axis of spherical coordinates. Note that the angle θ between $\boldsymbol{\xi}$ and $\boldsymbol{\beta}$ appears in Δ (in the definition of q), and thus this choice simplifies handling the expressions for the poles in $H_{j\ell}$. We will return to this point later.

From Eq. (4.31), the grading term $\mathcal{U}^g$ can be written as

$$U^g_{j\ell}(\boldsymbol{x} - \boldsymbol{x}') = -\frac{1}{4\pi\mu_0 r}\left(1 - \mathrm{e}^{-\beta r}\right)\delta_{j\ell} + A_{j\ell}(\boldsymbol{x} - \boldsymbol{x}'), \tag{4.32}$$

where

$$A_{j\ell} = \frac{1}{(2\pi)^3} \int H_{j\ell} \exp(-\mathrm{i}\boldsymbol{r} \cdot \boldsymbol{\xi})\, \mathrm{d}\boldsymbol{\xi}. \tag{4.33}$$

Although the derivation is omitted, writing the (intimidating) formula for $A_{j\ell}$ will hopefully supply ample argument for the difficulties mentioned in the introduction. For this purpose, choose vectors $\boldsymbol{n} = (n_1, n_2, n_3)$ and $\boldsymbol{m} = (m_1, m_2, m_3)$ so that $\{\boldsymbol{n}, \boldsymbol{m}, \hat{\boldsymbol{\beta}}\}$ (with $\hat{\boldsymbol{\beta}} = \boldsymbol{\beta}/\beta$) form an orthonormal system of coordinates. For the spherical coordinate system mentioned above it follows that

$$\boldsymbol{r} \cdot \boldsymbol{n} = r \sin\Theta \cos\Phi, \quad \boldsymbol{r} \cdot \boldsymbol{m} = r \sin\Theta \sin\Phi, \quad \boldsymbol{r} \cdot \hat{\boldsymbol{\beta}} = r \cos\Theta. \tag{4.34}$$

The expression for $A_{j\ell} = A_{j\ell}(r, \Theta, \Phi)$, as derived in Ref. 33 and corrected in Ref. 35, is

$$A_{j\ell} = -\frac{\beta}{4\pi\mu_0(1-\nu)} \mathcal{I}_1 - \frac{\beta}{2\pi^2\mu_0(1-\nu)} \left(\mathcal{I}_2 - \mathcal{I}_3 + \mathcal{I}_4 - \mathcal{I}_5\right), \tag{4.35}$$

where

$$\begin{aligned}
\mathcal{I}_1 &= \sum_{s=0}^{2} \sum_{n=0}^{2} \int_0^{\pi/2} \mathcal{R}_s^{(n)}(\theta)\, \mathrm{e}^{-|k| y_s}\, I_n(K y_s) \sin\theta \, \mathrm{d}\theta, \\
\mathcal{I}_2 &= \sum_{s=0}^{2} \int_{\theta_m}^{\pi/2} \mathcal{R}_s^{(0)} \sin\theta \int_{\eta_m}^{\pi/2} \sinh\Psi_s \, \mathrm{d}\eta \, \mathrm{d}\theta, \\
\mathcal{I}_3 &= \sum_{s=0}^{2} \int_{\theta_m}^{\pi/2} \mathcal{R}_s^{(2)} \sin\theta \int_{\eta_m}^{\pi/2} \sinh\Psi_s \cos 2\eta \, \mathrm{d}\eta \, \mathrm{d}\theta, \\
\mathcal{I}_4 &= \sum_{s=1}^{2} \int_{\theta_m}^{\pi/2} \mathcal{M}_s^{(1)} \sin\theta \int_{\eta_m}^{\pi/2} \cosh\Psi_s \sin\eta \, \mathrm{d}\eta \, \mathrm{d}\theta, \\
\mathcal{I}_5 &= \sum_{s=1}^{2} \int_{\theta_m}^{\pi/2} \widetilde{\mathcal{M}}_s^{(1)} \operatorname{sgn}(k) \sin\theta \int_{\eta_m}^{\pi/2} \sinh\Psi_s \sin\eta \, \mathrm{d}\eta \, \mathrm{d}\theta.
\end{aligned} \tag{4.36}$$

We will now proceed to define the notation in these expressions. First, $I_n(x)$ is the modified Bessel function of the first kind of order n, and the integration limits θ_m and $\eta_m(\Theta, \theta)$ are defined by

$$\begin{aligned}
\theta_m(\Theta) &= \left|\tfrac{1}{2}\pi - \Theta\right|, \\
|k(\Theta, \theta)| &= K(\Theta, \theta) \sin\eta_m,
\end{aligned} \tag{4.37}$$

where $k(\Theta, \theta) = \beta r \cos\theta \cos\Theta$ and $K(\Theta, \theta) = \beta r \sin\theta \sin\Theta$. The argument of the hyperbolic functions is

$$\Psi_s(\Theta, \theta, \eta) = K(\Theta, \theta) y_s(\theta) \left(\sin\eta_m(\Theta, \theta) - \sin\eta\right), \tag{4.38}$$

where the functions y_s are given by

$$y_0 = 1, \quad y_1 = \sqrt{q(\theta) + \sqrt{q^2(\theta) - 1}}, \quad y_2 = \sqrt{q(\theta) - \sqrt{q^2(\theta) - 1}}, \tag{4.39}$$

$q = q(\theta)$ being defined in Eq. (4.27). The functions $\mathcal{R}_s^{(n)}$ are

$$\begin{aligned} \mathcal{R}_s^{(0)} &= \mathcal{M}_s^{(0)}, \quad s = 0, 1, 2, \\ \mathcal{R}_s^{(2)} &= -\mathcal{M}_s^{(2)}, \quad s = 0, 1, 2, \\ \mathcal{R}_0^{(1)} &= 0, \\ \mathcal{R}_s^{(1)} &= -\left(\mathcal{M}_s^{(1)} + \widetilde{\mathcal{M}}_s^{(1)} \operatorname{sgn}(k(\Theta, \theta))\right), \quad s = 1, 2, \end{aligned} \tag{4.40}$$

where

$$\begin{aligned} \mathcal{M}_0^{(n)} &= \frac{f_n(1)}{2\, D(1)}, \quad n = 0, 2, \\ \mathcal{M}_s^{(n)} &= \frac{f_n(y_s)}{(1 - y_s^2)\, D'(y_s)}, \quad n = 0, 2 \text{ and } s = 1, 2, \\ \mathcal{M}_s^{(1)} &= \frac{f_1(y_s)}{D'(y_s)}, \quad s = 1, 2, \\ \widetilde{\mathcal{M}}_s^{(1)} &= \frac{\tilde{f}_1(y_s)}{D'(y_s)}, \quad s = 1, 2, \end{aligned} \tag{4.41}$$

and the final definitions needed to complete the expressions are the functions $D(x) = x^4 - 2x^2 q + 1$ and $D'(x) = -4x^3 + 4xq$ (the prime does not indicate derivative), and

$$\begin{aligned} f_0(x) &= \tfrac{1}{2}\{8\nu x^4 - (-x^2 + 1)(-2x^2 q + 1)\}(n_j n_\ell + m_j m_\ell)\sin^2\theta \\ &\quad + \left\{8\nu x^4 \sin^2\theta + (-x^2 + 1)[-x^2 - (-2x^2 q + 1)\cos^2\theta]\right\}\hat{\beta}_j\hat{\beta}_\ell, \\ f_1(x) &= x^3(4\nu - 1)(s_j\hat{\beta}_\ell - \hat{\beta}_j s_\ell)\sin\theta, \\ \tilde{f}_1(x) &= -\tfrac{1}{2}(s_j\hat{\beta}_\ell + \hat{\beta}_j s_\ell)(-2x^2 q + 1)\sin 2\theta, \\ f_2(x) &= -\tfrac{1}{2}[8\nu x^4 - (-x^2 + 1)(-2x^2 q + 1)]\left\{n_j(n_\ell\cos 2\Phi + m_\ell \sin 2\Phi)\right. \\ &\quad \left. + m_j(n_\ell \sin 2\Phi - m_\ell \cos 2\Phi)\right\}\sin^2\theta. \end{aligned} \tag{4.42}$$

It should be obvious at this point that evaluation of the Green's function is far from easy, and direct implementation will necessarily be computationally quite expensive. As a consequence, a boundary integral analysis, while possible, will involve long execution times, possibly long enough to wipe out any advantages of the method.

4.3. Traction Fundamental Solution $\mathcal{T}$

Computation time is not the only serious issue that the complexity of the fundamental solution poses for a boundary integral formulation of graded elasticity. The integral equation for surface displacement also requires the corresponding traction fundamental solution $\mathcal{T}(\boldsymbol{x}, \boldsymbol{x}')$, and this kernel is defined in terms of first order derivatives of $\mathcal{U}(\boldsymbol{x}, \boldsymbol{x}')$ with respect to $\boldsymbol{x}$. Moreover, as an ultimate goal is to develop a boundary integral capability for FGM fracture analysis, it is necessary to work with the corresponding equation for surface traction, and this involves a kernel function composed of second order derivatives of $\mathcal{U}$.

The first derivative generates formulas that are even lengthier and more expensive to compute than those for $\mathcal{U}(\boldsymbol{x}, \boldsymbol{x}')$, and equally important, differentiation of the continuous grading term results in a weakly singular contribution. The handling of these singularities is not difficult, even if this piece of the kernel function is evaluated numerically. However, a second derivative of this term will be strongly singular, and the boundary integration of this function using strictly numerical methods is much more problematic. A numerical treatment of the displacement boundary integral equation — based upon direct evaluation of the kernel expressions — is feasible.[34] Nevertheless, the prospects for extending this approach to develop a practical fracture analysis implementation are not particularly bright.

As in the previous section, the two objectives for presenting a brief summary of the traction kernel derivation[34] are, first, an attempt at completeness, and second, to justify the above comments. To obtain $\mathcal{T}(\boldsymbol{x}, \boldsymbol{x}')$, it is necessary to differentiate the expression for the fundamental displacement solution, these derivatives yielding the corresponding strains. Employing the constitutive law with the tensor of elastic stiffnesses given in Eq. (4.25), the corresponding stresses and tractions are then obtained.

Differentiation of Eq. (4.14) yields

$$\frac{\partial U_{j\ell}}{\partial x_k}(\boldsymbol{x},\boldsymbol{x}') = \exp\left(-\boldsymbol{\beta}\cdot(\boldsymbol{x}+\boldsymbol{x}')\right)\left(\frac{\partial U^0_{j\ell}}{\partial x_k}(\boldsymbol{x}-\boldsymbol{x}') + \frac{\partial U^g_{j\ell}}{\partial x_k}(\boldsymbol{x}-\boldsymbol{x}')\right)$$
$$-\beta_k U_{j\ell}(\boldsymbol{x},\boldsymbol{x}'), \tag{4.43}$$

where x_k denotes a component of $\boldsymbol{x}$. The derivative of the Kelvin term $U^0_{j\ell}$ will eventually become the traction kernel for a homogeneous material, and as relatively simple analytic expressions are well known,[36,37] the handling of the strong singularity is not a problem. The derivative of $U^g_{j\ell}$ is only weakly singular and can be expressed, in view of Eq. (4.32), as

$$\frac{\partial U^g_{j\ell}}{\partial x_k}(\boldsymbol{x}-\boldsymbol{x}') = -\frac{\delta_{j\ell}}{4\pi\mu_0}\left\{\beta\frac{\mathrm{e}^{-\beta r}}{r} - \frac{(1-\mathrm{e}^{-\beta r})}{r^2}\right\} r_{,k} + \frac{\partial A_{j\ell}}{\partial Q_k}(\boldsymbol{x}-\boldsymbol{x}'), \tag{4.44}$$

where

$$r_{,k} = \frac{\partial r}{\partial x_k} = \frac{x_k - x'_k}{r}. \tag{4.45}$$

Finally, from Eq. (4.35), the derivative of A_{jl} is the sum of the derivatives of the integrals $\mathcal{I}_i$,

$$\frac{\partial A_{jl}}{\partial x_k} = -\frac{\beta}{4\pi(1-\nu)\mu_0}\left\{\frac{\partial \mathcal{I}_1}{\partial x_k} + \frac{2}{\pi}\left(\frac{\partial \mathcal{I}_2}{\partial x_k} - \frac{\partial \mathcal{I}_3}{\partial x_k} + \frac{\partial \mathcal{I}_4}{\partial x_k} - \frac{\partial \mathcal{I}_5}{\partial x_k}\right)\right\}. \tag{4.46}$$

Using the standard result from calculus

$$\frac{d}{dx}\int_{A(x)}^{B} f(x,t)\mathrm{d}t = \int_{A(x)}^{B}\frac{\partial f(x,t)}{\partial x}\mathrm{d}t - f(x,A(x))\frac{dA}{dx}, \tag{4.47}$$

the process for differentiating the numerous formulas in the previous section is straightforward, albeit somewhat lengthy, and we again refer the reader to Ref. 34 for the details.

The strains $E_{ij\ell}(\boldsymbol{x},\boldsymbol{x}')$ corresponding to the fundamental displacements $U_{j\ell}(\boldsymbol{x},\boldsymbol{x}')$ are therefore

$$E_{ij\ell}(\boldsymbol{x},\boldsymbol{x}') = \frac{1}{2}\left(\frac{\partial U_{i\ell}}{\partial x_j}(\boldsymbol{x},\boldsymbol{x}') + \frac{\partial U_{j\ell}}{\partial x_i}(\boldsymbol{x},\boldsymbol{x}')\right), \tag{4.48}$$

and, incorporating the isotropic constitutive law Eq. (4.25), the stresses are

$$\Sigma_{ij\ell} = 2\mu(\boldsymbol{x})E_{ij\ell}(\boldsymbol{x},\boldsymbol{x}') + \lambda(\boldsymbol{x})E_{kk\ell}(\boldsymbol{x},\boldsymbol{x}')\delta_{ij}. \tag{4.49}$$

The stress can therefore be written as

$$\Sigma_{ij\ell}(\boldsymbol{x}-\boldsymbol{x}') = \exp\left(\boldsymbol{\beta}\cdot(\boldsymbol{x}-\boldsymbol{x}')\right)\left(\Sigma^0_{ij\ell}(\boldsymbol{x}-\boldsymbol{x}') + \Sigma^g_{ij\ell}(\boldsymbol{x}-\boldsymbol{x}')\right), \tag{4.50}$$

where the strongly singular term $\Sigma^0_{ij\ell}(\boldsymbol{x}-\boldsymbol{x}')$ is the homogeneous component analogous to the Kelvin displacements in the expression for the Green's function. It represents the stress tensor σ_{ij} at $\boldsymbol{x}$ originated by a unit point force in direction ℓ at $\boldsymbol{x}'$ in the homogeneous isotropic material with Lamé constants μ_0 and λ_0. The weakly singular grading contribution can be expressed in terms of $\mathcal{U}$ and its derivatives as

$$\begin{aligned}\Sigma^g_{ij\ell}(\boldsymbol{x}-\boldsymbol{x}') = \mu_0&\left(\frac{\partial U^g_{i\ell}}{\partial x_j} + \frac{\partial U^g_{j\ell}}{\partial x_i} - \beta_i\left(U^0_{j\ell}+U^g_{j\ell}\right) - \beta_j\left(U^0_{i\ell}+U^g_{i\ell}\right)\right)\\ &+\lambda_0\left(\frac{\partial U^g_{k\ell}}{\partial x_k} - \beta_k\left(U^0_{k\ell}+U^g_{k\ell}\right)\right)\delta_{ij}.\end{aligned} \tag{4.51}$$

It is worth pointing out the differences between the exponential prefactors in the expressions for the displacement Eq. (4.14) and stress kernels Eq. (4.50). Note that while $U_{j\ell}(\boldsymbol{x},\boldsymbol{x}')$ depends on the position of both the source and field points, $\boldsymbol{x}'$ and $\boldsymbol{x}$, $\Sigma^g_{ij\ell}(\boldsymbol{x}-\boldsymbol{x}')$ depends only on the difference of these points. However, the displacement kernel does maintain the usual reciprocity relation for a Green's function, and thus a symmetric-Galerkin boundary integral approximation would be possible. Note too the sign change in this prefactor. As a consequence, taking the source point $\boldsymbol{x}'$ as fixed, the exponential in the fundamental displacements $U_{j\ell}$ decreases in stiffer zones and increases in more compliant zones, whereas the stresses $\Sigma_{ij\ell}$, not surprisingly, behave in the opposite manner.

To complete this section, the Cauchy lemma defines the traction fundamental solution $T_{i\ell}(\boldsymbol{x},\boldsymbol{x}')$ associated with $\Sigma_{ij\ell}(\boldsymbol{x}-\boldsymbol{x}')$ by

$$\begin{aligned}T_{i\ell}(\boldsymbol{x},\boldsymbol{x}') &= \Sigma_{ij\ell}(\boldsymbol{x}-\boldsymbol{x}')\,n_j(\boldsymbol{x})\\ &= \exp\left(\boldsymbol{\beta}\cdot(\boldsymbol{x}-\boldsymbol{x}')\right)\left(T^0_{i\ell}(\boldsymbol{x},\boldsymbol{x}') + T^g_{i\ell}(\boldsymbol{x},\boldsymbol{x}')\right),\end{aligned} \tag{4.52}$$

where $\mathbf{n}(\boldsymbol{x})$ is the unit outward normal vector. The decomposition into homogeneous and gradient contributions follows as for the displacement and stress kernels.

4.4. Boundary Element Method

The boundary integral formulation for an isotropic, exponentially graded body $\Omega \subset \mathcal{R}^3$ with a bounded (Lipschitz and piecewise smooth) boundary $\partial\Omega = \Gamma$ will be briefly discussed in this section. The derivation follows the standard procedures for a homogeneous material.[36,37] Starting from the 2nd Betti Theorem of reciprocity of work for a graded material, one can derive the corresponding Somigliana identity:

$$C_{i\ell}(\boldsymbol{x}')u_i(\boldsymbol{x}') + \fint_{\Gamma} T_{i\ell}(\boldsymbol{x}, \boldsymbol{x}')u_i(\boldsymbol{x})dS(\boldsymbol{x}) = \int_{\Gamma} U_{i\ell}(\boldsymbol{x}, \boldsymbol{x}')t_i(\boldsymbol{x})dS(\boldsymbol{x}), \tag{4.53}$$

expressing the displacements $u_i(\boldsymbol{x}')$ at a domain or boundary point $\boldsymbol{x}' \in \Omega \cup \Gamma$ in terms of the boundary displacements $u_i(\boldsymbol{x})$ and tractions $t_i(\boldsymbol{x})$, $\boldsymbol{x} \in \Gamma$. Notice that zero body forces have been considered in Eq. (4.53). The strongly singular traction kernel integral is evaluated in the Cauchy principal value sense, and

$$C_{i\ell}(\boldsymbol{x}') = \lim_{\varepsilon \to 0_+} \int_{S_\varepsilon(\boldsymbol{x}')\cap\Omega} T_{i\ell}(\boldsymbol{x}, \boldsymbol{x}')dS(\boldsymbol{x}) \tag{4.54}$$

is the coefficient tensor of the free term, $S_\varepsilon(\boldsymbol{x}')$ being a spherical surface of radius ε centered at $\boldsymbol{x}'$. It is important to note that, despite the complexity of the $T_{i\ell}$ kernel expression, this evaluation is not a problem. The weakly singular grading term and the exponential prefactor in Eq. (4.53) will play no role in the limit procedure in Eq. (4.54). Thus, the value of $C_{i\ell}$ in Eq. (4.54) coincides with the value of $C_{i\ell}$ for the homogeneous isotropic material whose Poisson ratio equals $\nu = \lambda_0/2(\mu_0 + \lambda_0)$, *i.e.*,

$$C_{i\ell}(\boldsymbol{x}') = \lim_{\varepsilon \to 0_+} \int_{S_\varepsilon(\boldsymbol{x}')\cap\Omega} T^0_{i\ell}(\boldsymbol{x}, \boldsymbol{x}')dS(\boldsymbol{x}). \tag{4.55}$$

Hence, $C_{i\ell}(\boldsymbol{x}') = \delta_{i\ell}$ for $\boldsymbol{x}' \in \Omega$, $C_{i\ell}(\boldsymbol{x}') = \frac{1}{2}\delta_{i\ell}$ for $\boldsymbol{x}' \in \Gamma$ situated at a smooth part of Γ, and for an edge or corner point of Γ, $C_{i\ell}(\boldsymbol{x}')$ is given by the size, shape and spatial orientation of the interior solid angle at $\boldsymbol{x}'$. A general explicit expression of the symmetric tensor $C_{i\ell}(\boldsymbol{x}')$ in terms of the unit vectors tangential to the boundary edges and the unit outward normal vectors to the boundary surfaces at $\boldsymbol{x}'$ can be found in Ref. 38.

For an unbounded domain Ω, the form of the Somigliana identity in Eq. (4.53) holds if the following radiation condition[39] for the displacement solution u_i and the corresponding traction solution t_i is fulfilled for any fixed $\boldsymbol{x}' \in \Omega$:

$$\lim_{\rho \to \infty} \int_{\Gamma_\rho} (U_{i\ell}(\boldsymbol{x}, \boldsymbol{x}')t_i(\boldsymbol{x}) - T_{i\ell}(\boldsymbol{x}, \boldsymbol{x}')u_i(\boldsymbol{x}))\, dS(\boldsymbol{x}) = 0, \tag{4.56}$$

where Γ_ρ is the spherical surface of radius ρ centered at the origin of the coordinates. In fact, Eq. (4.56) implies that the integral in Eq. (4.56) vanishes for any sufficiently large Γ_ρ including Γ in its interior and any $\boldsymbol{x}' \in \Omega$, $\|\boldsymbol{x}'\| < \rho$. A further study of Eq. (4.56) using a detailed knowledge of the behaviour of $U^g_{jl}(\boldsymbol{x} - \boldsymbol{x}')$ for $r \to \infty$ (apparently not available at present due to the cumbersome expression of A_{jl} given in Eq. (4.35)) might be necessary in the deduction of a more explicit and equivalent expression of this radiation condition in the classical form (see Ref. 39 for some standard examples). This classical form of Eq. (4.56) would be very useful for practical applications of Eq. (4.53) to infinite domains.

The numerical implementation of Eq. (4.53) in this work employs standard approximation techniques. A collocation approximation based upon a nine-node continuous quadrilateral quadratic isoparametric element is employed to interpolate the boundary and the boundary functions. The evaluation of regular integrals is accomplished by Gaussian quadrature with 8×8 integration points, whereas an adaptive element subdivision[40] is utilized for nearly singular integrals. A standard polar coordinate transformation[40] is employed to handle the weakly singular integrals involving the kernel $U_{i\ell}$, and the rigid-body motion procedure is invoked for evaluating the sum of the coefficient tensor of the free term $C_{i\ell}$ and the Cauchy principal value integral with the kernel $T_{i\ell}$.

4.5. Numerical Examples

Two numerical examples are presented in order to illustrate the possibilities of the present 3-D BEM formulation and implementation for advanced stress analysis in an exponentially graded linearly elastic material (hereafter called the FGM-BEM). This code has been thoroughly described and tested in Refs. 34 and 35, and a parallel domain decomposition version was developed in Ref. 41.

The first example deals with the elastic solution for a pressurized thick-walled tube of a material that is exponentially graded in the radial direction. In the last decade, several analytic solutions for pressurized tubes with different radial grading were deduced.[41–46] Comparing the numerical solution obtained with an analytic solution available[41] allows a convenient check of the correctness of the FGM-BEM code. Moreover, this example shows the applicability of the FGM-BEM code to problems with spatial variation of the grading direction; this capability is clearly important, e.g. for the stress analysis of thermal barrier coatings of turbine blades.

In the second example, an exponentially graded material, having a constant grading vector $\boldsymbol{\beta}$ and with an interior spherical cavity, is subjected to a far field uniaxial tension. As an analytic solution for this problem is not available at present, it seems interesting to study the effect of the material grading on the values and positions of stress concentrations. The present analysis is focused on the dependency of the resulting stress concentration on the relative orientation of the load and grading direction, and on the grading gradient. Two extreme cases of parallel and perpendicular loading and grading directions are studied. It appears that the solution presented is one of the first attempts to study 3-D defects in FGMs of shapes other than cracks, e.g. cavities or inclusions of other materials having various geometries. In fact, the only results in this direction known to the authors are the approximate analytic solutions for problems of circular and elliptical holes in linearly graded materials presented recently.[47,48]

The numerical results obtained by the FGM-BEM code presented here show that it will be especially useful for more complex stress concentration problems, including multiple interacting voids and/or inclusions. The meshing of such defects is very easy in BEM and highly accurate results can be expected due to the analytic representation of the graded material inherent in the boundary integral kernel formulation.

4.5.1. *Pressurized thick-walled tube*

Consider an infinitely long thick-walled tube, inner radius r_0 and outer radius r_e, pressurized by an internal pressure $p > 0$ with grading in the radial direction,

$$\mu(\boldsymbol{x}) = \mu(r) = \mu_0 e^{2\beta r}, \quad \lambda(\boldsymbol{x}) = \lambda(r) = \lambda_0 e^{2\beta r}, \quad E(\boldsymbol{x}) = E(r) = E_0 e^{2\beta r}, \tag{4.57}$$

and where μ_0, λ_0, E_0 and β are constants. The Poisson ratio is taken as $\nu = 0.3$.

This tube problem can be modelled as a hollow cylinder with the boundary conditions defined by zero tractions on the outer ring $r = r_e$, a constant internal pressure $p > 0$ on the inner ring $r = r_0$, and symmetry boundary conditions (zero normal displacements and tangential stresses)

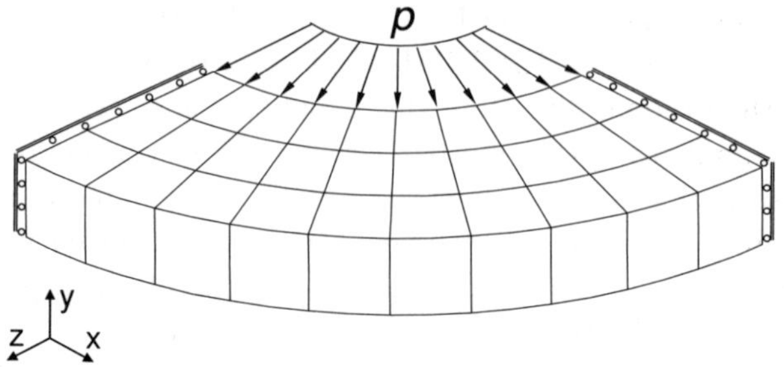

Fig. 4.1. Domain decomposition and boundary discretization of the hollow cylinder.

on the extreme faces (top and bottom) of the cylinder. Only a one-quarter section of the cylinder defining the elastic domain Ω is discretized, as shown in Fig. 4.1, with symmetry boundary conditions also applied on the surfaces $\theta = 0$ and $\theta = \pi/2$.

The domain decomposition of the quarter cylinder Ω employed in the calculations, with $r_0 = 1$ and $r_e = 2$ is shown in Fig. 4.1. This is a relatively coarse decomposition, as there are only $n = 10$ subdomains (polar segments) $\Omega_s \subset \Omega$ $(s = 1, \ldots, n)$, each one being discretized by three 9-node quadratic discontinuous boundary elements in the radial direction.*16 The (constant) grading direction for each subdomain Ω_s is approximated by the vector $\boldsymbol{\beta}^s$ in the radial direction associated with the mid-point in θ. For the values of β considered, $\beta = \pm 0.75$, the change in the elastic stiffness along the cylinder radius is $(E(r_e)/E(r_0))^{\pm 1} \cong 4.5$. Accurate results are nevertheless obtained with the crude discretization, a consequence of the exponential grading being handled exactly within the Green's functions.

The displacement and stress solutions shown in Figs. 4.2 and 4.3 agree excellently with the analytic solution in displacements and stresses.[41] To be more specific, define the percentages of the normalized errors as

$$\mathrm{err}(u_r) = \frac{u_r^{\mathrm{num.}} - u_r^{\mathrm{an.}}}{\max\limits_{r_0 \le r \le r_e} u_r^{\mathrm{an.}}} \times 100 \quad \text{and} \quad \mathrm{err}(\sigma_\theta) = \frac{\sigma_\theta^{\mathrm{num.}} - \sigma_\theta^{\mathrm{an.}}}{\max\limits_{r_0 \le r \le r_e} \sigma_\theta^{\mathrm{an.}}} \times 100, \tag{4.58}$$

where 'num.' and 'an.', respectively, refer to numerical and analytical values. By evaluating these errors at nodes in the radial direction, the following small error bounds are achieved for the radially graded cylinder (the superscripts $\pm$ refer to $\beta = \pm 0.75$): $0.2\% < \mathrm{err}(u_r^+) < 0.6\%$ and

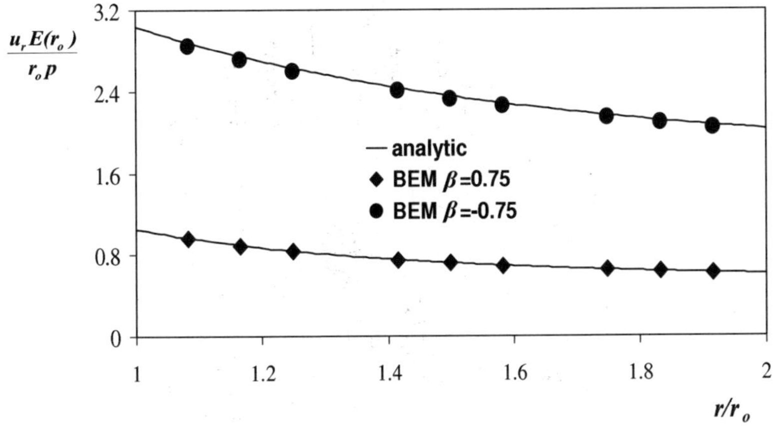

Fig. 4.2. Computed values of $u_r E(r_0)/r_0 p$ for $\beta = \pm 0.75$ in the hollow cylinder.

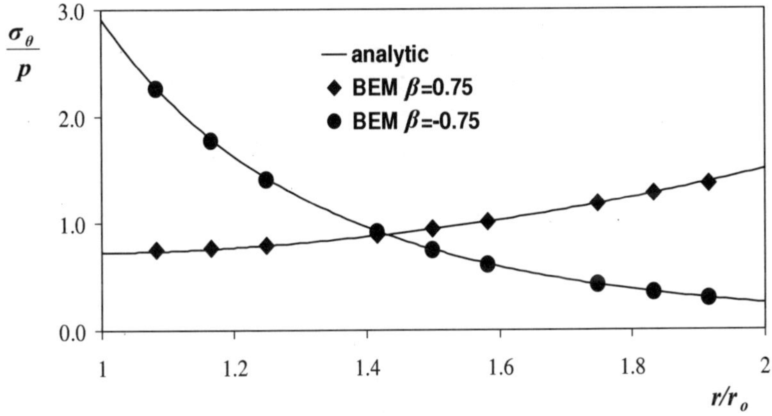

Fig. 4.3. Computed values of σ_θ/p for $\beta = \pm 0.75$ in the hollow cylinder.

$0.7\% < \mathrm{err}(u_r^-) < 1.3\%$ in the radial displacements and $0.01\% < \mathrm{err}(\sigma_\theta^+) < 1.3\%$ and $0.08\% < \mathrm{err}(\sigma_\theta^-) < 0.7\%$ in the hoop stresses.

4.5.2. *Spherical cavity in a cube subjected to uniaxial tension*

Consider an elastic domain Ω defined as a cube $\langle 0, 10\rangle^3$ of an exponentially graded elastic isotropic material having a spherical cavity of radius $r_0 = 1$ located at its center (see Fig. 4.4a). The origin $O = (0, 0, 0)$ of the global

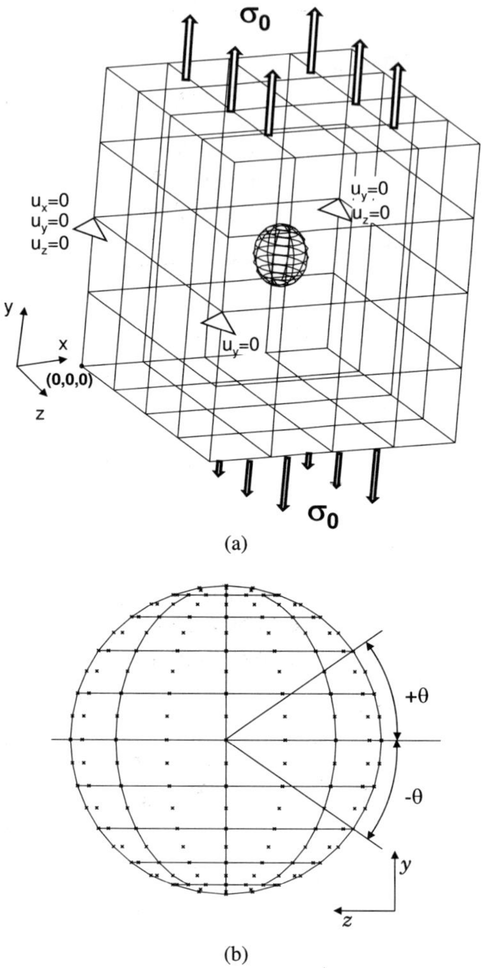

Fig. 4.4. (a) Geometry, boundary conditions and discretization of a cube with a spherical cavity, (b) Discretization of the cavity boundary and definition of the angle θ.

cartesian Coordinates is placed at a cube corner. The discretization of the domain boundary Γ by 9-node quadratic continuous boundary elements is also shown in Fig. 4.4a. There, $6 \times 16 = 96$ elements with 385 nodes are being used for the cube boundary, and 72 elements with 291 nodes for the cavity boundary. Fig. 4.4b shows in detail the discretization applied for the cavity boundary.

A tension load $\sigma_0 = 1$ in the y-direction is applied at the cube faces $y = 0$ (bottom) and $y = 10$ (top); the cavity boundary and the remaining cube faces are traction free. To avoid rigid-body motions of the domain Ω, six point supports are also specified.

Two cases for the grading direction are considered, along the y-axis (the grading and loading directions are parallel) and z-axis (the grading and loading directions are perpendicular), respectively, denoted as (a) and (b) hereafter. In case (a), the Young modulus varies as $E(y) = E_0 e^{2\beta y}$, whereas in case (b), $E(z) = E_0 e^{2\beta z}$, and values $\beta = 0.1, 0.2$ and 0.3 are employed. These values of β correspond to the following ratios between the maximum and minimum values of E in the cube, $E_{\max}/E_{\min} \cong 7.4$, 54.6 and 403.4, respectively. In addition, some results for the homogeneous material case defined by $\beta = 0$ are presented for comparison purposes. The Poisson ratio in all the configurations studied was taken as $\nu = 0.3$.

The displacement solutions obtained directly by the FGM-BEM code are illustrated in Fig. 4.5, where the deformed shapes of the cube and cavity boundaries are shown for cases (a) and (b) where $\beta = 0.3$; additionally, Fig. 4.6 shows in detail the corresponding deformed shapes of the spherical cavity.

With reference to the global deformation of the cube, according to Fig. 4.5a, the more compliant material in the bottom part of the domain is more elongated in the y-direction than the stiffer material in the top part. Consequently, due to the constant Poisson ratio, the bottom part of the domain is also more contracted in the transversal directions (x and z) than the top part. Similarly, in Fig. 4.5b, the more compliant material near the face $z = 0$ is more deformed than the stiffer material near the face $z = 10$. It can also be seen that both deformed shapes verify (at least approximately) the condition of the zero shear strains at cube corners, and the condition of zero shear strains in the planes perpendicular to cube edges.

Similar effects of the relative position of the grading and loading directions can also be observed on the detailed pictures of the deformed shapes of the spherical cavity shown in Fig. 4.6, although the variations (due to grading) of the deformed shape of the cavity are less pronounced here because of a smaller increment of stiffness along the cavity diameter. In Fig. 4.6a, the cavity is slightly more elongated and transversally contracted (in the x- and z-directions) in the bottom part in comparison with the top part, whereas in Fig. 4.6b, the cavity is more stretched on the left-hand side (corresponding to smaller values of z) than on the right-hand side (with larger values of z).

It should also be mentioned that all the deformed shapes presented, for both the cube and cavity boundaries, have the expected symmetries.

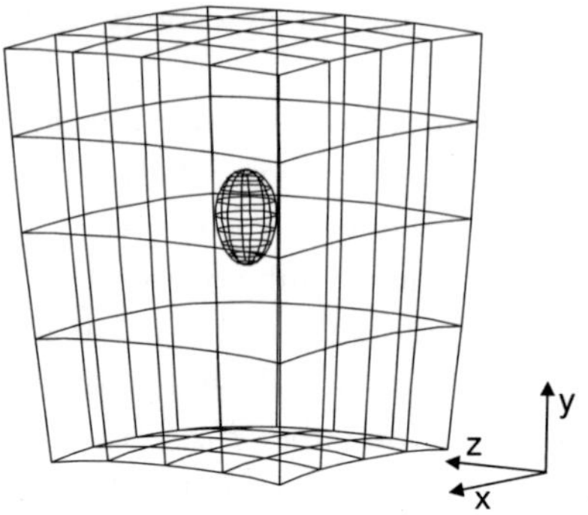

(a)

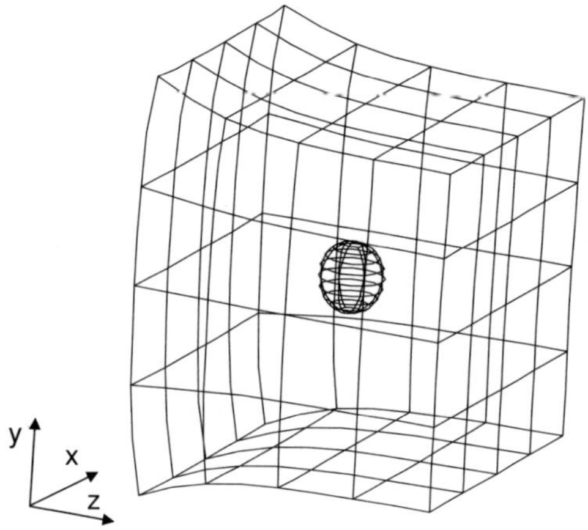

(b)

Fig. 4.5. Deformed shapes of the cube and spherical cavity boundaries for graded elastic domains (a) $E(y)$ and (b) $E(z)$. In both, $\beta = 0.3$.

Specifically, the following planes of symmetry are easily identified: in case (a), $x = 5$, $z = 5$, $z = x$ and $z = -x$, and in case (b), $x = 5$ and $y = 5$.

Summarizing the above analysis, we can say that the deformed shapes obtained reflect the expected behaviour caused by the material grading and add further (qualitative) evidence that the FGM-BEM code developed is producing correct solutions to exponentially graded material problems.

For strength assessment of FGMs, one of the key issues is the evaluation of stresses, and particularly stress concentration factors, at defects. Thus, to show the capabilities of the FGM-BEM code, the values of the normalized hoop stress σ_θ/σ_0 computed at nodes located in the plane $x = 5$ (recall

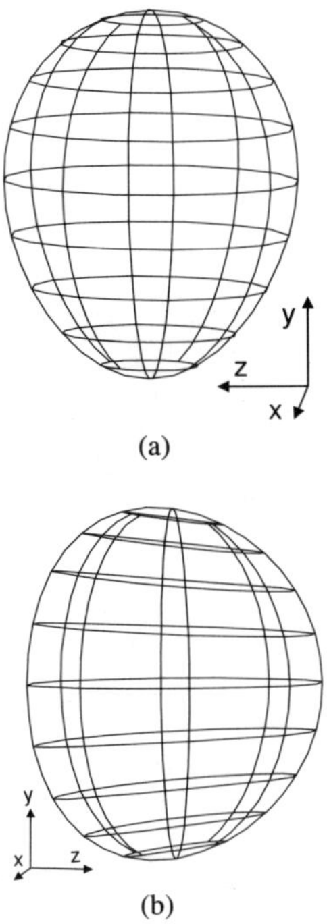

Fig. 4.6. Deformed shapes of the spherical cavity boundaries for graded elastic domains (a) $E(y)$ and (b) $E(z)$. In both, $\beta = 0.3$.

that this is a problem symmetry plane in both cases of grading considered) is plotted in radial plots, shown in Fig. 4.7, as a function of the angle θ (see Fig. 4.4b for θ definition). The zero stress value in the 'radial scale' is indicated by a circumference in these plots. The corresponding θ values are indicated at the extreme of each radius. In order to facilitate comparison of the present results with the results of other authors in future, the values of σ_θ/σ_0 at some particular points are shown in Table 4.1. These points are located at angles $\theta = 0°, \pm 90°$ and $180°$. The maximum value of σ_θ/σ_0, i.e. the stress concentration factor at the cavity boundary, is presented in a separate row, as in the case of the grading parallel to the load its position is different from these four particular angles.

The values of stresses presented have been computed, from the traction-free boundary conditions and the displacements at nodes located on the cavity boundary obtained directly in the BEM analysis, through a standard post-processing procedure commonly used in BEM for homogeneous materials.[37] In this procedure, tangential derivatives of the boundary

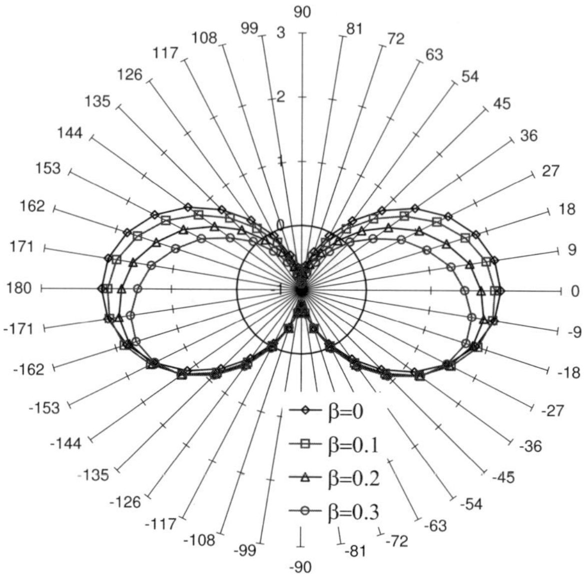

(a)

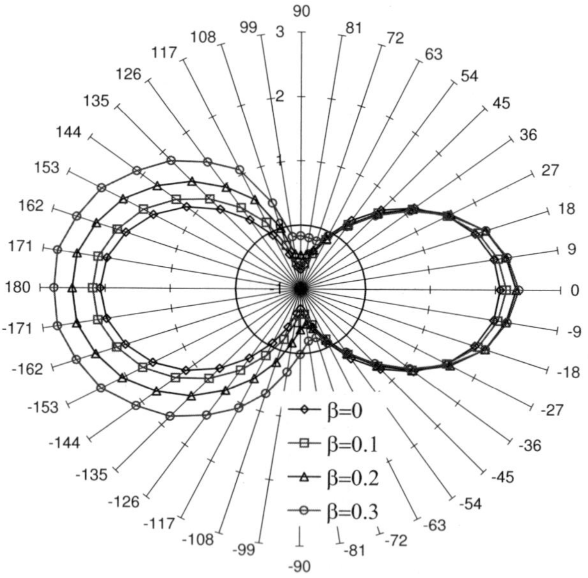

(b)

Fig. 4.7. Radial plots of the normalized hoop stresses σ_θ/σ_0 at nodes located on the cavity boundary in the symmetry plane $x = 5$, (a) $E(y)$ and (b) $E(z)$.

Table 4.1. Normalized hoop stress σ_θ/σ_0 evaluated along the spherical cavity boundary section at $x = 5$ for points $\theta = 0°, \pm 90°$ and $180°$.

E	E_0	$E_0 e^{2\beta y}$			$E_0 e^{2\beta z}$		
$(\theta \setminus \beta)$	(0)	0.1	0.2	0.3	0.1	0.2	0.3
0°	2.07	1.98	1.77	1.52	2.17	2.32	2.36
90°	−0.70	−0.73	−0.80	−0.91	−0.64	−0.47	−0.17
−90°	−0.70	−0.72	−0.76	−0.78	−0.61	−0.37	−0.02
180°	2.07	1.98	1.77	1.52	2.19	2.50	2.78
$\theta(\sigma_\theta^{\max})$	2.07	1.99	1.85	1.67	2.19	2.50	2.78

displacements are computed element-wise by differentiating boundary element shape functions, and used to evaluate strains in the plane tangent to the cavity boundary. These strains substituted into the constitutive law for the linear elastic graded material, Eq. (4.5), are combined with the traction-free boundary conditions to obtain nodal stresses defined element-wise. Finally, an average value of the nodal stresses associated with a node and defined on the boundary elements adjacent to this node is computed, obtaining in this way a reasonable approximation of boundary stresses at the nodes of the cavity discretization.

For the purpose of comparison, the results computed for the homogeneous case, where $\beta = 0$, are included in Table 4.1. When comparing these results with the analytic solution available,[49] $\sigma_\theta/\sigma_0 = 2.045$ for $\theta = 0°$ and $180°$, and $\sigma_\theta/\sigma_0 = -0.682$ for $\theta = \pm 90°$, the relative errors of the above described post-processing procedure for the direct BEM result in displacements then being 1.38% and 1.95%, respectively. Thus, we can expect that the percentage of the relative errors of the BEM results for graded materials shown in Table 4.1 will be similar or slightly worse due to the more complex constitutive law of the material.

Hoop stresses in case (a): parallel grading and loading directions. When looking at σ_θ/σ_0 values in Fig. 4.7a, it is observed that, while keeping the expected symmetry with respect to the line given by $\theta = -90°$ and $90°$, these values are strongly decreasing with increasing values of β for $\theta > 0$, whereas for $\theta < 0$ variations in σ_θ with increasing values of β are smaller and not monotonous. There is a shift of higher and positive stress values to the zone corresponding to more compliant material, defined by $\theta < 0$. This can be, at first sight, surprising, as, in general, higher stresses would be expected in the zone with the stiffer material. It seems that this shift is associated with a high level of material elongation in the zone of more compliant material, but an in-depth explanation of this phenomenon will require further study. It would be particularly interesting to analyze whether a similar shift appears in an analogous plane problem with a hole in a graded material subjected to a transverse uniaxial tension. A consequence of this shift is that the position of the stress concentration has moved from $\theta(\sigma_\theta^{\max}) = 0°$ for $\beta = 0$ to a negative value of $\theta(\sigma_\theta^{\max})$. The following estimations are obtained using the present boundary discretization: $\theta(\sigma_\theta^{\max}) = -9°$ for $\beta = 0.1,\ 0.2$, and $\theta(\sigma_\theta^{\max}) = -18°$ for $\beta = 0.3$. Values of σ_θ/σ_0 in Table 4.1 (case $E(y)$) show that the stress concentration factor is strongly decreasing with increasing values of β, whereas the maximum compressions achieved at $\theta = \pm 90°$ are increasing with increasing values of β, higher values of compressions being obtained at $\theta = 90°$ as could be expected due to stiffer material therein.

Hoop stresses in case (b): perpendicular grading and loading directions. When looking at σ_θ/σ_0 values in Fig. 4.7b, it is observed that these values keep the expected symmetry with respect to the line given by $\theta = 0°$ and 180° except for a few values at angles θ near $\pm 90°$. It is believed that an improvement in the cavity boundary discretization will prevent this asymmetry. As expected, stresses are strongly increasing with increasing values of β for $|\theta| > 90°$ in the zone corresponding to stiffer material, while their variations are much smaller for $|\theta| < 90°$ in the zone corresponding to more compliant material, although σ_θ at $\theta = 0$ is increasing with increasing values of β. Values of σ_θ/σ_0 in Table 4.1 (case $E(z)$) show that the stress concentration factor is strongly increasing with increasing values of β, the maximum value for $\beta = 0.3$ being 34% higher than in the homogeneous case. Although the expected symmetry in the maximum compressions achieved at $\theta = \pm 90°$ is perturbed for higher values of β (see Table 4.1), their tendency to decrease with increasing values of β is evident.

In concluding, we can say that the above preliminary study of stress concentrations at a spherical cavity in an FGM provides useful information for strength analysis of FGMs in the presence of voids. The FGM-BEM code developed can be directly applied to analyze related stress concentration problems for different shapes of voids (e.g. ellipsoidal), for interactions of multiple voids, and also for other kinds of defects like inclusions. Nevertheless, the critical issue is the computational efficiency of this code; thus, some proposals for its improvements will be discussed in the next section.

4.6. Analytic Expansion

The calculations in the previous section, along with those in the cited publications, establish that a boundary integral analysis for graded elasticity is possible. However, as discussed in the Introduction, the situation is far from ideal. Direct numerical evaluation of the formulas for the grading terms $\mathcal{U}^g$ and $\mathcal{T}^g$ is expensive, and nearly prohibitively so (it is debatable whether it is correct to say 'nearly' or not). Moreover, the extension of these techniques to fracture analysis calculations, necessitating writing and computing the kernel functions in the traction integral equation, is far from appealing.

It seems likely that further progress in this area will require a new, more computational friendly, representation of the graded Green's function and its derivatives. In this section, we outline an approach that appears to be successful; if so, the details will be published elsewhere.

Ignoring the exponential prefactor, the fundamental displacement tensor consists of two simple analytic pieces, the Kelvin solution and the diagonal contribution in Eq. (4.32), plus the complicated $A_{j\ell}$ term (Eq. (4.33))

$$\begin{aligned} A_{j\ell} &= \frac{1}{(2\pi)^3}\int H_{j\ell}\exp(-\mathrm{i}\boldsymbol{r}\cdot\boldsymbol{\xi})\,\mathrm{d}\boldsymbol{\xi} \\ &= -\frac{\kappa}{16\pi^3\mu_0}\int \frac{\Omega_{j\ell}}{(\xi^2+\beta^2)\xi^4\Delta}\exp(-\mathrm{i}\boldsymbol{r}\cdot\boldsymbol{\xi})\,\mathrm{d}\boldsymbol{\xi}, \end{aligned} \tag{4.59}$$

where $\Delta = \xi^4 + 2\xi^2\beta^2 q + \beta^4$ and $\Omega_{j\ell}$ is a polynomial in the components of $\boldsymbol{\xi}$. The difficulty in inverting the Fourier transform stems from the q term in Δ, whereby $q = 1 + 2\nu\kappa\sin^2\theta$. As θ is the angle between $\boldsymbol{\beta}$ and $\boldsymbol{\xi}$, the expressions for the poles in Eq. (4.59) are not easy to work with. Moreover, $\Omega_{j\ell}$ is a polynomial of degree 7 and the denominator is of degree 10, so the transform does not die off fast enough to allow an efficient and accurate brute force numerical evaluation of the three-dimensional Fourier integral.

In the method outlined below, additional (relatively) simple analytic expressions will be split off from $A_{j\ell}$, leaving a Fourier transform that decays faster at infinity. It is expected that numerical integration, or some other procedure, will then be able to handle this 'weakened remainder' term, left in the form of a Fourier integral. The key observation is to write

$$\begin{aligned} \frac{1}{\Delta} &= \frac{1}{(\xi^2+\beta^2)^2} + \left\{\frac{1}{\Delta} - \frac{1}{(\xi^2+\beta^2)^2}\right\} \\ &= \frac{1}{(\xi^2+\beta^2)^2} - 4\nu\kappa\sin^2\theta\frac{\xi^2\beta^2}{(\xi^2+\beta^2)^2\Delta}, \end{aligned} \tag{4.60}$$

allowing the inverse transform in Eq. (4.59) to be expressed as

$$\frac{\Omega_{j\ell}}{(\xi^2+\beta^2)\xi^4\Delta} = \frac{\Omega_{j\ell}}{(\xi^2+\beta^2)^3\xi^4} - 4\nu\kappa\beta^2\sin^2\theta\frac{\Omega_{j\ell}}{(\xi^2+\beta^2)^3\xi^2\Delta}. \tag{4.61}$$

Note that θ is absent in the first term on the right and that the poles are now easily determined; analytic inversion of this component should follow as in the treatment of the E term, Eq. (4.31), with r as the polar axis in the spherical coordinate system. Moreover, the 'remainder' term dies off faster at infinity than $A_{j\ell}$, as the denominator is now of order 12 in ξ.

Further analytic terms can be split off by iterating this entire process: replace the Δ^{-1} in the last term on the right by Eq. (4.60) to obtain

$$\frac{1}{\Delta} = \frac{1}{(\xi^2+\beta^2)^2} - 4\nu\kappa\sin^2\theta\frac{\xi^2\beta^2}{(\xi^2+\beta^2)^4} + (4\nu\kappa\sin^2\theta)^2\frac{\xi^4\beta^4}{(\xi^2+\beta^2)^4\Delta}. \tag{4.62}$$

For each iteration, more work is required to obtain the analytic expressions, but the decay in the remainder term will be faster.

4.7. Conclusions

The manufacture, characterization and application of functionally graded materials, or more generally *tailored materials*, is a relatively new field of study in materials science and engineering. The development of integral equation methods for the analysis and design of these *non-homogeneous materials* is similarly far from being a settled field. An important motivation for pursuing this work is the attempt to exploit the advantages that boundary integral simulations possess for specific problems, e.g. crack propagation and stress concentrations at voids and inclusions, for this new area.

Although other integral equation approaches for FGMs have been developed (see, for example, the cited works by J. and V. Sládek[31,32]), and specialized finite element methods exist as well,[50–54] this chapter has narrowly focused on (what is expected to be) the early work for one particular path: a 'standard' boundary integral equation formulation based upon the point load Green's function for an *exponentially graded* material. As reviewed herein, the exponential grading is mathematically tractable, in the sense that computable formulas for the fundamental displacement and traction tensors can be derived. This allows the writing of a boundary integral equation for exponential grading that is completely analogous to the standard equation for homogeneous isotropic materials based upon the Kelvin solution. Moreover, the new calculations presented in the previous section, together with past work, illustrate that the numerical implementation of this program can be successful.

However, as observed in Ref. 33 and 34, the forms for the fundamental solutions are certainly not convenient for numerical work. Direct numerical evaluation is very costly, and this necessitates employing a parallel algorithm,[41] even for relatively simple geometries. Equally important, an even more complicated function, the hypersingular kernel, is needed for fracture analysis. This function is defined in terms of second order derivatives of the displacement kernel, and the derivation and implementation utilizing the procedures employed for the displacement and traction kernels appears to be a rather daunting task.

If this approach is to lead to a viable technique for modelling FGMs, it will therefore be necessary to develop alternate representations for all kernel functions, ones that can form the basis for more efficient numerical algorithms. A mechanism for achieving this, that is currently under investigation, was outlined in Sec. 4.6. The key idea is to split the difficult inverse Fourier transform integral into two integrals: the first can be treated exactly, yielding a relatively simple analytic expression, while the integrand in the remainder dies off rapidly at infinity. It remains to be seen whether or not this new formulation will successfully overcome the difficulties of the original expressions.

Acknowledgements

The author thanks Federico París (University of Seville) for his motivation and continuous support of this work. Large parts of the work discussed herein were carried out by Paul Martin (Colorado School of Mines) and Rafael Criado (previously at the University of Seville and now at GHESA). L.J. Gray is grateful to Federico París for the hospitality during several visits at the University of Seville. This work was supported by the Spanish Ministry of Education and Science (Project TRA2005-06764), the Junta de Andalucía (Project of Excellence TEP1207) and by the Office of Advanced Scientific Computing Research, U.S. Department of Energy, under contract DE-AC05-00OR22725 with UT-Battelle, LLC. J.E. Ortiz was supported by a fellowship from the Spanish Ministry of Education and Science (Programa Ramón y Cajal). O. Sallah was supported by a fellowship from the Egyptian Ministry of Higher Education.

References

1. S. Rjasanow and O. Steinbach, *The Fast Solution of Boundary Integral Equations* (Springer, Berlin, 2007).
2. G. Barton, *Elements of Green's Functions and Propagation* (Oxford University Press, Oxford, 1999).
3. A. Pomp, The boundary-domain integral method for elliptic systems, Lect. Notes in Maths. **1683**. (Springer, Berlin, 1998).
4. R. P. Shaw, Green's functions for heterogeneous media potential problems, *Eng. Anal. Bound. Elem.* **13**, 219–221, (1994).
5. R. P. Shaw and N. Makris, Green's functions for Helmholtz and Laplace equations in heterogeneous media, *Eng. Anal. Bound. Elem.* **10**, 179–183, (1992).
6. R. P. Shaw and G. D. Manolis, A generalized Helmholtz equation fundamental solution using a conformal mapping and dependent variable transformation, *Eng. Anal. Bound. Elem.* **24**, 177–188, (2000).
7. W. T. Ang, J. Kusuma, and D. L. Clements, A boundary element method for a second order elliptic partial differential equation with variable coefficients, *Eng. Anal. Bound. Elem.* **18**, 311–316, (1996).
8. D. L. Clements, A boundary integral equation method for the numerical solution of a second order elliptic equation with variable coefficients, *J. Austral. Math. Soc. (Series B)* **22**, 218–228, (1980).
9. E. Divo and A. J. Kassab, Generalized boundary integral equation for heat conduction in non-homogeneous media: Recent developments on the sifting property, *Eng. Anal. Bound. Elem.* **22**, 221–234, (1998).
10. A. J. Kassab and E. Divo, A generalized boundary integral equation for isotropic heat conduction with spatially varying thermal conductivity, *Eng. Anal. Bound. Elem.* **18**, 273–286, (1996).
11. M. Bonnet and M. Guiggiani, Comments about the paper entitled 'A generalized boundary integral equation for isotropic heat conduction with spatially varying thermal conductivity' by A. J. Kassab and E. Divo, *Eng. Anal. Bound. Elem.* **22**, 235–240, (1998).
12. H. Power, On the existence of Kassab and Divo's generalized boundary integral equation formulation for isotropic heterogeneous steady state heat conduction problems, *Eng. Anal. Bound. Elem.* **20**, 341–345, (1997).
13. A. McKenney, L. Greengard, and A. Mayo, A fast Poisson solver for complex geometries, *J. Comput. Phys.* **118**, 348–355, (1995).
14. M. Ingber, A. Mammoli, and M. Brown, A comparison of domain integral evaluation techniques for boundary element methods, *Int. J. Numer. Meth. Engng.* **52**, 417–432, (2001).
15. J. Ding, W. Ye, and L. J. Gray, An accelerated surface discretization-based BEM approach for non-homogeneous linear problems in 3-D complex domains, *Int. J. Numer. Meth. Eng.* **63**, 1775–1795, (2005).
16. L. J. Gray, T. Kaplan, J. D. Richardson, and G. H. Paulino, Green's functions and boundary integral analysis for exponentially graded materials: Heat conduction, *J. Appl. Mech.* **70**, 543–549, (2003).

17. J. R. Berger, P. A. Martin, V. Mantič, and L. J. Gray, Fundamental solutions for steady-state heat transfer in an exponentially graded anisotropic material, *ZAMP* **56**, 293–303, (2005).
18. A. Sutradhar, G. H. Paulino, and L. J. Gray, Transient heat conduction in functionally graded materials, *Eng. Anal. Bound. Elem.* **26**, 119–132, (2002).
19. P. A. Martin, On functionally graded balls and cones, *J. Eng. Math.* **42**, 133–142, (2002).
20. E. C. N. Silva, M. C. Walters, and G. H. Paulino, Modeling bamboo as a functionally graded material: Lessons for the analysis of affordable materials, *J. Mater. Sci.* **41**, 6991–7004, (2006).
21. U. Leushake, T. Krell, and U. Schultz, Graded thermal barrier coating systems for gas turbine applications, *Materialwiss. Werkstofftech.* **28**, 391–394, (2004).
22. J. F. Li, K. Takagi, M. Ono, W. Pan, R. Wantanabe, A. Almajid, and M. Taya, Fabrication and evaluation of porous piezoelectric ceramics and porosity graded piezoelectric actuators, *J. Am. Ceram. Soc.* **86**, 1094–1098, (2003).
23. W. G. Cooley and A. Palazotto, Finite element analysis of functionally graded shell panels under thermal loading. *Proc. 2005 ASME International Congress and Exhibition*, (2005). Paper IMECE2005-85778.
24. V. Birman and L. W. Byrd, Modeling and analysis of functionally graded materials and structures, *Appl. Mech. Rev.* **60**, 195–212, (2007).
25. T. Hirai, Functionally Graded Materials, In: *Processing of Ceramics, Part 2*, R. J. Brook (ed.), In: *Materials Science and Technology, volume 17B*, R. W. Cahn, P. Haasen and E. J. Kramer (eds.), VCH Verlagsgesellschaft, Weinheim, 292–341, (1996).
26. A. J. Markworth, K. S. Ramesh, and W. P. Parks Jr., Modelling studies applied to functionally graded materials, *J. Mater. Sci.* **30**, 2183–2193, (1995).
27. S. Suresh and A. Mortensen, *Fundamentals of Functionally Graded Materials.* (The Institute of Materials, IOM Communications Ltd., London, 1998).
28. Y. Miyamoto, W. A. Kaysser, B. H. Rabin, A. Kawasaki, and R. G. Ford, *Functionally Graded Materials: Design, Processing and Applications.* (Kluwer Academic Publishers, Dordrecht, 1999).
29. C. Vrettos, In-plane vibrations of soil deposits with variable shear modulus: I. surface waves, *Int. J. Num. Anal. Meth. Geomech.* **14**, 209–222, (1990).
30. C. Vrettos, Time-harmonic Boussinesq problem for a continuously non-homogeneous soil, *Earthq. Eng. Struct. Dyn.* **20**, 961–977, (1991).
31. J. Sládek, V. Sládek, C. Zhang, P. Solek, and E. Pan, Evaluation of fracture parameters in continuously nonhomogeneous piezoelectric solids, *Int. J. Fracture* **145**, 313–326, (2007).
32. J. Sládek, V. Sládek, C. Hellmich, and J. Eberhardsteiner, Analysis of thick functionally graded plates by local integral equation method, *Commun. Numer. Meth. Eng.* **23**, 733–754, (2007).

33. P. A. Martin, J. D. Richardson, L. J. Gray, and J. Berger, On Green's function for a three-dimensional exponentially-graded elastic solid, *Proc. Royal Soc.* **458**, 1931–1948, (2002).
34. R. Criado, J. E. Ortiz, V. Mantič, L. J. Gray, and F. París, Boundary element analysis of three-dimensional exponentially graded isotropic elastic solids, *CMES - Comp. Model. Eng. Sci.* **22**, 151–164, (2007).
35. R. Criado, L. J. Gray, V. Mantič, and F. París, Green's function evaluation for three-dimensional exponentially-graded elasticity, *Int. J. Numer. Meth. Eng.* **74**, 1560–1591, (2008).
36. F. París and J. Cañas, *Boundary Element Method: Fundamentals and Applications* (Oxford University Press, Oxford, 1997).
37. M. H. Aliabadi, *The Boundary Element Method. Vol. II.* (John Wiley & Sons, Chichester, 2002).
38. V. Mantič, A new formula for the C-matrix in the Somigliana identity, *J. Elast.* **33**, 191–201, (1993).
39. M. Costabel and M. Dauge, On representation formulas and radiation conditions, *Math. Method. Appl. Sci.* **20**, 133–150, (1997).
40. J. C. Lachat and J. O. Watson, Effective numerical treatment of boundary integral equations: A formulation for three-dimensional elastostatics, *Int. J. Numer. Meth. Eng.* **10**, 991–1005, (1976).
41. J. E. Ortiz, W. A. Shelton, V. Mantič, R. Criado, L. J. Gray and F. París, A parallel domain decomposition BEM algorithm for three dimensional exponentially graded elasticity, *J. Appl. Mech.* **75**, 051108-1–051108-8, (2008).
42. C. O. Horgan and A. M. Chan, The pressurized hollow cylinder or disk problem for functionally graded isotropic linearly elastic materials, *J. Elast.* **55**, 43–59, (1999).
43. X. Zhang and N. Hasebe, Elasticity solution for a radially nonhomogeneous hollow circular cylinder, *J. Appl. Mech.* **66**, 598–606 (1999).
44. H. Xiang and Z. Shi, Elastic analyses of heterogeneous hollow cylinders, *Mech. Res. Commun.* **33**, 681–691, (2006).
45. H. Hatami-Marbini and H. M. Shodja, Thermoelastic fields of a functionally graded coated inhomogeneity with sliding/perfect interfaces, *J. Appl. Mech.* **74**, 389–398, (2007).
46. N. Tutuncu, Stresses in thick-walled FGM cylinders with exponentially-varying properties, *Eng. Struct.* **29**, 2032–2035, (2007).
47. X. Wang and N. Hasebe, Complex variable solution of plane problem for functionally graded materials, In: *Solid Mechanics and Its applications, Vol. 106: IUTAM Symposium on Dynamics of Advanced Materials and Smart Structures*, K. Watanabe, F. Ziegler (eds.). (Kluwer Academic Publishers, Dordrecht, 449–458, 2003).
48. X. Wang and N. Hasebe, On stress concentration for an arbitrarily oriented elliptical hole in functionally graded materials, In: *Developments in Mechanics of Structures and Materials*, A.J. Deeks, H. Hao (eds.). (Taylor & Francis, London, 305–310, 2005).
49. J. N. Goodier, Concentration of stress around spherical and cylindrical inclusions and flaws, *J. Appl. Mech.* **55**, 39–44, (1933).

50. J. H. Kim and G. H. Paulino, Finite element evaluation of mixed-mode stress intensity factors in functionally graded materials, *Int. J. Numer. Meth. Eng.* **53**, 1903–1935, (2002).
51. J. H. Kim and G. H. Paulino, Mixed-mode fracture of orthotropic functionally graded materials using the finite element method, *Eng. Frac. Mech.* **69**, 1769–1790, (2002).
52. R. Naghdabadi and S. A. H. Kordkheili, A finite element formulation for analysis of functionally graded plates and shells, *Arch. Appl. Mech.* **74**, 375–386, (2005).
53. G. Anlas, M. H. Santare and J. Lambros, Numerical calculation of stress intensity factors in functionally graded materials, *Int. J. Fracture* **104**, 131–143, (2000).
54. M. H. Santare and J. Lambros, Use of graded finite elements to model the behavior of nonhomogeneous materials, *J. Appl. Mech.* **67**, 819–822, (2000).

Chapter 5

FAST HIERARCHICAL BOUNDARY ELEMENT METHOD FOR LARGE-SCALE 3-D ELASTIC PROBLEMS

I. Benedetti and A. Milazzo
Dipartimento di Ingegneria Strutturale, Aerospaziale e Geotecnica
Università degli Studi di Palermo
Edificio 8, Viale delle Scienze, 90128, Palermo, Italy
i.benedetti@unipa.it, alberto.milazzo@unipa.it

M.H. Aliabadi
Department of Aeronautics, Imperial College London
South Kensington Campus, London, SW7 2AZ, United Kingdom
m.h.aliabadi@imperial.ac.uk

This chapter reviews recent developments in the fast solution of boundary element systems for large-scale 3-D elastic problems. Applications to both isotropic and anisotropic materials as well as to the analysis of cracked and uncracked solids are considered. An effective technique based on the combined use of a hierarchical representation of the boundary element collocation matrix and iterative solution procedures is described. The hierarchical representation of the collocation matrix is built starting from the generation of the cluster and block trees that take into account the nature of the considered problem, for example the possible presence of a crack. Low rank blocks are generated through adaptive cross approximation (ACA) algorithms and the final hierarchical matrix is further coarsened through suitable procedures also used for the generation of a *coarse* preconditioner, which is built taking full advantage of the hierarchical format. The final system is solved using a generalized minimum residual (GMRES) iterative solver. Applications show that the technique allows considerable savings in terms of storage memory, assembly time and solution time without loss of accuracy. Such features make the method appealing for large-scale applications.

5.1. Introduction

The boundary element method (BEM) has developed over the last three decades to become a powerful numerical tool for the analysis and solution of many physical and engineering problems.[1,2] Today, it represents a viable alternative to other numerical approaches, such as the finite element method (FEM).

The main advantage of boundary element techniques is the reduction in the degrees of freedom needed to model a given physical system. Such reduction is allowed by the underlying boundary integral formulation which requires, for its numerical solution, only the discretization of the boundary of the solution domain. This results not only in a reduction in size of the system matrix, but also in faster data preparation.

However, as the size of the problem increases, the time required to solve the final system of equations increases considerably. The system matrix obtained by the application of the BEM is fully populated and not symmetric. This results in increased storage memory requirements as well as increased solution time in comparison to the FEM. Since the matrix is fully populated, the memory required to store its coefficients is of order $O(N^2)$, where N denotes the number of unknowns. On the other hand, the solution of the system requires $O(N^3)$ operations if direct solvers are used or $O(M \times N^2)$ if iterative solvers are employed, where M denotes the number of iterations.

Much research has been devoted to the improvement of BE solution methods and many techniques using block-based solvers,[3,4] lumping technique[5] or iterative solvers[6] have been proposed. In the early 1980s, Rokhlin[7] developed an iterative strategy for the solution of the integral equations arising in classical potential theory. This work introduced the idea of coupling iterative solvers with the harmonic expansion of the kernels on suitable clusters of far field boundary elements, in order to reduce the computational cost of the solution process. In particular, the method was aimed at reducing the number of operations required to evaluate the matrix-vector products occurring in the application of iterative solvers and resulted in an $O(N)$ algorithm for the solution of the original equations.

Similar algorithms for the reduction of the computational complexity of the solution process were also developed in other fields of investigation that were not directly related to the boundary element method. Particularly interesting is the algorithm devised by Barnes and Hut[8] for the treatment of the gravitational N-body problem. They developed an $O(N \ln N)$ strategy based on the preliminary hierarchical subdivision of the space into cubic cells and on the following approximation of the mutual action between cells through a recursive scheme. A similar approach was presented by Greengard and Rokhlin,[9] who used multi-pole expansions to approximate potential and force fields of various nature generated by systems of many particles.

Although these algorithms were mainly developed for N-body problems, they can be extended to the treatment of boundary value problems, due to their similar underlying mathematical structure. The boundary element method is, in fact, based on the calculation of influence coefficients of the solution matrix by integration of the fundamental solution collocated on some source point, which represents a certain mutual influence between couples of points, over some surface elements. From this point of view, the approaches developed by Rokhlin[7] and Greengard and Rokhlin[9] are analogous, as already pointed out by the authors themselves.

Starting from these early works, fast multi-pole methods (FMMs) have been developed to efficiently solve boundary element formulations for different kinds of problems. Nishimura *et al.*[10] used FMMs in connection with a GMRES method[11] to solve 3-D crack problems for the Laplace equation. Fast algorithms have also been used for the treatment of elastic problems. Fu *et al.*[12] developed an FMM BEM for 3-D many-particle elastic problems based on spherical harmonic expansions of the kernel functions, while Popov and Power[13] used Taylor expansions to obtain an $O(N)$ algorithm for 3-D elasticity as well. Another interesting method intended for enhancing the matrix-vector multiplication was the panel clustering approach developed by Hackbusch and Nowak.[14]

Although the techniques noted above are very effective and are valuable tools for the fast solution of boundary element problems, their main disadvantage is that the knowledge of the kernel expansion is required in order to carry out the integration; all the terms of the series needed to reach a given accuracy must be computed in advance and then integrated, which can lead to a significant modification of the integration procedures in standard BEM codes.

From an algebraic point of view, however, the integration of a degenerate kernel (i.e. of a kernel expanded in series) over a cluster of elements corresponds to the approximation of the corresponding matrix block by a *low rank* block. This idea paved the way for the development of purely algebraic techniques for the approximation of large BEM matrices, like the mosaic-skeleton method.[15–17] Of particular interest was the observation, due to Tyrtyshnikov,[15] that low rank approximations could be built from only a few entries of the original block.

Subsequently, Bebendorf[18] proposed a method for the construction of such approximations, based on the computation of selected rows and columns from the original blocks. The technique was further developed by Bebendorf and Rjasanow[19] and was referred to as adaptive cross approximation (ACA). Such an algorithm allows a relatively simple generation of the approximation and enables both storage and matrix-vector multiplication in almost linear complexity.

The subdivision of the matrix into a hierarchical tree of sub-blocks and the blockwise approximation by low rank blocks is the basis for the *hierarchical representation* of the collocation matrix.[20,21] As in the case of FMMs, the use of the hierarchical format is aimed at reducing the storage requirement and the computational complexity arising in the BEM. Having represented the coefficient matrix in hierarchical format, the solution of the system can be obtained either directly, by inverting the matrix in hierarchical format, or indirectly, by using iterative schemes with or without preconditioners.[22,23] Both choices rely on the use of formatted matrix operations, i.e. on a suitable arithmetic for hierarchical matrices developed to take advantage of this special format.[24,25]

The use of iterative techniques, however, takes particular advantage of the employment of the hierarchical format. Different iterative solvers for algebraic systems of equations stemming from 2-D and 3-D BEM problems have been investigated. While early studies[26] had not shown good results, subsequent works[6,27] confirmed the applicability of iterative solution procedures to BEM systems and showed the potentiality for operations count reduction with respect to Gauss elimination, especially for large systems; however, they reported a serious lack of convergence for the worst ill-conditioned cases, when mixed boundary conditions are present, thus pointing out the need for preconditioning the system. Barra *et al.*[28] tested the performance of the GMRES algorithm, developed by Saad and Schultz,[11] for the solution of two-dimensional elasticity BEM equations, observing a more rapid convergence with respect to other iterative strategies, especially when preconditioning was used. They mentioned the possibility of developing new preconditioners based on the inherent nature of the BEM. Guru Prasad *et al.*[29] discussed the performance of several Krylov subspace methods and related such performance to the structure of the BEM matrices for some two- and three-dimensional thermal and elastic problems, highlighting the effect of the relative magnitude of the coefficients of the system matrix on the convergence of the algorithms. Moreover, they pointed out that the use of suitable preconditioning improves the eigenvalues clustering and the diagonal dominance of the matrix, thus resulting in enhanced convergence. Their analysis demonstrated that preconditioned Krylov methods, especially preconditioned GMRES, could be competitive or superior to direct methods. The importance of the diagonal dominance for the iterative solution of BEM equations has been shown by Urekew and Rencis,[30] while Merkel *et al.*[31] focused on eigenvalues clustering and its effect on the convergence of iterative solvers applied to the solution of some thermal and elastic industrial problems. Some issues in the analysis of larger three-dimensional BEM systems through preconditioned GMRES were evidenced by Leung and Walker,[32] who also proposed a strategy to overcome some limitations and extend the applicability of the algorithm to systems with some thousands of unknowns. Barra *et al.*[33] proposed a strategy for the construction of preconditioners for GMRES-solved BEM problems, while Wang *et al.*[34] investigated a class of preconditioners for fast multi-pole BEMs.

All the aforementioned studies have demonstrated the importance of preconditioners for an effective iterative solution. A general survey on preconditioners for improving the performance and reliability of Krylov subspace methods has recently been presented by Benzi,[35] who pointed out that the intense research on preconditioners has blurred the distinction between direct and iterative solvers. The importance of the subject has also been stressed by Saad and van der Vorst,[36] in their survey of iterative solvers for linear systems.

In this context, the use of the hierarchical format for BEM matrices, in conjunction with Krylov subspace methods, constitutes a recent and interesting development. The method proves to be efficient in dealing with large BEM systems and offers a quite natural approach to the construction of effective preconditioners.

Hierarchical matrices and their arithmetic have been extensively studied and assessed and their application has proved successful for the analysis of some interesting realistic problems. Apart from some benchmark tests reported in the papers devoted to the development of the technique (see, for example, the work of Bebendorf and Rjasanow[19]), interesting applications to various electromagnetic problems have been proposed by Kurz *et al.*,[37] Zhao *et al.*[38] and Ostrowski *et al.*[39] Bebendorf and Grzhibovskis[40] have recently extended the use of ACA to the analysis of elastic problems through Galerkin BEM, while Benedetti *et al.*[41] extended the use of hierarchical matrices to the treatment of 3-D elastic *crack* problems through dual boundary element method (DBEM). Some preliminary results about anisotropic elastic problems without crack have been presented by Benedetti *et al.*[42]

In the following sections, the development and the use of the fast hierarchical BEM for the analysis of three-dimensional elasticity problems is described. For the sake of generality, the formulation and the strategy solution are described referring to the DBEM, which can take into account the possible presence of cracks in the analysed domain. However, all the considerations can be directly applied to the straightforward BEM analysis of uncracked bodies. First, the basic formulation of the DBEM for 3-D fracture mechanics problems is briefly reviewed and the features of DBEM matrices are discussed. Next, the main steps for building the hierarchical representation of the solution matrix are discussed and some algorithms are reported. Some considerations about the application of the hierarchical format to boundary element formulations of 3-D crack problems are pointed out. Some applications to both isotropic and anisotropic problems complete the work and demonstrate the capability of the method.

5.2. The 3-D Dual Boundary Element Method

The DBEM is a general and efficient technique for modelling both two-dimensional[43,44] and three-dimensional[45,46] isotropic and anisotropic[47–51] crack problems in the framework of the BEMs.[52,53] The DBEM is based on the use of two independent boundary integral equations, namely the displacement integral equation, collocated on the external boundary and on one of the crack surfaces, and the traction integral equations, collocated on the other crack surface and introduced to overcome the problems originating from the coincidence of the crack nodes.

Assuming continuity of displacements at the boundary nodes, the boundary integral representation for the displacements u_j is given by

$$c_{ij}(\boldsymbol{x}_0)u_j(\boldsymbol{x}_0) + \fint_{\Gamma} T_{ij}(\boldsymbol{x}_0, \boldsymbol{x})u_j(\boldsymbol{x})\, d\Gamma = \int_{\Gamma} U_{ij}(\boldsymbol{x}_0, \boldsymbol{x})t_j(\boldsymbol{x})\, d\Gamma \qquad (5.1)$$

where U_{ij} and T_{ij} represent the displacement and traction fundamental solutions at the boundary point $\boldsymbol{x}$ when collocating at the point $\boldsymbol{x}_0$; c_{ij} are coefficients depending on the boundary geometry and computed through rigid body considerations; and the symbol $\fint$ stands for Cauchy principal value integral, whose presence is the consequence of the $O(r^{-2})$ strength of the T_{ij} integrands.

The displacement equation (5.1) is collocated on the boundary Γ and on one of the crack surfaces. When collocated at the crack node $\boldsymbol{x}_0^-$, it assumes the form

$$c_{ij}(\boldsymbol{x}_0^-)u_j(\boldsymbol{x}_0^-) + c_{ij}(\boldsymbol{x}_0^+)u_j(\boldsymbol{x}_0^+) + \fint_\Gamma T_{ij}(\boldsymbol{x}_0^-, \boldsymbol{x})u_j(\boldsymbol{x})\, d\Gamma$$

$$= \int_\Gamma U_{ij}(\boldsymbol{x}_0^-, \boldsymbol{x})t_j(\boldsymbol{x})\, d\Gamma \tag{5.2}$$

where $\boldsymbol{x}_0^-$ and $\boldsymbol{x}_0^+$ are the two coincident crack nodes. For smooth crack surfaces at the point $\boldsymbol{x}_0^-$, it is $c_{ij}(\boldsymbol{x}_0^-) = c_{ij}(\boldsymbol{x}_0^+) = (1/2)\delta_{ij}$.

The traction integral equation collocated at the point $\boldsymbol{x}_0^+$, where continuity of strains is assumed, is given by

$$c_{ij}(\boldsymbol{x}_0^+)t_j(\boldsymbol{x}_0^+) - c_{ij}(\boldsymbol{x}_0^-)t_j(\boldsymbol{x}_0^-) + n_j(\boldsymbol{x}_0^+) \,=\!\!\!\!\!\!\int_\Gamma T_{ijk}(\boldsymbol{x}_0^+, \boldsymbol{x})u_k(\boldsymbol{x})\, d\Gamma$$

$$= n_j(\boldsymbol{x}_0^+) \fint_\Gamma U_{ijk}(\boldsymbol{x}_0^+, \boldsymbol{x})t_k(\boldsymbol{x})\, d\Gamma \tag{5.3}$$

where the kernels U_{ijk} and T_{ijk} contain derivatives of U_{ij} and T_{ij} respectively; n_j are the components of the outward normal at the point $\boldsymbol{x}_0^+$; and $=\!\!\!\!\!\!\int$ stands for the Hadamard principal value integral, originating from the presence of the $O(r^-3)$ kernels T_{ijk}.

Equations (5.1), (5.2) and (5.3) provide the boundary integral model for the analysis of general crack problems. If no cracks are involved, then only equation (5.1) is needed for the analysis.

5.2.1. *DBEM systems of equations*

In this section, the main features of DBEM systems of equations are discussed in view of the construction of a hierarchical representation of the collocation matrix. All the considerations, however, apply to BEM systems, after suitable simplification.

The DBEM leads to a system of equations that can be written in the form

$$\begin{bmatrix} \mathbf{H}_{bb} & \mathbf{H}_{bd} & \mathbf{H}_{bt} \\ \mathbf{H}_{db} & \mathbf{H}_{dd} & \mathbf{H}_{dt} \\ \mathbf{S}_{tb} & \mathbf{S}_{td} & \mathbf{S}_{tt} \end{bmatrix} \begin{bmatrix} \boldsymbol{u}_b \\ \boldsymbol{u}_d \\ \boldsymbol{u}_t \end{bmatrix} = \begin{bmatrix} \mathbf{G}_{bb} & \mathbf{G}_{bd} & \mathbf{G}_{bt} \\ \mathbf{G}_{db} & \mathbf{G}_{dd} & \mathbf{G}_{dt} \\ \mathbf{D}_{tb} & \mathbf{D}_{td} & \mathbf{D}_{tt} \end{bmatrix} \begin{bmatrix} \boldsymbol{t}_b \\ \boldsymbol{t}_d \\ \boldsymbol{t}_t \end{bmatrix} \tag{5.4}$$

where the subscripts denote the boundary of the domain (b), the crack surface where the displacement integral equation is collocated (d) and the crack surface where the traction integral equation (t) is collocated. The blocks $\mathbf{H}_{bd}$ and $\mathbf{G}_{bd}$, for example, contain the coefficients computed by integrating T_{ij} and U_{ij} over the elements lying on the displacement crack surface (d) while collocating the integral equation on the boundary nodes (b). The blocks $\mathbf{S}$ and $\mathbf{D}$ are obtained from the integration of the kernels T_{ijk} and U_{ijk}, when the traction integral equation is collocated on the related crack surface.

After the application of the boundary conditions, assuming free crack surfaces, the final system is written

$$\begin{bmatrix} \mathbf{A}_{bb} & \mathbf{H}_{bd} & \mathbf{H}_{bt} \\ \mathbf{A}_{db} & \mathbf{H}_{dd} & \mathbf{H}_{dt} \\ \mathbf{B}_{tb} & \mathbf{S}_{td} & \mathbf{S}_{tt} \end{bmatrix} \begin{bmatrix} \boldsymbol{X}_b \\ \boldsymbol{u}_d \\ \boldsymbol{u}_t \end{bmatrix} = \begin{bmatrix} \boldsymbol{Y}_b \\ \boldsymbol{Y}_d \\ \boldsymbol{Y}_t \end{bmatrix} \tag{5.5}$$

where the vector $\boldsymbol{X}_b$ contains unknown boundary displacements or tractions and the vectors $\boldsymbol{Y}$ are obtained after applying the boundary conditions, as a linear combination of the columns of the blocks $\mathbf{H}$, $\mathbf{G}$, $\mathbf{D}$ and $\mathbf{S}$ corresponding to the prescribed displacement and traction nodal values. The blocks $\mathbf{A}_{bb}$ and $\mathbf{A}_{db}$ contain a mix of columns from the corresponding blocks $\mathbf{H}_{bb}$ and $\mathbf{G}_{bb}$, and $\mathbf{H}_{db}$ and $\mathbf{G}_{db}$, while $\mathbf{B}_{tb}$ mixes columns from $\mathbf{D}_{tb}$ and $\mathbf{S}_{tb}$.

The coefficient matrix in equation (5.5) has some important features stemming from the inherent BEM characteristics and the special computational strategy used. The matrix is, in fact, fully populated, non-symmetric and not definite. Such features are common to the matrices generally produced by the BEM. Moreover, the off-diagonal blocks $\mathbf{H}_{dt}$ and $\mathbf{S}_{td}$ are characterized by the presence of high-strength terms. Such terms originate from the geometrical coincidence of the two crack surfaces; that implies the presence of singular terms in the related blocks. When collocating on the displacement crack surface, for example, both the collocation point $\boldsymbol{x}_0^-$ and the geometrically coincident point $\boldsymbol{x}_0^+$ are singular, thus generating high-strength terms in both $\mathbf{H}_{dd}$ and $\mathbf{H}_{dt}$. Moreover, it is to be noted that the blocks $\mathbf{D}$ and $\mathbf{S}$, originating from the collocation of the hypersingular boundary integral equation (i.e. the traction equation) contain terms whose strength is considerably higher with respect to those contained in the blocks $\mathbf{H}$ and $\mathbf{G}$.

The solution of fully populated, non-symmetric and not definite systems, especially when accuracy and reliability are of primary concern, is usually tackled by direct methods, such as Gauss elimination, as they are easy to implement, robust and tend to require predictable time and storage resources. However, when dealing with large three-dimensional systems, involving several thousands of equations, the use of direct solvers becomes too computationally expensive, both in terms of operations count and memory requirements. In such cases, iterative solvers may represent a preferable choice, becoming essential for very large systems.[35]

The application of hierarchical matrices in conjunction with iterative solvers to DBEM matrices of the form given in equation (5.5) is discussed in the next section.

5.3. Hierarchical Dual Boundary Element Method

In this section, the use of hierarchical matrices for the approximation and solution of systems of equations produced by the DBEM is discussed. Before going into the details of the method, it is useful to give a summary of the conditions that must be met and the steps that must be carried out to obtain the hierarchical representation.

The overall objective of this special format is to reduce the storage requirements as well as to speed up the operations involving the matrix, by representing the matrix itself as a collection of blocks, some of which admit a particular approximated representation that can be obtained by computing only a few entries from the original matrix. These special blocks are called low rank blocks. The blocks that cannot be represented in this way must be computed and stored entirely and are called full rank blocks.

Low rank blocks constitute an approximation of suitably selected blocks of the discrete integral operator based, from the analytical point of view, on a suitable expansion of the kernel of the continuous integral operator.[15,16,18,19] This expansion, and consequently the existence of low rank approximants, is based on the *asymptotic smoothness* of the kernel functions, i.e. on the fact that the kernels $U_{ij}(\boldsymbol{x}_0, \boldsymbol{x})$ and $T_{ij}(\boldsymbol{x}_0, \boldsymbol{x})$ are singular only when $\boldsymbol{x}_0 = \boldsymbol{x}$. For more precise information about asymptotic smoothness, the interested reader is referred to the works of Bebendorf,[18] Bebendorf and Rjasanow[19] and especially to the paper of Bebendorf and Grzhibovskis,[40] where the application to elastic solids is considered, laying the grounds for the applicability of ACA to the class of problems considered in the present work. Here, it is only mentioned that the asymptotic smoothness represents a sufficient condition for the existence of low rank approximants and that it does not exclude strongly or hypersingular kernels, like $U_{ijk}(\boldsymbol{x}_0, \boldsymbol{x})$ and $T_{ijk}(\boldsymbol{x}_0, \boldsymbol{x})$. Moreover, the regularity of the boundary over which the approximation is carried out is not requested.

Once the conditions for the approximants' existence have been assessed, the subdivision of the matrix into low and full rank blocks is based on geometrical considerations. Every block in the matrix is characterized by two subsets of indices, corresponding respectively to the row and column indices. In the standard collocation BEM, every row index is associated to a degree of freedom of a collocation node, while every column index is associated to a degree of freedom of a discretization node, whose coefficient is computed by integrating those elements to which the node itself belongs. Every block is then related to two sets of boundary elements: the one containing the collocation points corresponding to the row indices, here denoted by Ω_{x_0}; and the one grouping the elements over which the integration is carried out, denoted by Ω_x, that contains the nodes corresponding to the columns. If these two sets of boundary elements are *separated*, then the block will be represented and stored in low rank format, while it will be entirely generated and stored in full rank format otherwise. The blocks meeting the requirement of separation are deemed *admissible*. A schematic diagram of the process leading to the boundary subdivision, and the correspondence with the suitable matrix block, is illustrated in Fig. 5.1.

In Fig. 5.1, both the cluster of collocation points, related to a set of rows of the collocation matrix, and the cluster of integration elements,

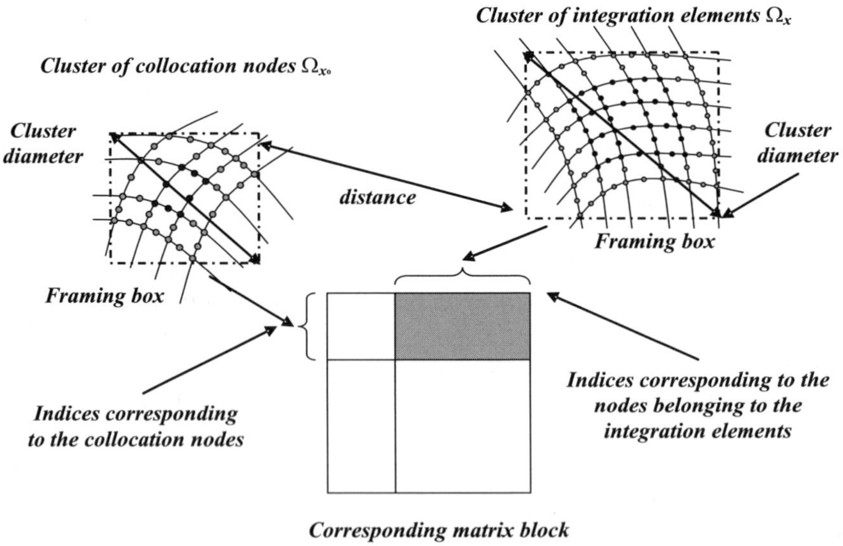

Fig. 5.1. Schematic of the boundary subdivision process.

which contain the nodes related to the columns of the matrix, are shown. As schematically illustrated, the admissibility condition is checked choosing suitable boxes framing the two clusters. This strategy is dictated by the need to reduce the computational costs of the boundary subdivision and particularly the subsequent admissibility check.

The entire process leading to the subdivision in sub-blocks and to their further classification is based on a previous *hierarchical partition* of the matrix index set aimed at grouping subsets of indices corresponding to contiguous nodes and elements on the basis of some computationally efficient geometrical criterion. This partition is stored in a binary tree of index subsets, or *cluster tree*, that constitutes the basis for the following construction of the hierarchical block subdivision that will be stored in a quaternary *block tree*. Such a process of hierarchical subdivision and tree generation will be further illustrated in the following sections. Here, it is important to focus on the fact that the block (quaternary) tree stems from the index (binary) tree.

As the admissible blocks have been located, their approximation can be computed. While in fast multi-pole or panel clustering methods the knowledge of the explicit form of the kernel expansion is required in order to approximate the integral operator, low rank blocks can be generated directly by computing some entries from the original blocks, through ACA algorithms. It is important to highlight that the ACA allows the approximated block to reach adaptively the *a priori* selected accuracy. These features make such a technique particularly appealing, as it is not necessary to modify or rewrite the routines for the boundary integration in previously developed codes.

Once the basic hierarchical representation has been set up, the collocation matrix can be treated in different ways to obtain the system solution. It is worth noting that the representation obtained by ACA is not yet optimal in terms of storage requirements. The low rank blocks can be, in fact, recompressed, taking advantage of the reduced singular value decomposition (SVD)[54] that allows a further storage reduction without accuracy penalties. Moreover, since the initial matrix partition is generally not optimal,[55] once the blocks have been generated and recompressed, the entire structure of the hierarchical block tree can be modified by a suitable *coarsening* procedure.[22] These consecutive manipulations have the objective of further reducing the storage requirements and speeding up the solution, maintaining the preset accuracy. It is important to note that such recompression schemes can be applied sequentially immediately after the blocks' generation. When a block has been generated, it can be immediately recompressed. Afterwards, a collection of four contiguous recompressed blocks can be tested for coarsening. This fact implies that the needed memory is less than that required for storing the ACA-generated matrix.

As an optimal coarsened representation is obtained, the solution of the system can be tackled either by direct solvers or iterative methods. In both cases, the efficiency of the solution relies on the use of a special arithmetic, i.e. a set of algorithms that implement the operations on matrices represented in hierarchical format, such as addition, multiplication and inversion.[24] For a direct solution, the computation of the hierarchical LU decomposition of the collocation matrix is needed to carry out an effective hierarchical inversion.[23] Iterative solutions, on the other hand, are mainly based on the efficiency of the matrix-vector product, but can be noticeably sped up by the use of suitable preconditioners. An effective preconditioner for BE matrices based on hierarchical LU decomposition has recently been proposed by Bebendorf.[23]

In the following section, these points will be further developed and the main algorithms involved will be discussed. The modifications required to take into account the presence of cracks will be pointed out.

5.3.1. *Boundary subdivision and cluster tree*

The construction of a partition of the matrix index set is the basis for the following definition of the hierarchical block tree. The objective of the partition is to subdivide the index set into subsets (or clusters) of indices corresponding to contiguous boundary element nodes. This process leads to the identification of separate or not separate couples of boundary element groups. In the case of three-dimensional elastic problems, as three different indices are associated to each discretization point, it is preferable to partition the set of the boundary nodes indices itself. The process starts from the complete set of indices $I = \{1, 2, \ldots, n\}$, where n denotes the number of collocation points. This initial set constitutes the root of the tree. Each cluster in the tree, called the tree node, not to be confused with the geometrical discretization node, is split into two subsets, called sons, on the basis of a selected criterion. The common tree node from which two sets originate is called the parent. The tree nodes that cannot be further split are the leaves of the tree. Usually a node cannot be further split when it contains a number of indices equal to or less than a minimum number n_{min}, called the cardinality of the tree, a previously fixed value.

However, the procedure must be slightly modified for the analysis of cracked configurations through the DBEM. The crack can be either embedded in or emanate from a surface, but in either case it is located *inside* the boundary surface and its geometry is usually clearly distinguishable from the geometry of the boundary. This circumstance naturally induces a first distinction, dictated by the geometry, between boundary and crack nodes. Moreover, as already mentioned, crack modeling requires special considerations: the crack is discretized by using discontinuous elements, collocating displacement equations on the nodes belonging to one crack surface and traction equations on the other. As different integral operators, or kernels, correspond to displacement and traction integral equations, it is then appropriate to further distinguish between the nodes on one crack surface and the nodes on the other one; this fact results in the two sons of the *crack nodes cluster* being subdivided into the cluster containing the nodes on which displacement equations are collocated and the one corresponding to the nodes on which the traction equations are collocated. This subdivision and the sets of node indices stemming from this process are graphically represented in Fig. 5.2.

The algorithm used for the construction of the cluster tree is also reported (Algorithm 5.1). It is similar to the algorithm introduced by Giebermann[56] and used in Grasedyck[22] and is aimed at generating a geometrically balanced cluster tree, after the initial subdivision between boundary nodes, displacement crack nodes and traction crack nodes. It requires, as input, a set of indices associated to a set of collocation points, the set of the coordinates of such points and the minimum number of points allowed in a subset, i.e. the cardinality. The output of the procedure is the entire structure of the binary tree, from the root to the leaves. Note that

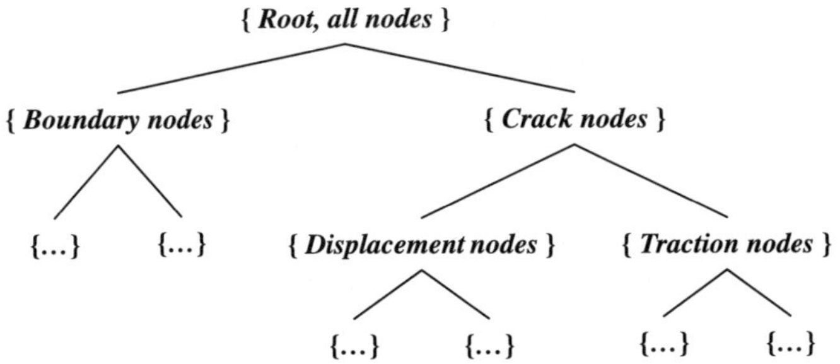

Fig. 5.2. Index sets induced by presence of cracks.

Algorithm 5.1 Recursive SplitCluster($s, \mathbf{x}, n_{min}$)

if s is the tree root cluster **then**
  define $s_1 = \{i \in s : \mathbf{x}_i \in$ set of *boundary* nodes $\}$
  define $s_2 = \{i \in s : \mathbf{x}_i \in$ set of *crack* nodes $\}$
else if s is the cluster of *all* the crack nodes **then**
  define $s_1 = \{i \in s : \mathbf{x}_i \in$ set of *displacement* crack nodes $\}$
  define $s_2 = \{i \in s : \mathbf{x}_i \in$ set of *traction* crack nodes $\}$
else if $\#s \leq n_{min}$ **then**
  set $sons(s) = \{\emptyset\}$
  return
else
  for i=1,3 **do**
    $M_i = max\{(\mathbf{x}_j)_i : j \in s\}$
    $m_i = min\{(\mathbf{x}_j)_i : j \in s\}$
  end for
  find j such that $M_j - m_j$ is the largest
  define $s_1 = \{i \in s : (\mathbf{x}_i)_j \leq (M_j + m_j)/2\}$
  define $s_2 = s - s_1$
end if
set $sons(s) = \{s_1, s_2\}$
for i=1,2 **do**
  call SplitCluster($s_i, \mathbf{x}, n_{min}$)
end for

$(\mathbf{x}_j)_i$ indicates the *i-th* coordinate of the *j-th* collocation point, while the operator $\#(\cdot)$ gives the number of elements in a set.

5.3.2. *Block tree and admissibility condition*

The block tree is built recursively, starting from the complete index set $I \times I$ (both rows and columns) of the collocation matrix and the previously-found cluster tree. The objective of this process is to split, hierarchically, the matrix into sub-blocks and to classify the leaves of the tree into admissible (low rank) or non admissible (full rank) blocks. The classification is based on a geometrical criterion that assesses the separation of the clusters of boundary elements associated with the considered block. Such a criterion takes into consideration the features of the boundary mesh. For 3-D DBEM, eight-noded continuous and discontinuous quadrilateral elements are used. Let Ω_{x_0} denote the cluster of elements containing the discretization nodes corresponding to the row indices of the considered block and Ω_x denote the set of elements over which the integration is carried out to compute the coefficient corresponding to the column indices. The admissibility condition can be written

$$min(diam\ \Omega_{x_0}, diam\ \Omega_x) \leq \eta \cdot dist(\Omega_{x_0}, \Omega_x) \quad (5.6)$$

where $\eta > 0$ is a parameter influencing the number of admissible blocks on one hand and the convergence speed of the adaptive approximation of low rank blocks on the other hand.[25]

Since the actual diameters and the distance between two clusters are generally time-consuming to be exactly computed, the condition is usually assessed with respect to bounding boxes parallel to the reference axes,[22,56] as already mentioned in reference to Fig. 5.1. In this case, Ω_{x_0} and Ω_x in equation (5.6) are replaced by the boxes Q_{x_0} and Q_x. The bounding box clustering technique adopted in the present work is generally used for its simplicity, although it produces non-optimal partitions that can be improved by suitable procedures, as will soon be illustrated. Other clustering techniques able to produce better initial partitions have been proposed in the literature. The construction using the *principal component analysis*[57] should significantly improve the quality of the initial partition, but such a scheme is not tested in this work.

The algorithm is graphically illustrated in Fig. 5.3. Starting from the root (the entire matrix), each block is subdivided into four sub-blocks until either the admissibility condition is satisfied or the block is sufficiently small that it cannot be further subdivided. The light grey boxes represent low rank blocks while the dark grey boxes are the full rank ones.

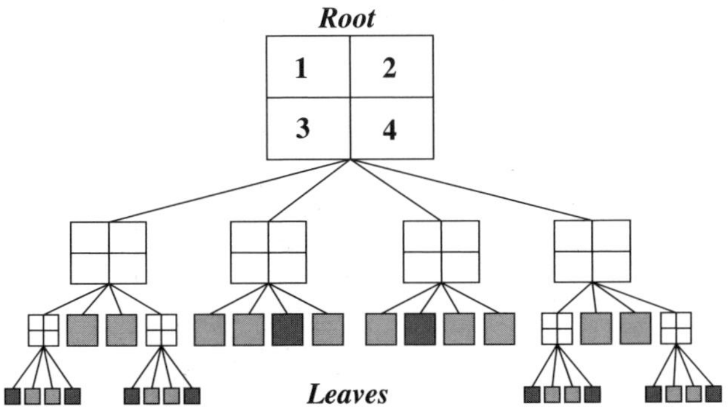

Fig. 5.3. Graphical illustration of the block splitting procedure.

The presence of cracks requires some extra considerations. As illustrated in Fig. 5.2, when a crack is present, the second level of the cluster tree has two nodes: the first corresponding to the discretization nodes on the boundary and the other corresponding to those on the crack. While it is absolutely acceptable to check the admissibility condition for the matrix block corresponding to collocation on the boundary nodes and integration on the two coincident crack surfaces *considered as a whole*, the reverse is not true. When generally collocating on the crack nodes, two different kinds of boundary integral equations are being evoked, namely displacement equations and traction equations. These correspond to different integral operators that should be approximated separately. Besides the admissibility condition (5.6), the supplementary constraint that the block corresponding to collocation on the nodes of the crack cluster *as a whole* and integration on the boundary *is inadmissible* must be considered. In other words, referring to equation (5.5), if the condition (5.6) is satisfied, it is admissible to approximate the two submatrices $\mathbf{H}_{bd}$ and $\mathbf{H}_{bt}$ through a single approximate low rank block, while it is not admissible to check the condition (5.6) for the sub-matrix comprised of $\mathbf{A}_{db}$ and $\mathbf{B}_{tb}$. This further condition may induce a characteristic asymmetric structure on the hierarchical block tree, as shown in Fig. 5.4. Notice that the dashed lines appearing in the white block do not separate it into four sub-blocks, but are drawn only to point out the lost block tree symmetry.

Finally, it is important to consider specifically the assessment of the condition (5.6) when clusters of boundary elements and crack elements are involved at the same time. As mentioned above, the admissibility

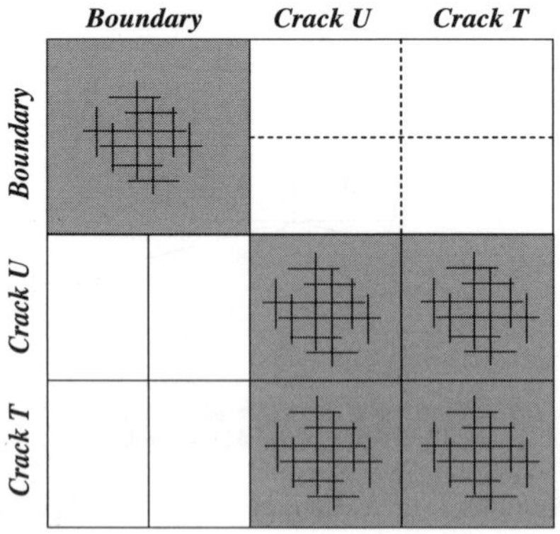

Fig. 5.4. Structure of the hierarchical matrix including crack surfaces.

condition is generally checked by considering framing boxes; however, cracks are always *contained* by the external boundary and the common procedure, since the boxes are one inside the other, could lead to the premature classification of some perfectly admissible blocks as inadmissible. To avoid such a circumstance, a special procedure has been devised. When the distance between a cluster of crack elements and a cluster of boundary elements is being computed, only the crack elements are framed by a box, and the distance between the two clusters is computed by considering the boundary cluster element by element. This procedure avoids the premature skipping of admissible blocks, especially in the very favourable case of embedded cracks. The element by element distance check, if carried out only when crack elements are involved, is not too expensive computationally.

The extended admissibility condition is used in the algorithm for the generation of the block tree reported here (Algorithm 5.2). It takes into account the possible presence of cracks, following the considerations developed above. The output of the procedure, when it is initially called passing the block corresponding to the entire matrix, is the block tree from the root to the leaves that are classified in low and full rank.

Algorithm 5.2 Recursive SplitBlock($B_{s \times t}, n_{min}$)

```
if s is the cluster of all the crack nodes then
  set sons(B_{s×t}) = {B_{σ×τ} : σ ∈ sons(s), τ ∈ sons(t)}
else if B_{s×t} is admissible then
  set B_{s×t} as a low rank block
  set sons(B_{s×t}) = {∅}
else if #s < n_min or #t < n_min then
  set B_{s×t} as a full rank block
  set sons(B_{s×t}) = {∅}
else
  set sons(B_{s×t}) = {B_{σ×τ} : σ ∈ sons(s), τ ∈ sons(t)}
end if
if sons(B_{s×t}) ≠ {∅} then
  for all B_{σ×τ} ∈ sons(B_{s×t}) do
    call SplitBlock(B_{σ×τ})
  end for
end if
```

5.3.3. *Low rank blocks and ACA algorithms*

Let M be an $m \times n$ admissible block in the collocation matrix. It admits the low rank representation

$$M \simeq M_k = A \cdot B^T = \sum_{i=1}^{k} a_i \cdot b_i^T \tag{5.7}$$

where A is of order $m \times k$ and B is of order $n \times k$, where k is the *rank* of the new representation. The approximating block M_k satisfies the relation $\|M - M_k\|_F \leq \varepsilon \|M\|_F$, where $\| \cdot \|_F$ represents the Frobenius norm and ε is the set accuracy. Sometimes it is useful to represent the matrix using the alternative sum representation, where a_i and b_i are the *i-th* columns of A and B, respectively. The approximate representation allows storage savings with respect to the full rank representation and speeds up the matrix-vector product.[24]

Different ACA algorithms can be used to generate the approximate blocks. The original algorithm was proposed by Bebendorf[18] and was further developed by Bebendorf and Rjasanow.[19] Grasedyck[22] proposed the so-called ACA+ algorithm and compared its performances to those of the standard scheme. Bebendorf and Grzhibovskis[40] proposed a strategy for overcoming some problems that may arise when populating some low rank blocks involving the interaction, through double layer kernels, of sets of coplanar elements. Useful illustrations of the basic ACA scheme can be found in the works of Kurz *et al.*[37] and Bebendorf and Kriemann,[58] while the reader is referred to the work of Grasedyck[22] for ACA+. Algorithm 5.3 describes the partially pivoted ACA scheme. In this work also, a slightly different scheme has been used to circumvent some problems originating when computing some particular blocks, as illustrated in the following.

The previously mentioned schemes allow the block to reach *adaptively* the *a priori* set accuracy ε and are substantially based on the computation of selected columns and rows of the original block that, suitably manipulated, furnish exactly the columns a_i and b_i appearing in equation (5.7). Different schemes often differ for the choice of the pivots; that can have a noticeable effect on the convergence and quality of the approximation.[58] Once a new column a_i and row b_i^T have been generated and added to those previously computed, the convergence toward the required accuracy is checked against a suitable stopping criterion. If the criterion is satisfied, the computation is stopped, else a new column and a new row are generated and stored.

Algorithm 5.3 Partially Pivoted ACA for $B_{\sigma\times\tau}$

$k = 1$; $Z = \{\emptyset\}$
repeat
  if k=1 **then**
    $i_k = \min(\sigma - Z)$
  else
    $i_k = \text{argmax}_{i\in\sigma-Z}(|(a_{k-1})_i|)$
  end if
  $Z = Z \cup \{i_k\}$
  $\overline{\mathbf{b}}_k = \mathbf{M}(i_k, :) - \sum_{r=1}^{k-1}(\mathbf{a}_r)_{i_k}\mathbf{b}_r^T$
  if $\overline{\mathbf{b}}_k \neq \mathbf{0}$ **then**
    $j_k = \text{argmax}_{j\in\sigma}(|(\overline{\mathbf{b}}_k)_j|)$; $\mathbf{b}_k = (\overline{\mathbf{b}}_k)_{j_k}^{-1}\overline{\mathbf{b}}_k$
    $\mathbf{a}_k = \mathbf{M}(:, j_k) - \sum_{r=1}^{k-1}(\mathbf{b}_r)_{j_k}\mathbf{a}_r$
    $k = k + 1$
  end if
until *The stop criterion is satisfied or* $Z = \sigma$

The stopping criterion is based on the assessment of the convergence of the approximating block in terms of the Frobenius norm.[18,37] Since the original blocks are not accessible, only the partial approximations M_k, with k running, are used to check the convergence. The Frobenius norm can be computed by the following recursive formula

$$\|M_k\|_F^2 = \|M_{k-1}\|_F^2 + 2\sum_{i=1}^{k-1}(a_k^T a_i)(b_i^T b_k) + \|a_k\|_F^2\|b_k\|_F^2 \tag{5.8}$$

where a_k and b_k represent the column and row computed at the $k-th$ iteration. A suitable stopping criterion can be expressed as

$$\|a_k\|_F\|b_k\|_F \leq \varepsilon\|M_k\|_F \tag{5.9}$$

that prescribes a stop to the iteration when the inequality is satisfied for a required preset accuracy ε.

The construction of the approximating block not only reduces the storage needed to represent an admissible block, but also reduces the assembly time for the set-up of the collocation matrix, as the integration is carried out only on those elements that allow the population of the required columns and rows.

With reference to the form of the DBEM system of equations (5.5), some additional consideration on the construction of the approximating

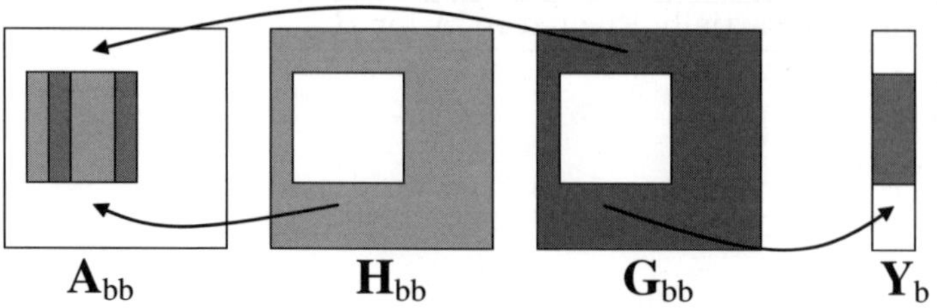

Fig. 5.5. Construction of an A block.

blocks is convenient. The blocks contained in $\mathbf{A}_{bb}$ and $\mathbf{A}_{db}$ include a mix of columns from the blocks $\mathbf{H}_{bb}$ and $\mathbf{G}_{bb}$ and $\mathbf{H}_{db}$ and $\mathbf{G}_{db}$, respectively, and require some attention. Moreover, some of the columns from the corresponding **H** and **G** blocks may give a contribution to the right-hand side. Let us suppose that, following the application of the boundary conditions, the low rank block M belonging to the sub-matrix $\mathbf{A}_{bb}$ is comprised of many **H** columns and few **G** columns, and that few **G** columns contribute to the right-hand side (see Fig. 5.5).

In this case, it may not be convenient to generate the approximation of the entire **G** block (the white block belonging to $\mathbf{G}_{bb}$ in Fig. 5.5). On the contrary, it is more effective to generate the approximation of the entire **H** block through ACA, then to annihilate the terms in the rows b_i^T corresponding to the columns to be replaced with the **G** columns, and eventually to compute exactly the few needed **G** columns and add them to the representation (5.7), using row vectors filled with zeros and ones placed in the positions corresponding to the columns to be replaced. The other **G** columns, among the few exactly computed, contribute to the right-hand side, through appropriate coefficients.

The choice between computing a block through ACA and computing few columns exactly depends on the number of columns from a block **H** and **G** that are actually needed for the construction of a specific **A** block. If the number of needed columns is much less than the average rank, it is convenient to compute the columns exactly, since the ACA representation of the few columns would require more columns (and rows) than their exact representation. Analogous considerations hold for the block $\mathbf{B}_{tb}$ that mixes columns from $\mathbf{S}_{tb}$ and $\mathbf{D}_{tb}$.

The blocks contained in $\mathbf{H}_{bd}$ and $\mathbf{H}_{bt}$ do not require special consideration and can be computed through standard ACA. However, it is interesting to observe that, if the crack is sufficiently far from the boundary, both these sub-matrices belong to a single big low rank block: this is the block for which the numerical compression works better.

The blocks contained in $\mathbf{H}_{dd}$, $\mathbf{H}_{dt}$, $\mathbf{S}_{td}$ and $\mathbf{S}_{tt}$ often stem from the integration of T_{ij} and T_{ijk} over clusters of coplanar elements (case of plane crack). In such circumstance the standard ACA can fail, as shown by Grasedyck[22] with a counterexample for partial pivoting. To avoid this potential problem, and the consequent loss of accuracy, ACA+ can be used for the approximation of these blocks. In this work, however, a different strategy, based on the computation of more than one row at each ACA iteration, has been used. The strategy is referred to as *big-volume search*[17,58] and, although slightly more expensive then standard ACA, has proven to be very reliable in the performed numerical experiments. Briefly, the search of the pivot is not limited to a single row per iteration, but is extended to more rows that are generated and stored at every ACA step. The efficiency of the scheme is, hence, based on the availability of more matrix entries.

5.3.4. *Block recompression and tree coarsening*

After the blocks have been generated, a further reduction of the required memory can be achieved by suitable recompression schemes,[22,54] so that the amount of memory required for the final storage is lower than that needed for the ACA generated matrix. Moreover, the recompression schemes further speed up the computation, maintaining the desired preset accuracy ε.

Two different recompression schemes can be applied: one that acts on the single leaves; and one that modifies the entire tree structure, through a process of reabsorption of the leaves.

The block recompression, i.e. the first scheme, is sometimes referred to as *truncation*[24,25] and is based on the SVD of low and full rank blocks. Since the size of the full rank leaves is bounded by n_{min}, their SVD can be efficiently computed by suitable available efficient algorithms, like the `LAPACK dgesvd.f`; for the low rank leaves, on the other hand, it is possible to perform a reduced SVD.[54]

Let $M_k = A \cdot B^T$ be a low rank block. The reduced decomposition can be computed by

$$M_k = A \cdot B^T = Q_A R_A R_B^T Q_B^T = Q_A U \Sigma V^T Q_B^T \tag{5.10}$$

where $Q_X R_X$ is the QR decomposition of the matrix X, while $U\Sigma V^T$ represents the SVD of $R_A R_B^T$. If $A \in \mathbb{R}_{m\times k}$ then $Q_A \in \mathbb{R}_{m\times k}$ while $R_A \in$

$\mathbb{R}_{k\times k}$. The SVD of the reduced matrix $R_A R_B^T \in \mathbb{R}_{k\times k}$ can then be computed by using `dgesvd.f`.

In equation (5.10) there is still no approximation. If the singular values $\sigma_i < \varepsilon\sigma_1$ contained in the diagonal of Σ are discarded, then the following approximation is obtained

$$M_k \simeq Q_A \widetilde{U}\widetilde{\Sigma}\widetilde{V}^T Q_B^T = \widetilde{A}\widetilde{B}^T \tag{5.11}$$

where $\widetilde{A} = Q_A\widetilde{U}\widetilde{\Sigma}$ and $\widetilde{B} = \widetilde{V}^T Q_B^T$. In equation (5.11) $\widetilde{U}, \widetilde{V} \in \mathbb{R}_{k\times\widetilde{k}}$ are obtained from the original matrices discarding the last $k - \widetilde{k}$ columns, while $\widetilde{\Sigma} \in \mathbb{R}_{\widetilde{k}\times\widetilde{k}}$ is the diagonal matrix $diag\{\sigma_1, \sigma_2, \ldots, \sigma_{\widetilde{k}}\}$. An analogous procedure is followed for the full rank leaves that, after discarding the smallest singular values, are stored in low rank format if the total size in low rank is less than the size requested in full rank representation. Note that the scheme can be applied immediately after a block has been generated.

Such procedure operates on single blocks, reducing the size of low rank blocks and converting the full rank blocks satisfying the previous condition into low rank blocks. It is important to emphasize the role of the accuracy ε that appears in all the computations. The required accuracy could also be lowered if a less accurate representation of the matrix would be needed for special purposes. The importance of this fact will become apparent in the construction of a preconditioner for iterative solvers.

Besides the blockwise SVD, it is possible to apply a further compression, referred to as *tree coarsening*, aimed at modifying the entire tree structure. The coarsening tries to unify groups of four leaves that are sons of the same block tree node. If some conditions are met, the SVD decomposition of the unification of the four blocks is computed and, if after discarding the smaller singular values along with the corresponding columns and rows, the result requires less storage, the four blocks are unified, or reabsorbed, in the parent tree node. This scheme can be applied when four adjacent sub-blocks have been generated and recompressed. The procedure that performs the coarsening has been presented by Grasedyck[22] and is reported in Algorithm 5.4. Note that the unification of the four low rank sons of a block is performed by expanding their columns and rows to the parent's block dimensions by suitably placed zeros and then adding such expanded vectors to provide the parent's low rank representation.

Algorithm 5.4 Recursive CoarsenTree($B_{s \times t}$)

call CoarsenTree($B_{\sigma \times \tau}$) for all $B_{\sigma \times \tau} \in Sons(B_{s \times t})$
if ($Sons(B_{s \times t})$ are all low rank leaves) **then**
  Unify the four leaves;
  Compute the SVD of the unification;
  Check the storage requirements of the SVD;
  if (storage of the SVD < storage of the four sub-blocks) **then**
    Store the SVD and set $Sons(s \times t) = \{\emptyset\}$
  end if
end if

Two points are stressed here: a) the block recompression is applied immediately after the block generation and not after the generation of the entire matrix which allows an actual reduction of the storage requirements; b) the recompression schemes provide a valuable tool for the construction of an effective preconditioner[23] based on the computation of a coarse approximation of the collocation matrix, as described in the following sections.

5.3.5. *Hierarchical arithmetic*

To carry out operations on hierarchical matrices, it is necessary to define a suitable hierarchical arithmetic. The operations involving the entire hierarchical matrices (operations between matrices, trees or sub-trees) are based on elementary operations between low or full rank blocks (operations between blocks or leaves). Often such operations, involving the SVD, require a truncation with respect to a previously set accuracy, which thus acquires a fundamental importance in all the considered algorithms. *All the hierarchical operations are defined with respect to a set accuracy.*

The main difficulty when implementing such algorithms is distinguishing between the full range of possible different cases. The multiplication, for example, is defined with respect to the two block factors and the so-called target block. Each of these blocks could be subdivisible or not subdivisible and, if not subdivisible, i.e. leaf, could be low or full rank. Each of these cases requires separate treatment to reduce the overall complexity of the operation.

Rigorous information on addition between hierarchical matrices, truncation with respect to a given accuracy, matrix-vector multiplication, matrix-matrix multiplication and hierarchical inversion can be found in the works of Hackbusch[20] and Grasedyck and Hackbusch,[24] where some algorithms are also given and their arithmetic complexity is analysed. A collection of useful algorithms for practical implementation is given by Börm *et al.*[25] The hierarchical LU decomposition, based on the previous arithmetic, is discussed by Bebendorf.[23] The interested reader is referred to the mentioned works for analytic details. In the following sections, the basic algorithms are illustrated for the sake of completeness.

5.3.5.1. *Addition*

The addition of two trees with the same structure is performed by adding the leaves of the two trees in full or low rank format. Such an operation preserves the structure of the two added matrices and the result is then a hierarchical matrix with the same structure. The addition of the two trees (or branches) H_1 and H_2 is denoted sometimes by $H = H_1 \oplus H_2$.

The addition of two full rank blocks is performed by simply adding the two blocks in full matrix format. On the other hand, the addition of two low rank leaves, is given by

$$M_1 + M_2 = A_1 \cdot B_1^T + A_2 \cdot B_2^T = [A_1, A_2] \cdot [B_1, B_2]^T \tag{5.12}$$

As is evident, the block resulting from the exact addition in low rank format has a higher rank with respect to the two added blocks. To maintain the advantages given by the low rank representation, an SVD truncation to the set accuracy is eventually performed on the resulting block. The importance of this truncation will appear more clearly when the multiplication is discussed.

5.3.5.2. *Multiplication*

The hierarchical multiplication is performed between dimensionally compatible hierarchical matrices or blocks with the same or different structure. According to the literature on the subject,[24,25] the multiplication between the two trees (or sub-trees) H_1 and H_2 is denoted sometimes by $H = H_1 \odot H_2$. It is worth saying that the hierarchical multiplication and its implementation are noticeably more involved than the hierarchical addition. Such complexity is a consequence of the many different cases that must be taken into account. Let A and B be the two dimensionally compatible blocks to be multiplied. Besides the two factors, the so-called target matrix

T and its structure must also be taken into account. Three main cases are possible:

(i) The two factors and the target matrix are subdivided. In this case, the following recursion is invoked

$$\begin{pmatrix} A_{11} & A_{12} \\ A_{21} & A_{22} \end{pmatrix} \begin{pmatrix} B_{11} & B_{12} \\ B_{21} & B_{22} \end{pmatrix} = \begin{pmatrix} A_{11}B_{11} + A_{12}B_{21} & A_{11}B_{12} + A_{12}B_{22} \\ A_{21}B_{11} + A_{22}B_{21} & A_{21}B_{12} + A_{22}B_{22} \end{pmatrix} \quad (5.13)$$

where also the addition in low rank format is called (note that all the additions and multiplications are intended in hierarchical format, although we used the standard operation symbols instead of $\oplus$ and $\odot$ for brevity). From these equations, the importance of defining a target matrix is clear. The entire operation is performed by multiplying blocks and subsequently adding the result to the right target block. It is to be noted that the compatibility between the dimensions of the two factors to be multiplied and those of the target matrix plays a crucial role and must then be taken carefully into account when implementing the relative algorithms. Moreover, it becomes evident that the structure of the target block itself, i.e. if the target block is a leaf or is subdivided, is of primary importance in controlling the flow of the operation, as emerges in the following two cases.

(ii) At least one of the two factors is a leaf, i.e. is not subdivided. This case has several sub-cases. It could either happen that only one of the factors or both of them are leaves. Moreover the leaves can be low or full rank. Each of these circumstances deserves a separate implementation, but basically all these cases are dealt with by considering that if a leaf is full rank, its dimensions are bounded by n_{min}, while if it is low rank, k itself is low. It is then convenient to compute the product block by taking into account that

$$M_1M_2 = A_1B_1^TM_2 = A_1\widetilde{B}_1{}^T \quad (5.14)$$

$$M_1M_2 = M_1A_2B_2^T = \widetilde{A}_2B_2{}^T \quad (5.15)$$

$$M_1M_2 = A_1B_1^TA_2B_2^T = A_1\widetilde{B}_1{}^T = \widetilde{A}_2B_2{}^T \quad (5.16)$$

These represent the multiplication between a low rank and a full rank leaf and between two low rank leaves. The result block must then be split suitably and added to the target matrix. It is worthwhile

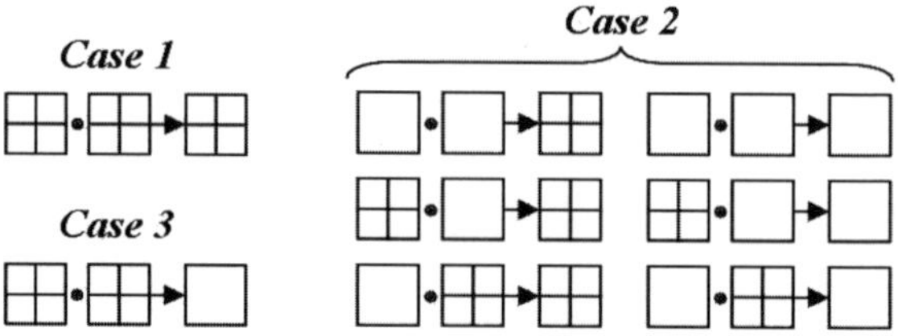

Fig. 5.6. Different cases occurring in hierarchical multiplication.

remembering that every addition in low rank format must be followed by a truncation needed to maintain the low rank of the resulting block. Skipping this passage could lead to an explosion of the average rank of the product matrix that would destroy the advantages of the low rank representation.

(iii) The target block is a leaf and the two factors are subdivided. In this case, the multiplication is performed in the sub-blocks and the result blocks are then expanded to the dimensions of the target, truncated and added to it. After every addition, a new truncation is performed.

The classification of the many different cases and the respective action is graphically illustrated in Fig. 5.6. It is to be noted that when a block is not subdivisible, it can be low or full rank.

5.3.5.3. *LU decomposition*

The hierarchical *LU* decomposition is based on the previously developed arithmetics.[23] If a block H is divisible, then its decomposition can be written as

$$\begin{pmatrix} H_{11} & H_{12} \\ H_{21} & H_{22} \end{pmatrix} = \begin{pmatrix} L_{11} & 0 \\ L_{21} & L_{22} \end{pmatrix} \begin{pmatrix} U_{11} & U_{12} \\ 0 & U_{22} \end{pmatrix} \tag{5.17}$$

which turns into the following subproblems

$$L_{11}U_{11} = H_{11} \tag{5.18}$$

$$L_{11}U_{12} = H_{12} \tag{5.19}$$

$$L_{21}U_{11} = H_{21} \tag{5.20}$$

$$L_{22}U_{22} = H_{22} - L_{21}U_{12} \tag{5.21}$$

Equation 5.18 represents the *LU* decomposition of the first sub-block and can be tackled by the LAPACK subroutine `dgetrf.f`, if the block is not further divisible. Otherwise, the scheme is reapplied recursively. Once L_{11} and U_{11} have been computed, equation 5.19 gives U_{12}, while equation 5.20 allows the calculation of L_{21}. In both cases, it is necessary to distinguish between different cases and apply different suitable schemes. Equation 5.21 can be tackled once the previous equations have been solved.

5.3.5.4. *Inversion*

Once the basic hierarchical arithmetics, i.e. the addition and the multiplication in hierarchical format, have been implemented, the algorithm for the hierarchical inversion is relatively straightforward.[20] If H is the matrix to be inverted, then

$$H^{-1} = \begin{pmatrix} H_{11}^{-1} + H_{11}^{-1} H_{12} S^{-1} H_{21} H_{11}^{-1} & -H_{11}^{-1} H_{12} S^{-1} \\ -S^{-1} H_{21} H_{11}^{-1} & S^{-1} \end{pmatrix} \tag{5.22}$$

where $S = H_{22} - H_{21} H_{11}^{-1} H_{12}$. It is to be noted that the order of the computations is important. Once H_{11}^{-1} is obtained, it is possible to compute S and then all the other factors. Moreover, the original matrix can be overwritten.

5.3.6. *System solution*

The solution of the system can be obtained either directly, through hierarchical matrix inversion,[24] or indirectly, through iterative solvers that exploit the efficient matrix-vector product in low rank format.[23,57]

The iterative solvers can be used with or without preconditioners. When the condition number is high and slows down the convergence rate, as is often the case when dealing with BEM systems, a preconditioner can be computed taking full advantage of the representation in hierarchical format. If $Ax = b$ is the system to be solved, then a left preconditioner is an easily invertible matrix P, such that the condition number of the system $P^{-1}Ax = P^{-1}b$ is lower than the original one, thus improving the convergence rate of the iterative solver.

The hierarchical representation offers the opportunity to naturally build an effective preconditioner.[22,23] A coarse preconditioner can be obtained by first generating a coarse approximation $A(\varepsilon_p)$ of the original collocation matrix $A(\varepsilon_c)$, where the relationship $\varepsilon_p > \varepsilon_c$ holds and where ε denotes the set accuracy for hierarchical representation. This coarse approximation, with reduced memory storage, can then be decomposed through the hierarchical LU decomposition to give the preconditioner P. The resulting system

$$(LU)^{-1}Ax = (LU)^{-1}b \tag{5.23}$$

has a lower condition number and the convergence rate of iterative solvers is noticeably improved. It should be noted that in equation (5.23) there is no need to compute the matrix product $(LU)^{-1}A$, as it is more efficient to directly use the forward and backward substitution for the inversion of the matrix LU in iterative solution schemes. Backward and forward substitutions take full advantage of the hierarchical matrix-vector multiplication and this is the reason why the preconditioner is LU decomposed.

In this work, the GMRES[11] with a coarse hierarchical left preconditioner has been used as a solver.

5.3.7. *Some details about code implementation*

In this work, subroutines and functions for the treatment of hierarchical matrices have been implemented in `FORTRAN 90`. Different modules have been implemented to deal with the diverse tasks involved in the hierarchical treatment of the boundary element elastostatic problems. The basic module `Hdata` implements all the needed data structures; the module `Htree` contains all the procedures that build the cluster and block trees starting from the boundary information; the module `Hsetup` implements all the subroutines that set up the hierarchical matrix computing low and full rank blocks, also forcing the boundary conditions; the module `Hblocks` implements all the procedures that work on single blocks, like the full and reduced SVD; finally, the module `Harithmetics` implements the arithmetics on the trees (addition, multiplication, LU decomposition and inversion).

The basic data structure is called `BlockNode` (Fig. 5.7) and it allows the natural treatment of either the full or low rank blocks of the hierarchical matrix.

The field `IndexSet` identifies the position of the block in the full matrix, identifying rows and columns on the basis of a previously defined index partition; the field `Identifier` can be assigned three different values that allow one to establish the nature of the block and specifically if the block is a leaf or not and, in the case it is a leaf, if it is a low or full rank block;

```
TYPE BlockNode
  INTEGER, DIMENSION(6) :: IndexSet
  INTEGER :: Identifier
  INTEGER :: Rank
  TYPE(BlockNode), POINTER :: parent
  TYPE(BlockNode), POINTER :: son11
  TYPE(BlockNode), POINTER :: son12
  TYPE(BlockNode), POINTER :: son21
  TYPE(BlockNode), POINTER :: son22
  TYPE(vector), POINTER :: HeadColumns
  TYPE(vector), POINTER :: TailColumns
  TYPE(vector), POINTER :: HeadRows
  TYPE(vector), POINTER :: TailRows
  DOUBLE PRECISION, DIMENSION(:,:), POINTER :: FBL
END TYPE BlockNode
```

Fig. 5.7. Basic data structure for hierarchical matrices.

the pointers `parent`, `son11`, `son12`, `son21` and `son22` are needed to build and maintain the structure of the quaternary hierarchical tree. It is worth stressing again that some operations, like the coarsening of the block tree, rely on the possibility of moving through the tree in both directions, i.e. from root to leaves and from leaves to root.

If the block is a leaf, on the basis of the value taken by `Identifier`, the suitable data structure is allocated to store the information. If the block is full rank, then the array `FBL` (full rank block) is allocated, its dimension being inferred by `IndexSet`. On the other end, if the block is low rank, then the pointers `HeadColumns`, `TailColumns`, `HeadRows` and `TailRows` are associated: they point to the head and tail of two different lists of vectors collecting, respectively, the columns and rows of the low rank representation.

5.4. Numerical Experiments

In this section, results obtained by applying the fast hierarchical BEM are presented. Both isotropic and anisotropic problems are analysed and the performance of the developed scheme is assessed. The computations involving isotropic material structures have been performed using an Intel® Core™ 2 Duo Processor T5500 (1.66 GHz) and 2 GB of RAM. The tests involving anisotropic materials have been performed on an Intel® Core™ 2 Duo Processor T9300 (2.5 GHz) and 2 GB of RAM.

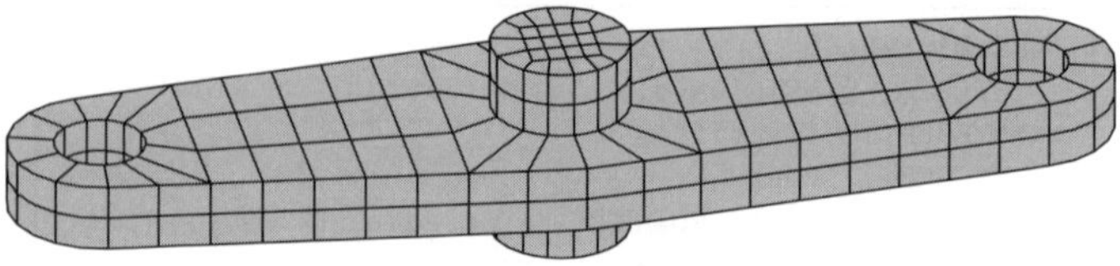

Fig. 5.8. Analysed 3-D configuration.

5.4.1. *Uncracked isotropic elastic bracket*

A mechanical element, Fig. 5.8, is first analysed to obtain some insight into the structure of the hierarchical collocation matrix for structures without cracks. The mechanical bracket has the two central cylinders clamped and is anti-symmetrically loaded at the holes in such a way that the resulting load is a moment lying along the central axis. A mesh with 1,032 elements and 3,094 nodes has been considered. The standard method required 350 seconds for the generation of the collocation matrix and 1,484 seconds for the solution of the system through Gauss elimination. These times have been compared with the related times required by the fast method to obtain both the assembly speed-up ratio and the solution speed-up ratio at different parameter settings. In all the following tables, the speed-up ratio is defined as

$$\textit{Speed-up ratio} = \frac{\textit{Hierarchical time}}{\textit{Standard time}} \tag{5.24}$$

where the times are those required by the standard and hierarchical schemes to assemble the matrix or solve the system. In particular, the standard assembly time has been compared with the time needed for the ACA generation of the collocation matrix, while the standard solution time has been compared with the fast solution time, which is comprised of the collocation matrix compression and coarsening time, the preconditioner generation, coarsening and *LU* decomposition time and the GMRES iteration time.

A first set of analyses has been performed to investigate the effect of the preconditioner accuracy on the convergence of the iterative solution. For this purpose, the accuracy of the collocation matrix has been set to $\varepsilon_c = 10^{-5}$, the admissibility parameter has been chosen as $\eta = \sqrt{2}$ and the minimal block size has been set to $n_{min} = 36$. The GMRES relative accuracy is 10^{-8}.

With these settings, the hierarchical collocation matrix is stored using only 35.61% (after coarsening) of the memory required for the allocation of the original matrix. The matrix is generated through ACA in 198 seconds, i.e. 57% of the standard assembly time, and it is recompressed and coarsened in 216 seconds. It may be of interest to mention that, in this specific application, 60% of the collocation matrix has been generated through ACA in 55 seconds (28% of the hierarchical assembly time). The remaining 40% of the collocation matrix is in full rank format and its evaluation requires the remaining 72% of the hierarchical assembly time, mainly due to the computation of the singular integrals whose evaluation is needed to populate such blocks. The approximate solution accuracy is $\|x-\tilde{x}\|_{L^2}/\|x\|_{L^2} = 2.9 \times 10^{-4}$. Both the collocation matrix coarsening time and the solution accuracy are independent of the preconditioner accuracy. The GMRES converges towards the same approximate solution, whose accuracy depends only on the accuracy of $\tilde{A}$ (the collocation matrix) and $\tilde{b}$ (the right-hand side).

Table 5.1 reports the storage memory needed to store the preconditioner, expressed as a percentage of the full collocation matrix, the time required to set up and decompose the preconditioner, the time and the number of iterations required by the GMRES to converge and, finally, the solution speed-up ratio, defined as the ratio between the fast solution time and the standard solution time. Times are expressed in seconds. It is interesting to note that the time required to set up the preconditioner and for its LU decomposition grows when the preconditioner required accuracy

Table 5.1. Storage, time and speed-up ratio for different preconditioner accuracies.

ε_p	Storage	Setup (s)	LU (s)	GMRES (s)	Iterations	Speed-up
No Prec.	0.00%	0.0	0.0	3,470.8[a]	5,000[a]	2.49[a]
Jacobi	0.00%	0.0	0.0	3,474.6[b]	5,000[b]	2.49[b]
$5 \cdot 10^{-1}$	1.45%	23.9	12.7	3,906.0[c]	5,000[c]	2.80[c]
$1 \cdot 10^{-1}$	4.95%	28.4	45.5	10.0	29	0.19
$5 \cdot 10^{-2}$	6.54%	31.5	63.2	8.7	25	0.21
$1 \cdot 10^{-2}$	10.59%	47.3	117.7	4.0	10	0.25
$5 \cdot 10^{-3}$	12.62%	58.5	153.2	3.9	9	0.28
$1 \cdot 10^{-3}$	17.31%	82.1	252.6	3.0	6	0.36
$5 \cdot 10^{-4}$	19.38%	90.8	303.4	2.6	5	0.40
$1 \cdot 10^{-4}$	25.20%	104.2	476.7	2.5	4	0.53
$5 \cdot 10^{-5}$	27.89%	97.9	586.2	2.1	3	0.60
$1 \cdot 10^{-5}$	35.61%	3.1[d]	988.2	2.3	3	0.81

[a]Reached GMRES relative accuracy: 3.0×10^{-6} (no convergence).
[b]GMRES relative accuracy: $3.0 \cdot 10^{-2}$.
[c]GMRES relative accuracy: $6.1 \cdot 10^{-8}$.
[d]Only the time for copying the collocation matrix.

grows (as ε_p becomes smaller the required accuracy increases). On the other hand, quite naturally, the number of GMRES iterations and the iterative solution time decrease when ε_p decreases. In the extreme case of the preconditioner retaining the same accuracy as that of the collocation matrix, the preconditioner LU decomposition can be used for a direct solution. It is worth noting, however, that a very coarse preconditioner ($\varepsilon_p = 10^{-1}$) provides the fastest solution, as is evident from the reported solution speed-up ratios, while both the unpreconditioned GMRES and the Jacobi preconditioned GMRES fail to converge.

Figure 5.9 shows the block-wise structure of the collocation matrix as generated by ACA, the coarsened collocation matrix and the structure of the coarsest effective preconditioner ($\varepsilon_p = 10^{-1}$). The number of blocks obtained for the selected minimal block size goes from 3,547 in the ACA generated matrix to 2,545 in the coarsened matrix, while the preconditioner counts 1,063 blocks.

Every block is filled with a tone of grey proportional to the ratio between the memory required for low rank representation and the memory in full rank format. Full rank blocks are black while almost white blocks are those for which the numerical compression works better. It is worth noting the reduction in the number of blocks obtained going from the ACA-generated matrix to the coarsened matrix. The big difference in the number of blocks is due to the suboptimal choice of the admissibility parameter η, as will be discussed soon. It is also worthwhile to remember that the required memory is not that for storing the ACA-generated matrix, but it is that, lower, required for storing the coarsened matrix. Here, the two different compressions are shown to illustrate the mechanism of coarsening, but the operation can be performed recursively while populating the blocks.

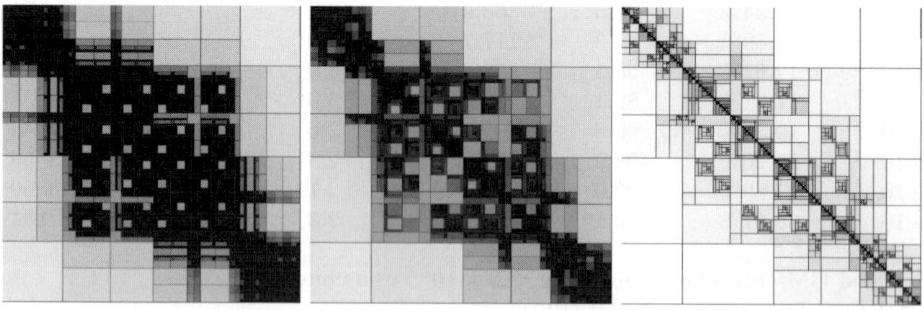

Fig. 5.9. Block-wise representation of the ACA generated matrix, the coarsened matrix and the preconditioner.

Table 5.2. Influence of the prescribed accuracy for the collocation matrix on storage and speed-up ratios.

ε_c	Stor. A	Stor. B	Assembly	Speed-up	Sol.	Speed-up	$\frac{\|x-\tilde{x}\|_{L^2}}{\|x\|_{L^2}}$
10^{-6}	57.49%	45.59%	218.3	0.62	319.1	0.21	$6.0 \cdot 10^{-6}$
10^{-5}	53.62%	35.61%	197.6	0.57	300.2	0.20	$2.9 \cdot 10^{-4}$
10^{-4}	50.09%	25.17%	180.5	0.52	269.6	0.18	$5.6 \cdot 10^{-3}$
10^{-3}	47.15%	17.28%	169.0	0.48	201.9	0.14	$3.4 \cdot 10^{-2}$
10^{-2}	44.17%	10.56%	159.1	0.45	150.8	0.10	$1.6 \cdot 10^{-1}$

Table 5.2 reports on memory requirements before and after coarsening, assembly time and assembly speed-up ratio, solution time (compression and coarsening, preconditioner generation and LU decomposition, GMRES) and solution speed-up ratio and the accuracy of the solution for different values of the collocation matrix prescribed accuracy. The tests have been performed by setting $\varepsilon_p = 1.0 \cdot 10^{-1}$ and $\eta = \sqrt{2}$. Memory requirements, assembly times and solution times decrease when the preset accuracy decreases as the average rank of the approximation is reduced. However, reducing the requested accuracy obviously also reduces the approximation quality of the final solution. From an engineering point of view, the selection of a suitable criterion for selecting the collocation matrix accuracy is of fundamental importance. Note that the L^2 norm used in Table 5.2 does not give insight into the quality of the approximation for engineering purposes. A node by node check of the solution has confirmed, however, that, for a selected accuracy $\varepsilon_c = 1.0 \times 10^{-5}$, the average errors are to the order of $0.1 \div 1.0\%$. Bigger percentage errors can occur for degrees of freedom whose standard solution values are smaller than the requested accuracy. This consideration suggests that it is advisable to set the accuracy at the same order of magnitude as that of the smaller quantities of interest in the analysis.

Table 5.3 reports the memory storage before and after coarsening, the number of blocks before and after coarsening, the assembly time and

Table 5.3. Influence of the admissibility parameter.

η	Stor. A	Stor. B	N. of blocks	Assembly	Speed-up	Coars.	Speed-up
$\sqrt{2}$	53.62%	35.61%	3,547 - 2,545	197.6	0.57	216.2	0.20
3	45.04%	34.61%	2,581 - 2,284	209.2	0.60	117.5	0.13
4	43.73%	34.17%	2,341 - 2,152	218.3	0.62	98.9	0.12
5	43.52%	33.63%	2,257 - 2,044	220.5	0.63	111.2	0.13

assembly speed-up ratio, the coarsening time and the solution speed-up ratio at different values of the admissibility parameter. The other parameters have been set to $\varepsilon_c = 10^{-5}$ and $\varepsilon_p = 10^{-1}$. The time for generating and manipulating the preconditioner is independent of η. On the contrary, the time required for coarsening the matrix strongly depends on it. The choice of η directly affects the quality of the ACA-generated matrix and a good choice results in a matrix closer to the optimal matrix produced by the coarsening procedure, as can also be noted from the reduction in the number of blocks. This is the reason for the influence on the coarsening time. Note that the matrices obtained after coarsening require almost the same memory, regardless of the initial ACA-generated matrix storage: the coarsening produces an almost optimal hierarchical matrix, reducing the differences related to the choice of η. However, though a larger part of the matrix is generated through ACA, which should lead to a reduction in the assembly time, the average rank of the approximation increases, as the new admissible blocks converge more slowly to the preset accuracy. This aspect may have a negative effect on the assembly time, but that is balanced by a more relevant reduction in the solution time. Figure 5.10 shows the structure of the ACA matrix for two different values of η. It is evident that for the $\eta = 4$ value more blocks become admissible than for the $\eta = \sqrt{2}$ value.

5.4.2. *Embedded crack in isotropic bar*

A second configuration, a cylinder with an embedded crack, is considered (see Fig. 5.11). The cylinder is subjected to uniaxial stress acting on the two bases, so as to produce a mode I crack load.

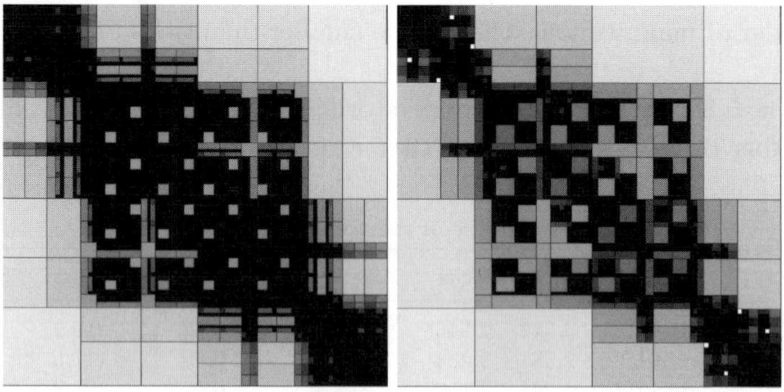

Fig. 5.10. Block-wise structure of the ACA matrix for $\eta = \sqrt{2}$ and $\eta = 4$.

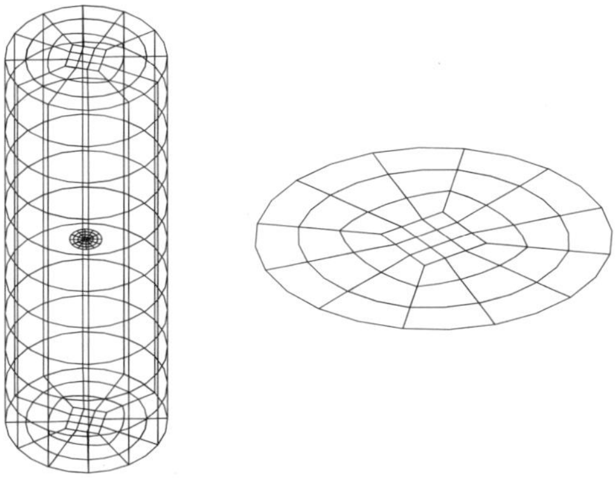

Fig. 5.11. Cylinder with embedded crack and crack basic mesh.

A mesh with 800 elements and 3,652 nodes is considered. The standard technique requires 805 seconds to assemble the collocation matrix and 2,394 seconds to solve the system. It is worth noting that the numerical integration of the singular, strongly singular and hypersingular kernels occurring during the assembly of the collocation matrix requires, in this case, 60% of the standard integration time. Since singular integrals occur near the diagonal blocks, which are full rank in the hierarchical representation, this percentage represents the lower bound for the hierarchical assembly time.

First, a parametric analysis on the influence of the preconditioner accuracy is performed, as in the previous case. For this first set of computations, the collocation matrix accuracy is set to $\varepsilon_c = 10^{-5}$, the admissibility parameter is $\eta = 6$, while $n_{min} = 36$. With these parameters, the collocation matrix is generated in 689 seconds, i.e. 85% of the standard assembly time, it is compressed and coarsened in 168 seconds and is stored using 23.70% of the original space. The accuracy of the final solution is $\|x - \tilde{x}\|_{L^2}/\|x\|_{L^2} = 1.2 \times 10^{-4}$, which is very good in terms of point by point accuracy. Table 5.4 reports the preconditioner storage, the preconditioner setup and LU decomposition times (in seconds), the GMRES solution time, the number of GMRES iterations and the solution speed-up ratio.

Again, it can be observed that the required memory, the setup time and the LU decomposition time grow as the required preconditioner accuracy grows. On the contrary, the number of GMRES iterations, and the GMRES time as consequence, decrease as the accuracy grows. Moreover, it is important to note that the best speed-up ratio is obtained with the coarsest

Table 5.4. Storage and times for different values of prescribed accuracy for the preconditioner.

ε_p	Storage	Setup (s)	LU (s)	GMRES (s)	Iterations	Speed-up
No Prec.	0.00%	0.0	0.0	3,642.1[a]	5,000[a]	1.52[a]
Jacobi	0.00%	0.0	0.0	3,644.3[b]	5,000[b]	1.52[b]
$1 \cdot 10^{-0}$	0.86%	15.2	0.4	92.4	306	0.12
$5 \cdot 10^{-1}$	2.42%	15.7	115.2	60.1	197	0.15
$1 \cdot 10^{-1}$	5.00%	17.9	163.9	12.9	40	0.15
$1 \cdot 10^{-2}$	9.06%	26.5	273.2	5.9	16	0.20
$1 \cdot 10^{-3}$	13.39%	47.4	448.6	3.8	8	0.28
$1 \cdot 10^{-4}$	18.32%	68.6	802.8	2.5	4	0.43
$1 \cdot 10^{-5}$	23.69%	2.7[c]	1,371.6	2.3	3	0.64

[a]The relative GMRES accuracy was $5.9 \cdot 10^{-5}$ (no convergence reached).
[b]The GMRES relative accuracy was $8.8 \cdot 10^{-2}$.
[c]Only the time for copying the collocation matrix.

Table 5.5. Influence of the admissibility parameter.

η	Stor. A	Stor. B	Blocks	A. speed-up	Coarsening	S. speed-up
2	43.94%	23.72%	4,735 - 2,380	0.85	229.7	0.14
4	38.10%	23.72%	4,003 - 2,380	0.85	185.4	0.12
6	35.85%	23.70%	3,643 - 2,338	0.86	168.2	0.12
8	35.15%	23.67%	3,433 - 2,314	0.86	162.4	0.11

preconditioner ($\varepsilon_p = 1.0$), while a fine preconditioner could be used as a direct solver. Finally, the construction of the hierarchical preconditioner is actually necessary, as the unpreconditioned GMRES as well as the Jacobi preconditioned GMRES fail to converge.

The influence of the admissibility parameter has been investigated and the results are reported in Table 5.5. The analysis is performed setting $\varepsilon_c = 10^{-5}$ and $\varepsilon_p = 1.0$. The same considerations as in the previous case hold.

Finally, three different meshes are analysed to obtain some insight into the behaviour of the solver at varying numbers of degrees of freedom. The first mesh has 66 elements and 400 nodes, the second 300 elements and 1,352 nodes and the third uses 800 elements and 3,652 nodes. The settings are $\varepsilon_c = 10^{-5}$, $\varepsilon_p = 1.0$, $\eta = 6$. Table 5.6 reports the results obtained for the three different meshes in terms of memory ratio, assembly speed-up ratio, solution speed-up ratio and number of GMRES iterations. Also, the times for standard assembly and standard solution are reported

Table 5.6. Memory savings and speed-up ratios.

Elem.	Nodes	Storage	St. Ass.	Speed-up	St. Sol.	Speed-Up	Iterations
66	400	69.78%	79.4	1.07	3.2	1.23	49
300	1,352	40.74%	215.5	1.04	122.8	0.28	113
800	3,652	23.70%	805.5	0.86	2,398.8	0.12	306

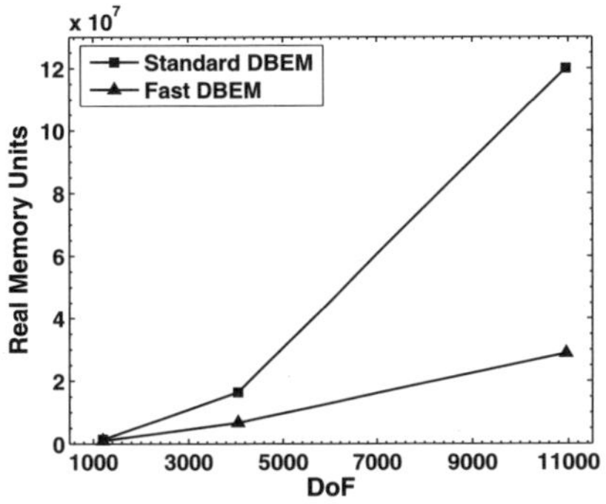

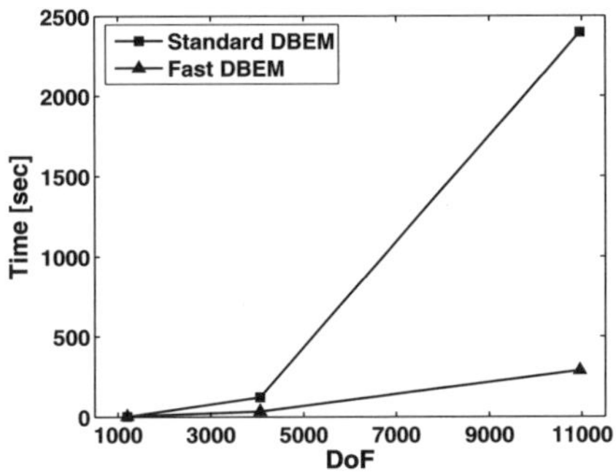

Fig. 5.12. Comparison between standard and fast DBEM.

and are expressed in seconds. It appears evident that the advantages of the described technique become more relevant with larger meshes. While memory savings are always obtained also for coarse meshes, the assembly and solution speed-up ratios are less than one only beyond certain threshold, under which the direct solver performs better.

Figure 5.12 shows the comparison in terms of required memory and solution time between standard and fast DBEM. It is worth noting the almost linear behaviour of the fast DBEM with respect to the number of degrees of freedom.

Figure 5.13 shows the block-wise structure of the collocation matrix for the finest mesh as generated by ACA, the coarsened collocation matrix and the structure of the preconditioner. The number of blocks goes from 3,643 in the ACA generated matrix to 2,338 in the coarsened matrix and 430 blocks in the preconditioner. It is interesting to point out how the low rank blocks corresponding to collocation on the boundary and integration on the crack and vice-versa are clearly distinguishable. The geometry and mesh features have a numerical counterpart in the block-wise structure of the hierarchical matrices.

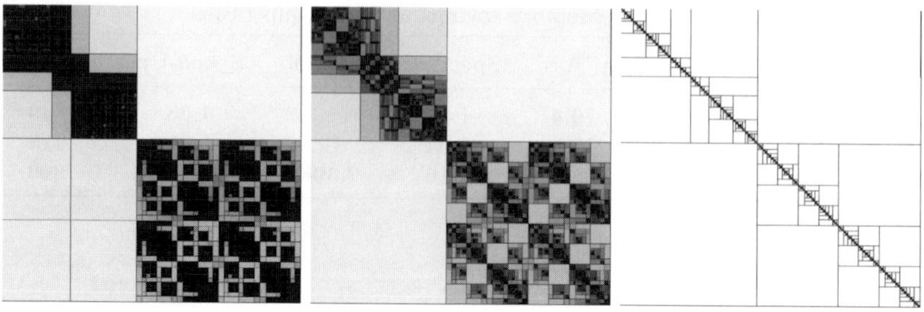

Fig. 5.13. Block-wise representation of the ACA generated matrix, the coarsened matrix and the preconditioner.

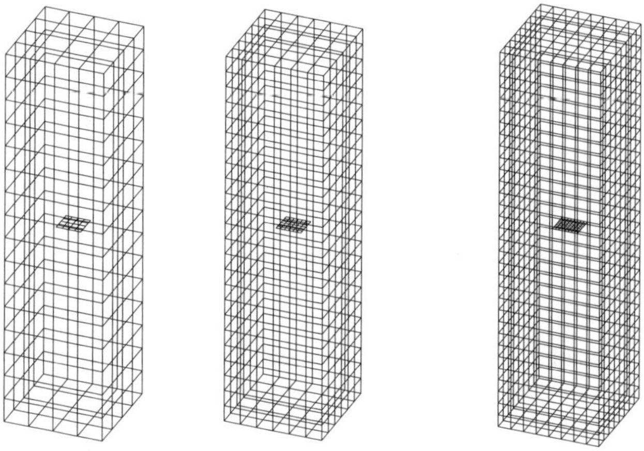

Fig. 5.14. Square crack embedded in anisotropic bar. Three uniformly refined meshes.

5.4.3. *Embedded crack in anisotropic bar*

The performance of the fast hierarchical DBEM has also been tested for anisotropic crack problems, to assess the applicability of the proposed scheme to such class of problems and to evaluate the sensitivity of the method to different materials.[59] The configuration shown in Fig. 5.14 has been analysed (three sample meshes are depicted).

The prismatic bar is subjected to an opening load acting on the bases and different uniformly refined meshes are considered for analysing the performances of the solver at varying degrees of freedom.

Three different materials are analysed. The considered material properties are listed in Table 5.7. It is worth noting that, in the following reported results, the isotropic material configurations have been analysed

Table 5.7. Material properties [GPa].

C_{11}	C_{12}	C_{13}	C_{14}	C_{22}	C_{23}	C_{24}	C_{33}	C_{44}	C_{55}	C_{56}	C_{66}
Aluminum Polycrystal (isotropic)											
111	61	61	0	111	61	0	111	25	25	0	25
Barium sodium niobate (orthorhombic)											
239	104	50	0	247	52	0	135	65	66	0	76
Sapphire (trigonal)											
494	158	114	−23	494	114	23	496	145	145	−23	168

NB: All the non-reported constants are zero.

Table 5.8. Influence of the admissibility parameter.

η	Stor. A	Stor. B	Blocks	A. Speed-up	Coarsening	S. Speed-up
Aluminum Polycrystal (isotropic)						
1	53.0 %	31.7%	3,757 - 1,951	0.81	92.4	0.19
2	45.8 %	31.7%	3,013 - 1,951	0.81	66.8	0.15
4	41.6 %	31.6%	2,347 - 1,906	0.89	47.6	0.12
6	41.6 %	31.5%	2,305 - 1,867	0.90	48.0	0.12
8	41.5 %	31.5%	2,287 - 1,849	0.90	48.0	0.12
Barium sodium niobate (orthorhombic)						
1	52.9 %	31.9 %	3,757 - 1,978	0.79	90.8	0.18
2	45.5 %	31.9 %	3,013 - 1,978	0.79	66.0	0.15
4	41.2 %	31.8 %	2,347 - 1,936	0.87	47.3	0.12
6	41.2 %	31.8 %	2,305 - 1,900	0.87	47.1	0.12
8	41.1 %	31.8 %	2,287 - 1,882	0.88	47.2	0.12
Sapphire (trigonal)						
1	53.5 %	33.8%	3,757 - 2,152	0.82	99.5	0.19
2	46.4 %	33.8%	3,013 - 2,146	0.83	65.8	0.15
4	42.5 %	33.5%	2,347 - 2,035	0.91	47.8	0.12
6	42.5 %	33.5%	2,305 - 1,999	0.92	47.8	0.12
8	42.5 %	33.4%	2,287 - 1,981	0.93	47.6	0.12

for uniformity by using the numerical scheme for anisotropy described in Appendix B. The studies on the influence of the admissibility parameter, the collocation matrix accuracy and the preconditioner accuracy have been performed on a mesh with 720 elements and 2,522 nodes.

Table 5.8 reports the influence of the admissibility parameter for the three materials. The analysis is performed setting $\varepsilon_c = 10^{-5}$ and $\varepsilon_p = 10^{-1}$. The minimal block size has been set to $n_{min} = 36$ and the GMRES relative accuracy is 10^{-8}. The same considerations as those reported for isotropic materials apply and no noticeable sensitivity with respect to the degree of

Table 5.9. Storage and times for different preconditioner accuracies.

ε_p	Storage	Setup (s)	LU (s)	GMRES (s)	Iterations	Speed-up
Aluminum Polycrystal (isotropic)						
No Prec.	0.0%	0.0	0.0	828.6[a]	3,000[a]	1.27[a]
Jacobi	0.0%	0.0	0.0	750.3[b]	3,000[a]	1.23[a]
$1 \cdot 10^{-0}$	1.2%	6.0	0.1	23.8	188	0.12
$1 \cdot 10^{-1}$	5.3%	6.9	17.5	3.4	23	0.12
$1 \cdot 10^{-2}$	10.3%	11.0	35.6	2.2	13	0.15
$1 \cdot 10^{-3}$	15.8%	22.8	63.9	1.2	6	0.21
$1 \cdot 10^{-4}$	22.4%	32.2	123.4	1.0	4	0.31
Barium sodium niobate (orthorhombic)						
No Prec.	0.00%	0.0	0.0	803.2[a]	3,000[a]	1.31[a]
Jacobi	0.00%	0.0	0.0	764.3[a]	3,000[a]	1.25[a]
$1 \cdot 10^{-0}$	1.2%	6.9	0.1	37.2	260	0.14
$1 \cdot 10^{-1}$	5.6%	7.5	17.8	3.6	24	0.12
$1 \cdot 10^{-2}$	10.4%	12.0	35.4	2.9	17	0.15
$1 \cdot 10^{-3}$	16.0%	25.1	67.9	1.1	5	0.22
$1 \cdot 10^{-4}$	22.7%	33.3	125.7	0.8	3	0.32
Sapphire (trigonal)						
No Prec.	0.00%	0.0	0.0	825.1[a]	3,000[a]	1.37[a]
Jacobi	0.00%	0.0	0.0	800.0[a]	3,000[a]	1.33[a]
$1 \cdot 10^{-0}$	1.2%	7.2	0.1	35.3	236	0.14
$1 \cdot 10^{-1}$	5.6%	8.2	16.3	4.2	26	0.12
$1 \cdot 10^{-2}$	10.6%	13.0	40.8	2.0	11	0.16
$1 \cdot 10^{-3}$	16.6%	27.8	72.5	1.1	5	0.23
$1 \cdot 10^{-4}$	24.2%	24.8	120.3	1.1	4	0.30

[a]No convergence reached.

anisotropy is shown. In particular, it is apparent that some values of the admissibility parameter originate almost optimal block trees, thus reducing the coarsening time.

Table 5.9 reports storage and times for different preconditioner accuracies. For this purpose, the accuracy of the collocation matrix has been set to $\varepsilon_c = 10^{-5}$, the admissibility parameter has been chosen as $\eta = 4$ and the minimal block size has been set to $n_{min} = 36$. The GMRES relative accuracy is 10^{-8}. As can be observed, the coarsest preconditioners provide faster solutions. Also for the anisotropic case, it is shown that preconditioning is needed and both unpreconditioned systems and Jacobi preconditioned systems fail to converge within the maximum number of iterations chosen. The preconditioner setup time increases at increasing requested accuracies, while the number of GMRES iterations decreases. It is worth highlighting once again that more accurate preconditioners are

Table 5.10. Memory savings and speed-up ratios.

Elem.	Nodes	Storage	St. Ass.	Speed-up	St. Sol.	Speed-up	Iterations
Aluminum Polycrystal (isotropic)							
320	1,122	49.5%	39.9	1.09	56.0	0.30	19
500	1,752	41.2%	79.1	1.01	213.1	0.17	18
720	2,522	31.6%	149.0	0.89	638.3	0.12	23
980	3,432	25.4%	283.7	0.71	1,606.4	0.12	27
1,280	4,482	20.0%	515.0	0.61	3,582.6	0.09	24
Barium sodium niobate (orthorhombic)							
320	1,122	49.5%	39.9	1.08	56.2	0.30	25
500	1,752	41.3%	79.0	1.01	213.3	0.18	18
720	2,522	31.9%	152.1	0.87	647.3	0.12	24
980	3,432	26.0%	284.7	0.70	1,674.1	0.11	28
1,280	4,482	20.5%	518.2	0.59	3,633.2	0.08	26
Sapphire (trigonal)							
320	1,122	50.8%	39.8	1.11	55.9	0.31	20
500	1,752	43.1%	79.3	1.02	214.0	0.17	16
720	2,522	33.5%	149.0	0.92	638.3	0.12	26
980	3,432	27.6%	287.6	0.72	1,601.7	0.11	26
1,280	4,482	21.8%	515.0	0.64	3,582.9	0.09	25

closer to the inverse of the collocation matrix, making then the iterative technique closer to the direct solution of the system.

Table 5.10 reports memory savings and speed-up ratios versus the number of degrees of freedom. The settings are $\varepsilon_c = 10^{-5}$, $\varepsilon_p = 10^{-1}$, $\eta = 4$, $n_{min} = 36$ and the GMRES relative accuracy is 10^{-8}. It is apparent that storage memory, assembly time and solution time vary almost linearly with respect to the number of degrees of freedom, thus providing considerable savings for large systems in comparison to direct solvers. Such considerations apply to all the considered materials, highlighting the versatility of the technique and the positive independence from the degree of anisotropy. This aspect is particularly appealing as anisotropic problems are usually quite time-consuming. The presented results also compare well with those computed using the pure isotropic scheme reported in the previous sections, thus confirming the robustness of the approach.

Figures 5.15 and 5.16 depict the assembly times and solution times at varying degrees of freedom for the three different materials. Only the assembly time shows a slight sensitivity to the material, while the solution time appears to be quite insensitive to it.

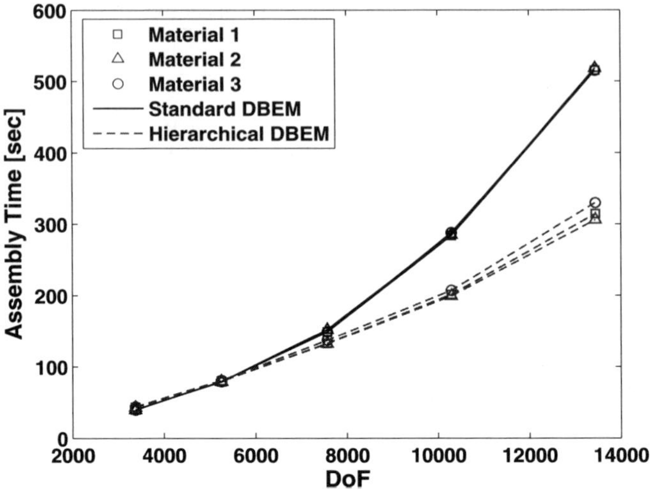

Fig. 5.15. Comparison between standard and Hierarchical DBEM assembly times.

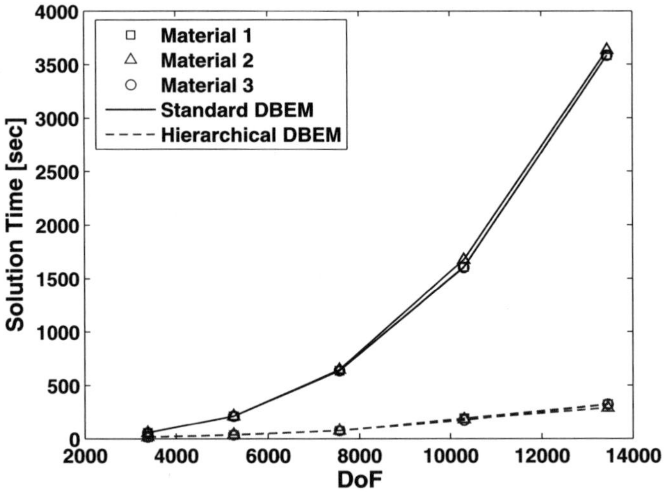

Fig. 5.16. Comparison between standard and Hierarchical DBEM solution times.

5.5. Conclusions

In this chapter, recent developments in the fast solution of BEM and DBEM systems of equations have been presented. The development of a fast hierarchical DBEM has been described and applications to both isotropic and anisotropic 3-D crack problems have been presented. The method was shown to be very effective and reliable in terms of accuracy. Moreover, it allows a considerable reduction in both the amount of memory needed for storing the system coefficients and the time required for the system solution. In particular, it has been shown that both storage memory and solution time vary almost linearly with respect to the number of degrees of freedom, thus providing considerable savings for large systems in comparison to direct solvers. Such considerations apply to both isotropic and anisotropic problems as the study demonstrated negligible sensitivity with respect to the degree of anisotropy. The results show that hierarchical matrices are particularly suitable for crack problems. This is because cracks are normally isolated surfaces which are far from most of the remaining boundary, hence they correspond to large low rank blocks. As demonstrated, this allows one to increase the number of elements on the crack surfaces with only modest increases in storage memory and solution time.

Appendix A. Fundamental Solutions

For isotropic problems, the fundamental solution kernels U_{ij}, T_{ij}, U_{ijk} and T_{ijk} are given by the Kelvin displacements and tractions and a combination of their derivatives as follows

$$U_{ij}(\boldsymbol{x}_0, \boldsymbol{x}) = \frac{1}{16\pi(1-\nu)\mu\, r}[(3-4\nu)\delta_{ij} + r_{,i}r_{,j}] \tag{A.1}$$

$$\begin{aligned} T_{ij}(\boldsymbol{x}_0, \boldsymbol{x}) = &-\frac{1}{8\pi(1-\nu)r^2}\left\{[(1-2\nu)\delta_{ij} + 3r_{,i}r_{,j}]\frac{\partial r}{\partial n}\right. \\ &\left. -(1-2\nu)(r_{,i}n_j - r_{,j}n_i)\right\} \end{aligned} \tag{A.2}$$

$$U_{ijk}(\boldsymbol{x}_0, \boldsymbol{x}) = \frac{1}{8\,\pi(1-\nu)r^2}[(1-2\nu)(\delta_{ij}r_{,k} + \delta_{ik}r_{,j} - \delta_{jk}r_{,i}) + 3r_{,i}r_{,j}r_{,k}] \tag{A.3}$$

$$\begin{aligned} T_{ijk}(\boldsymbol{x}_0, \boldsymbol{x}) = &\frac{\mu}{4\,\pi(1-\nu)r^3}\left\{3\frac{\partial r}{\partial n}[(1-2\nu)\delta_{ij}r_{,k}\right. \\ &+\nu(\delta_{ik}r_{,j} + \delta_{jk}r_{,i}) - 5r_{,i}r_{,j}r_{,k}] + 3\nu(n_i r_{,j}r_{,k} + n_j r_{,i}r_{,k}) \\ &\left.+(1-2\nu)(3n_k r_{,i}r_{,j} + n_j\delta_{ik} + n_i\delta_{jk}) - (1-4\nu)n_k\delta_{ij}\right\} \end{aligned} \tag{A.4}$$

where n_i are the components of the outward normal at the point $\boldsymbol{x}$ and the the following relations hold

$$r_k = x_k - x_{0\,k} \tag{A.5}$$

$$r = \|\boldsymbol{x} - \boldsymbol{x}_0\| = (r_k\, r_k)^{1/2} \tag{A.6}$$

$$r_{,k} = \frac{\partial r}{\partial x_k} = \frac{r_k}{r} \tag{A.7}$$

For the general 3-D anisotropic elasticity problem, no closed form fundamental solutions are available and fundamental solution kernels U_{ij}, T_{ij}, U_{ijk} and T_{ijk} are numerically obtained from the Green's function for general anisotropic domains.[60] For the special case of transversely isotropic materials, the fundamental solution kernels can be expressed in analytic form as reported by Pan and Chou[61] and Ariza and Dominguez.[50]

Different techniques for such numerical evaluation have been presented by various authors. Vogel and Rizzo[62] derived and applied the integral representation of the static fundamental solution for a general anisotropic three-dimensional continuum. Following that, Wilson and Cruse[63] proposed an efficient numerical implementation to speed up the BEM solution of 3-D anisotropic problems. Since these pioneering works, much research has been focused on the derivation of fundamental solutions for three-dimensional anisotropic elasticity. Indeed, the lack of closed form fundamental solutions has hindered the use of the method in anisotropic applications. BEM approaches for general anisotropic problems have been developed using either the integral expression of the fundamental solution[60,62–64] or its explicit approximate expressions.[65–69] While the former are computationally expensive, the latter involve tedious calculations of the kernels' derivatives. For such reasons, the BEM for the analysis of three-dimensional anisotropic structures has been quite slow and difficult to implement with respect to isotropic formulations.

The technique used in this chapter is based on the scheme proposed by Wilson and Cruse[63] and is briefly reviewed in the following.[60] Let $\mathbf{b}$ be a unit vector having the same direction as that of the line connecting the collocation point $\boldsymbol{x}_0$ with the integration point $\boldsymbol{x}$. The anisotropic Green function can be written as

$$G_{ij} = \frac{1}{8\pi^2 r}\int_0^{2\pi} M_{ij}^{-1}(\varphi)d\varphi \tag{A.8}$$

where r is the distance between the observed and source points, and the integration is performed on the plane perpendicular to $\mathbf{b}$. The integrand function M_{ij}^{-1} is composed of the inverse of the matrix defined by

$$M_{ij} = C_{ipjq} z_p z_q \tag{A.9}$$

where C_{ipjq} are the components of the stiffness tensor and z_i are the components of a unit vector representative of the vectors that lie on the integration plane (perpendicular to $\mathbf{b}$). Such components are expressed in terms of the anomaly φ. It is worth noting that the Green's function comprises a singular part depending only on r; and a finite part which depends on the material properties and the direction $\boldsymbol{x}_0\boldsymbol{x}$.

To evaluate the boundary integral equation (BIE) kernels of the DBEM, the derivatives of the Green's function are also needed. They are given by

$$\frac{\partial G_{ij}}{\partial x_p} = -\frac{1}{4\pi^2 r^2}\int_0^{\pi} (-b_p M_{ij}^{-1} + z_p F_{ij})d\varphi \tag{A.10}$$

$$\frac{\partial^2 G_{ij}}{\partial x_p \partial x_q} = \frac{1}{4\pi^2 r^2}\int_0^{\pi} [2b_p b_q M_{ij}^{-1} - 2(z_p b_q + z_q b_q)F_{ij} + z_p z_q A_{ij}]d\varphi \tag{A.11}$$

where, once again, the integration is performed on a plane perpendicular to $\mathbf{b}$ and

$$F_{ij} = C_{mnpq} M_{im}^{-1} M_{pj}^{-1}(z_n b_q + z_q b_n) \tag{A.12}$$

$$A_{ij} = C_{mnpq}[(z_n b_q + z_q b_n)(F_{im} M_{pj}^{-1} + M_{im}^{-1} F_{pj}) - 2M_{im}^{-1} M_{pj}^{-1} b_n b_q] \tag{A.13}$$

Also for the Green's function derivatives the same considerations as those for the Green's function hold. In particular, such functions are comprised of a singular part and a finite part depending on material properties.

Once the general anisotropic Green's function and its derivatives have been introduced, the DBEM kernels are given by

$$U_{ij} = G_{ij} \tag{A.14}$$

$$T_{ij} = C_{kjmn}\frac{\partial G_{mi}}{\partial x_n} n_k \tag{A.15}$$

$$U_{ijk} = -C_{ijpq}\left(\frac{\partial G_{pk}}{\partial x_q} + \frac{\partial G_{qk}}{\partial x_p}\right) \tag{A.16}$$

$$T_{ijk} = -C_{ijpq} C_{lkmn}\left(\frac{\partial^2 G_{mp}}{\partial x_q \partial x_n} + \frac{\partial^2 G_{mq}}{\partial x_p \partial x_n}\right) n_l \tag{A.17}$$

Appendix B. Numerical Scheme

The discrete boundary element model is built starting from the equations (5.1–5.3) and sub-dividing the external boundary and the crack surfaces into a set of boundary elements over which the displacements and the tractions, as well as the geometry, are expressed by means of suitable shape functions and nodal values.[53]

Care must be taken in the choice of suitable boundary elements, in order to fulfil the conditions for the existence of the singular integrals. In particular, the existence of Cauchy and Hadamard principal values requires Hölder continuity of the displacements and their derivatives at the collocation points. Such restrictions can be satisfied through the use of special boundary elements. In this work, the same modelling strategy as that adopted by Mi and Aliabadi[45,46] is used. The continuity of the displacement derivatives, which is the stronger constraint required for the existence of the integrals in the traction equation, is guaranteed by using *discontinuous* eight-node quadratic elements for the modelling of *both* the crack surfaces. The boundary, on which only the displacement equation is collocated, is modelled by using *continuous* eight-node quadratic elements. *Semi-discontinuous* eight-noded elements are used to model those portions of the external boundary intersecting the crack surfaces when the crack reaches the external boundary. Further information on slightly different discretization procedures can be found in the works of Cisilino and Aliabadi.[70,71]

The isotropic kernels are integrated following the procedures used in the aforementioned references. For anisotropic kernels, the integrals are computed by using standard Gauss quadrature for regular elements and accounting for suitable integration schemes to improve the accuracy of nearly singular integrals.[53] Strongly and hypersingular kernels must be computed as finite part integrals (Cauchy and Hadamard principal values). In this work, a technique based on polar coordinate transformations[60] and Kutt's numerical quadrature[72–74] is employed to compute both strongly singular and hypersingular integrals.

The computation of the integrals involved in the BIE requires the evaluation of the fundamental solution kernels at the integration points. The form of the general anisotropic Green's function and its derivatives suggests an efficient method to speed up their computation.[60,63] The modulation functions pertaining to the Green's function and its derivatives are regular and can be easily computed through standard schemes. They can be calculated in advance for different directions of the unit vector **b** and the results can be stored in a database. When a particular couple of source and observed points are considered, the direction of the corresponding unit vector **b** is determined and the related value of the modulation functions can be computed by interpolation from the stored database. The Green's function and its derivatives are then given by the product of the modulation functions by the inverse powers of the source/observed point distance.

References

1. L. C. Wrobel, *The Boundary Element Method, vol. 1: Applications in Thermo-Fluids and Acoustics* (John Wiley & Sons, Ltd Chichester, 2002).
2. M. H. Aliabadi, *The Boundary Element Method, vol. 2: Applications in Solids and Structures* (John Wiley & Sons, Ltd Chichester, 2002).
3. J. M. Crotty, A block equation solver for large unsymmetric matrices arising in the boundary integral equation method, *International Journal for Numerical Methods in Engineering* **18**, 997–1017, (1982).
4. R. H. Rigby and M. H. Aliabadi, Out-of-core solver for large, multi-zone boundary element matrices, *International Journal for Numerical Methods in Engineering* **38**, 1507–1533, (1995).
5. J. H. Kane, B. L. Kashava Kumar, and S. Saigal, An arbitrary condensing, noncondensing solution strategy for large scale, multi-zone boundary element analysis, *Computer Methods in Applied Mechanics and Engineering* **79**, 219–244, (1990).
6. W. J. Mansur, F. C. Araujo, and E. B. Malaghini, Solution of BEM systems of equations via iterative techniques, *International Journal for Numerical Methods in Engineering* **33**, 1823–1841, (1992).
7. H. Rokhlin, Rapid solution of integral equations of classical potential theory, *Journal of Computational Physics* **60**, 187–207, (1985).
8. J. Barnes and P. Hut, A hierarchical $O(N \ln N)$ force-calculation algorithm, *Nature* **324**, 446–449, (1986).
9. L. Greengard and V. Rokhlin, A fast algorithm for particle simulations, *Journal of Computational Physics* **73** (2), 325–348, (1987).
10. N. Nishimura, K. I. Yoshida, and S. Kobayashi, A fast multipole boundary integral equation method for crack problems in 3D, *Engineering Analysis with Boundary Elements* **23**, 97–105, (1999).
11. Y. Saad and M. H. Schultz, GMRES: A generalized minimal residual algorithm for solving nonsymmetric linear systems, *SIAM J. Sci. Stat. Comput.* **7** (3), 856–869, (1986).
12. Y. Fu, K. J. Klimkowski, G. J. Rodin, E. Berger, J. C. Browne, J. K. Singer, R. A. Van de Geijn, and K. S. Vemaganti, A fast solution method for three-dimensional many-particle problems of linear elasticity, *International Journal for Numerical Methods in Engineering* **42**, 1215–1229, (1998).
13. V. Popov and H. Power, An $O(N)$ Taylor series multipole boundary element method for three-dimensional elasticity problems, *Engineering Analysis with Boundary Elements* **25**, 7–18, (2001).

14. W. Hackbusch and Z. P. Nowak, On the fast matrix multiplication in the boundary element method by panel clustering, *Numerische Mathematik* **73**, 207–243, (1989).
15. E. E. Tyrtyshnikov, Mosaic-skeleton approximations, *Calcolo* **33**, 47–57, (1996).
16. S. A. Goreinov, E. E. Tyrtyshnikov, and N. L. Zamarashkin, A theory of pseudoskeleton approximations, *Linear Algebra and its Applications* **261**, 1–21, (1997).
17. E. E. Tyrtyshnikov, Incomplete cross approximation in the Mosaic-Skeleton Method, *Computing* **64**, 367–380, (2000).
18. M. Bebendorf, Approximation of boundary element matrices, *Numerische Mathematik* **86**, 565–589, (2000).
19. M. Bebendorf and S. Rjasanow, Adaptive low-rank approximation of collocation matrices, *Computing* **70**, 1–24, (2003).
20. W. Hackbusch, A sparse matrix arithmetic based on H-matrices. Part I: Introduction to H-matrices, *Computing* **62**, 89–108, (1999).
21. W. Hackbusch and B. N. Khoromskij, A sparse H-matrix arithmetic. Part II: Application to multidimensional problems, *Computing* **64**, 21–47, (2000).
22. L. Grasedyck, Adaptive recompression of H-matrices for BEM, *Computing* **74**, 205–223, (2005).
23. M. Bebendorf, Hierarchical *LU* decomposition-based preconditioners for BEM, *Computing* **74**, 225–247, (2005).
24. L. Grasedyck and W. Hackbusch, Construction and arithmetics of H-matrices, *Computing* **70**, 295–334, (2003).
25. S. Börm, L. Grasedyck, and W. Hackbusch, Introduction to hierarchical matrices with applications, *Engineering Analysis with Boundary Elements* **27**, 405–422, (2003).
26. R. L. Mullen and J. J. Rencis, Iterative methods for solving boundary element equations, *Computers & Structures* **25** (5), 713–723, (1987).
27. F. P. Valente and H. L. G. Pina, Iterative solvers for BEM algebraic systems of equations, *Engineering Analysis with Boundary Elements* **22**, 117–124, (1998).
28. L. P. S. Barra, A. L. G. A. Coutinho, W. J. Mansur, and J. F. C. Telles, Iterative solution of BEM equations by GMRES algorithm, *Computers & Structures* **44** (6), 1249–1253, (1992).
29. K. Guru Prasad, J. H. Kane, D. E. Keyes, and C. Balakrishna, Preconditioned Krylov solvers for BEA, *International Journal for Numerical Methods in Engineering* **37**, 1651–1672, (1994).
30. T. J. Urekew and J. J. Rencis, The importance of diagonal dominance in the iterative solution of equations generated from the boundary element method, *International Journal for Numerical Methods in Engineering* **36**, 3509–3527, (1993).
31. M. Merkel, V. Bulgakov, R. Bialecki, and G. Kuhn, Iterative solution of large-scale 3D-BEM industrial problems, *Engineering Analysis with Boundary Elements* **22**, 183–197, (1998).

32. C. Y. Leung and S. P. Walker, Iterative solution of large three-dimensional BEM elastostatic analyses using the GMRES technique, *International Journal for Numerical Methods in Engineering* **40**, 2227–2236, (1997).
33. L. P. S. Barra, A. L. G. A. Coutinho, J. F. C. Telles, and W. J. Mansur, Multi-level hierarchical preconditioners for boundary element systems, *Engineering Analysis with Boundary Elements* **12**, 103–109, (1993).
34. H. Wang, Z. Yao, and P. Wang, On the preconditioners for fast multipole boundary element methods for 2D multi-domain elastostatics, *Engineering Analysis with Boundary Elements* **29**, 673–688, (2005).
35. M. Benzi, Preconditioning techniques for large linear systems: A survey, *Journal of Computational Physics* **182**, 418–477, (2002).
36. Y. Saad and H. A. van der Vorst, Iterative solution of linear systems in the 20th century, *Journal of Computational and Applied Mathematics* **123**, 1–33, (2000).
37. S. Kurz, O. Rain, and S. Rjasanow, The adaptive cross approximation technique for the 3-D boundary element method, *IEEE Transactions on Magnetics* **38** (2), 421–424, (2002).
38. K. Zhao, M. N. Vouvakis, and J. F. Lee, The adaptive cross approximation algorithm for accelerated method of moments computation of EMC problems, *IEEE Transaction on Electromagnetic Compatibility* **47**, 763–773, (2005).
39. J. Ostrowski, K. I. Andjelíc, M. Bebendorf, B. Crangânu-Creţu, and J. Smajić, Fast BEM-solution of Laplace problems with H-matrices and ACA, *IEEE Transaction on Magnetics* **42** (4), 627–630, (2006).
40. M. Bebendorf and R. Grzhibovskis, Accelerating Galerkin BEM for linear elasticity using adaptive cross approximation, *Mathematical Methods in the Applied Sciences* **29**, 1721–1747, (2006).
41. I. Benedetti, M. H. Aliabadi, and G. Daví, A fast 3D dual boundary element method based on hierarchical matrices, *International Journal of Solids and Structures* **45** (7–8), 2355–2376, (2008).
42. I. Benedetti, A. Milazzo, and M. H. Aliabadi, A fast 3D BEM for anisotropic elasticity based on hierarchical matrices. In: M.H. Aliabadi and R. Abascal (eds.), *Advances in Boundary Element Techniques IX, Proceedings of the International Conference on Boundary Element Techniques (BETEQ IX)*, EC, Ltd, UK, pp. 433–438, (July, 2008).
43. A. Portela, M. H. Aliabadi, and D. P. Rooke, The dual boundary element method: effective implementation for crack problems, *International Journal for Numerical Methods in Engineering* **33**, 1269–1287, (1992).
44. A. Portela, M. H. Aliabadi, and D. P. Rooke, Dual boundary element incremental analysis of crack propagation, *Computers & Structures* **46** (2), 237–247, (1993).
45. Y. Mi and M. H. Aliabadi, Dual boundary element method for three-dimensional fracture mechanics analysis, *Engineering Analysis with Boundary Elements* **10**, 161–171, (1992).
46. Y. Mi and M. H. Aliabadi, Three-dimensional crack growth simulation using BEM, *Computers & Structures* **52** (5), 871–878, (1994).

47. V. Sladek and J. Sladek, Three dimensional crack analysis for anisotropic body, *Applied Mathematical Modelling* **6**, 374–382, (1982).
48. A. Le Van and J. Royer, Boundary formulation for three-dimensional anisotropic crack problems, *Applied Mathematical Modelling* **20**, 662–674, (1996).
49. E. Pan and F. G. Yuan, Boundary element analysis of three-dimensional cracks in anisotropic solids, *International Journal for Numerical Methods in Engineering* **48**, 211–237, (2000).
50. M. P. Ariza and J. Dominguez, Boundary element formulation for 3D transversly isotropic cracked bodies, *International Journal for Numerical Methods in Engineering* **60**, 719–753, (2004).
51. Z. Q. Yue, H. T. Xiao, and E. Pan, Stress intensity factors of square crack inclined to interface of transversly isotropic bimaterial, *Engineering Analysis with Boundary Elements* **31**, 50–65, (2007).
52. M. H. Aliabadi, Boundary element formulations in fracture mechanics, *Applied Mechanics Reviews* **50** (2), 83–96, (1997).
53. M. H. Aliabadi, A new generation of boundary element methods in fracture mechanics, *International Journal of Fracture* **86**, 91–125, (1997).
54. M. Bebendorf. *Effiziente numerische Losung von Randintegralgleichungen unter Verwendung von Niedrigrang-Matrizen.* PhD thesis, Universitat Saarbrucken, (2001).
55. W. Hackbusch, B. N. Khoromskij, and R. Kriemann, Hierarchical matrices based on weak admissibility criterion, *Computing* **73**, 207–243, (2004).
56. K. Giebermann, Multilevel approximation of boundary integral operators, *Computing* **67**, 183–207, (2001).
57. M. Bebendorf, Approximate inverse preconditioning of FE systems for elliptic operators with non-smooth coefficients, *SIAM Journal on Matrix Analysis and Applications* **27** (4), 909–929, (2006).
58. M. Bebendorf and R. Kriemann, Fast parallel solution of boundary integral equations and related problems, *Computing and Visualization in Science* **8**, 121–135, (2005).
59. I. Benedetti, A. Milazzo, and M. H. Aliabadi, A fast dual boundary element method for 3D anisotropic crack problems, *International Journal for Numerical Methods in Engineering.* (2009). DOI: 10.1002/nme.2666.
60. N. A. Schlar, *Anisotropic Analysis Using Boundary Elements.* (Computational Mechanics Publications, Southampton UK, 1994).
61. Y. C. Pan and T. W. Chou, Point force solution for an infinite transversely isotropic solid, *Journal of Applied Mechanics* **43**, 608–612, (1976).
62. S. M. Vogel and F. J. Rizzo, An integral equation formulation of threedimensional anisotropic elastostatic boundary value problems, *Journal of Elasticity* **3**, 203–216, (1973).
63. R. B. Wilson and T. A. Cruse, Efficient implementation of anisotropic three dimensional boundary-integral equation stress analysis, *International Journal for Numerical Methods in Engineering* **12**, 1383–1397, (1978).
64. C. Y. Wang and M. Denda, 3D BEM for general anisotropic elasticity, *International Journal of Solids & Structures* **44**, 7073–7091, (2007).

65. T. Chen and F. Z. Lin, Boundary integral formulations for three-dimensional anisotropic solids, *Computational Mechanics* **15**, 485–496, (1995).
66. T. C. T. Ting and V. G. Lee, The three-dimensional elastostatic Green's function for general anisotropic linear elastic solids, *Quarterly Journal of Mechanics and Applied Mathematics* **50**, 407–426, (1997).
67. G. Nakamura and K. Tanuma, A formula for the fundamental solution of anisotropic elasticity, *Quarterly Journal of Mechanics and Applied Mathematics* **50**, 179–194, (1997).
68. M. A. Sales and L. J. Gray, Evaluation of the anisotropic Green's function and its derivatives, *Computers & Structures* **69**, 247–254, (1998).
69. F. Tonon, E. Pan, and B. Amadei, Green's functions and boundary element method formulation for 3D anisotropic media, *Computers & Structures* **79**, 469–482, (2001).
70. A. P. Cisilino and M. H. Aliabadi, Three-dimensional BEM analysis for fatigue crack growth in welded components, *International Journal of Pressure Vessels & Piping* **70**, 135–144, (1997).
71. A. P. Cisilino and M. H. Aliabadi, Dual boundary element assessment of three-dimensional fatigue crack growth, *Engineering Analysis with Boundary Elements* **28**, 1157–1173, (2004).
72. H. R. Kutt, The numerical evaluation of principal value integrals by finite-part integration, *Numerische Mathematik* **24** 205–210, (1975).
73. H. R. Kutt. Quadrature formulae for finite-part integrals. Technical Report WSK 178, National Research Institute for Mathematical Sciences, CSIR, Pretoria, (1975).
74. H. R. Kutt, On the numerical evaluation of finite-part integrals involving an algebraic singularity. Technical Report WSK 179, National Research Institute for Mathematical Sciences, CSIR, Pretoria, (1975).

Chapter 6

MODELLING OF PLATES AND SHALLOW SHELLS BY MESHLESS LOCAL INTEGRAL EQUATION METHOD

J. Sladek and V. Sladek
Institute of Construction and Architecture,
Slovak Academy of Sciences, 84503 Bratislava, Slovakia
jan.sladek@savba.sk
Vladimir.sladek@savba.sk

P.H. Wen
School of Engineering and Materials Science
Queen Mary, University of London
Mile End, London E1 4NS, U.K.
p.h.wen@qmul.ac.uk

Ch. Zhang
Department of Civil Engineering, University of Siegen
Paul-Bonatz-Str. 9-11, D-57076 Siegen, Germany
c.zhang@uni-siegen.de

This chapter presents a meshless method based on the local Petrov-Galerkin weak-form for static and dynamic analyses of orthotropic thick plates and shallow shells. The analyzed planar domain of the considered plate/shell is divided into circular sub-domains. The nodal points are randomly distributed over the analyzed domain. Each node is the center of a circular sub-domain surrounding this node. For dynamic problems, the Laplace-transform technique is applied. A unit test function is used in the local weak-form of the governing equations for the transformed quantities. Applying the Gauss divergence theorem to the weak-form, local boundary-domain integral equations are derived. A meshless approximation based on the Moving Least-Squares (MLS) method is utilized. An inverse Laplace-transform method is employed to obtain time-dependent solutions.

6.1. Introduction

In recent years the demand for construction of huge and lightweight plate and shell structures has been increasing. To minimize the weight of spatial structures a layered profile of the plate and shell is utilized frequently. In such a case a delaminating of individual layers may occur due to a jump change of the material properties and subsequently physical fields too. To remove this phenomenon the functionally graded materials (FGMs) can be utilized.[1,2] FGMs are multi-phase materials with a pre-determined property profile, where the phase volume fractions are varying gradually in space. This results in continuously nonhomogenous material properties at the macroscopic structural scale. Often, these spatial gradients in the material behavior render FGMs as superior to conventional composites because of their continuously graded structures and properties. FGMs may exhibit isotropic or anisotropic material properties, depending on the processing technique and the practical engineering requirements. Due to the high mathematical complexity of the boundary or initial-boundary value problems, analytical approaches for FGMs are restricted to simple geometry and boundary conditions. It is well known that the classical thin plate theory of Kirchhoff gives rise to certain non-physical simplifications mainly related to the omission of the shear deformation and the rotary inertia, which become more significant for plates with increasing their thickness. The choice of an appropriate mathematical model together with a consistent computational method is important for such a kind of structures. Previous research results showed that the transverse shear effects are more significant for orthotropic plates and shells than for isotropic ones due to a high ratio of in-plane Young's modulus to transverse shear modulus.[3,4] The effects of shear deformation and rotary inertia are taken into account in the Reissner–Mindlin plate bending theory.[5,6] The Reissner–Mindlin theory reduces the original three-dimensional (3-D) thick plate and shell problem to a two-dimensional (2-D) problem. Governing equations for thick orthotropic plates and shells are quite complicated.

Most significant advances in spatial structures have been made using the finite element method (FEM). Elimination of shear locking in thin walled structures by FEM is difficult and techniques developed are less accurate. It is well known that standard displacement-based type shell element is too stiff and thus suffers from locking phenomena. Locking problem arises due to inconsistencies in discrete representation of the transverse shear energy and membrane energy.[7] Much effort has been devoted to deriving shell elements which account for the out-of-plane shear deformation in thick shells that are free from locking in the thin shell limit. In view of the increasing reliability on the use of computer modelling to replace or reduce experimental works in the context of a virtual testing, it is therefore imperative that alternative computer-based methods are developed to allow for independent verification of the finite element solutions. Recently, Arciniega and Reddy[8] consistently derived shell model using absolute tensor notation for a nonlinear analysis of shell structures.

The boundary element method (BEM) has emerged as an alternative numerical method to solve plate and shell problems. The first application of the boundary integral equation method to Reissner's plate model was given by Van der Ween.[9] Dynamic analysis of elastic Reissner–Mindlin plates has been performed by the direct BEM in the frequency domain.[10] All previous BEM applications are dealing with isotropic Reissner–Mindlin plates. Wang and Huang[2] were the first who applied BEM to orthotropic thick plates. A review on early applications of BEM to shells is given by Beskos.[11] The first application of BEM to shells is given by Newton and Tottenham,[12] where they presented a method based on the decomposition of the fourth-order governing equation into a set of the second-order ones. Lu and Huang[13] derived a direct BEM formulation for shallow shells involving shear deformation. For elastodynamic shell problems it is appropriate to use the weighted residual method with the static fundamental solution as a test function.[14,15] Dirgantara and Aliabadi[16] applied the domain-boundary element method for shear deformable shells under a static load. They used a test function corresponding to the thick plate bending problem. Lin and Long[17] used this method for geometrically non-linear analysis of shallow shells. All previous BEM applications are dealing with isotropic shallow shells. Only Wang and Schweizerhof[4,18] applied a boundary integral equation method for moderately thick laminated orthotropic shallow shells. They used the static fundamental solution corresponding to a shear deformable orthotropic shell and applied it for free vibration analysis of thick shallow shells. The fundamental solution for a thick orthotropic shell under a dynamic load is not available according to the best of the author's knowledge.

Meshless approaches for problems of continuum mechanics have attracted much attention during the past decade.[19,20] In spite of the great success of the FEM and the BEM as accurate and effective numerical tools for the solution of boundary value problems with complex domains, there is still a growing interest in developing new advanced numerical methods. Elimination of shear locking in thin walled structures by standard FEM is difficult and techniques developed are not accurate. However, the FEM based on the mixed approximations with the reduced integration performs quite well. Meshless methods with continuous approximation of stresses are more convenient for such kinds of structures.[21] In recent years, meshfree or meshless formulations are becoming to be popular due to their high adaptivity and low costs to prepare input data for numerical analyses. Many of meshless methods are derived from a weak-form formulation on global domain or a set of local sub-domains. In the global formulation background cells are required for the integration of the weak-form. It should be noticed that integration is performed only on those background cells with a nonzero shape function. In methods based on local weak-form formulation no cells are required. The first application of a meshless method to plate/shell problems was given by Krysl and Belytschko,[22,23] where they applied the element-free Galerkin method. The moving least-square (MLS) approximation yields C^1 continuity which satisfies the Kirchhoff hypotheses. The continuity of the MLS approximation is given by the minimum between the continuity of the basis functions and that of the weight function. So continuity can be tuned to a desired value. Their results showed excellent convergence, however, their formulation is not applicable to shear deformable plate/shell problems. Recently, Noguchi *et al.*[24] used a mapping technique to transform the curved surface into flat two-dimensional space. Then, the element-free Galerkin method can be applied also to thick plates or shells including the shear deformation effects. The reproducing kernel particle method (RKPM)[25] has been successfully applied for large deformations of thin isotropic shells.[26]

The meshless local Petrov-Galerkin (MLPG) method[20] is a fundamental base for the derivation of many meshless formulations, since the trial and test functions can be chosen from different functional spaces. The method has been successfully applied also to plate problems.[27–30] In the present chapter a meshless method based on the local Petrov-Galerkin weak-form is applied to static and dynamic problems for orthotropic thick plates and shallow shells. The nodal points are randomly distributed over the plane domain of the considered plate/shell. Each node is the center of a circle surrounding this node. The Laplace-transform technique is applied to the set of governing differential equations for elastodynamic Reissner bending theory. A unit test function is used in the local weak-form of the governing equations for transformed fields. Applying the Gauss divergence theorem to the weak-form, the local boundary-domain integral equations are derived. The numerical integration of the domain integrals arising from the inertial term and the initial values on a simple domain does not give rise to difficulties if the meshless approximation based on the Moving Least-Squares (MLS) method is utilized. The quasi-static boundary value problems must be solved for several values of the Laplace-transform parameter selected for each considered time instant. The Stehfest's inversion method[31] is employed to obtain the time-dependent solution.

6.2. The MLS Approximation

In general, a meshless method uses a local interpolation to represent the trial function with the values (or the fictitious values) of the unknown variable at some randomly spread nodes. The moving least squares approximation may be considered as one of such schemes, and is used in the current work. Consider a sub-domain Ω_x, the neighborhood of a point x and denoted as the domain of definition of the MLS approximation for the trial function at $\boldsymbol{x}$, which is located in the problem domain Ω. To approximate the distribution of the function u in Ω_x, over a number of randomly located nodes $\{\mathbf{x}^a\}$, $a = 1, 2, \ldots, n$, the MLS approximant $u^h(\mathbf{x})$ of u, $\forall \mathbf{x} \in \Omega_x$, can be defined by

$$u^h(\mathbf{x}) = \mathbf{p}^T(\mathbf{x})\mathbf{a}(\mathbf{x}) \quad \forall \mathbf{x} \in \Omega_x, \tag{6.1}$$

where $\mathbf{p}^T(\mathbf{x}) = [p^1(\mathbf{x}), p^2(\mathbf{x}), \ldots, p^m(\mathbf{x})]$ is a complete monomial basis of order m; and $\mathbf{a}(\mathbf{x})$ is a vector containing coefficients $a^j(\mathbf{x})$, $j = 1, 2, \ldots, m$ which are functions of the space co-ordinates $\mathbf{x} = [x_1, x_2, x_3]^T$. For example, for a 2-D problem

$$\mathbf{p}^T(\mathbf{x}) = [1, x_1, x_2], \quad \text{linear basis } m = 3 \tag{6.2a}$$

$$\mathbf{p}^T(\mathbf{x}) = [1, x_1, x_2, (x_1)^2, x_1 x_2, (x_2)^2], \quad \text{quadratic basis } m = 6 \tag{6.2b}$$

The coefficient vector $\mathbf{a}(\mathbf{x})$ is determined by minimizing a weighted discrete L_2 norm, defined as

$$J(\mathbf{x}) = \sum_{a=1}^{n} \nu^a(\mathbf{x})[p^T(\mathbf{x}^a)a(\mathbf{x}) - \hat{u}^a]^2, \tag{6.3}$$

where $v^a(\mathbf{x})$ is the weight function associated with the node a, with $v^a(\mathbf{x}) > 0$. Recall that n is the number of nodes in Ω_x for which the weight functions $v^a(\mathbf{x}) > 0$ and $\hat{u}^a$ are the fictitious nodal values, and not the nodal values of the unknown trial function $u^h(\mathbf{x})$ in general (see Fig. 6.1 for simple one dimensional case for the distinction between u_i and $\hat{u}_i$). The

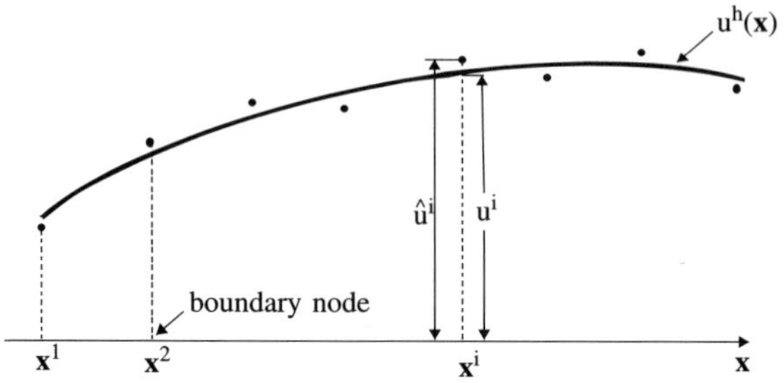

Fig. 6.1. The moving least-square approximation.

stationarity of J in Eq. (6.3) with respect to $\mathbf{a}(\mathbf{x})$ leads to the following linear relation between $\mathbf{a}(\mathbf{x})$ and $\hat{\mathbf{u}}$

$$\mathbf{A}(\mathbf{x})\mathbf{a}(\mathbf{x}) = \mathbf{B}(\mathbf{x})\hat{\mathbf{u}}, \tag{6.4}$$

where

$$\mathbf{A}(\mathbf{x}) = \sum_{a=1}^{n} w^a(\mathbf{x})\mathbf{p}(\mathbf{x}^a)\mathbf{p}^T(\mathbf{x}^a)$$
$$\mathbf{B}(\mathbf{x}) = [v^1(\mathbf{x})\mathbf{p}(\mathbf{x}^1), v^2(\mathbf{x})\mathbf{p}(\mathbf{x}^2), \ldots, v^n(\mathbf{x})\mathbf{p}(\mathbf{x}^n)]. \tag{6.5}$$

The MLS approximation is well defined only when the matrix $\mathbf{A}$ in Eq. (6.4) is non-singular. A necessary condition to be satisfied that requirement is that at least m weight functions are non-zero (i.e. $n \geq m$) for each sample point $\mathbf{x} \in \Omega$ and that the nodes in Ω_x will not be arranged in a special pattern such as on a straight line.

Solving for $\mathbf{a}(\mathbf{x})$ from Eq. (6.4) and substituting it into Eq. (6.1) give a relation

$$\mathbf{u}^h(\mathbf{x}) = \mathbf{\Phi}^T(\mathbf{x}) \cdot \hat{\mathbf{u}} = \sum_{a=1}^{n} \phi^a(\mathbf{x})\hat{u}^a; \quad \hat{u}^a \neq u(\mathbf{x}^a) \wedge \hat{u}^a \neq u^h(\mathbf{x}^a), \tag{6.6}$$

where

$$\mathbf{\Phi}^T(\mathbf{x}) = \mathbf{p}^T(\mathbf{x})\mathbf{A}^{-1}(\mathbf{x})\mathbf{B}(\mathbf{x}). \tag{6.7}$$

$\phi^a(\mathbf{x})$ is usually called the shape function of the MLS approximation corresponding to the nodal point $\mathbf{x}^a$. From Eqs. (6.5) and (6.7), it may be seen that $\phi^a(\mathbf{x}) = 0$ when $v^a(\mathbf{x}) = 0$. In practical applications, $v^a(\mathbf{x})$ is generally chosen such that it is non-zero over the support of nodal points $\mathbf{x}^a$. The support of the nodal point $\mathbf{x}^a$ is usually taken to be circle of radius

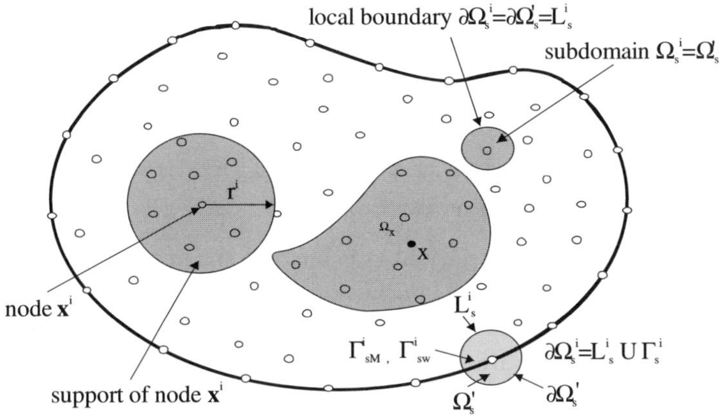

Fig. 6.2. Local boundaries for weak formulation, the domain Ω_x for MLS approximation, and support area of weight function around node $\mathbf{x}^i$.

r^a centered at $\mathbf{x}^a$ (see Fig. 6.2). This radius is an important parameter of the MLS approximation because it determines the range of interaction (coupling) between degrees of freedom defined at nodes.

Both the Gaussian and spline weight functions with compact supports can be considered in a numerical analysis. The Gaussian weight function can be written as

$$v^a(\mathbf{x}) = \begin{cases} \exp[-(d^a/c^a)^2] - \exp[-(r^a/c^a)^2] & 0 \leq d^a \leq r^a \\ 0 & d^a \geq r^a \end{cases}, \tag{6.8}$$

where $d^a = |\mathbf{x} - \mathbf{x}^a|$; c^a is a constant controlling the shape of the weight function v^a and r^a is the size of support. The size of support r^a should be large enough to have a sufficient number of nodes covered in the domain of definition to ensure the regularity of matrix $\mathbf{A}$.

A 4th-order spline-type weight function

$$v^a(\mathbf{x}) = \begin{cases} 1 - 6\left(\frac{d^a}{r^a}\right)^2 + 8\left(\frac{d^a}{r^a}\right)^3 - 3\left(\frac{d^a}{r^a}\right)^4, & 0 \leq d^a \leq r^a, \\ 0, & d^a \geq r^a, \end{cases} \tag{6.9}$$

is more convenient for modelling since the C^1-continuity of the weight function is ensured over the entire domain. Then, the continuity condition of the bending moments, the shear forces and the normal forces is satisfied.

The partial derivatives of the MLS shape functions are obtained as[19]

$$\phi^a_{,k} = \sum_{j=1}^{m} [p^j_{,k}(\mathbf{A}^{-1}\mathbf{B})^{ja} + p^j(\mathbf{A}^{-1}\mathbf{B}_{,k} + \mathbf{A}^{-1}_{,k}\mathbf{B})^{ja}], \tag{6.10}$$

wherein $\mathbf{A}_{,k}^{-1} = (\mathbf{A}^{-1})_{,k}$, represents the derivative of the inverse of $\mathbf{A}$ with respect to x_k, given by

$$\mathbf{A}_{,k}^{-1} = -\mathbf{A}^{-1}\mathbf{A}_{,k}\mathbf{A}^{-1}.$$

6.3. Analyses of Orthotropic FGM Plates Under Mechanical and Thermal Load

Consider an elastic orthotropic plate of constant thickness h, with the mean surface occupying the domain Ω in the plane (x_1, x_2). The Reissner–Mindlin plate bending theory[5,6] is used to describe the plate deformation. The transverse shear strains are represented as constant throughout the plate thickness and some correction coefficients are required for computation of transverse shear forces in that theory. Then, the spatial displacement field, expressed in terms of displacement components u_1, u_2, and u_3, has the following form[32]

$$u_1(\mathbf{x}, t) = x_3 w_1(\mathbf{x}, t), \quad u_2(\mathbf{x}, t) = x_3 w_2(\mathbf{x}, t), \quad u_3(\mathbf{x}, t) = w_3(\mathbf{x}, t), \tag{6.11}$$

where w_α and w_3 represent the rotations around the x_α-direction and the out-of-plane deflection, respectively (Fig. 6.3). The linear strains are given by

$$\begin{aligned}
\varepsilon_{11}(\mathbf{x}, t) &= x_3 w_{1,1}(\mathbf{x}, t), \quad \varepsilon_{22}(\mathbf{x}, t) = x_3 w_{2,2}(\mathbf{x}, t),\\
\varepsilon_{12}(\mathbf{x}, t) &= x_3 (w_{1,2}(\mathbf{x}, t) + w_{2,1}(\mathbf{x}, t))/2,\\
\varepsilon_{13}(\mathbf{x}, t) &= (w_1(\mathbf{x}, t) + w_{3,1}(\mathbf{x}, t))/2.\\
\varepsilon_{23}(\mathbf{x}, t) &= (w_2(\mathbf{x}, t) + w_{3,2}(\mathbf{x}, t))/2.
\end{aligned} \tag{6.12}$$

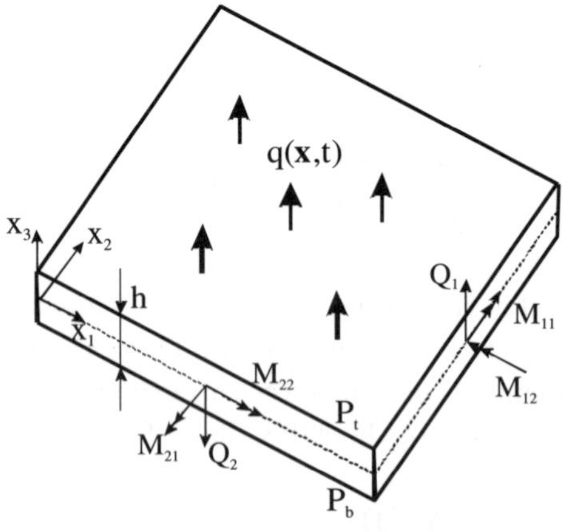

Fig. 6.3. Sign convention of bending moments and forces for FGM plate.

Both mechanical and thermal loads are analyzed simultaneously. For a pure mechanical load it is sufficient to state vanishing value of the temperature difference in the constitutive equation. Since uncoupled thermoelasticity is considered here, the thermal field is not influenced by deformations. Then, the temperature distribution can be analyzed separately. The computational method to solve heat conduction equations in plates is given at the end of this paragraph.

6.3.1. *Meshless local integral equations for a Reissner–Mindlin plate*

In the case of orthotropic materials, the relation between the stress σ_{ij} and the strain ε_{ij} when temperature θ changes are considered, is governed by the well known Duhamel-Neumann constitutive equations for the stress tensor

$$\sigma_{ij}(\mathbf{x}, t) = c_{ijkl}\varepsilon_{kl}(\mathbf{x}, t) - \gamma_{ij}\theta(\mathbf{x}, x_3, t), \tag{6.13}$$

where c_{ijkl} are the material stiffness coefficients. The stress-temperature modulus can be expressed through the stiffness coefficients and the coefficients of linear thermal expansion α_{kl}

$$\gamma_{ij} = c_{ijkl}\alpha_{kl}. \tag{6.14}$$

For plane problems the constitutive equation (6.13) is frequently written in terms of the second-order tensor of elastic constants.[33] The constitutive equation for orthotropic materials and plane stress problem has the following form

$$\begin{bmatrix} \sigma_{11} \\ \sigma_{22} \\ \sigma_{12} \\ \sigma_{13} \\ \sigma_{23} \end{bmatrix} = \mathbf{G}(\mathbf{x}) \begin{bmatrix} \varepsilon_{11} \\ \varepsilon_{22} \\ 2\varepsilon_{12} \\ 2\varepsilon_{13} \\ 2\varepsilon_{23} \end{bmatrix} - \begin{bmatrix} \gamma_{11} \\ \gamma_{22} \\ 0 \\ 0 \\ 0 \end{bmatrix} \theta(\mathbf{x}, x_3, t), \tag{6.15}$$

where

$$\mathbf{G}(\mathbf{x}) = \begin{bmatrix} E_1/e & E_1\nu_{21}/e & 0 & 0 & 0 \\ E_2\nu_{12}/e & E_2/e & 0 & 0 & 0 \\ 0 & 0 & G_{12} & 0 & 0 \\ 0 & 0 & 0 & G_{13} & 0 \\ 0 & 0 & 0 & 0 & G_{23} \end{bmatrix} \quad \text{with } e = 1 - \nu_{12}\nu_{21},$$

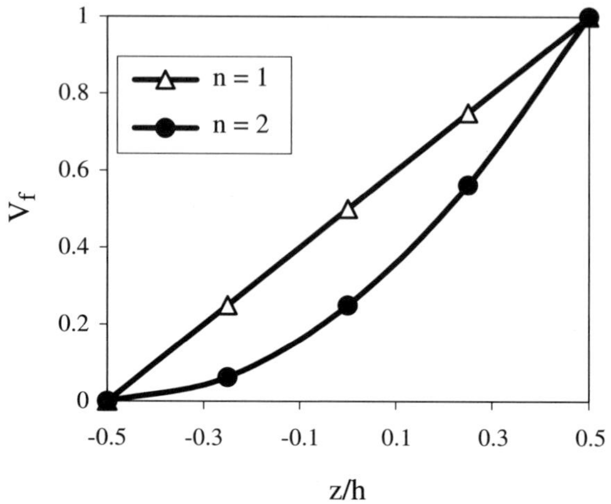

Fig. 6.4. Variation of volume fraction over the plate thickness, for linear and quadratic power-law index.

E_α are the Young's moduli refering to the axes x_α, $\alpha = 1, 2$, G_{12}, G_{13} and G_{23} are shear moduli, $\nu_{\alpha\beta}$ are Poisson's ratios.

Next, we assume that the material properties are graded along the plate thickness, and we represent the profile for volume fraction variation by

$$P(x_3) = P_b + (P_t - P_b)V \quad \text{with } V = \left(\frac{x_3}{h} + \frac{1}{2}\right)^n, \tag{6.16}$$

where P denotes a generic property like modulus, P_t and P_b denote the property of the top and bottom faces of the plate, respectively, and n is a parameter that dictates the material variation profile (Fig. 6.4). Poisson's ratios are assumed to be uniform.

The bending moments $M_{\alpha\beta}$ and the shear forces Q_α are defined as

$$\begin{bmatrix} M_{11} \\ M_{22} \\ M_{12} \end{bmatrix} = \int_{-h/2}^{h/2} \begin{bmatrix} \sigma_{11} \\ \sigma_{22} \\ \sigma_{12} \end{bmatrix} x_3 dx_3 \quad \text{and} \quad \begin{bmatrix} Q_1 \\ Q_2 \end{bmatrix} = \kappa \int_{-h/2}^{h/2} \begin{bmatrix} \sigma_{13} \\ \sigma_{23} \end{bmatrix} dx_3, \tag{6.17}$$

where $\kappa = 5/6$ in the Reissner plate theory.

Substituting Eqs. (6.15) and (6.12) into moment and force resultants (6.17) allows the expression of the bending moments $M_{\alpha\beta}$ and shear forces Q_α for α, $\beta = 1, 2$, in terms of rotations, lateral displacements of the orthotropic plate and temperature. In the case of considered continuous gradation of material properties through the plate thickness, one obtains

$$\begin{aligned} M_{\alpha\beta} &= D_{\alpha\beta}(w_{\alpha,\beta} + w_{\beta,\alpha}) + C_{\alpha\beta} w_{\gamma,\gamma} - H_{\alpha\beta} \\ Q_\alpha &= C_\alpha(w_\alpha + w_{3,\alpha}), \end{aligned} \tag{6.18}$$

where

$$H_{\alpha\beta} = \int_{-h/2}^{h/2} x_3 \gamma_{\alpha\beta} \theta(\mathbf{x}, x_3, t) dx_3.$$

In Eq. (6.18), repeated indices α, β do not imply summation, and the material parameters $D_{\alpha\beta}$ and $C_{\alpha\beta}$ are given as

$$\begin{aligned} D_{11} &= \frac{D_1}{2}(1 - \nu_{21}), \quad & D_{22} &= \frac{D_2}{2}(1 - \nu_{12}), \quad & D_{12} &= D_{21} = \frac{\bar{G}_{12} h^3}{12}, \\ C_{11} &= D_1 \nu_{21}, & C_{22} &= D_2 \nu_{12}, & C_{12} &= C_{21} = 0, \\ D_\alpha &= \frac{\bar{E}_\alpha h^3}{12e}, & D_1 \nu_{21} &= D_2 \nu_{12}, & C_\alpha &= \kappa h \bar{G}_{\alpha 3}, \end{aligned} \tag{6.19}$$

where

$$\bar{E}_\alpha \equiv \begin{cases} E_{\alpha t} = E_{\alpha b}, & n = 0 \\ (E_{\alpha b} + E_{\alpha t})/2, & n = 1 \\ (3E_{\alpha b} + 2E_{\alpha t})/5, & n = 2 \end{cases},$$

$$\bar{G}_{12} \equiv \begin{cases} G_{12t} = G_{12b}, & n = 0 \\ (G_{12b} + G_{12t})/2, & n = 1 \\ (3G_{12b} + 2G_{12t})/5, & n = 2 \end{cases},$$

$$\bar{G}_{\alpha 3} \equiv \begin{cases} G_{\alpha 3t} = G_{\alpha 3b}, & n = 0 \\ (G_{\alpha 3b} + G_{\alpha 3t})/2, & n = 1 \\ (2G_{\alpha 3b} + G_{\alpha 3t})/3, & n = 2 \end{cases},$$

with the same meaning of subscripts b and t as in Eq. (6.16).

For a general variation of material properties through the plate thickness

$$D_{11} = \int_{-h/2}^{h/2} x_3^2 E_1(x_3) \frac{1 - \nu_{21}}{e} dx_3, \quad D_{22} = \int_{-h/2}^{h/2} x_3^2 E_2(x_3) \frac{1 - \nu_{12}}{e} dx_3,$$

$$D_{12} = \int_{-h/2}^{h/2} x_3^2 G_{12}(x_3) dx_3, \qquad C_{11} = \int_{-h/2}^{h/2} x_3^2 E_1(x_3) \frac{\nu_{21}}{e} dx_3,$$

$$C_{22} = \int_{-h/2}^{h/2} x_3^2 E_2(x_3) \frac{\nu_{12}}{e} dx_3, \qquad C_\alpha = \kappa \int_{-h/2}^{h/2} G_{\alpha 3}(x_3) dx_3.$$

Using the Reissner's linear theory of thick plates,[34] the equations of motion may be written as

$$M_{\alpha\beta,\beta}(\mathbf{x},t) - Q_\alpha(\mathbf{x},t) = \frac{\rho h^3}{12}\ddot{w}_\alpha(\mathbf{x},t),$$
$$Q_{\alpha,\alpha}(\mathbf{x},t) + q_3(\mathbf{x},t) = \rho h\ddot{w}_3(\mathbf{x},t), \quad \mathbf{x} \in \Omega, \tag{6.20}$$

where ρ is the mass density, and throughout the paper the Greek indices vary from 1 to 2. The plate is subjected to a transverse dynamic load $q_3(\mathbf{x},t)$. The dots over a quantity indicate differentiations with respect to time t.

To eliminate the time variable t in the governing equations (6.20), the Laplace-transform is applied

$$L[f(\mathbf{x},t)] = \bar{f}(\mathbf{x},s) = \int_0^\infty f(\mathbf{x},t)e^{-st}dt.$$

Then, one obtains

$$\bar{M}_{\alpha\beta,\beta}(\mathbf{x},s) - \bar{Q}_\alpha(\mathbf{x},s) = \frac{\rho h^3}{12}s^2\bar{w}_\alpha(\mathbf{x},s) - \bar{R}_\alpha(\mathbf{x},s), \tag{6.21}$$
$$\bar{Q}_{\alpha,\alpha}(\mathbf{x},s) = \rho h s^2\bar{w}_3(\mathbf{x},s) - \bar{R}_3(\mathbf{x},s), \tag{6.22}$$

where s is the Laplace-transform parameter, while $\bar{R}_\alpha$ and $\bar{R}_3$ are given by

$$\bar{R}_\alpha(\mathbf{x},s) = \frac{\rho h^3}{12}[sw_\alpha(\mathbf{x}) + \dot{w}_\alpha(\mathbf{x})],$$
$$\bar{R}_3(\mathbf{x},s) = \bar{q}_3(\mathbf{x},s) + \rho h s w_3(\mathbf{x}) + \rho h\dot{w}_3(\mathbf{x}),$$

with $w_\alpha(\mathbf{x})$, $w_3(\mathbf{x})$, $\dot{w}_\alpha(\mathbf{x})$ and $\dot{w}_3(\mathbf{x})$ being the initial values and the initial velocities of the generalized displacement field.

Instead of writing the global weak-form for the above governing equations, the MLPG methods construct the weak-form over local sub-domains such as Ω_s, which is a small region taken for each node inside the global domain.[20] The local sub-domains overlap each other and cover the whole global domain Ω (Fig. 6.2). The local sub-domains could be of any geometrical shape and size. In the current paper, the local sub-domains are taken to be of circular shapes. The local weak-form of the governing equations (6.21) and (6.22) for $\mathbf{x}^i \in \Omega_s^i$ can be written as

$$\int_{\Omega_s^i}\left[\bar{M}_{\alpha\beta,\beta}(\mathbf{x},s) - \bar{Q}_\alpha(\mathbf{x},s) - \frac{\rho h^3}{12}s^2\bar{w}_\alpha(\mathbf{x},s) + \bar{R}_\alpha(\mathbf{x},s)\right]$$
$$\times\, w^*_{\alpha\gamma}(\mathbf{x})d\Omega = 0, \tag{6.23}$$

$$\int_{\Omega_s^i} [\bar{Q}_{\alpha,\alpha}(\mathbf{x},s) - \rho h s^2 \bar{w}_3(\mathbf{x},s) + \bar{R}_3(\mathbf{x},s)] w_3^*(\mathbf{x}) d\Omega = 0, \tag{6.24}$$

where $w_{\alpha\beta}^*(\mathbf{x})$ and $w^*(\mathbf{x})$ are weight or test functions.

Applying the Gauss divergence theorem to Eqs. (6.23) and (6.24), one obtains

$$\begin{aligned}
&\int_{\partial\Omega_s^i} \bar{M}_\alpha(\mathbf{x},s) w_{\alpha\gamma}^*(\mathbf{x}) d\Gamma - \int_{\Omega_s^i} \bar{M}_{\alpha\beta}(\mathbf{x},s) w_{\alpha\gamma,\beta}^*(\mathbf{x}) d\Omega \\
&\quad - \int_{\Omega_s^i} \bar{Q}_\alpha(\mathbf{x},s) w_{\alpha\gamma}^*(\mathbf{x}) d\Omega - \int_{\Omega_s^i} \frac{\rho h^3}{12} s^2 \bar{w}_\alpha(\mathbf{x},s) w_{\alpha\gamma}^*(\mathbf{x}) d\Omega \\
&\quad + \int_{\Omega_s^i} \bar{R}_\alpha(\mathbf{x},s) w_{\alpha\gamma}^*(\mathbf{x}) d\Omega = 0,
\end{aligned} \tag{6.25}$$

$$\begin{aligned}
&\int_{\partial\Omega_s^i} \bar{Q}_\alpha(\mathbf{x},s) n_\alpha(\mathbf{x}) w^*(\mathbf{x}) d\Gamma - \int_{\Omega_s^i} \bar{Q}_\alpha(\mathbf{x},s) w_{,\alpha}^*(\mathbf{x}) d\Omega \\
&\quad - \int_{\Omega_s^i} \rho h s^2 \bar{w}_3(\mathbf{x},s) w^*(\mathbf{x}) d\Omega + \int_{\Omega_s^i} \bar{R}_3(\mathbf{x},s) w^*(\mathbf{x}) d\Omega = 0,
\end{aligned} \tag{6.26}$$

where $\partial\Omega_s^i$ is the boundary of the local sub-domain and

$$\bar{M}_\alpha(\mathbf{x},s) = \bar{M}_{\alpha\beta}(\mathbf{x},s) n_\beta(\mathbf{x})$$

is the Laplace-transform of the normal bending moment and n_α is the unit outward normal vector to the boundary $\partial\Omega_s^i$. The local weak-forms (6.25) and (6.26) are the starting point for deriving local boundary integral equations on the basis of appropriate test functions. Unit step functions are chosen for the test functions $w_{\alpha\beta}^*(\mathbf{x})$ and $w^*(\mathbf{x})$ in each sub-domain

$$\begin{aligned}
w_{\alpha\gamma}^*(\mathbf{x}) &= \begin{cases} \delta_{\alpha\gamma} & \text{at} \quad \mathbf{x} \in (\Omega_s \cup \partial\Omega_s) \\ 0 & \text{at} \quad \mathbf{x} \notin (\Omega_s \cup \partial\Omega_s) \end{cases}, \\
w^*(\mathbf{x}) &= \begin{cases} 1 & \text{at} \quad \mathbf{x} \in (\Omega_s \cup \partial\Omega_s) \\ 0 & \text{at} \quad \mathbf{x} \notin (\Omega_s \cup \partial\Omega_s) \end{cases}.
\end{aligned} \tag{6.27}$$

Then, the local weak-forms (6.25) and (6.26) are transformed into the following local boundary integral equations

$$\begin{aligned}
&\int_{\partial\Omega_s^i} \bar{M}_\alpha(\mathbf{x},s) d\Gamma - \int_{\Omega_s^i} \bar{Q}_\alpha(\mathbf{x},s) d\Omega - \int_{\Omega_s^i} \frac{\rho h^3}{12} s^2 \bar{w}_\alpha(\mathbf{x},s) d\Omega \\
&\quad + \int_{\Omega_s^i} \bar{R}_\alpha(\mathbf{x},s) d\Omega = 0,
\end{aligned} \tag{6.28}$$

$$\int_{\partial\Omega_s^i} \bar{Q}_\alpha(\mathbf{x}, s) n_\alpha(\mathbf{x}) d\Gamma - \int_{\Omega_s^i} \rho h s^2 \bar{w}_3(\mathbf{x}, s) d\Omega + \int_{\Omega_s^i} \bar{R}_3(\mathbf{x}, s) d\Omega = 0. \tag{6.29}$$

In the above local integral equations, the trial functions $\bar{w}_\alpha(\mathbf{x}, s)$, related to rotations, and $\bar{w}_3(\mathbf{x}, s)$, related to deflections, are chosen as the moving least-squares (MLS) approximations over a number of nodes randomly spread within the domain of influence.

According to Sec. 2, one can write the approximation formula for the generalized displacements (two rotations and deflection)

$$\bar{\mathbf{w}}^h(\mathbf{x}, s) = \mathbf{\Phi}^T(\mathbf{x}) \cdot \hat{\mathbf{w}}(s) = \sum_{a=1}^{n} \phi^a(\mathbf{x}) \hat{\mathbf{w}}^a(s), \tag{6.30}$$

Substituting the approximation (6.30) into the definition of the normal bending (6.17), one obtains for $\bar{\mathbf{M}}(\mathbf{x}, s) = [\bar{M}_1(\mathbf{x}, s), \bar{M}_2(\mathbf{x}, s)]^T$

$$\begin{aligned} \bar{\mathbf{M}}(\mathbf{x}, s) &= \mathbf{N_1} \sum_{a=1}^{n} \mathbf{B}_1^a(\mathbf{x}) \mathbf{w}^{*a}(s) + \mathbf{N_2} \sum_{a=1}^{n} \mathbf{B}_2^a(\mathbf{x}) \mathbf{w}^{*a}(s) - \bar{\mathbf{H}}(\mathbf{x}, s) \\ &= \mathbf{N}_\alpha(\mathbf{x}) \sum_{a=1}^{n} \mathbf{B}_\alpha^a(\mathbf{x}) \mathbf{w}^{*a}(s) - \bar{\mathbf{H}}(\mathbf{x}, s), \end{aligned} \tag{6.31}$$

where the vector $\mathbf{w}^{*a}(s)$ is defined as a column vector $\mathbf{w}^{*a}(s) = [\hat{w}_1^a(s), \hat{w}_2^a(s)]^T$, the vector $\bar{\mathbf{H}}(\mathbf{x}, s) = [\bar{H}_{11} n_1, \bar{H}_{22} n_2]^T$, the matrices $\mathbf{N}_\alpha(\mathbf{x})$ are related to the normal vector $\mathbf{n}(\mathbf{x})$ on $\partial\Omega_s$ by

$$\mathbf{N}_1(\mathbf{x}) = \begin{bmatrix} n_1 & 0 & n_2 \\ 0 & n_2 & n_1 \end{bmatrix} \quad \text{and} \quad \mathbf{N}_2(\mathbf{x}) = \begin{bmatrix} C_{11} & 0 \\ 0 & C_{22} \end{bmatrix} \begin{bmatrix} n_1 & n_1 \\ n_2 & n_2 \end{bmatrix},$$

and the matrices $\mathbf{B}_\alpha^a$ are represented by the gradients of the shape functions as

$$\mathbf{B}_1^a(\mathbf{x}) = \begin{bmatrix} 2D_{11}\phi_{,1}^a & 0 \\ 0 & 2D_{22}\phi_{,2}^a \\ D_{12}\phi_{,2}^a & D_{12}\phi_{,1}^a \end{bmatrix}, \quad \mathbf{B}_2^a(\mathbf{x}) = \begin{bmatrix} \phi_{,1}^a & 0 \\ 0 & \phi_{,2}^a \end{bmatrix}.$$

The influence on the material gradation is incorporated in $C_{\alpha\beta}$ and $D_{\alpha\beta}$ defined in Eq. (1.19).

Similarly one can obtain the approximation for the shear forces

$$\bar{\mathbf{Q}}(\mathbf{x},s) = \mathbf{C}(\mathbf{x}) \sum_{a=1}^{n} [\phi^a(\mathbf{x})\mathbf{w}^{*a}(s) + \mathbf{F}^a(\mathbf{x})\hat{w}_3^a(s)], \tag{6.32}$$

where $\bar{\mathbf{Q}}(\mathbf{x},s) = [\bar{Q}_1(\mathbf{x},s), \bar{Q}_2(\mathbf{x},s)]^T$ and

$$\mathbf{C}(\mathbf{x}) = \begin{bmatrix} C_1(\mathbf{x}) & 0 \\ 0 & C_2(\mathbf{x}) \end{bmatrix}, \quad \mathbf{F}^a(\mathbf{x}) = \begin{bmatrix} \phi^a_{,1} \\ \phi^a_{,2} \end{bmatrix}.$$

Then, insertion of the MLS-discretized force fields (6.31) and (6.32) into the local boundary integral equations (6.28) and (6.29) yields the discretized local integral equations (LIEs)

$$\begin{aligned} &\sum_{a=1}^{n} \Bigg[\int_{L_s^i + \Gamma_{sw}^i} \mathbf{N}_\alpha(\mathbf{x})\mathbf{B}_\alpha^a(\mathbf{x})d\Gamma \\ &\quad - \int_{\Omega_s^i} \left(\mathbf{C}(\mathbf{x}) + \mathbf{E}\frac{\rho h^3(\mathbf{x})}{12}s^2 \right) \phi^a(\mathbf{x})d\Omega \Bigg] \mathbf{w}^{*a}(s) \\ &\quad - \sum_{a=1}^{n} \hat{w}_3^a(s) \int_{\Omega_s^i} \mathbf{C}(\mathbf{x})\mathbf{F}^a(\mathbf{x})d\Omega = \int_{L_s^i + \Gamma_{sM}^i} \bar{\mathbf{H}}(\mathbf{x},s)d\Gamma \\ &\quad - \int_{\Gamma_{sM}^i} \tilde{\bar{\mathbf{M}}}(\mathbf{x},s)d\Gamma - \int_{\Omega_s^i} \bar{\mathbf{R}}(\mathbf{x},s)d\Omega, \end{aligned} \tag{6.33}$$

$$\begin{aligned} &\sum_{a=1}^{n} \left(\int_{\partial\Omega_s^i} \mathbf{C}_n(\mathbf{x})\phi^a(\mathbf{x})d\Gamma \right) \mathbf{w}^{*a}(s) \\ &\quad + \sum_{a=1}^{n} \hat{w}_3^a(s) \left(\int_{\partial\Omega_s^i} \mathbf{C}_n(\mathbf{x})\mathbf{F}^a(\mathbf{x})d\Gamma - \int_{\Omega_s^i} \rho s^2 h(\mathbf{x})\phi^a(\mathbf{x})d\Omega \right) \\ &= - \int_{\Omega_s^i} \bar{R}_3(\mathbf{x},s)d\Omega, \end{aligned} \tag{6.34}$$

in which

$$\mathbf{E} = \begin{pmatrix} 1 & 0 \\ 0 & 1 \end{pmatrix}, \quad \mathbf{C}_n(\mathbf{x}) = (n_1, n_2) \begin{pmatrix} C_1 & 0 \\ 0 & C_2 \end{pmatrix} = (C_1 n_1, C_2 n_2).$$

Equations (6.33) and (6.34) are considered on the sub-domains adjacent to the interior nodes $\mathbf{x}^i$ as well as to the boundary nodes on Γ_{sM}^i. For the source point $\mathbf{x}^i$ located on the global boundary Γ the boundary of the sub-domain $\partial\Omega_s^i$ is decomposed into L_s^i and Γ_{sM}^i (part of the global boundary with prescribed bending moment) according to Fig. 6.2.

It should be noted here that there are neither Lagrange multipliers nor penalty parameters introduced into the local weak forms (6.23) and (6.24) because the essential boundary conditions on Γ_{sw}^i (part of the global boundary with prescribed rotations or displacements) can be imposed directly, using the interpolation approximation (6.30)

$$\sum_{a=1}^{n} \phi^a(\mathbf{x}^i)\hat{\mathbf{w}}^a(s) = \tilde{\bar{\mathbf{w}}}(\mathbf{x}^i, \mathbf{s}) \quad \text{for } \mathbf{x}^i \in \mathbf{\Gamma}_{sw}^i, \tag{6.35}$$

where $\tilde{\bar{w}}(\mathbf{x}^i, s)$ is the Laplace-transform of the generalized displacement vector prescribed on the boundary Γ_{sw}^i. For a clamped plate all three vector components (rotations and deflection) are vanishing on the fixed edge, and Eq. (6.35) is used at all the boundary nodes in such a case. However, for a simply supported plate only the third component of the displacement vector (deflection) is prescribed, while the rotations are unknown. Then, the entire equation (6.33) and the third component of Eq. (6.35) are applied to the nodes lying on the global boundary. On those parts of the global boundary where no displacement boundary conditions are prescribed both local integral equations (6.33) and (6.34) are applied.

The time-dependent values of rotations, displacements, moments, and shear forces are obtained from the corresponding Laplace-transformed quantities by performing inverse transform. Thereby, great attention is paid to the numerical inversion of the Laplace-transform, since it is an ill-posed problem, in general, and small truncation errors can be greatly magnified in the inversion process with yielding poor numerical results. In the present analysis, the sophisticated Stehfest's algorithm[31] is used for the numerical inversion. If $\bar{f}(s)$ is the Laplace-transform of $f(t)$, an approximate value f_a of $f(t)$ for a specific time t is given by

$$f_a(t) = \frac{\ln 2}{t} \sum_{i=1}^{N} v_i \bar{f}\left(\frac{\ln 2}{t} i\right), \tag{6.36}$$

where

$$v_i = (-1)^{N/2+i} \sum_{k=[(i+1)/2]}^{\min(i,N/2)} \frac{k^{N/2}(2k)!}{(N/2-k)!k!(k-1)!(i-k)!(2k-i)!}. \tag{6.37}$$

In numerical analyses, we have considered $N = 10$ for single precision arithmetic. It means that for each time t it is needed to solve N boundary value problems for the corresponding Laplace-transform parameters $s = i \ln 2/t$, with $i = 1, 2, \ldots, N$. If M denotes the number of the time instants in which we are interested to know $f(t)$, the number of the Laplace-transform solutions $\bar{f}(s_j)$ is then $M \times N$.

6.3.2. *Meshless local integral equations for heat conduction problems in plates*

Consider a boundary value problem for the heat conduction problem in a continuously nonhomogeneous anisotropic medium, which is described by the governing equation:

$$\rho(\mathbf{x})c(\mathbf{x})\frac{\partial\theta}{\partial t}(\mathbf{x}, t) = [k_{ij}(x)\theta_{,j}(\mathbf{x}, t)]_{,i} + Q(\mathbf{x}, t), \qquad (6.38)$$

where $\theta(\mathbf{x}, t)$ is the temperature field, $Q(\mathbf{x}, t)$ is the density of body heat sources, k_{ij} is the thermal conductivity tensor, $\rho(\mathbf{x})$ is the mass density and $c(\mathbf{x})$ the specific heat.

Let the analyzed plate is denoted by Ω with the top and bottom surfaces S^+ and S^-, respectively. Arbitrary temperature or heat flux boundary conditions can be prescribed on all considered surfaces. The initial condition is assumed

$$\theta(\mathbf{x}, t)|_{t=0} = \theta(\mathbf{x}, 0)$$

in the analyzed domain Ω.

The above stated problem has been recently solved by Qian and Batra[35] by an approximate computational technique. The temperature field is expanded in the plate thickness direction by using Legendre polynomials as basis functions. The original 3-D problem is transformed into a set of 2-D problems there. In the present paper, a more general 3-D analysis based on the MLPG method is applied. The MLS approximation is used here. The approximations described in Sec. 2 for 2-D problems are still valid with only modification of basis polynomials as

$$\begin{aligned}
\mathbf{p}^T(\mathbf{x}) &= [1, x_1, x_2, x_3], \text{ linear basis } m = 4,\\
\mathbf{p}^T(\mathbf{x}) &= [1, x_1, x_2, x_3, x_1^2, x_2^2, x_3^2, x_1x_2, x_2x_3, x_3x_1],\\
&\qquad \text{quatratic basis } m = 10.
\end{aligned} \qquad (6.39)$$

Applying the Laplace transformation to the governing Eq. (6.38), one obtains

$$[k_{ij}(\mathbf{x})\bar{\theta}_{,j}(\mathbf{x}, s)]_{,i} - \rho(\mathbf{x})c(\mathbf{x})s\,\bar{\theta}(\mathbf{x}, s) = -\bar{F}(\mathbf{x}, s), \qquad (6.40)$$

where

$$\bar{F}(\mathbf{x}, s) = \bar{Q}(\mathbf{x}, s) + \theta(\mathbf{x}, 0)$$

is the redefined body heat source in the Laplace-transform domain with initial boundary condition for temperature and s is the Laplace-transform parameter.

Again the weak form is constructed over local sub-domains Ω_s, which is a small sphere taken for each node inside the global domain. The local weak form of the governing Eq. (6.39) for $\mathbf{x}^a \in \Omega_s^a$ can be written as

$$\int_{\Omega_s^a} \left[(k_{lj}(\mathbf{x})\bar{\theta}_{,j}(\mathbf{x}, s))_{,l} - \rho(\mathbf{x})c(\mathbf{x})s\,\bar{\theta}(\mathbf{x}, s) + \bar{F}(\mathbf{x}, s) \right] \theta^*(\mathbf{x}) d\Omega = 0, \quad (6.41)$$

where $\theta^*(\mathbf{x})$ is a weight (test) function.

Applying the Gauss divergence theorem to Eq. (6.41) we obtain

$$\int_{\partial\Omega_s^a} \bar{q}(\mathbf{x}, s)\theta^*(\mathbf{x}) d\Gamma - \int_{\Omega_s^a} k_{lj}(\mathbf{x})\bar{\theta}_{,j}(\mathbf{x}, s)\theta^*_{,l}(\mathbf{x}) d\Omega$$
$$- \int_{\Omega_s^a} \rho(\mathbf{x})c(\mathbf{x})s\bar{\theta}(\mathbf{x}, s)\theta^*(\mathbf{x}) d\Omega + \int_{\Omega_s^a} \bar{F}(\mathbf{x}, s)\theta^*(\mathbf{x}) d\Omega = 0, \quad (6.42)$$

where $\partial\Omega_s^a$ is the boundary of the local sub-domain and

$$\bar{q}(\mathbf{x}, s) = k_{lj}(\mathbf{x})\bar{\theta}_{,j}(\mathbf{x}, s) n_l(\mathbf{x}).$$

The local weak form (6.42) is a starting point to derive local boundary integral equations providing an appropriate test function selection. If a Heaviside step function is chosen as the test function $\theta^*(\mathbf{x})$ in each sub-domain

$$\theta^*(\mathbf{x}) = \begin{cases} 1 & \text{at } \mathbf{x} \in \Omega_s^a \\ 0 & \text{at } \mathbf{x} \notin \Omega_s^a \end{cases}$$

the local weak form (6.42) is transformed into the following simple local boundary integral equation

$$\int_{\partial\Omega_s^a} \bar{q}(\mathbf{x}, s) d\Gamma - \int_{\Omega_s^a} \rho(\mathbf{x})c(\mathbf{x})s\bar{\theta}(\mathbf{x}, s) d\Omega = - \int_{\Omega_s^a} \bar{F}(\mathbf{x}, s) d\Omega. \quad (6.43)$$

Equation (6.43) is recognized as the flow balance condition of the sub-domain. In stationary case there is no domain integration involved in the left hand side of this local boundary integral equation. If an assumption of zero body heat sources is made, a pure boundary integral formulation will be obtained.

The MLS is used for approximation of the heat flux $\bar{q}(\mathbf{x}, s)$

$$\bar{q}^h(\mathbf{x}, s) = k_{ij} n_i \sum_{a=1}^{n} \phi^a_{,j}(\mathbf{x}) \hat{\theta}^a(s).$$

Substituting the MLS-approximations into the local integral Eq. (6.43) the system of algebraic equations is obtained

$$\sum_{a=1}^{n} \left(\int_{L_s + \Gamma_{sp}} \mathbf{n}^T \mathbf{K} \mathbf{P}^a(\mathbf{x}) d\Gamma - \int_{\Omega_s} \rho c s \phi^a(\mathbf{x}) d\Gamma \right) \hat{\theta}^a(s)$$
$$= - \int_{\Gamma_{sq}} \tilde{\bar{q}}(\mathbf{x}, s) d\Gamma - \int_{\Omega_s} \bar{R}(\mathbf{x}, s) d\Omega, \tag{6.44}$$

at interior nodes as well as to the boundary nodes with prescribed heat flux on Γ_{sq}. In Eq. (6.44), we have used the notations

$$\mathbf{K} = \begin{bmatrix} k_{11} & k_{12} & k_{13} \\ k_{12} & k_{22} & k_{23} \\ k_{13} & k_{23} & k_{33} \end{bmatrix}, \quad \mathbf{P}^a(\mathbf{x}) = \begin{bmatrix} \phi^a_{,1} \\ \phi^a_{,2} \\ \phi^a_{,3} \end{bmatrix}, \quad \mathbf{n}^T = (n_1, n_2, n_3). \tag{6.45}$$

The time dependent values of the transformed quantities can be obtained by an inverse Laplace-transform. In the present analysis, the Stehfest's inversion algorithm[31] is used.

6.4. Analyses of Orthotropic FGM Shells Under Mechanical and Thermal Load

Consider a linear elastic orthotropic shallow shell of constant thickness h and with its mid-surface being described by $x_3 = f(x_1, x_2)$ in a domain Ω with the boundary contour Γ in the base plane $x_1 - x_2$. The shell is subjected to a transient dynamic load $q_i(\mathbf{x}, t)$. The Reissner–Mindlin bending theory[34] is used to describe the shell deformation. The total displacements of the shell are given by the superposition of the bending deformations and the membrane deformations. Then, the spatial displacement field u'_i and strains ε'_{ij} caused by bending are the same as for a plate and they are given by Eqs. (6.11) and (6.12), respectively. The constitutive Eq. (6.15) is valid for bending stresses σ'_{ij} also here.

The membrane strains have the form[35]

$$\varepsilon_{\alpha\beta} = \frac{1}{2}(u_{\alpha,\beta} + u_{\beta,\alpha}) + \delta_{\alpha\beta} k_{\alpha\beta} w_3, \tag{6.46}$$

where $k_{\alpha\beta}$ are the principal curvatures of the shell in x_1- and x_2-directions with the assumption $k_{12} = k_{21} = 0$, and u_α are the in-plane displacements. The summation convention is not assumed in Eq. (6.46).

Then, the membrane stresses are given as

$$\begin{bmatrix}\sigma_{11}\\ \sigma_{22}\\ \sigma_{12}\end{bmatrix} = \mathbf{D}(\mathbf{x})\begin{bmatrix}\varepsilon_{11}\\ \varepsilon_{22}\\ 2\varepsilon_{12}\end{bmatrix} - \begin{bmatrix}\gamma_{11}\\ \gamma_{22}\\ 0\end{bmatrix}\theta(x_1, x_2, 0, t), \tag{6.47}$$

where

$$\mathbf{D}(\mathbf{x}) = \begin{bmatrix} E_1/e & E_1\nu_{21}/e & 0\\ E_2\nu_{12}/e & E_2/e & 0\\ 0 & 0 & G_{12}\end{bmatrix}.$$

The bending moments $M_{\alpha\beta}$, the shear forces Q_α are defined by Eqs. (6.17) and the normal forces $N_{\alpha\beta}$ by

$$\begin{bmatrix}N_{11}\\ N_{22}\\ N_{12}\end{bmatrix} = \int_{-h/2}^{h/2}\begin{bmatrix}\sigma_{11}\\ \sigma_{22}\\ \sigma_{12}\end{bmatrix} dx_3. \tag{6.48}$$

Substituting Eqs. (6.46) and (6.47) into the force resultants (6.48) yields the expression of the normal forces $N_{\alpha\beta}$ $(\alpha, \beta = 1, 2)$ in terms of the deflection and the lateral displacements of the orthotropic shell

$$\begin{bmatrix}N_{11}\\ N_{22}\\ N_{12}\end{bmatrix} = \mathbf{P}\begin{bmatrix}u_{1,1}\\ u_{2,2}\\ u_{1,2}+u_{2,1}\end{bmatrix} + \begin{bmatrix}Q_{11}\\ Q_{22}\\ 0\end{bmatrix} w_3 - \begin{bmatrix}\theta_{11}\\ \theta_{22}\\ 0\end{bmatrix}, \tag{6.49}$$

where

$$\theta_{\alpha\beta} = \int_{-h/2}^{h/2}\gamma_{\alpha\beta}\theta(\mathbf{x}, x_3, t)dx_3,$$

$$\mathbf{P} = \begin{bmatrix} E_1^*/e & E_1^*\nu_{12}/e & 0\\ E_2^*\nu_{12}/e & E_2^*/e & 0\\ 0 & 0 & G_{12}^*\end{bmatrix},$$

$$\begin{bmatrix}Q_{11}\\ Q_{22}\\ 0\end{bmatrix} = \begin{bmatrix}(k_{11}+k_{22}\nu_{12})E_1^*/e\\ (k_{11}\nu_{12}+k_{22})E_2^*/e\\ 0\end{bmatrix}, \tag{6.50}$$

$$E_\alpha^* \equiv \int_{-h/2}^{h/2} E_\alpha(x_3)dx_3 = \begin{cases} E_{\alpha t} = E_{\alpha b}, & n = 0 \\ (E_{\alpha b} + E_{\alpha t})/2, & n = 1 \\ (2E_{\alpha b} + E_{\alpha t})/3, & n = 2 \end{cases},$$

$$G_{12}^* \equiv \int_{-h/2}^{h/2} G_{12}(x_3)dx_3 = \begin{cases} G_{12t} = G_{12b}, & n = 0 \\ (G_{12b} + G_{12t})/2, & n = 1 \\ (2G_{12b} + G_{12t})/3, & n = 2 \end{cases}.$$

Using the Reissner's linear theory of shallow shells,[34] the equations of motion may be written as

$$\begin{aligned} &M_{\alpha\beta,\beta}(\mathbf{x},t) - Q_\alpha(\mathbf{x},t) = \frac{\rho h^3}{12}\ddot{w}_\alpha(\mathbf{x},t), \\ &Q_{\alpha,\alpha}(\mathbf{x},t) - k_{\alpha\beta}N_{\alpha\beta}(\mathbf{x},t) + q_3(\mathbf{x},t) = \rho h\ddot{w}_3(\mathbf{x},t), \\ &N_{\alpha\beta,\beta}(\mathbf{x},t) + q_\alpha(\mathbf{x},t) = \rho\ddot{u}_\alpha(\mathbf{x},t), \quad \mathbf{x} \in \Omega. \end{aligned} \tag{6.51}$$

To eliminate the time variable t in the governing equations, the Laplace-transform is applied to Eq. (6.51). Then, one obtains

$$\bar{M}_{\alpha\beta,\beta}(\mathbf{x},s) - \bar{Q}_\alpha(\mathbf{x},s) = \frac{\rho h^3}{12}s^2\bar{w}_\alpha(\mathbf{x},s) - \bar{R}_\alpha(\mathbf{x},s), \tag{6.52}$$

$$\bar{Q}_{\alpha,\alpha}(\mathbf{x},s) - k_{\alpha\beta}\bar{N}_{\alpha\beta}(\mathbf{x},s) = \rho hs^2\bar{w}_3(\mathbf{x},s) - \bar{R}_3(\mathbf{x},s), \tag{6.53}$$

$$\bar{N}_{\alpha\beta,\beta}(\mathbf{x},s) = \rho s^2\bar{u}_\alpha(\mathbf{x},s) - \bar{R}'_\alpha(\mathbf{x},s), \tag{6.54}$$

where s is the Laplace-transform parameter, and $\bar{R}_\alpha$, $\bar{R}_3$ and $\bar{R}'_\alpha$ are given by

$$\begin{aligned} &\bar{R}_\alpha(\mathbf{x},s) = \frac{\rho h^3}{12}[sw_\alpha(\mathbf{x}) + \dot{w}_\alpha(\mathbf{x})], \\ &\bar{R}_3(\mathbf{x},s) = \bar{q}_3(\mathbf{x},s) + \rho hsw_3(\mathbf{x}) + \rho h\dot{w}_3(\mathbf{x}), \\ &\bar{R}'_\alpha(\mathbf{x},s) = \bar{q}_\alpha(\mathbf{x},s) + \rho su_\alpha(\mathbf{x}) + \rho\dot{u}_\alpha(\mathbf{x}), \end{aligned}$$

with $w_k(\mathbf{x})$, $u_\alpha(\mathbf{x})$, $\dot{w}_k(\mathbf{x})$ and $\dot{u}_\alpha(\mathbf{x})$ representing the initial values of the bending and the membrane displacements and their respective velocities.

Again the MLPG method constructs the weak-form over local sub-domains Ω_s

$$\begin{aligned} \int_{\Omega_s^i} \bigg[&\bar{M}_{\alpha\beta,\beta}(\mathbf{x},s) - \bar{Q}_\alpha(\mathbf{x},s) - \frac{\rho h^3}{12}s^2\bar{w}_\alpha(\mathbf{x},s) \\ &+ \bar{R}_\alpha(\mathbf{x},s)\bigg] w_{\alpha\gamma}^*(\mathbf{x})d\Omega = 0, \end{aligned} \tag{6.55}$$

$$\int_{\Omega_s^i} [\bar{Q}_{\alpha,\alpha}(\mathbf{x},s) - k_{\alpha\beta}\bar{N}_{\alpha\beta}(\mathbf{x},s) - \rho h s^2 \bar{w}_3(\mathbf{x},s) + \bar{R}_3(\mathbf{x},s)] w_3^*(\mathbf{x}) d\Omega = 0, \tag{6.56}$$

$$\int_{\Omega_s^i} [\bar{N}_{\alpha\beta,\beta}(\mathbf{x},s) - \rho s^2 \bar{u}_\alpha(\mathbf{x},s) + \bar{R'}_\alpha(\mathbf{x},s)] u_{\alpha\gamma}^*(\mathbf{x}) d\Omega = 0. \tag{6.57}$$

Applying the Gauss divergence theorem to local weak forms and choosing the test functions as a unit step function with support in the current subdomain, one obtains local boundary-domain integral equations

$$\int_{\partial\Omega_s^i} \bar{M}_\alpha(\mathbf{x},s) d\Gamma - \int_{\Omega_s^i} \bar{Q}_\alpha(\mathbf{x},s) d\Omega - \int_{\Omega_s^i} \frac{\rho h^3}{12} s^2 \bar{w}_\alpha(\mathbf{x},s) d\Omega + \int_{\Omega_s^i} \bar{R}_\alpha(\mathbf{x},s) d\Omega = 0, \tag{6.58}$$

$$\int_{\partial\Omega_s^i} \bar{Q}_\alpha(\mathbf{x},s) n_\alpha(\mathbf{x}) d\Gamma - \int_{\Omega_s^i} k_{\alpha\beta}(\mathbf{x}) \bar{N}_{\alpha\beta}(\mathbf{x},s) d\Omega - \int_{\Omega_s^i} \rho h s^2 \bar{w}_3(\mathbf{x},s) d\Omega + \int_{\Omega_s^i} \bar{R}_3(\mathbf{x},s) d\Omega = 0, \tag{6.59}$$

$$\int_{\partial\Omega_s^i} \bar{T}_\alpha(\mathbf{x},s) d\Gamma - \int_{\Omega_s^i} \rho s^2 \bar{u}_\alpha(\mathbf{x},s) d\Omega + \int_{\Omega_s^i} \bar{R}'_\alpha(\mathbf{x},s) d\Omega = 0. \tag{6.60}$$

where

$$\bar{T}_\alpha(\mathbf{x},s) = \bar{N}_{\alpha\beta}(\mathbf{x},s) n_\beta(\mathbf{x}). \tag{6.61}$$

Similar to approximation of rotations and deflection, one can utilize the approximation for the in-plane displacements as

$$\bar{\mathbf{u}}^h(\mathbf{x},s) = \mathbf{\Phi}^T(\mathbf{x}) \cdot \hat{\mathbf{u}}(s) = \sum_{a=1}^{n} \phi^a(\mathbf{x}) \hat{\mathbf{u}}^a(s). \tag{6.62}$$

The bending moment and shear forces are approximated by Eqs. (6.31) and (6.32), respectively. The traction vector can be approximated similarly

$$\bar{\mathbf{T}}(\mathbf{x},s) = \mathbf{N}_1(\mathbf{x}) \mathbf{P}(\mathbf{x}) \sum_{a=1}^{n} \mathbf{B}^a(\mathbf{x}) \mathbf{u}^{*a}(s) + \mathbf{J}(\mathbf{x}) \sum_{a=1}^{n} \phi^a(\mathbf{x}) \hat{w}_3^a(s) - \bar{\mathbf{\Theta}}(\mathbf{x},s), \tag{6.63}$$

where the vector $\mathbf{u}^{*a}(s)$ is defined as a column vector

$$\mathbf{u}^{*a}(s) = [\hat{u}_1^a(s), \hat{u}_2^a(s)]^{\mathrm{T}}, \quad \bar{\boldsymbol{\Theta}}(\mathbf{x}, s) = [\bar{\theta}_{11}n_1, \bar{\theta}_{22}n_2]^T,$$

$$\mathbf{J}(\mathbf{x}) = [Q_{11}n_1, Q_{22}n_2]^T,$$

$$\mathbf{B}^a(\mathbf{x}) = \begin{bmatrix} \phi_{,1}^a & 0 \\ 0 & \phi_{,2}^a \\ \phi_{,2}^a & \phi_{,1}^a \end{bmatrix}.$$

We need to approximate also

$$\begin{aligned} k_{\alpha\beta}(\mathbf{x})\bar{N}_{\alpha\beta}(\mathbf{x}, s) = \mathbf{K}(\mathbf{x})^{\mathrm{T}}\mathbf{P}(\mathbf{x}) \sum_{a=1}^{n} \mathbf{B}^a(\mathbf{x})\mathbf{u}^{*a}(s) \\ + O(\mathbf{x}) \sum_{a=1}^{n} \phi^a(\mathbf{x})\hat{w}_3^a(s) \\ - \bar{\theta}_{11}(\mathbf{x})k_{11}(\mathbf{x}) - \bar{\theta}_{22}(\mathbf{x})k_{22}(\mathbf{x}), \end{aligned} \tag{6.64}$$

where

$$\mathbf{K}(\mathbf{x}) = \begin{bmatrix} k_{11} \\ k_{22} \\ 2k_{12} \end{bmatrix},$$

$$O(\mathbf{x}) = Q_{11}(\mathbf{x})k_{11}(\mathbf{x}) + Q_{22}(\mathbf{x})k_{22}(\mathbf{x}).$$

Furthermore, in view of the MLS approximations (6.31), (6.32) and (6.63) for the unknown fields in the local boundary-domain integral equations (6.58)–(6.60), we obtain their discretized forms as

$$\begin{aligned} &\sum_{a=1}^{n} \left[\int_{L_s^i + \Gamma_{sw}^i} \mathbf{N}_\alpha(\mathbf{x})\mathbf{B}_\alpha^a(\mathbf{x})d\Gamma \right. \\ &\quad \left. - \int_{\Omega_s^i} \left(\mathbf{C}(\mathbf{x}) + \mathbf{I}\frac{\rho h^3(\mathbf{x})}{12}s^2 \right) \phi^a(\mathbf{x})d\Omega \right] \mathbf{w}^{*a}(s) \\ &\quad - \sum_{a=1}^{n} \left[\int_{\Omega_s^i} \mathbf{C}(\mathbf{x})\mathbf{F}^a(\mathbf{x})d\Omega \right] \hat{w}_3^a(s) \\ &= \int_{L_s^i + \Gamma_{sM}^i} \bar{\mathbf{H}}(\mathbf{x}, s)d\Gamma - \int_{\Gamma_{sM}^i} \tilde{\bar{\mathbf{M}}}(\mathbf{x}, s)d\Gamma - \int_{\Omega_s^i} \bar{\mathbf{R}}(\mathbf{x}, s)d\Omega, \end{aligned} \tag{6.65}$$

$$
\begin{aligned}
&\sum_{a=1}^{n}\left[\int_{\partial\Omega_s^i} \mathbf{C}_n(\mathbf{x})\phi^a(\mathbf{x})d\Gamma\right]\mathbf{w}^{*a}(s)\\
&\qquad - \mathbf{K}(\mathbf{x})^{\mathrm{T}}\mathbf{P}(\mathbf{x})\sum_{a=1}^{n}\left[\int_{\Omega_s^i}\mathbf{B}^a(\mathbf{x})d\Omega\right]\mathbf{u}^{*a}(s)\\
&\qquad + \sum_{a=1}^{n}\left[\int_{\partial\Omega_s^i}\mathbf{C}_n(\mathbf{x})\mathbf{F}^a(\mathbf{x})d\Gamma - \int_{\Omega_s^i} O(\mathbf{x})\phi^a(\mathbf{x})d\Omega\right.\\
&\qquad \left. - \rho h s^2\int_{\Omega_s^i}\phi^a(\mathbf{x})d\Omega\right]\hat{w}_3^a(s)\\
&\quad = \int_{\Omega_s^i}[\bar{\theta}_{11}(\mathbf{x},s)k_{11} + \bar{\theta}_{22}(\mathbf{x},s)k_{22}]d\Omega - \int_{\Omega_s^i}\bar{R}_3(\mathbf{x},s)d\Omega,
\end{aligned}
\tag{6.66}
$$

$$
\begin{aligned}
&\sum_{a=1}^{n}\left[\int_{L_s^i+\Gamma_{su}^i}\mathbf{N}_1(\mathbf{x})\mathbf{P}(\mathbf{x})\mathbf{B}^a(\mathbf{x})d\Gamma - \rho s^2\mathbf{I}\int_{\Omega_s^i}\phi^a(\mathbf{x})d\Omega\right]\mathbf{u}^{*a}(s)\\
&\qquad + \sum_{a=1}^{n}\left[\int_{\Omega_s^i}\mathbf{J}(\mathbf{x})\phi^a(\mathbf{x})d\Omega\right]\hat{w}_3^a(s)\\
&\quad = \int_{L_s^i+\Gamma_{sM}^i}\bar{\mathbf{\Theta}}(\mathbf{x},s)d\Gamma - \int_{\Gamma_{sP}^i}\tilde{\bar{\mathbf{T}}}(\mathbf{x},s)d\Gamma - \int_{\Omega_s^i}\bar{\mathbf{R}}'(\mathbf{x},s)d\Omega.
\end{aligned}
\tag{6.67}
$$

Recall that the discretized local boundary-domain integral equations (6.65)–(6.67) are considered on the sub-domains adjacent to the interior nodes $\mathbf{x}^i$ as well as to the boundary nodes on Γ_{sM}^i and Γ_{sP}^i. For the source point $\mathbf{x}^i$ located on the global boundary Γ the boundary of the sub-domain $\partial\Omega_s^i$ is composed of the interior and the boundary portions L_s^i and Γ_{sM}^i, respectively, or alternatively of L_s^i and Γ_{sP}^i, with the portions Γ_{sM}^i and Γ_{sP}^i lying on the global boundary with prescribed bending moments or stress vector, respectively. Equations (6.65) and (6.67) are vector equations for the two components of rotations and in-plane displacements, respectively. Then, the set of Eqs. (6.65)–(6.67) represents 5 equations at each node for five unknown components, namely, two rotations, one out-of-plane deflection and two in-plane displacements.

It should be noted here that there are neither Lagrange-multipliers nor penalty parameters introduced into the local weak-forms (6.55)–(6.57) because the essential boundary conditions on Γ_{sw}^i or Γ_{su}^i can be imposed directly by using the MLS approximations (6.30) and (6.62)

$$\sum_{a=1}^{n} \phi^a(\mathbf{x})\hat{\mathbf{w}}^a(s) = \tilde{\bar{\mathbf{w}}}(\mathbf{x}^i, s) \quad \text{for } \mathbf{x}^i \in \Gamma^i_{sw}, \tag{6.68}$$

$$\sum_{a=1}^{n} \phi^a(\mathbf{x})\hat{\mathbf{u}}^a(s) = \tilde{\bar{\mathbf{u}}}(\mathbf{x}^i, s) \quad \text{for } \mathbf{x}^i \in \Gamma^i_{su}, \tag{6.69}$$

where $\tilde{\bar{\mathbf{w}}}(\mathbf{x}^i, s)$ and $\tilde{\bar{\mathbf{u}}}(\mathbf{x}^i, s)$ are the Laplace-transforms of the generalized displacement vector prescribed on the boundary Γ^i_{sw} and Γ^i_{su}, respectively. The MLS approximation does not possess Kronecker-delta property in the present form. If a singular weight function were introduced into the MLS approximation, the Kronecker-delta property would be recovered.[37] In such a case instead of fictitious nodal values one would use the nodal values of the generalized displacements in the approximations (6.30) and (6.62) with assuming such nodal values being prescribed on Γ^i_{sw} and Γ^i_{su}, respectively. For essential boundary conditions only one column in the matrix form of Eqs. (6.68) or (6.69) has prescribed quantities and other ones are zero. For a clamped shell all three vector components (two rotations and one deflection) and two components of the in-plane displacements are vanishing at the fixed edge and only Eqs. (6.68) and (6.69) are used at the boundary nodes in such a case. On the other hand, for a simply supported shell only the third component of the generalized displacement vector (i.e. the deflection) is prescribed and the rotations are unknown. Then, equations (6.65) and (6.67) together with Eq. (6.68) for the third vector component are applied for a point on the global boundary. If no geometrical boundary conditions are prescribed on the part of the boundary, all three local integral equations (6.65)–(6.67) are applied.

6.5. Numerical Examples

In this section, numerical results are presented for plates and shells under mechanical or thermal loads. Both stationary and transient boundary conditions with the Heaviside time dependence are considered here. In order to test the accuracy, the numerical results obtained by the present method are compared with the results provided by the FEM-NASTRAN code using a very fine mesh. Clamped and simply supported boundary conditions are considered. In all numerical calculations, FGM plates and shells with isotropic and orthotropic material properties are analyzed.

6.5.1. *Plates*

Consider a clamped square plate with a side-length $a = 0.254\,\mathrm{m}$ and plate thicknesses $h/a = 0.05$. The plate is subjected to a uniformly distributed static load. Firstly, isotropic material properties of the FGM plate are considered. The following material parameters on top side of the plate are used in numerical analysis: Young's moduli $E_{2t} = 0.6895 \cdot 10^{10}\,\mathrm{N/m}^2$, $E_{1t} = E_{2t}$, Poisson's ratios $\nu_{21} = 0.3$, $\nu_{12} = 0.3$. Linear and quadratic variations of volume fraction V defined in equation (4) are considered here, and Young's moduli on the bottom side are: $E_{1b} = E_{1t}/2$ and $E_{1b} = E_{2b}$.

In our numerical calculations, 441 nodes with a regular distribution were used for the approximation of the rotations and the deflection (Fig. 6.5). If s is the distance of two neighboring nodes, then the radius of the circular sub-domain is chosen as $r_{loc} = 0.4s$ and the radius of the support domain for node a is $r^a = 4r_{loc}$. Smaller values of the support domain lead to lower approximation accuracy, and larger values of the support domain prolong the computational time for the evaluation of the shape functions. More detailed analyses on the influence of the support domain on computational accuracy can be found in the monograph of Atluri.[20]

The variation of the deflection with the x_1-coordinate at $x_2 = a/2$ of homogeneous and FGM plates are presented in Fig. 6.6.

The deflection value is normalized by the corresponding central deflection of an isotropic plate with homogeneous material properties identical to the ones on the top of the FGM plate. For a uniformly

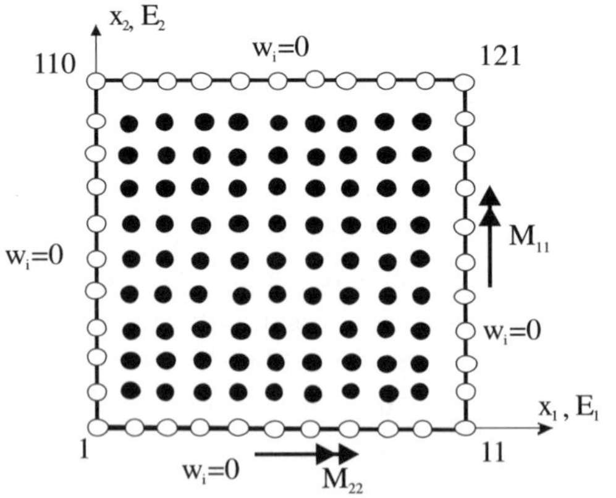

Fig. 6.5. Node distribution for numerical analyses of a clamped square plate.

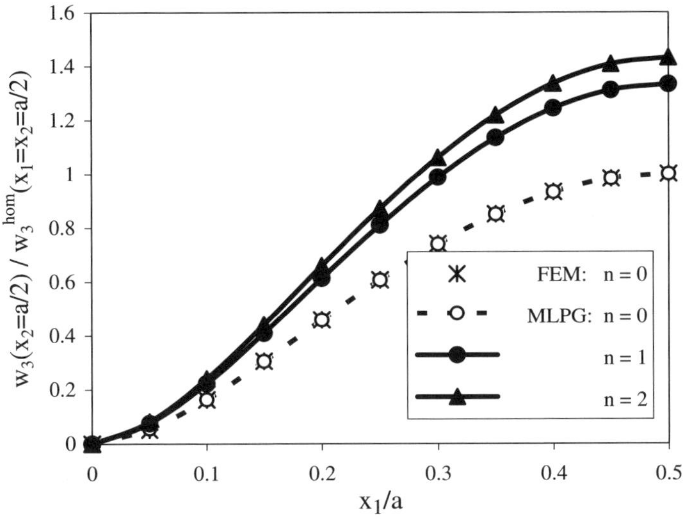

Fig. 6.6. Isotropic clamped square plates under a uniform static load: Variation of the deflection with the x_1-coordinate at $x_2 = a/2 = \text{const}$.

distributed load $q_0 = 300$ psi($2.07 \times 10^6\,\text{Nm}^{-2}$) we have $w_3^{\text{hom}}(a/2) = 8.842 \cdot 10^{-3}\,\text{m}$. Since Young's modulus on bottom side is considered to be smaller than on the top one, deflection for FGM plate is larger than for homogenous plate with material properties corresponding to the top side, $E_{2t} = 0.6895 \cdot 10^{10}\,\text{N/m}^2$. The numerical results are compared with the results obtained by the FEM-ANSYS code with a fine mesh of 400 quadrilateral eight-node shell elements for a quarter of the plate. The variation of the bending moment M_{11} is presented in Fig. 6.7.

Here, the bending moments are normalized by the central bending moment value corresponding to a homogeneous isotropic plate $M_{11}^{\text{hom}}(a/2) = 3064\,\text{Nm}$. The variation of material properties through the plate thickness has practically no influence on the bending moment. Bending moments in homogeneous and FGM plates are almost the same. Minimal differences between them are caused by numerical inaccuracies.

Next, orthotropic FGM clamped plate is analyzed. The following material parameters are used in numerical analysis: Young's moduli $E_{2t} = 0.6895 \cdot 10^{10}\,\text{N/m}^2$, $E_{1t} = 2E_{2t}$, Poisson's ratios $\nu_{21} = 0.15$, $\nu_{12} = 0.3$. The used shear moduli correspond to Young's modulus E_2, namely, $G_{12} = G_{13} = G_{23} = E_2/2(1+\nu_{12})$. A quadratic variation of the volume fraction V is considered with Young's moduli on the bottom side are: $E_{1b} = E_{1t}/2$ and

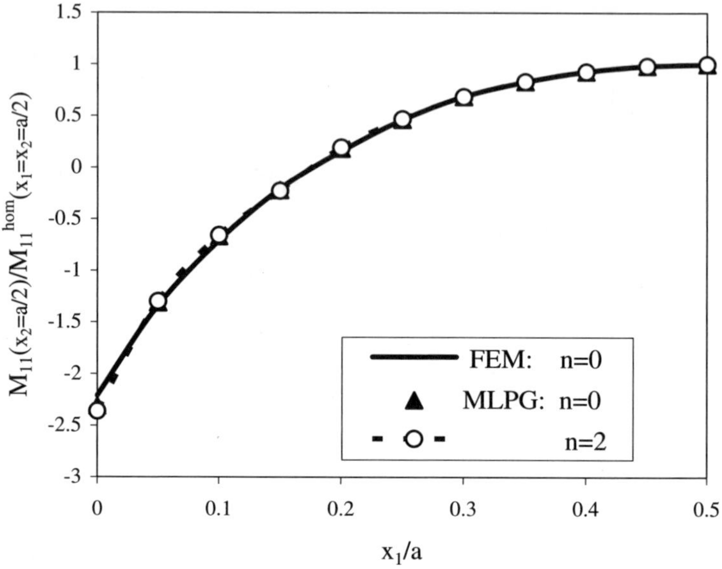

Fig. 6.7. Isotropic clamped square plates under a uniform static load: Variation of the bending moment with the x_1-coordinate at $x_2 = a/2 = \text{const}$.

$E_{1b} = 2E_{2b}$. Variation of the deflection with the x_1-coordinate at $x_2 = a/2$ of homogeneous (dashed line) and FGM (solid line) plates are presented in Fig. 6.8.

The deflection value is normalized by the corresponding deflection at the center of an isotropic homogeneous plate with material constants given above and $E_{1t} = E_{2t}$. One can observe that the deflection value is reduced for an orthotropic plate if one of the Young's moduli is increased. However, the deflection value of FGM plate is larger than that of the homogeneous plate with material properties identical to those on the top of the FGM plate, characterized by larger Young's moduli. Variation of the bending moment along $(x_1, a/2)$ for an orthotropic FGM plate is presented in Fig. 6.9. In an orthotropic plate, the absolute values of the bending moment at the plate center and at the center of the clamped side are higher than those occurring in an isotropic plate, for both homogeneous and FGM plates.

A simply supported isotropic thick square plate under an impact load with Heaviside time variation is analyzed. The used geometrical and material parameters are the same as in the static case. For the numerical modelling, we have used again 441 nodes with a regular distribution.

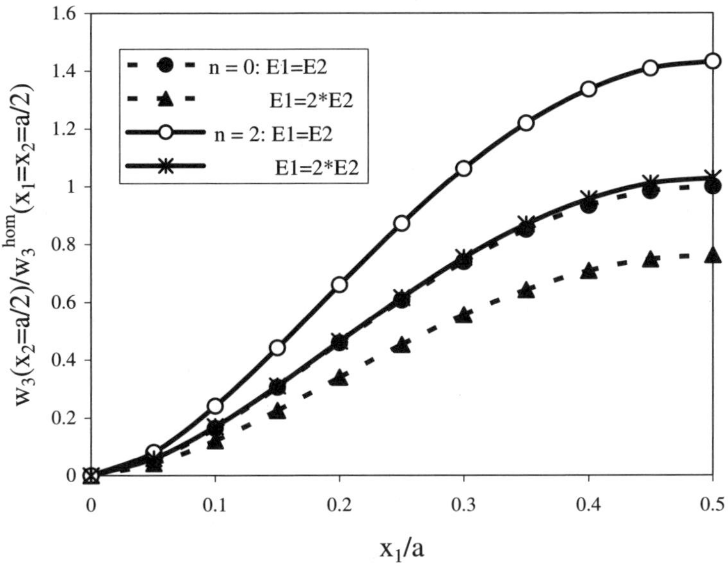

Fig. 6.8. Orthotropic clamped square plates under a uniform static load: Variation of the deflection with the x_1-coordinate at $x_2 = a/2 = \text{const.}$

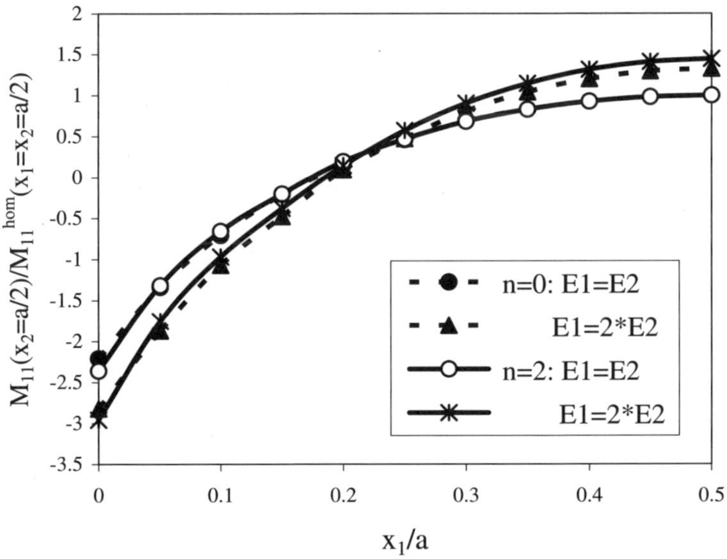

Fig. 6.9. Orthotropic clamped square plates under a uniform static load: Variation of the bending moment with the x_1-coordinate at $x_2 = a/2 = \text{const.}$

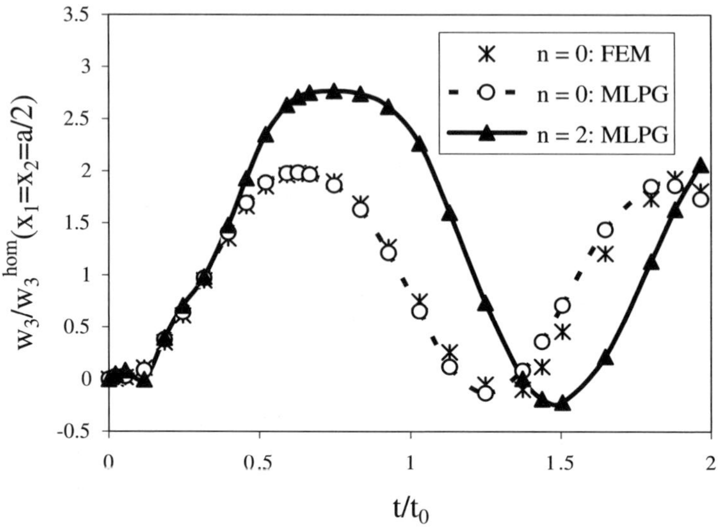

Fig. 6.10. Time variation of the deflection at the center of isotropic simply supported square plates subjected to a suddenly applied load.

The time variation of the central deflection at $x_1 = x_2 = a/2$ for both homogeneous and FGM plates are presented in Fig. 6.10. The deflection value is normalized by the central deflection of a homogeneous plate under a static load, $w_3^{\text{hom}}(a/2) = 28.29 \cdot 10^{-3}\,\text{m}$. The time variable is normalized by a characteristic time[37] $t_0 = a^2/4\sqrt{\rho h/D} = 1.35 \cdot 10^{-2} s$ corresponding to homogeneous plate, too. The MLPG results for a homogeneous plate are compared with those obtained by FEM-ANSYS computer code, where 400 quadrilateral eight-node shell elements with 1000 time increments have been used. A good agreement between both the numerical results is achieved. A quadratic variation of the volume fraction V is considered for FGM plate with Young's moduli on the bottom side as: $E_{1b} = E_{1t}/2$ and $E_{1b} = E_{2b}$. The oscillation of the plate deflection for FGM plate is moderated relatively to a homogeneous plate with higher flexural rigidity. The absolute value of the central deflection of FGM plate is larger than that of the homogeneous plate.

The time variation of the central bending moment M_{11} is given in Fig. 6.11. The bending moment at the center of the plate is normalized by the quantity corresponding to a homogeneous plate under a static load, $M_{11}^{\text{hom}}(a/2) = 6480\,\text{Nm}$. In this case, the amplitudes of the bending moments for both FGM and homogeneous plates are almost the same.

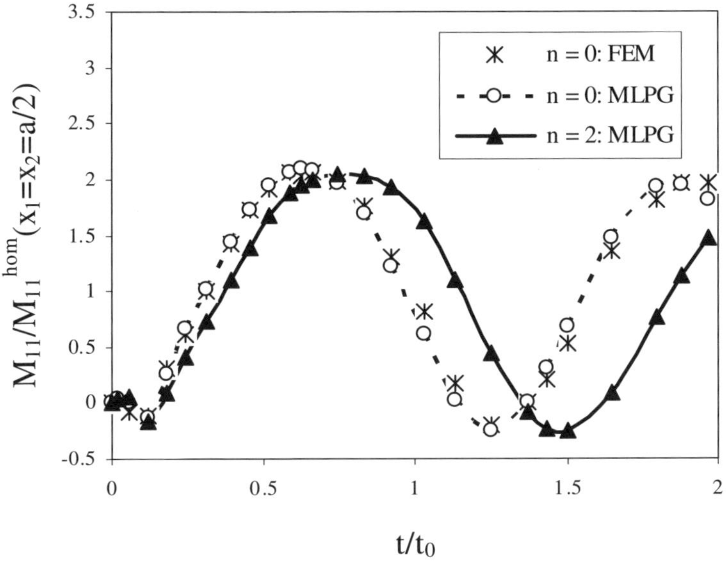

Fig. 6.11. Time variation of the bending moment at the center of isotropic simply supported square plates subjected to a suddenly applied load.

However, in the FGM plate with smaller flexural rigidity, the peaks of the moment amplitudes occur at later time instants.

Next, the orthotropic FGM clamped plate under a uniform impact load with a Heaviside time dependence is analyzed. The material properties and geometry parameters are the same as in the static case analyzed above and mass density $\rho = 0.2589\,\mathrm{lb\,in}^{-3}(7.166 \times 10^3\,\mathrm{kgm}^{-3})$. Again a quadratic variation of the volume fraction V is considered for the FGM plate. The time variation of the deflection at $x_1 = x_2 = a/2$ for both the isotropic and orthotropic FGM plates are presented in Fig. 6.12. The deflections are normalized by the static equivalent for the isotropic and homogeneous plate, $w_3^{\mathrm{hom}}(a/2) = 8.842 \cdot 10^{-3}\,\mathrm{m}$. The amplitudes of the deflections for the orthotropic plate are reduced with respect to the isotropic one, due to higher Young's modulus E_1 for the orthotropic plate. However, the oscillations are more rapid for the orthotropic FGM plate.

The comparison of the time variations of bending moments for the isotropic and orthotropic plates is given in Fig. 6.13. Here, the bending moments are normalized by the bending moment at the center of the homogeneous isotropic plate under a static load, $M_{11}^{\mathrm{hom}}(a/2) = 3064\,\mathrm{Nm}$. One can observe that the amplitudes of the bending moments in the

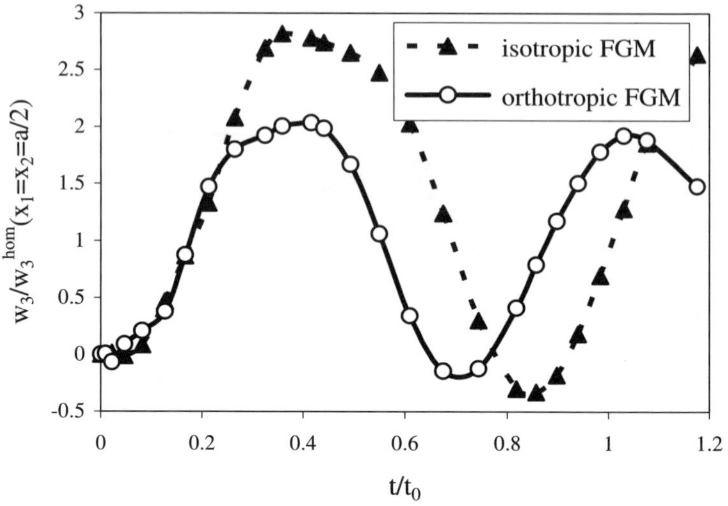

Fig. 6.12. Influence of orthotropic material properties on time variation of the deflection for a clamped square plates under a uniform impact load.

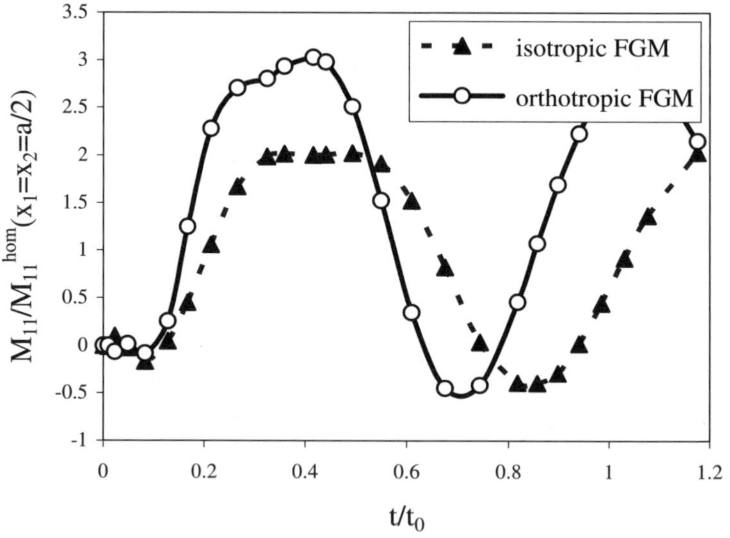

Fig. 6.13. Influence of orthotropic material properties on time variation of the bending moment for a clamped square plates under a uniform impact load.

orthotropic plate are larger than those in the isotropic plate. Again, more rapid oscillations are encountered in the orthotropic plate with higher flexural rigidity.

Next, a simply supported square plate under a thermal load is analyzed. On the top surface of the plate a uniformly distributed temperature $\theta = 1$ deg is prescribed. The bottom surface is kept at vanishing temperature. Orthotropic mechanical properties of the plate are considered with Young's moduli like in a mechanical load case and the thermal expansion coefficients $\alpha_{11} = \alpha_{22} = 1 \cdot 10^{-5}$ deg^{-1}.

The variation of the deflection with the x_1-coordinate at $x_2 = a/2$ of the plate is presented in Fig. 6.14 with assuming isotropic thermal expansion coefficients. Opposite to mechanical load case (Fig. 6.8) the deflection is not reduced in the orthotropic plate as compared with the isotropic plate. It is due to increasing equivalent load for orthotropic plate at the same temperature distributions in both cases.

The variations of the bending moment M_{11} for orthotropic plate are presented in Fig. 6.15. We consider orthotropic properties for Young's moduli and for thermal expansion coefficients either isotropic or orthotropic. The bending moments are normalized by the central value for an isotropic plate. One can observe that orthotropic material properties of thermal expansion coefficient have a strong influence on the bending moment values.

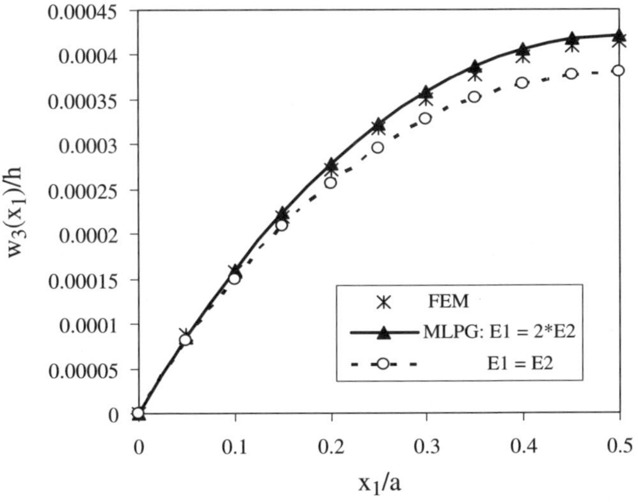

Fig. 6.14. Influence of orthotropic material properties on the plate deflection.

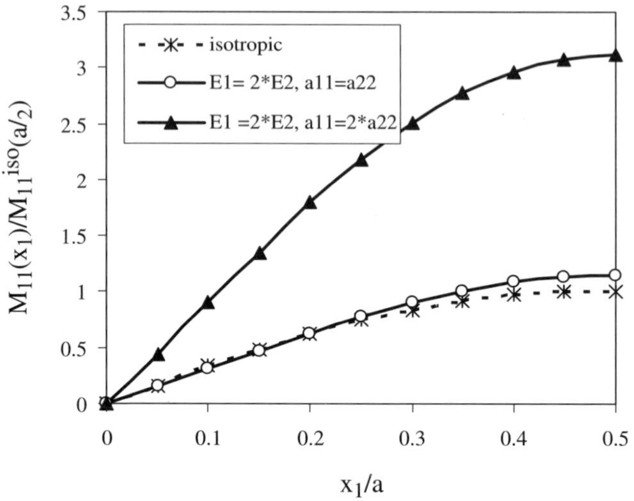

Fig. 6.15. Variation of the bending moment with the x_1-coordinate for a simply supported orthotropic square plate.

A simply supported square plate with a side-length $a = 0.254\,\text{m}$ and the plate thicknesses $h/a = 0.05$ is considered. On the top surface of the plate a thermal shock $\theta = H(t-0)$ with Heaviside time variation is applied. If the ends of the plate are thermally insulated, a uniform temperature distribution on plate surfaces is given. The bottom surface is thermally insulated too. In such a case the temperature distribution is given by Carslaw and Jaeger[39]

$$\theta(x_3, t) = 1 - \frac{4}{\pi}\sum_{n=0}^{\infty}\frac{(-1)^n}{2n+1}\exp\left[-\frac{(2n+1)^2\pi^2\kappa t}{4h^2}\right] \times \cos\frac{(2n+1)\pi x_3}{2h}, \tag{6.70}$$

where diffusivity coefficient $\kappa = k/\rho c$, with thermal conductivity $k = 100\,\text{W/m deg}$, mass density $\rho = 7500\,\text{kg/m}^3$ and specific heat $c = 400\,\text{Ws/kg deg}$.

Isotropic material parameters are considered: Young's moduli $E_1 = E_2 = 0.6895 \cdot 10^{10}\,\text{N/m}^2$, Poisson's ratios $\nu_{21} = \nu_{12} = 0.3$, the thermal expansion coefficients $\alpha_{11} = \alpha_{22} = 1 \cdot 10^{-5}\,\text{deg}^{-1}$. The used shear moduli correspond to Young's modulus E_2, namely,

$$G_{12} = G_{13} = G_{23} = E_2/2(1+\nu_{12}).$$

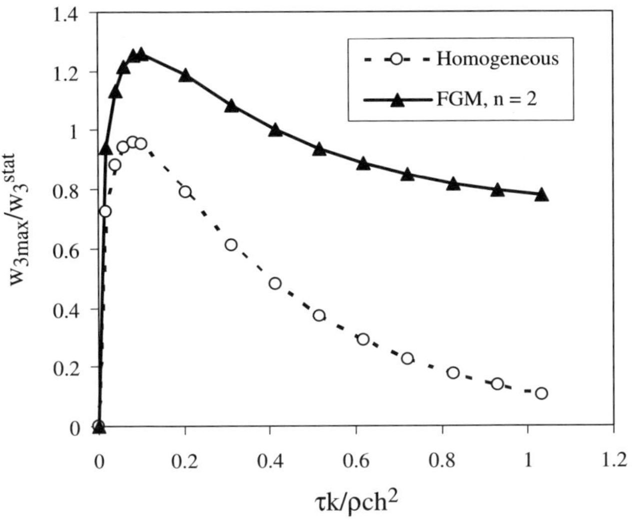

Fig. 6.16. Time variation of the central deflection in the plate with vanishing heat flux.

Homogeneous and FGM properties with n= 2 are considered here. Numerical results for the central plate deflection are presented in Fig. 6.16. Deflections are normalized by the central deflection corresponding to stationary thermal distribution with $\theta = 1$ deg on the top plate surface and vanishing temperature on the bottom surface. For homogeneous material properties the corresponding stationary deflection is $w_3^{stat} = 0.4829 \cdot 10^{-5}$ m. One can observe that in the whole time interval deflection for a homogeneous plate is lower than in a stationary case. The stiffness of the FGM plate is lower than for a homogeneous one.

For homogeneous plate the deflection is approaching to zero for late time instants. It is opposite to FGM plate. The bending moment at the center of the plate M_{11}^{stat} =0.4699 Nm is used as a normalized parameter in Fig. 6.17. The peak values of the bending moments for both homogeneous and FGM plates are almost the same. Since the deflection is not vanishing for late time instants at FGM plate, the bending moment is finite too.

6.5.2. *Shells*

A shallow spherical shell with a square contour and orthotropic material properties is investigated in the first example here (see Fig. 6.18). The shell is subjected to a uniformly distributed static load. The following orthotropic material parameters are used in our numerical analysis: Young's moduli

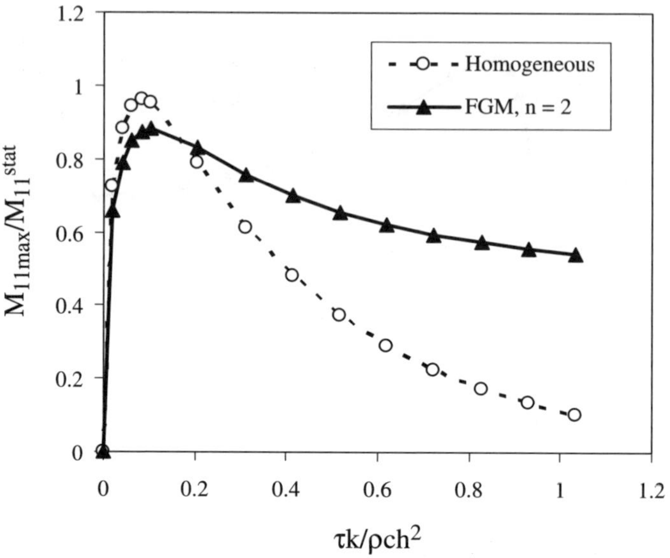

Fig. 6.17. Time variation of the bending moment.

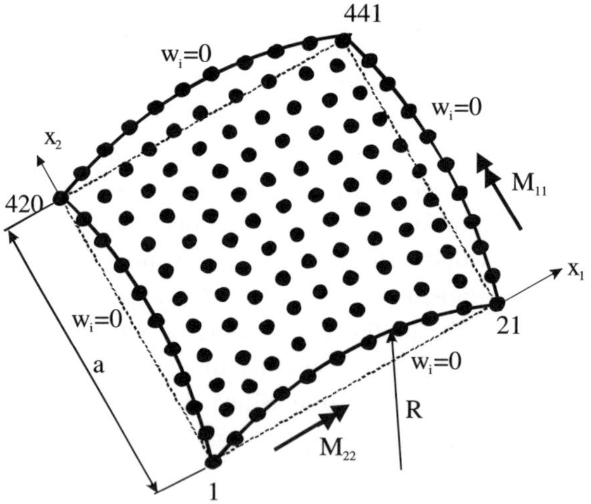

Fig. 6.18. Geometry and boundary conditions of the square shallow spherical shell.

$E_{2t} = 0.6895 \cdot 10^{10}\,\mathrm{N/m^2}$ and $E_{1t} = 2E_{2t}$, Poisson's ratios $\nu_{21} = 0.15$ and $\nu_{12} = 0.3$. The used shear moduli correspond to Young's modulus E_2 and Poisson's ratio ν_{12}, namely,

$$G_{12} = G_{13} = G_{23} = E_2/2(1 + \nu_{12}).$$

A quadratic variation of the volume fraction is considered with Young's moduli on the bottom side being $E_{1b} = E_{1t}/2$ and $E_{1b} = 2E_{2b}$. The mass density is taken as $\rho = 7.166 \times 10^3\,\mathrm{kgm^{-3}}$. The following geometrical parameters are assumed: projected side-length of the shell $a = 0.254\,\mathrm{m}$ and thickness of the shell $h = 0.0127\,\mathrm{m}$.

In our numerical calculations, 441 nodes with a regular distribution are used for the approximation of the rotations and the deflection. The variation of the deflection with the x_1-coordinate at $x_2 = a/2$ of homogeneous (dashed line) and FGM (solid line) shells are presented in Fig. 6.19. The deflection is normalized by the corresponding deflection at the center of an isotropic homogeneous plate ($R = \infty$) with the material constants given above and $E_{1t} = E_{2t}$. For a uniformly distributed load $q_0 = 300\,\mathrm{psi}$ (or $2.07 \times 10^6\,\mathrm{N/m^2}$) we have $w_3^{phom}(a/2) = 8.842 \cdot 10^{-3}\,\mathrm{m}$.

One can observe that the deflection value is reduced for an orthotropic shell if one of the Young's moduli is increased. However, the deflection value of the FGM shell is larger than that of the homogeneous shell with the material properties identical to those on the top-surface of the FGM shell, characterized by larger Young's moduli.

In the next example, a uniform loading with a Heaviside time variation is applied on the clamped square shallow spherical shell with isotropic material properties. The geometrical and material parameters are the same as in the static case. The time variations of the central deflection for both the homogeneous shell and the FGM shell are presented in Fig. 6.20. The deflection value is normalized by the static central deflection of an isotropic plate ($R = \infty$) with homogeneous material properties identical to those prevailing on the top-surface of the FGM shell. The time is normalized by $t_0 = a^2/4\sqrt{\rho h/D} = 1.35 \cdot 10^{-2}\,\mathrm{s}$. For the FEM analysis we have used 400 quadrilateral eight-node shell elements with 1000 time increments. One can observe a quite good agreement between both the MLPG and the FEM results. The central deflection is larger and its oscillation is slower for an FGM shell than for a homogeneous one with a higher flexural rigidity.

The time-variation of the central bending moment M_{11} of the clamped square shallow spherical shell with curvatures $R/a = 10$ is presented in

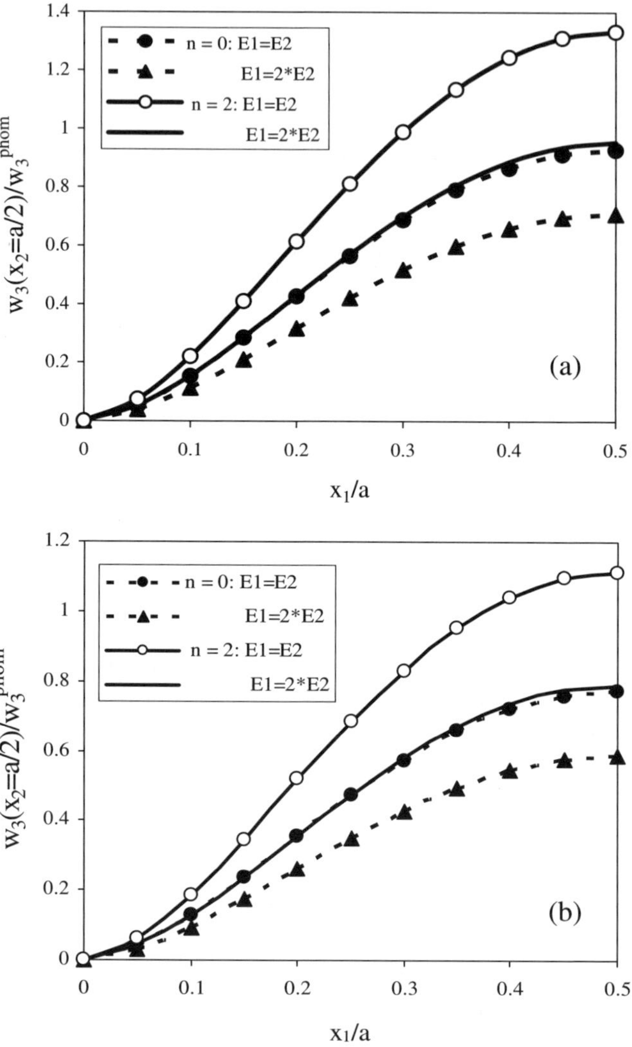

Fig. 6.19. Variation of the deflection with the x_1-coordinate at $x_2 = a/2$ for a clamped orthotropic square shallow spherical shell: a) $R/a = 10$, b) $R/a = 5$.

Fig. 6.21. Small discrepancies of the present MLPG and the FEM results are observed here mainly for larger time instants. Presumably they are caused by the inaccuracy of the Laplace-inversion technique. However, the periodicity of the oscillations and the maximum values are almost the same for both the FEM and the MLPG results. The maximum values of the

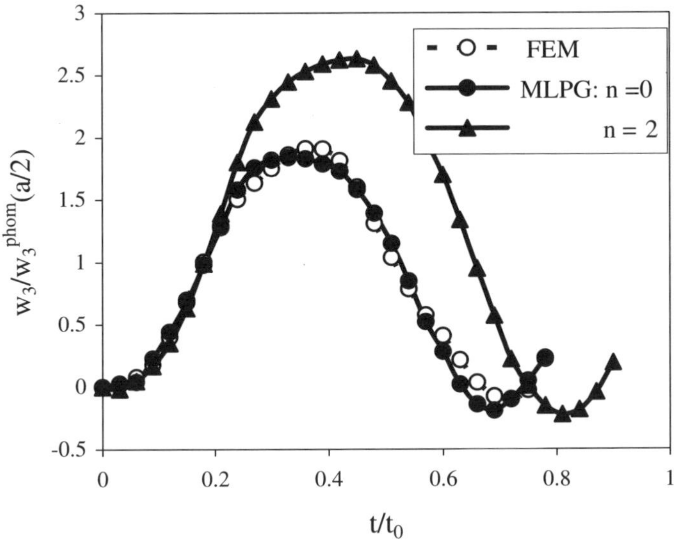

Fig. 6.20. Time variation of the deflection at the center of a clamped isotropic square shallow spherical shell with $R/a = 10$ and subjected an impact loading.

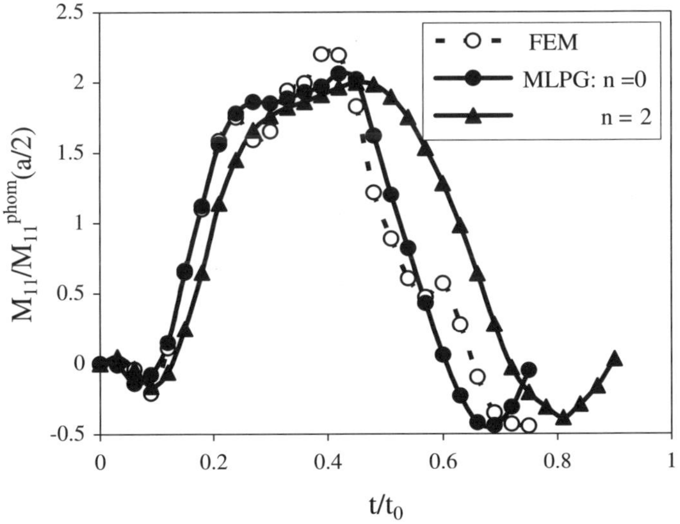

Fig. 6.21. Time variation of the bending moment at the center of a clamped square shallow spherical shell with $R/a = 10$ and subjected to an impact loading.

bending moment M_{11} for both the homogeneous shell and the FGM shell are almost the same. However, that value is lower for more curved shells with a larger curvature.

6.6. Conclusions

The following conclusions can be drawn from the present chapter:

1. A meshless local Petrov-Galerkin method is applied to orthotropic plates and shallow shells under static, dynamic and thermal loads. The behavior of the plates and shells is described by the Reissner–Mindlin theory, which takes the shear deformation into account.
2. The Laplace-transform technique is applied to eliminate the time variable in the coupled governing partial differential equations of the Reissner–Mindlin shell theory. The use of the Laplace-transform in the dynamic analysis converts the time-dependent problem to a set of quasi-static problems.
3. The analyzed domain is divided into small overlapping circular sub-domains. A unit step function is used as the test function in the local weak-form. The derived local boundary-domain integral equations are nonsingular. The Moving Least-Squares (MLS) scheme is adopted for approximating the physical quantities.
4. The main advantage of the present method is its simplicity and generality. The used test function here is simpler than the fundamental solution for orthotropic shells. Therefore, the method seems to be promising for problems, which cannot be solved by the conventional BEM due to unavailable fundamental solutions. However, the computational time in the proposed method is larger since the shape function in the MLS approximation is much more complicated than in FEM using simple polynomials. It is due to the fact that the meshless approximation usually involves more nodes and the corresponding shape functions are more complex. However, by using a mixed formulation[40] the radius of the support domain can be reduced to obtain the same accuracy as in the traditional approximation. A smaller size of the support domain dcreases the bandwidth of the system matrix and the computational time is significantly reduced. Therefore, in future a mixed formulation will be applied to the present plate and shell analyses.
5. The proposed method can be further extended to nonlinear problems, where meshless approximations may have certain advantages over the conventional domain-type disctretization approaches.

Acknowledgments

This work is supported by the Slovak Science and Technology Assistance Agency registered under number APVV-51-021205, the Slovak Grant Agency VEGA-2/6109/27, the EPSRC research grant (U.K.) EP/E050573/1, and the German Research Foundation (DFG).

References

1. S. Suresh and A. Mortensen, *Fundamentals of Functionally Graded Materials* (Institute of Materials, London, 1998).
2. Y. Miyamoto, W.A. Kaysser, B.H. Rabin, A. Kawasaki, and R.G. Ford, *Functionally Graded Materials; Design, Processing and Applications* (Kluwer Academic Publishers, Dordrecht, 1999).
3. J. Wang and M. Huang, *Acta Mechanica Sinica* **7**, (1991).
4. J. Wang and K. Schweizerhof, *Int. J. Solids and Structures* **33**, (1996).
5. E. Reissner, *Journal of Applied Mechanics ASME* **12**, (1945).
6. R. D. Mindlin, *Journal of Applied Mechanics ASME* **18**, (1951).
7. E. Dvorkin and K. J. Bathe, *Engn. Comput.* **1**, (1984).
8. R. A. Arciniega and J. N. Reddy, *Computer Methods in Applied Mechanics and Engineering* **196**, (2007).
9. F. Van der Ween, *International Journal for Numerical Methods in Engineering* **18**, (1982).
10. C. P. Providakis and D. E. Beskos, *Engineering Analysis with Boundary Elements* **14**, (1994).
11. D. E. Beskos, in *Boundary Element Analysis of Plates and Shells,* Ed. D. E. Beskos (Springer-Verlag, Berlin, 1991), p. 93.
12. D. A. Newton and H. Tottenham, *Journal of Engineering Mathematics* **2**, (1968).
13. P. Lu and M. Huang, *Int. J. Solids and Structures* **29**, (1992).
14. J. D. Zhang and S. N. Atluri, *Computers and Structures* **24**, (1986).
15. C. P. Providakis and D. Beskos, *Computational Methods in Applied Mechanical Engineering* **92**, (1991).
16. T. Dirgantara and M. H. Aliabadi, *International Journal for Numerical Methods in Engineering* **45**, (1999).
17. J. Lin and S. Long, *Engn. Analysis with Boundary Elements* **18**, (1996).
18. J. Wang and K. Schweizerhof, *Computers & Structures* **58**, (1996).
19. T. Belytschko, Y. Krogauz, D. Organ, M. Fleming, and P. Krysl, *Comp. Meth. Appl. Mech. Engn.* **139**, (1996).
20. S. N. Atluri, *The Meshless Method (MLPG) For Domain & BIE Discretizations* (Tech Science Press, Norcross, 2004).
21. B. M. Donning and K. M. Liu, *Computer Methods in Applied Mechanics and Engineering* **152**, (1998).
22. P. Krysl and T. Belytschko, *Computational Mechanics* **17**, (1996).
23. P. Krysl and T. Belytschko, *Int. J. Solids and Structures* **33**, (1996).

24. H. Noguchi, T. Kawashima and T. Miyamura, *International Journal for Numerical Methods in Engineering* **47**, (2000).
25. W. K. Liu, S. Jun, and S. Zhang, *International Journal for Numerical Methods in Fluids* **20**, (1995).
26. S. Li, W. Hao, and W. K. Liu, *Computational Mechanics* **25**, (2000).
27. J. Sladek, V. Sladek, and H. A. Mang, *International Journal for Numerical Methods in Engineering* **55**, (2002).
28. J. Sladek, V. Sladek, and H. A. Mang, *Computers & Structures* **81**, (2003).
29. S. Y. Long and S. N. Atluri, *CMES: Computer Modeling in Engineering & Sciences* **3**, (2002).
30. J. Soric, Q. Li, T. Jarak and S. N. Atluri, *CMES: Computer Modeling in Engineering & Sciences* **6**, (2004).
31. H. Stehfest, *Comm. Assoc. Comput. Mach.* **13**, (1970).
32. J. N. Reddy, *Mechanics of Laminated Composite Plates: Theory and Analysis* (CRC Press, Boca Raton, 1997).
33. S. G. Lekhnitskii, *Theory of Elasticity of an Anisotropic Body* (Holden Day, 1963).
34. E. Reissner, *Journal Math. Physics* **25**, (1946).
35. L. F. Qian and R. C. Batra, *Computational Mechanics* **35**, (2005).
36. S. Lukasiewicz, *Local Loads in Plates and Shells* (Noordhoff, London, 1979).
37. J. S. Chen and H. P. Wang, *Comp. Meth. Appl. Mech. Engn.* **187**, (2000).
38. C. P. Providakis and D. E. Beskos, *International Journal for Numerical Methods in Engineering* **49**, (2000).
39. H. S. Carslaw and J. C. Jaeger, *Conduction of Heat in Solids* (Clarendon, Oxford, 1959).
40. S. N. Atluri, H. T. Liu, and Z. D. Han, *CMES: Computer Modeling in Engineering & Sciences* **15**, (2006).

Chapter 7

BOUNDARY ELEMENT TECHNIQUE FOR SLOW VISCOUS FLOWS ABOUT PARTICLES

A. Sellier
LadHyX, Ecole Polytechnique. 91128. Palaiseau Cédex
sellier@ladhyx.polytechnique.fr

This chapter introduces the boundary element technique for slow viscous and steady flows about N-particule clusters made of solid bodies and/or spherical bubbles. It aims at providing a guide to adequately handle the steps encountered when applying this technique to the Stokes equations. In addition to general basic properties of these equations (reciprocal identity, energy dissipation, unicity or not), key integral representations of the velocity and pressure fields are established together with the associated relevant boundary-integral equations. A numerical strategy appealing to so-called boundary elements is examined and a few benchmark tests against exact or very accurate semi-analytical solutions are presented to highlight the abilities of the method.

7.1. Introduction

As illustrated in Fig. 7.1, many basic applications in fluid mechanics today involve either unbounded or bounded multiphase flows for which a collection of so-called particles (solid bodies, bubbles or drops) moves in a suspending *Newtonian* (see Section 7.2.1) *and homogeneous* liquid under prescribed external effects (gravity, capillarity) and/or flow field. The liquid velocity and pressure fields are governed by the unsteady and non-linear Navier-Stokes equations. It is thus tremendously involved to determine each particle motion using direct numerical simulations. However, particles are often of small enough typical size so that inertial effects are negligible (see Section 7.2.1) whereas the characteristic time of diffusion of vorticity is much smaller than the characteristic time of the liquid flow. The suspending *homogeneous Newtonian* liquid, with uniform viscosity μ and density ρ, then admits *quasi-steady* velocity field $\mathbf{u}_e + \mathbf{u}$ and pressure field $p + p_e$ with

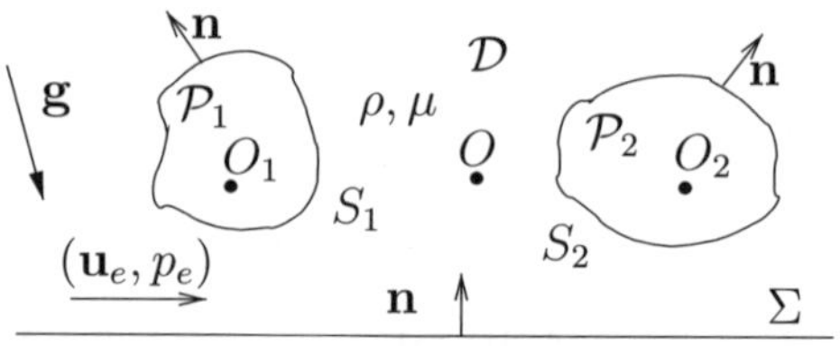

Fig. 7.1. A 2-particle cluster ($N = 2$) suspended in a Newtonian liquid near a solid plane wall Σ and subject to the gravity $\mathbf{g}$ and the ambient (steady Stokes) flow $(\mathbf{u}_e, p_e)$.

$(\mathbf{u}_e, p_e)$ prevailing in the absence of the cluster and the disturbances $(\mathbf{u}, p)$ obeying the steady Stokes equations (see Section 1.2.1)

$$\mu\nabla^2\mathbf{u}_e = \nabla p_e - \rho\mathbf{g} \quad \text{and} \quad \nabla.\mathbf{u}_e = 0 \text{ in } \mathcal{D}, \tag{7.1}$$

$$\mu\nabla^2\mathbf{u} = \nabla p \quad \text{and} \quad \nabla.\mathbf{u} = 0 \text{ in } \mathcal{D},\ (\mathbf{u}, p) \to (\mathbf{0}, 0) \text{ at infinity} \tag{7.2}$$

with $\mathbf{g}$ the gravity and $\mathcal{D}$ the steady liquid domain. The far-field behaviour in (7.2) is required far from the cluster in any direction where $\mathcal{D}$ is not bounded. The flow $(\mathbf{u}_e, p_e)$ is prescribed and one looks at $(\mathbf{u}, p)$ subject to (7.2) and additional boundary conditions on the cluster's surface (which depend on each particle's nature as illustrated by (7.3) and (7.5)–(7.6)) and, if any, on other surfaces bounding the liquid. This has been achieved in the considerable literature[1,2] using either analytical or numerical methods depending upon the external flow $(\mathbf{u}_e, p_e)$ and the number, nature and shapes of the particles and boundaries. If analytical results are available for solid sphere and ellipsoids[3–8] experiencing a given rigid-body translation or rotation, this is not the case for non-ellipsoidal solid particles. The disturbed flow $(\mathbf{u}, p)$ about a moving solid sphere with centre O and radius a immersed in a *general* external steady Stokes flow $(\mathbf{u}_e, p_e)$ has been solved using spherical harmonics.[8] The sphere with surface S moving at prescribed translational velocity $\mathbf{U}$ (the velocity of the centre O) and angular velocity $\mathbf{\Omega}$ is found by solving (7.2) together with the no-slip condition

$$\mathbf{u}(\mathbf{x}) = -\mathbf{u}_e(\mathbf{x}) + \mathbf{U} + \mathbf{\Omega} \wedge \mathbf{x} \quad \text{for } M \text{ on S with } \mathbf{x} = \mathbf{OM}, \tag{7.3}$$

to experience the following net force $\mathbf{F}$ and torque $\mathbf{C}$, with respect to O,

$$\mathbf{F} = 6\pi\mu a[\mathbf{u}_e(O) + \frac{a^2}{6}\nabla^2\mathbf{u}_e(O) - \mathbf{U}], \quad \mathbf{C} = 4\pi\mu a^3[\nabla \wedge \mathbf{u}_e(O) - 2\mathbf{\Omega}]. \tag{7.4}$$

A single spherical bubble with radius a and centre O does not rotate and thus may solely translate at the velocity $\mathbf{U}$. One this time solves (7.2) for other boundary conditions on the bubble surface S. Denoting by $\boldsymbol{\sigma}$ and $\boldsymbol{\sigma}_e$ the stress tensors (defined by (7.12)) of the flows $(\mathbf{u}, p)$ and $(\mathbf{u}_e, p_e)$, it is then required that

$$\mathbf{u}.\mathbf{n} = [\mathbf{U} - \mathbf{u}_e].\mathbf{n} \text{ on } S \tag{7.5}$$

$$\boldsymbol{\sigma}.\mathbf{n} - [\mathbf{n}.\boldsymbol{\sigma}.\mathbf{n}]\mathbf{n} = [\mathbf{n}.\boldsymbol{\sigma}_e.\mathbf{n}]\mathbf{n} - \boldsymbol{\sigma}_e.\mathbf{n} \text{ on } S. \tag{7.6}$$

Solving (7.2) with (7.5)–(7.6) shows[9,10] that the spherical bubble experiences a zero net torque about O and a net force $\mathbf{F}$ given by

$$\mathbf{F} = 4\pi\mu a[\mathbf{u}_e(O) - \mathbf{U}]. \tag{7.7}$$

The Stokes flow $(\mathbf{u}, p)$ about one spherical solid body or bubble experiencing a given rigid-body motion or translation, respectively in the vicinity of a plane and solid or fluid-fluid interface in absence of external flow $(\mathbf{u}_e, p_e)$ has been obtained by using bipolar coordinates.[11] Those coordinates also made it possible to deal with two spherical solid bodies[12–17] or one spherical solid immersed in a liquid bounded by a plane solid wall[18–20] or a spherical cavity.[21,22] The case of a bubble near a fluid-fluid interface has been treated in a similar fashion in Ref. 23. Like (7.4) and (7.7) the obtained results provide excellent benchmark tests for other numerical approaches to be employed for different boundaries and/or for a collection of arbitrarily-shaped solid bodies and/or spherical bubbles. One possible and so-called singularity procedure[24,25] puts *inside* each particle some singularities obeying (7.2) but lacks oa theroretical guide to efficiently select those singularities (type and location) and estimate the achieved accuracy. The closely-connected collocation and multipole methods[26–29] share the same drawbacks. By contrast, the boundary element method[30] offers considerable merits. It spreads suitable singularities on the cluster's boundary whose density are obtained by inverting a few boundary-integral equations on this surface and therefore solely requires to mesh the cluster's surface. On theoretical grounds it goes back to Ref. 31 whilst its first efficient numerical implementation for a solid particle seemingly is due to Ref. 32. Since, the ability of the technique to nicely handle the case of clusters of solid bodies, bubbles or drops has been emphasized in many works.[30]

This chapter smoothly introduces the boundary element technique by sucessively addressing each step usually encountered when running this approach. More precisely, basic assumptions, solutions and properties of steady Stokes flows $(\mathbf{u}, p)$ governed by (7.2) are first examined in Section 7.2 and Section 7.3. The key integral representations of the velocity and pressure disturbances $\mathbf{u}$ and p together with the resulting boundary-integral equations are established in Section 7.4. A possible numerical strategy is proposed in Section 7.5 whereas a few illustrating benchmark tests are reported in Section 7.6. Finally, Section 7.7 closes the chapter by adding a few concluding remarks.

7.2. Stokes Equations. First Properties and Simple Solutions

This section introduces the steady Stokes equations and a few properties and basic solutions.

7.2.1. *Assumptions. Navier-Stokes equations. Reynolds number and Stokes approximation*

We consider, as already sketched in Fig. 7.1, $N \geq 1$ particles $\mathcal{P}_n$ immersed in a *Newtonian and homogeneous* liquid occupying the domain $\mathcal{D}$. Each particle $\mathcal{P}_n$ is either a solid body with arbitrary shape or a spherical bubble. It has attached point O_n and smooth boundary S_n on which the unit normal vector $\mathbf{n}$ is directed into $\mathcal{D}$. The particles and fluid motions are examined with respect to a given laboratory framework with attached Cartesian coordinates (O, x_1, x_2, x_3). The boundary $\partial\mathcal{D}$ of $\mathcal{D}$ reduces to $S = \cup_{n=1}^{N} S_n$ or $S \cup \Sigma$ if the liquid is not bounded or bounded by a surface Σ, respectively. On Σ, which is either closed (liquid bounded by a cavity) or not (open fluid domains), the normal $\mathbf{n}$ is also directed into the liquid.

We henceforth use the usual tensor summation convention with $\mathbf{x} = \mathbf{OM} = x_i \mathbf{e}_i$ for unit vectors $\mathbf{e}_i$ being such that $\mathbf{e}_i.\mathbf{e}_j = \delta_{ij}$ with δ being the Kronecker symbol. At time t and at $\mathbf{x}$ in $\mathcal{D}$ the *homogeneous Newtonian* liquid subject to the gravity $\mathbf{g}$ has velocity $\mathbf{v}(\mathbf{x}, t) = v_i(\mathbf{x}, t)\mathbf{e}_i$ and pressure $P(\mathbf{x}, t)$ governed by the unsteady Navier-Stokes equations [1]

$$\nabla.\mathbf{v} = 0 \quad \text{and} \quad \rho\left\{\frac{\partial \mathbf{v}}{\partial t} + (\mathbf{v}.\nabla)\mathbf{v}\right\} = -\nabla P + \rho\mathbf{g} + \mu\nabla^2\mathbf{v} \text{ in } \mathcal{D}. \qquad (7.8)$$

Those equations, to be supplemented with far-field behaviours for $(\mathbf{v}, P)$ in directions where $\partial\mathcal{D}$ is not bounded and initial and boundary conditions on $\partial\mathcal{D}$, are *time-dependent and non-linear* and thus intricate to solve.

If the particles have length scale a and the velocity $\mathbf{v}$ has typical magnitude V one can introduce, in absence of external forcing, a typical time scale $\tau = a/V$ and the dimensionless Reynolds number $\mathrm{Re} = \rho a V/\mu$.

Under these notations, one readily obtains

$$\rho|(\mathbf{v}.\nabla)\mathbf{v}| = O(\mathrm{Re})|\mu\nabla^2\mathbf{v}|, \quad \rho\left|\frac{\partial\mathbf{v}}{\partial t}\right| = O(\mathrm{Re})|\mu\nabla^2\mathbf{v}|. \tag{7.9}$$

As Re $\to 0$, the inertial term $\rho(\mathbf{v}.\nabla)\mathbf{v}$ and $\rho\partial\mathbf{v}/\partial t$ are neglected in (7.8) which results in the steady low-Reynolds-number or Stokes equations

$$\nabla.\mathbf{v} = 0 \text{ and } \mu\nabla^2\mathbf{u} = \nabla P - \rho\mathbf{g} \text{ in } \mathcal{D}. \tag{7.10}$$

Assuming that $(\mathbf{u}_e, p_e)$ is a steady Stokes flow, i.e. it obeys (7.1), the linearity of (7.10) suggests that the velocity $\mathbf{u} = \mathbf{v} - \mathbf{u}_e$ and pressure $p = P - p_e$ should be introduced. The perturbation flow $(\mathbf{u}, p)$ fulfills the steady Stokes equations

$$\nabla.\mathbf{u} = 0 \quad \text{and} \quad \mu\nabla^2\mathbf{v} = \nabla p \text{ in } \mathcal{D} \tag{7.11}$$

and vanishes far from the cluster where the liquid is not bounded. Using the notation $f_{,i} = \partial f/\partial x_i$, we cast (7.11) into another form by introducing the rate-of-strain and stress tensors $\mathbf{e}$ and $\boldsymbol{\sigma}$ associated with $(\mathbf{u}, p)$. Setting $u_i = \mathbf{u}.\mathbf{e}_i$, these tensors admit the following Cartesian components

$$e_{ij} = (u_{i,j} + u_{j,i})/2, \quad \sigma_{ij} = -p\delta_{ij} + 2\mu e_{ij}. \tag{7.12}$$

Using (7.12), the Stokes equations (7.11) indeed become

$$\nabla.\mathbf{u} = 0 \quad \text{and} \quad \nabla.\boldsymbol{\sigma} = \mathbf{0} \text{ in } \mathcal{D} \quad \text{or} \quad u_{j,j} = 0 \quad \text{and} \quad \sigma_{ij,j} = 0 \text{ in } \mathcal{D}. \tag{7.13}$$

7.2.2. *First properties and noticeable solutions*

As it stands, (7.13) does not impose any far-field or boundary condition for $(\mathbf{u}, p)$ and thus admits many solutions. Moreover, it is *linear*, therefore permitting one to build solutions by combining auxiliary ones. Combining (7.13) with the definitions (7.12) also shows that p is harmonic in the liquid domain, i.e. that $\Delta p = p_{,ii} = 0$ in $\mathcal{D}$ under the notation $f_{,ij} = (f_{,i})_{,j}$. Finally, three classes of trivial solutions with uniform pressure arise:

(1) Solution 1: p uniform and $\mathbf{u} = \nabla\phi$ with ϕ harmonic. Indeed, (7.13) is readily satisfied under the property $\Delta\phi = 0$.
(2) Solution 2: p uniform and $\mathbf{u} = \mathbf{a} \wedge \nabla\phi$ with arbitrary harmonic function ϕ and (constant) vector $\mathbf{a} = a_j\mathbf{e}_j$. Indeed, $u_i = \mathbf{u}.\mathbf{e}_i = \epsilon_{ijk}a_j\phi_{,k}$ with ϵ the usual permutation tensor. Hence, $\nabla.\mathbf{u} = \Delta u_i = 0$ for ϕ harmonic.
(3) Solution 3: p uniform and $\mathbf{u} = \mathbf{a} \wedge \mathbf{x}$ with a constant vector $\mathbf{a}$.

The Solutions 1 and 3 immediately provide the key illustrating examples detailed below:

(i) A rigid-body motion $\mathbf{u}(\mathbf{x}) = \mathbf{U} + \mathbf{\Omega} \wedge \mathbf{x}$ with uniform vectors $\mathbf{U}$ and $\mathbf{\Omega}$ reads $\mathbf{u}(\mathbf{x}) = \nabla\phi + \mathbf{\Omega} \wedge \mathbf{x}$ with $\phi = \mathbf{U}.\mathbf{x}$ *harmonic*. Thus, *a rigid-body motion is a steady Stokes flow with uniform pressure.*
(ii) The steady Stokes flow $(\mathbf{u}, p)$ about a sphere, with radius a and center O, rotating at the angular velocity $\mathbf{\Omega}$ in a quiescent and unbounded liquid is $p = 0$ and $\mathbf{u} = a^3\mathbf{\Omega} \wedge \mathbf{x}/r^3$ for $r = |\mathbf{x}| > a$ ($\mathbf{u} = \mathbf{\Omega} \wedge \nabla\phi_3$ with $\phi_3 = -a^3/r$ *harmonic* for $r > a$). On the sphere's surface, the resulting traction $\mathbf{f} = \boldsymbol{\sigma}.\mathbf{n}$ is tangent and reads

$$\mathbf{f} = \boldsymbol{\sigma}.\mathbf{n} = -3\mu(\mathbf{\Omega} \wedge \mathbf{x})/a \quad \text{on } S. \tag{7.14}$$

The net force $\mathbf{F}$ and torque $\mathbf{C}$ (about center O) then read

$$\mathbf{F} = \int_S \mathbf{f} dS = \mathbf{0}, \quad \mathbf{C} = \int_S \mathbf{x} \wedge \mathbf{f} dS = -8\pi\mu a^3\mathbf{\Omega}. \tag{7.15}$$

7.3. General Solution. Singularities. Singularity Method and Illustrating Examples

This section gives the general form of solutions to (7.13) and a few additional solutions. It also pays attention to the singularity method which is employed to treat the case of a translating spherical solid sphere or bubble.

7.3.1. *General form of a steady Stokes flow*

The exibited Solutions 1 and 2 appeal to one *harmonic* function ϕ. As established in Ref. 33, a general solution $(\mathbf{u}, p)$ to (7.13) actually depends upon seven *harmonic* functions $\varphi, \Phi_1, \Phi_2, \Phi_3, \psi_1, \psi_2$ and ψ_3. More precisely, under the notations $\mathbf{\Phi} = \Phi_i\mathbf{e}_i$ and $\mathbf{\Psi} = \Psi_i\mathbf{e}_i$ it reads

$$\mathbf{u} = \nabla(\mathbf{x}.\mathbf{\Phi}) - 2\mathbf{\Phi} + \nabla\varphi + \nabla \wedge \mathbf{\Psi}, \quad p = 2\mu\nabla.\mathbf{\Phi} \quad \text{in } \mathcal{D}. \tag{7.16}$$

The derivation of (7.16) is available in Ref. 33. The solutions 1, 2 and 3 are retrieved for the only non-zero contributions $\varphi = \phi$, $\mathbf{\Psi} = -\phi\mathbf{a}$ and $\mathbf{\Phi} = -\mathbf{\Omega}\wedge\mathbf{x}/2$. Moreover, less than seven *harmonic* functions are needed as soon as symmetries exist (two *harmonic* functions are sufficient for axisymmetric

Stokes flows[33]). Finally, (7.16) yields the following basic properties:

(1) Property 1: Like any harmonic function, the flow $(\mathbf{u}, p)$ is smooth and admits derivatives of any order in the entire liquid domain.
(2) Property 2: The pressure p is *harmonic* and of specific form (7.16). This latter property is further exploited in Section 7.3.2.2.
(3) Property 3: Both p and $|\mathbf{u}|$ vanish as $r = |\mathbf{x}|$ becomes large in directions where $\mathcal{D}$ is not bounded. Using the behaviour of harmonic functions (7.16) then shows that p and $|\mathbf{u}|$ at least behave in those directions like $1/r^2$ and $1/r$, respectively. Such far-field behaviours are exploited in Section 7.3.2.2.

7.3.2. *Examples of singularity solutions in an unbounded domain*

We examine solutions defined for $r > 0$ that vanish as r becomes large. These solutions, singular at the origin O and called singularities, are employed in the singularity method introduced in Section 7.3.3.

7.3.2.1. *Potential velocity fields*

We look in the domain $r > 0$ at Solutions 1 vanishing far from O. We take $p = 0$ and $\mathbf{u} = \nabla\varphi$ in (7.16) and discard the trivial case φ uniform. Since $\Delta\varphi = 0$ a possible choice is $\varphi = -d/r$ with d being a constant. The resulting pressure, velocity and stress components read

$$p = 0, \quad u_i = dx_i/r^3, \quad \sigma_{ij} = 2\mu d[\delta_{ij}/r^3 - 3x_i x_j/r^5]. \tag{7.17}$$

The potential flow (7.17) is singular at $r = 0$ and induced by a so-called source point at O. The flow rate q through and the net force $\mathbf{F}$ and torque $\mathbf{C}$ exerted on a closed and arbitrary (due to (7.13)) fluid surface that encloses O and has a unit normal vector $\mathbf{n}$ directed outside are

$$q := \int_S \mathbf{u}.\mathbf{n}dS = s, \quad \mathbf{F} := \int_S \boldsymbol{\sigma}.\mathbf{n}dS = \mathbf{0}, \quad \mathbf{C} := \int_S \mathbf{x} \wedge \boldsymbol{\sigma}.\mathbf{n}dS = \mathbf{0}. \tag{7.18}$$

By linearity, one obtains other potential Stokes flows (Solutions 1 with $p = 0$) by combining derivatives of this flow. As an example, let us introduce a *potential dipole* located at O for which the potential function is

$\varphi = -d_k x_k/r^3$. For this dipole, the flow and stress tensor obey

$$u_i = d_k \left[\frac{3x_i x_k}{r^5} - \frac{\delta_{ik}}{r^3} \right], \quad \frac{\sigma_{ij}}{6\mu} = d_k \left\{ \frac{\delta_{ik}x_j + \delta_{ij}x_k + \delta_{jk}x_i}{r^5} - 5\frac{x_i x_j x_k}{r^7} \right\}. \tag{7.19}$$

Further differentiating the source (7.17) provides potential quadrupole, octupole and general multipole singularities located at the pole O producing more and more quickly decaying velocity and stress tensor far from O. For those potential dipoles and flows, it is clear that $q = 0$ and $\mathbf{F} = \mathbf{C} = \mathbf{0}$.

7.3.2.2. *Rotational velocity fields*

The previous singularities yield potential Stokes flows but, of course, rotational singularities also exist. One can think about Solution 2 with p *uniform* and $\mathbf{u} = \mathbf{a} \wedge \nabla\phi$ for ϕ given by differentiating the harmonic function $1/r$ as many times as desired. By virtue of (7.16), another class of solutions with a non-uniform pressure is obtained by setting $p = 2\mu[\Phi_{1,1} + \Phi_{2,2} + \Phi_{3,3}]$ for three harmonic functions Φ_i. Since p is required to vanish far from O it is necessary to first select $p = 2\mu s_k x_k/r^3$ with $\mathbf{s} = s_k \mathbf{e}_k$ a constant vector. Anticipating further results (see Section 7.4.1.1), *one*[a] associated Stokes flow is given by

$$p = 2\mu s_k x_k/r^3, \quad u_i = s_k \left[\frac{\delta_{ik}}{r} + \frac{x_i x_k}{r^3} \right], \quad \sigma_{ij} = -6\mu x_i x_j x_k s_k/r^5. \tag{7.20}$$

The velocity and stress are singular at the pole O and this singularity is called a Stokeslet with intensity $\mathbf{s}$. The mass flux q, net force $\mathbf{F}$ and torque $\mathbf{C}$ *about the pole* O for a closed fluid surface $\mathbf{S}$ enclosing O read

$$q = 0, \quad \mathbf{F} = \int_S \boldsymbol{\sigma}.\mathbf{n}dS = -8\pi\mu\mathbf{s}, \quad \mathbf{C} = \int_S \mathbf{x} \wedge \boldsymbol{\sigma}.\mathbf{n}dS = \mathbf{0}. \tag{7.21}$$

Again, other singularities are obtained by differentiating the Stokeslet solution resulting at the Stokeslet doublet, the Stokeslet quadrupole and, in general, Stokeslet multipoles.[30]

[a]Other solutions are readily obtained by adding any divergence-free velocity field $\mathbf{w}$ obeying $\nabla^2\mathbf{w} = \mathbf{0}$ for $r > 0$ and quiescent far from O.

7.3.3. Singularity method and examples

Each previous singularity induces at $\mathbf{x}$ a pressure $p_s(\mathbf{x})$, a velocity $\mathbf{u}_s(\mathbf{x})$ and a stress tensor $\boldsymbol{\sigma}_s(\mathbf{x})$. When located at another pole $\mathbf{y}$, the singularity produces the fields $p_s(\mathbf{x}-\mathbf{y}), \mathbf{u}_s(\mathbf{x}-\mathbf{y})$ and $\boldsymbol{\sigma}_s(\mathbf{x}-\mathbf{y})$ at $\mathbf{x}$, such that $|\mathbf{x}-\mathbf{y}| > 0$. The singularity method obtains the exterior fluid flow $(\mathbf{u}, p)$ governed by (7.11) by putting a collection of singularities outside the liquid domain $\mathcal{D}$, i.e. inside the particles $\mathcal{P}_n$ and elsewhere outside $\mathcal{D}$ if there exists a boundary Σ (recall Fig. 7.1). The intensity of given (number, nature and location) singularities is obtained by enforcing boundary conditions (similar to (7.3) or (7.5)–(7.6)) at points on the cluster's surface S and on Σ. The flow $(\mathbf{u}, p)$ is subsequently approximated in the liquid by using the determined singularities. Unfortunately, there is no recipe to select the number, nature and location of the singularities to adequately control the error for $(\mathbf{u}, p)$. For a single solid sphere or prolate spheroid, more and more singularities are required for complex external flows $(\mathbf{u}_e, p_e)$,[25] and studies[24,30] also reveal the sensitivity to the choice of singularities.

Sometimes the singularity method is exact. This is the case for a spherical particle (solid or bubble) with center O and radius a translating at the velocity $\mathbf{U}$ in a quiescent liquid. The liquid flow $(\mathbf{u}, p)$ vanishing far from O is produced by putting, at O, a Stokeslet and a potential dipole with unknown intensity $\mathbf{s}$ and $\mathbf{d}$, respectively. In other words, one sets

$$p = 2\mu s_k x_k / r^3, \quad u_i = s_k \left[\frac{\delta_{ik}}{r} + \frac{x_i x_k}{r^3}\right] + d_k \left[\frac{3x_i x_k}{r^5} - \frac{\delta_{ik}}{r^3}\right] \text{ for } r > 0. \tag{7.22}$$

For a *solid* sphere with surface S, the boundary condition $\mathbf{u} = \mathbf{U}$ at $r = a$ and (7.21) immediately give

$$\mathbf{s} = 3a\mathbf{U}/4, \quad \mathbf{d} = -a^3\mathbf{U}/4, \quad \mathbf{F} = \int_S \boldsymbol{\sigma}.\mathbf{n} dS = -6\pi\mu a\mathbf{U}. \tag{7.23}$$

Using (7.22) also gives the surface traction $\mathbf{f} = \boldsymbol{\sigma}.\mathbf{n}$ on S, and the exerted hydrodynamic $\mathbf{F}$ and torque $\mathbf{C}$ about O. One easily obtains

$$\mathbf{f} = \boldsymbol{\sigma}.\mathbf{n} = -3\mu\mathbf{U}/[2a] \text{ on } S, \quad \mathbf{F} = -6\pi\mu a\mathbf{U}, \quad \mathbf{C} = \int_C \mathbf{x} \wedge \mathbf{f} dS = \mathbf{0}. \tag{7.24}$$

For a spherical *bubble* with boundary S, using the conditions (7.5)–(7.6) for $\mathbf{u}_e = \mathbf{0}$ and $\boldsymbol{\sigma}_e = \mathbf{0}$ this times provides $\mathbf{d} = \mathbf{0}$ and

$$\mathbf{s} = \frac{a\mathbf{U}}{2}, \quad \mathbf{f} = \boldsymbol{\sigma}.\mathbf{n} = -\frac{3\mu(\mathbf{U}.\mathbf{x})\mathbf{x}}{a^4}, \quad \mathbf{F} = \int_S \boldsymbol{\sigma}.\mathbf{n}dS = -4\pi\mu a\mathbf{U}. \tag{7.25}$$

7.3.4. *General properties*

This section presents additional general and key properties of the solutions to (7.13). These properties will be exploited in Section 7.4.

7.3.4.1. *Reciprocal identity*

Let us consider two arbitrary steady Stokes flows $(\mathbf{u}, p)$ and $(\mathbf{u}', p')$ governed by (7.13). The following so-called reciprocal identity then holds

$$\int_{\partial\mathcal{D}} \mathbf{u}.\boldsymbol{\sigma}'.\mathbf{n}dS = \int_{\partial\mathcal{D}} \mathbf{u}'.\boldsymbol{\sigma}.\mathbf{n}dS. \tag{7.26}$$

Here, we recall that $\mathcal{D} = S \cup \Sigma$ with S being the cluster's surface. If $\mathcal{D}$ is bounded whatever the direction one obtains for $\mathbf{n}$ directed into the liquid and using the divergence theorem

$$\int_{\partial\mathcal{D}} [\mathbf{u}.\boldsymbol{\sigma}'.\mathbf{n} - \mathbf{u}'.\boldsymbol{\sigma}.\mathbf{n}]dS = \int_{\mathcal{D}} [u_i'\sigma_{ij} - u_i\sigma_{ij}']_{,j}d\Omega. \tag{7.27}$$

The identity (7.26) holds since, using (7.12)–(7.13), one arrives at $[u_i'\sigma_{ij} - u_i\sigma_{ij}']_{,j} = u_{i,j}'\sigma_{ij} - u_{i,j}\sigma_{ij}' = \mu[u_{i,j}'u_{j,i} - u_{i,j}u_{j,i}'] = 0$. If $\mathcal{D}$ is not bounded in some directions, then Σ is not bounded. For $R > 0$ large enough, one selects a closed cavity $\Sigma_R \cup S_R$ containing the cluster and made of a bounded subset of Σ, here denoted by Σ_R, and a subset S_R of the spherical surface $r = R$. The liquid domain in the cavity is bounded by $S \cup \Sigma_R \cup S_R$ for which the relation (7.26) thus holds. As R becomes large, the contribution over S_R vanishes due to the far-field behaviours of the Stokes flows $(\mathbf{u}, p)$ and $(\mathbf{u}', p')$ (see Property 3 given in Section 7.3.1) and Σ_R tends to Σ. Hence, (7.26) also holds for $\mathcal{D}$ not totally bounded.

7.3.4.2. *Energy dissipation and unicity of the solution*

For a steady Stokes flow $(\mathbf{u}, p)$ solution to (7.13), one introduces the rate of dissipation of mechanical energy E at the flow boundary as

$$E = \int_{\partial\mathcal{D}} \mathbf{u}.\boldsymbol{\sigma}.\mathbf{n}dS. \tag{7.28}$$

Using the divergence theorem, (7.12)–(7.13), and the auxiliary surface S_R when the liquid domain is not totally bounded yields

$$
\begin{aligned}
E &= -\int_{\mathcal{D}} [u_i \sigma_{ij}]_{,j} d\Omega = -\int_{\mathcal{D}} (u_{i,j}\sigma_{ij}) d\Omega \\
&= -\frac{\mu}{2}\int_{\mathcal{D}} (u_{i,j} + u_{j,i})(u_{i,j} + u_{j,i}) d\Omega \leq 0. \qquad (7.29)
\end{aligned}
$$

The result (7.29) permits one to discuss the unicity of a solution $(\mathbf{u}, p)$ to (7.13) when supplemented with boundary conditions on $S \cup \Sigma$ and, if need, vanishing far from the cluster (unbounded flow domain $\mathcal{D}$). We assume these additional boundary conditions on $S \cup \Sigma$ to read $\mathcal{L}(\mathbf{u}, p) = d$ with $\mathcal{L}$ a linear operator and d a prescribed data. Upon introducing another flow $(\mathbf{u}', p')$ governed by (7.13) and those boundary conditions, the difference $\mathbf{u}'' = \mathbf{u}' - \mathbf{u}, p'' = p' - p$ obeys the Stokes equations and the homogenous conditions $\mathcal{L}(\mathbf{u}, p) = 0$. Whenever these latter conditions yield $\mathbf{u}''.\boldsymbol{\sigma}''.\mathbf{n} = 0$ on $S \cup \Sigma$, the relation (7.29) shows that the continuous[b] rate-of-strain tensor $\mathbf{e}''$ (recall (7.12)) is zero in $\mathcal{D}$. Since only rigid-body motions have a zero rate-of-strain tensor, $\mathbf{u}''$ is written as $\mathbf{u}'' = \mathbf{U}'' + \boldsymbol{\Omega}'' \wedge \mathbf{x}$ and the pressure p'' is uniform in $\mathcal{D}$. Several different cases, dictated by the boundary conditions required on $S \cup \Sigma$ for the flow $(\mathbf{u}, p)$, then arise:

(1) Case 1: Not totally bounded domain $\mathcal{D}$. The homogeneous far-field condition shows that $\mathbf{u}''$ and p'' vanish in $\mathcal{D}$ and $(\mathbf{u}, p)$ is thus unique.
(2) Case 2: Domain $\mathcal{D}$ enclosed by a cavity. *The solution* $(\mathbf{u}, p)$ *is then not-necessarily unique,* as shown below:
   (i) If each velocity component $\mathbf{u}''.\mathbf{e}_i$ vanishes on a subset A_i of $S \cup \Sigma$, it vanishes in $\mathcal{D}$ with $\mathbf{u}$ unique and p defined up to an arbitrary constant.
   (ii) If the previous conditions break down, $(\mathbf{u}, p)$ is unique up to an arbitrary rigid-body flow. For instance, this arises if one provides the traction $\mathbf{f} = \boldsymbol{\sigma}.\mathbf{n}$ for the flow $(\mathbf{u}, p)$ on the *entire* boundary $S \cup \Sigma$.

7.3.4.3. *Basic application*

Both (7.26) and (7.29) admit straightforward applications. One example is given by the motion of a single particle $\mathcal{P}$ (solid body or spherical bubble) with attached point O and surface S immersed in the external steady Stokes

[b]Note that $\mathbf{e}''$ is regular in the liquid domain $\mathcal{D}$ (see Property 1 in Section 7.3.1).

flow $(\mathbf{u}_e, p_e)$ bounded by a solid and motionless surface Σ (for instance a plane solid wall or a closed and solid cavity) where $\mathbf{u}_e$ vanishes. The disturbed flow $(\mathbf{u}_e + \mathbf{u}, p_e + p)$ fulfills a no-slip condition on Σ and tends to $(\mathbf{u}_e, p_e)$ far from the particle. Accordingly, the disturbance $(\mathbf{u}, p)$ obeys

$$\nabla.\mathbf{u} = 0, \quad \nabla^2\mathbf{u} = \nabla p \quad \text{in } \mathcal{D}; \quad (\mathbf{u}, p) \to (\mathbf{0}, 0) \text{ at infinity}, \quad \mathbf{u} = \mathbf{0} \text{ on } \Sigma. \tag{7.30}$$

It exerts on the particle the net hydrodynamic force $\mathbf{F}$ and torque $\mathbf{C}$

$$\mathbf{F} = \int_S \boldsymbol{\sigma}.\mathbf{n}dS, \quad \mathbf{C} = \int_S \mathbf{x} \wedge \boldsymbol{\sigma}.\mathbf{n}dS \tag{7.31}$$

and obeys on S conditions related to the particle's nature. For a solid body moving at the translational and angular velocities $\mathbf{U}$ and $\boldsymbol{\Omega}$, one prescribes

$$\mathbf{u} = -\mathbf{u}_e + \mathbf{U} + \boldsymbol{\Omega} \wedge \mathbf{x} \text{ on } S \tag{7.32}$$

and it is worth introducing six auxiliary steady Stokes flows $(\mathbf{u}_T^{(i)}, p_T^{(i)})$ and $(\mathbf{u}_R^{(i)}, p_R^{(i)})$ with stress tensors $\boldsymbol{\sigma}_T^{(i)}$ and $\boldsymbol{\sigma}_R^{(i)}$, respectively. Those flows obey (7.30) and (7.32) for $\mathbf{u}_e = \mathbf{0}$ with $(\mathbf{U}, \boldsymbol{\Omega}) = (\mathbf{e}_i, \mathbf{0})$ for $(\mathbf{u}_T^{(i)}, p_T^{(i)})$ and $(\mathbf{U}, \boldsymbol{\Omega}) = (\mathbf{0}, \mathbf{e}_i)$ for $(\mathbf{u}_R^{(i)}, p_R^{(i)})$. The auxiliary flows are produced by a translation or a rotation of the particle. Setting $\mathbf{f}_L^{(i)} = \boldsymbol{\sigma}_L^{(i)}.\mathbf{n}$ on S for $L = T, R$ and exploiting (7.26) for those flows and the flow $(\mathbf{u}, p)$ then yields

$$\mathbf{F}.\mathbf{e}_i = \int_S \mathbf{u}.\mathbf{f}_T^{(i)} dS = \mathbf{U}. \int_S \mathbf{f}_T^{(i)} dS + \boldsymbol{\Omega}. \int_S [\mathbf{x} \wedge \mathbf{f}_T^{(i)}] dS - \int_S \mathbf{u}_e.\mathbf{f}_T^{(i)} dS, \tag{7.33}$$

$$\mathbf{C}.\mathbf{e}_i = \int_S \mathbf{u}.\mathbf{f}_R^{(i)} dS = \mathbf{U}. \int_S \mathbf{f}_R^{(i)} dS + \boldsymbol{\Omega}. \int_S [\mathbf{x} \wedge \mathbf{f}_R^{(i)}] dS - \int_S \mathbf{u}_e.\mathbf{f}_R^{(i)} dS. \tag{7.34}$$

For a spherical bubble translating at the velocity $\mathbf{U}$ without rotating, the flow $(\mathbf{u}, p)$ this time obeys the boundary conditions

$$\mathbf{u}.\mathbf{n} = [\mathbf{U} - \mathbf{u}_e].\mathbf{n} \quad \text{on } S \tag{7.35}$$

$$\boldsymbol{\sigma}.\mathbf{n} - [\mathbf{n}.\boldsymbol{\sigma}.\mathbf{n}]\mathbf{n} = [\mathbf{n}.\boldsymbol{\sigma}_e.\mathbf{n}]\mathbf{n} - \boldsymbol{\sigma}_e.\mathbf{n} \quad \text{on } S \tag{7.36}$$

which suggests that three Stokes flows $(\mathbf{u}_T^{(i)}, p_T^{(i)})$ governed by (7.30) and (7.35)–(7.36) for $\mathbf{u}_e = \boldsymbol{\sigma}_e.\mathbf{n} = \mathbf{0}$ and $\mathbf{U} = \mathbf{e}_i$ should be introduced.

The applied torque is zero and the net force $\mathbf{F}$ is obtained by using the decompositions

$$\mathbf{F}.\mathbf{e}_i = \int_S [\mathbf{e}_i - (\mathbf{e}_i.\mathbf{n})\mathbf{n}].[\boldsymbol{\sigma}.\mathbf{n} - (\mathbf{n}.\boldsymbol{\sigma}.\mathbf{n})\mathbf{n}]dS + \int_S (\mathbf{e}_i.\mathbf{n})(\mathbf{n}.\boldsymbol{\sigma}.\mathbf{n})dS, \tag{7.37}$$

$$A_i := \int_S \mathbf{u}_T^{(i)}.\boldsymbol{\sigma}.\mathbf{n}dS = \int_S (\mathbf{e}_i.\mathbf{n})(\mathbf{n}.\boldsymbol{\sigma}.\mathbf{n})dS + \int_S [\mathbf{u}_T^{(i)} - (\mathbf{u}_T^{(i)}.\mathbf{n})\mathbf{n}] \\ .[\boldsymbol{\sigma}.\mathbf{n} - (\mathbf{n}.\boldsymbol{\sigma}.\mathbf{n})\mathbf{n}]dS, \tag{7.38}$$

$$B_i := \int_S \mathbf{u}.\boldsymbol{\sigma}_T^{(i)}.\mathbf{n}dS = \int_S [(\mathbf{U} - \mathbf{u}_e).\mathbf{n}][\mathbf{n}.\boldsymbol{\sigma}_T^{(i)}.\mathbf{n}]dS. \tag{7.39}$$

Indeed, (7.26) shows that $A_i = B_i$. Combining (7.37), (7.38) and (7.39) then provides

$$\mathbf{F}.\mathbf{e}_i = \int_S (\mathbf{U}.\mathbf{n})(\mathbf{n}.\boldsymbol{\sigma}_T^{(i)}.\mathbf{n})dS \\ + \int_S \{[\mathbf{u}_T^{(i)} - \mathbf{e}_i].[\boldsymbol{\sigma}_e.\mathbf{n} - (\mathbf{n}.\boldsymbol{\sigma}_e.\mathbf{n})\mathbf{n} - (\mathbf{u}_e.\mathbf{n})(\mathbf{n}.\boldsymbol{\sigma}_T^{(i)}.\mathbf{n})\}dS. \tag{7.40}$$

Using $\mathbf{f}_T^{(i)}$ given by (7.25) provides (7.7) and, for a solid sphere, (7.4) is gained by injecting (7.24) into (7.33)–(7.34). In the absence of external flow (7.33)–(7.34) and (7.40) also read $(\mathbf{F}, \mathbf{C}) = -\mu\mathbf{M}_s.(\mathbf{U}, \boldsymbol{\Omega})$ and $\mathbf{F} = -\mu\mathbf{M}_b.\mathbf{U}$ with 6×6 and 3×3 square matrices $\mathbf{M}_s$ and $\mathbf{M}_b$. Exploiting (7.26) and (7.29) also shows that such matrices are symmetric and definite-positive.

7.4. Integral Representations for the Pressure and Velocity Fields. Resulting Boundary Integral Equations

We introduce in this section Green's solutions and integral representations of the velocity and pressure fields for a steady flow governed by (7.13). This permits one to subsequently establish key boundary-integral equations.

7.4.1. *Green solution*

Henceforth we denote by $\mathcal{D}'$ the domain consisting of *both the particles and* $\mathcal{D}$. As seen in Fig. 7.1, this domain is either the entire space or bounded by a surface Σ. The steady Stokes flow $(\mathbf{v}, q)$ induced by a point force of intensity $\mathbf{t}$ placed at the pole $\mathbf{y}$ in $\mathcal{D}'$ obeys

$$\mu\nabla^2\mathbf{v} = \nabla q - \delta_{3d}(\mathbf{x}-\mathbf{y})\mathbf{t}, \quad \nabla.\mathbf{v} = 0 \text{ in } \mathcal{D}' \tag{7.41}$$

where $\delta_{3d}(\mathbf{x}-\mathbf{y}) = \delta_d(x_1-y_1)\delta_d(x_2-y_2)\delta_d(x_3-y_3)$ if $y_i = \mathbf{y}.\mathbf{e}_i$ with δ_d the usual Dirac pseudo-function. The so-called *Green* or *fundamental solution* $(\mathbf{v}, q)$ is *not unique*: one can add a constant pressure and a divergence-free velocity $\mathbf{w}$ such that $\nabla^2\mathbf{w} = \mathbf{0}$. It is, however, possible to retain a solution *linearly* depending upon the vector $\mathbf{t}$. More precisely, for such a solution $(\mathbf{v}, q)$, with stress tensor $\boldsymbol{\sigma}$, there exist so-called Green pressure vector $\mathbf{P}$, second-rank Green velocity tensor $\mathbf{G}$ and third-rank Green stress tensor $\mathbf{T}$ with Cartesian components P_j, G_{ij} and T_{ijk} such that

$$q(\mathbf{x}) = \frac{P_j(\mathbf{x},\mathbf{y})t_j}{8\pi}, \quad v_i(\mathbf{x}) = \frac{G_{ij}(\mathbf{x},\mathbf{y})t_j}{8\pi\mu}, \quad \sigma_{ik}(\mathbf{x}) = \frac{T_{ijk}(\mathbf{x},\mathbf{y})t_j}{8\pi} \tag{7.42}$$

with, by virtue of the definition (7.12), the link

$$T_{ijk}(\mathbf{x},\mathbf{y}) = -\delta_{ik}P_j(\mathbf{x},\mathbf{y}) + G_{ij,k}(\mathbf{x},\mathbf{y}) + G_{kj,i}(\mathbf{x},\mathbf{y}). \tag{7.43}$$

In view of (7.43), it is then sufficient in practice to obtain the vector $\mathbf{P}$ and the velocity tensor $\mathbf{G}$.

7.4.1.1. *Free-space Green tensor*

When $\mathcal{D}'$ is the entire space, it becomes easy to obtain the Green tensor. Recalling that $\Delta(\phi) + \delta_{3d}(\mathbf{x}-\mathbf{y}) = 0$ for $\phi(\mathbf{x},\mathbf{y}) = [4\pi|\mathbf{x}-\mathbf{y}|]^{-1}$, one gets

$$\Delta q - \delta_{3d,i}(\mathbf{x}-\mathbf{y})t_i = 0, \quad \Delta\phi_{,i}(\mathbf{x},\mathbf{y}) + \delta_{3d,i}(\mathbf{x}-\mathbf{y}) = 0. \tag{7.44}$$

Thus, the pressure q obeys

$$\Delta[q + \mathbf{t}.\nabla\phi] = 0. \tag{7.45}$$

For the desired flow $(\mathbf{v}, q)$ *linear in* $\mathbf{t}$, defined in the entire space $|\mathbf{x}-\mathbf{y}| > 0$ and vanishing far from the pole $\mathbf{y}$, one accordingly selects $q = -\mathbf{t}.\nabla\phi$. Consequently, for $\mathbf{x} \neq \mathbf{y}$ the velocity component v_i satisfies

$$4\pi\mu\Delta v_i = \left[\frac{\delta_{ij}}{|\mathbf{x}-\mathbf{y}|} - \frac{3(x_i-y_i)(x_j-y_j)}{|\mathbf{x}-\mathbf{y}|^3}\right] t_j. \tag{7.46}$$

Inspecting (7.46) then suggests seeking the solution *linear in* $\mathbf{t}$ in the form

$$v_i(\mathbf{x}) = \frac{1}{4\pi}\left[\frac{b\delta_{ij}}{|\mathbf{x}-\mathbf{y}|} + \frac{c(x_i-y_i)(x_j-y_j)}{|\mathbf{x}-\mathbf{y}|^3}\right] t_j \tag{7.47}$$

where the constants b and c are determined by enforcing the divergence-free condition $u_{i,i} = 0$ and (7.46). This provides the widely-employed and so-called free-space Oseen–Burgers[7] solution

$$G_{ij}^{\infty}(\mathbf{x},\mathbf{y}) = \frac{\delta_{ij}}{|\mathbf{x}-\mathbf{y}|} + \frac{(x_i-y_i)(x_j-y_j)}{|\mathbf{x}-\mathbf{y}|^3}, \tag{7.48}$$

$$P_j^{\infty}(\mathbf{x},\mathbf{y}) = \frac{2(x_j-y_j)}{|\mathbf{x}-\mathbf{y}|^3}, \quad T_{ijk}^{\infty}(\mathbf{x},\mathbf{y}) = -\frac{6(x_i-y_i)(x_j-y_j)(x_k-y_k)}{|\mathbf{x}-\mathbf{y}|^5}. \tag{7.49}$$

The associated fundamental solution $(\mathbf{v}, q)$ to (7.41), which might be also obtained by Fourier transform,[34] agrees with the previously-proposed Stokeslet singularity (7.20). As revealed by (7.48)–(7.49), its pressure vector and velocity or stress tensors are *singular* as $\mathbf{x}$ tends to $\mathbf{y}$. Moreover, this singularity flow admits the following basic properties:

(i) Two symmetry properties. For any pair (i,j) one has

$$G_{ij}^{\infty}(\mathbf{x},\mathbf{y}) = G_{ji}^{\infty}(\mathbf{y},\mathbf{x}), \quad G_{ij}^{\infty}(\mathbf{x},\mathbf{y}) = G_{ji}^{\infty}(\mathbf{x},\mathbf{y}). \tag{7.50}$$

(ii) For a prescribed pair (i,k) the flow $T_{ijk}^{\infty}(\mathbf{x},\mathbf{y})\mathbf{e}_j$ is a steady Stokes flow with a pressure $\Pi_{ik}^{\infty}(\mathbf{x},\mathbf{y})$ given by[c]

$$\Pi_{ik}^{\infty}(\mathbf{x},\mathbf{y}) = 4\left\{\frac{3(x_i-y_i)(x_k-y_k)}{|\mathbf{x}-\mathbf{y}|^5} - \frac{\delta_{ik}}{|\mathbf{x}-\mathbf{y}|^3}\right\}. \tag{7.51}$$

7.4.1.2. *Case of a bounded fluid domain*

When $\mathcal{D}'$ is bounded by Σ it is, of course, still possible to use the previous and simple Oseen–Burgers solution since (7.41) is free from boundary conditions. It is, however, worth (see Section 7.4.2) introducing another Green solution $(\mathbf{P}^c, \mathbf{G}^c, \mathbf{T}^c)$ for which the associated flow $(\mathbf{v}, q)$ satisfies (7.41) *and* additional boundary conditions on Σ (the ones imposed on the liquid flow $(\mathbf{u}, p)$ solution to (7.2)). By superposition, another such Green

[c]Since obtained by elementary algebra, the proof is left to the reader.

solution reads

$$P_j^c(\mathbf{x},\mathbf{y}) = P_j^\infty(\mathbf{x},\mathbf{y}) + R_j(\mathbf{x},\mathbf{y}), \tag{7.52}$$

$$G_{ij}^c(\mathbf{x},\mathbf{y}) = G_{ij}^\infty(\mathbf{x},\mathbf{y}) + R_{ij}(\mathbf{x},\mathbf{y}), \tag{7.53}$$

$$T_{ijk}^c(\mathbf{x},\mathbf{y}) = T_{ijk}^\infty(\mathbf{x},\mathbf{y}) + R_{ijk}(\mathbf{x},\mathbf{y}) \tag{7.54}$$

with functions $R_j(\mathbf{x},\mathbf{y}), R_{ij}(\mathbf{x},\mathbf{y})$ and $R_{ijk}(\mathbf{x},\mathbf{y})$ selected so that $(\mathbf{v},q)$ satisfies the additional boundary conditions on Σ. The steady Stokes flow associated with these functions obeys (7.13) and Property 1 given in Section 7.3.1 shows that R_j, R_{ij} and R_{ijk} are *smooth and bounded* in the entire domain $\mathcal{D}'$. Accordingly, the Green solution $(\mathbf{P}^c, \mathbf{G}^c, \mathbf{T}^c)$ exhibits, as the observation point $\mathbf{x}$ tends to the pole $\mathbf{y}$, the singular behaviours (7.48)–(7.49). In addition, in any direction where the liquid domain is not bounded, the flow $(\mathbf{v},q)$ vanishes and, as $|\mathbf{x}-\mathbf{y}|$ becomes large, the functions R_j, R_{ij} and R_{ijk} at least decay as $|\mathbf{x}-\mathbf{y}|^{-1}$ for R_{ij} or $|\mathbf{x}-\mathbf{y}|^{-2}$ for R_j and R_{ijk}. Whenever the flow $(\mathbf{v},q)$, with stress tensor $\boldsymbol{\sigma}$, satisfies $\mathbf{v}.\boldsymbol{\sigma}.\mathbf{n} = 0$ on the entire surface Σ, it is also possible to introduce another second-rank pressure tensor $\Pi_{ik}^c(\mathbf{x},\mathbf{y})$, and the symmetry property $G_{ij}^c(\mathbf{x},\mathbf{y}) = G_{ji}^c(\mathbf{y},\mathbf{x})$ holds.[d]

In practice, it's worth using the Green solution $(\mathbf{P}^c, \mathbf{G}^c, \mathbf{T}^c)$ when at least the tensor $\mathbf{G}^c$ is known in a closed form permitting a straightforward numerical implementation. Except for a few simple surfaces Σ with specific boundary conditions it is, unfortunately, not possible to analytically determine $\mathbf{G}^c$. The most encountered case is the motionless and solid boundary Σ where one requires the velocity to vanish, i.e that $G_{ij}^c(\mathbf{x},\mathbf{y}) = 0$ for $\mathbf{x}$ located on Σ. The case of a solid and motionless plane wall was treated in Ref. 35. For a liquid in the $x_3 > 0$ domain, setting $c_1 = c_2 = 1, c_3 = -1$ and denoting by $\mathbf{y}'(y_1, y_2, -y_3)$ the symmetric of $\mathbf{y}(y_1, y_2, y_3)$ with respect to the $x_3 = 0$ plane wall Σ,[35] not only obtained the second-order velocity Green tensor components

$$\begin{aligned} R_{ij}(\mathbf{x},\mathbf{y}) = &-\left\{\frac{\delta_{ij}}{|\mathbf{y}'-\mathbf{x}|} + \frac{[(\mathbf{y}'-\mathbf{x}).\mathbf{e}_i][(\mathbf{y}'-\mathbf{x}).\mathbf{e}_j]}{|\mathbf{y}'-\mathbf{x}|^3}\right\} \\ &-\frac{2c_j(\mathbf{y}.\mathbf{e}_3)}{|\mathbf{y}'-\mathbf{x}|^3}\left\{\delta_{i3}(\mathbf{y}'-\mathbf{x}).\mathbf{e}_j - \delta_{j3}(\mathbf{y}'-\mathbf{x}).\mathbf{e}_i\right. \\ &\left.+(\mathbf{x}.\mathbf{e}_3)\left[\delta_{ij} - \frac{3[(\mathbf{y}'-\mathbf{x}).\mathbf{e}_i][(\mathbf{y}'-\mathbf{x}).\mathbf{e}_j]}{|\mathbf{y}'-\mathbf{x}|^2}\right]\right\}, \end{aligned} \tag{7.55}$$

[d] A derivation for the simple case $\mathbf{v} = \mathbf{0}$ on Σ is available in Ref. 30 and the announced extended results are obtained in a similar fashion.

but also, after gaining the pressure, computed the function R_{ijk} arriving at

$$
\begin{aligned}
R_{ijk}(\mathbf{x},\mathbf{y}) = {} & -\frac{6[(\mathbf{y}'-\mathbf{x}).\mathbf{e}_i][(\mathbf{y}'-\mathbf{x}).\mathbf{e}_j][(\mathbf{y}'-\mathbf{x}).\mathbf{e}_j]}{|\mathbf{y}'-\mathbf{x}|^5} - \frac{12c_j(\mathbf{y}.\mathbf{e}_3)}{|\mathbf{y}'-\mathbf{x}|^5} \\
& \times \Bigg\{ \delta_{ij}[(\mathbf{y}'-\mathbf{x}).\mathbf{e}_j][(\mathbf{y}'-\mathbf{x}).\mathbf{e}_3] - \delta_{j3}[(\mathbf{y}'-\mathbf{x}).\mathbf{e}_i][(\mathbf{y}'-\mathbf{x}).\mathbf{e}_j] \\
& - (\mathbf{x}.\mathbf{e}_3)\Bigg(\delta_{ij}[(\mathbf{y}'-\mathbf{x}).\mathbf{e}_j] + \delta_{ij}[(\mathbf{y}'-\mathbf{x}).\mathbf{e}_j] + \delta_{jj}[(\mathbf{y}'-\mathbf{x}).\mathbf{e}_i] \\
& + \frac{5[(\mathbf{y}'-\mathbf{x}).\mathbf{e}_i][(\mathbf{y}'-\mathbf{x}).\mathbf{e}_j][(\mathbf{y}'-\mathbf{x}).\mathbf{e}_j]}{|\mathbf{y}'-\mathbf{x}|^5} \Bigg) \Bigg\}.
\end{aligned} \tag{7.56}
$$

Result (7.55) has been exploited in Ref. 36 to deal with the migration of solid spheres in the vicinity of a plane solid wall under the action of an external shear flow.

When the boundary Σ is a spherical, solid and motionless cavity with center O and radius R it is also possible to calculate[7] the velocity tensor $\mathbf{G}^c(\mathbf{x},\mathbf{y})$ that vanishes for $\mathbf{x}$ on Σ. Unfortunately, the analytical result obtained in[7] is not suitable for a convenient and accurate numerical implementation. We therefore display below another equivalent formula, recently obtained and employed in Ref. 37 which is quite suitable for numerics. Introducing the notations

$$
\mathbf{y}' = \frac{R^2\mathbf{y}}{|\mathbf{y}|^2}, \quad \mathbf{t} = \frac{\mathbf{y}}{|\mathbf{y}|}, \quad \mathbf{a} = \mathbf{x} - (\mathbf{x}.\mathbf{t})\mathbf{t}, \quad h = \frac{|\mathbf{x}-\mathbf{y}'||\mathbf{y}|}{R}, \tag{7.57}
$$

and the real constant A and vector $\mathbf{A}$ such as

$$
A = \left\{ |\mathbf{x}| \pm \frac{2R^2\mathbf{x}.\mathbf{t}}{R^2 + Rh \mp |\mathbf{x}||\mathbf{y}|} \right\} / \{|\mathbf{x}| \pm \mathbf{x}.\mathbf{t}\}, \tag{7.58}
$$

$$
\begin{aligned}
\mathbf{A} = {} & \mp |\mathbf{y}|\mathbf{x} + [|\mathbf{y}||\mathbf{x}| \pm (1 \mp 2)R^2]\mathbf{t} \\
& \pm 2\left[\frac{2R^2|\mathbf{y}|\mathbf{x} + [R^3 h \mp R^2|\mathbf{y}||\mathbf{x}|]\mathbf{t}}{R^2 + Rh \mp |\mathbf{y}||\mathbf{x}|} \right]
\end{aligned} \tag{7.59}
$$

where superscripts or subscripts in (7.58)–(7.59) are used for $\mathbf{x}.\mathbf{t} \geq 0$ or $\mathbf{x}.\mathbf{t} < 0$, respectively. It has then been found that

$$
\begin{aligned}
R_{ij}(\mathbf{x},\mathbf{y}) = {} & -\frac{\delta_{ij}}{h} - \frac{(\mathbf{x}.\mathbf{e}_i)(\mathbf{x}.\mathbf{e}_j)}{h^3} + \frac{(\mathbf{t}.\mathbf{e}_i)(\mathbf{t}.\mathbf{e}_j)}{h}\left[\frac{|\mathbf{x}|^2}{h^2} - 1\right] \\
& - \left[\frac{2|\mathbf{y}|\mathbf{t}.\mathbf{x}}{h^3}\right](\mathbf{t}.\mathbf{e}_i)(\mathbf{t}.\mathbf{e}_j) + |\mathbf{y}|\left[\frac{(\mathbf{t}.\mathbf{e}_i)(\mathbf{x}.\mathbf{e}_j) + (\mathbf{t}.\mathbf{e}_j)(\mathbf{x}.\mathbf{e}_i)}{h^3}\right]
\end{aligned}
$$

$$- \frac{[|\mathbf{x}|^2 - R^2][|\mathbf{y}|^2 - R^2]}{2} \left\{ \frac{\delta_{ij}}{R^3 h^3} - \frac{3}{R^2} \left[\frac{(\mathbf{h}.\mathbf{e}_i)(\mathbf{h}.\mathbf{e}_j)}{h^5} \right] \right.$$

$$- 2\frac{\mathbf{t}.\mathbf{e}_j}{R^2} \left[\frac{\mathbf{t}.\mathbf{e}_i}{h^3} - \frac{3(\mathbf{h}.\mathbf{e}_i)(\mathbf{h}.\mathbf{t})}{h^5} \right] + \frac{3A}{R^4 h} [\delta_{ij} - (\mathbf{t}.\mathbf{e}_j)(\mathbf{t}.\mathbf{e}_i)]$$

$$+ \frac{3\mathbf{a}.\mathbf{e}_j}{R} \left[-\frac{E}{R^3 h} \left\{ \frac{|\mathbf{y}|\mathbf{h}.\mathbf{e}_i}{Rh^2} + \frac{2\mathbf{a}.\mathbf{e}_i}{|\mathbf{a}|^2} \right\} + \frac{\mathbf{E}.\mathbf{e}_i}{R^4 h^2 [|\mathbf{x}|_-^+ (\mathbf{x}.\mathbf{t}))]} \right.$$

$$\left. \left. + \mathbf{a}.\mathbf{e}_i \left[\frac{(2R^2)_-^+ |\mathbf{y}||\mathbf{x}|}{R^4 h^2 |\mathbf{a}|^2} \right] \right] \right\}. \tag{7.60}$$

As announced, and can be checked for (7.55) and (7.60), for a solid and motionless boundary Σ the Green velocity tensor $\mathbf{G}^c$ satisfies the symmetry property $G^c_{ij}(\mathbf{x}, \mathbf{y}) = G^c_{ji}(\mathbf{y}, \mathbf{x})$.

In summary, one should note that, depending upon the nature of the boundary Σ, the symmetry properties (7.50) might be not satisfied. However, the retained Green solution (7.42) exhibits in any case (bounded or not bounded liquid domain) the key behaviours

$$P_j(\mathbf{x}, \mathbf{y}) \sim \frac{2(x_j - y_j)}{|\mathbf{x} - \mathbf{y}|^3} \quad \text{as } \mathbf{x} \to \mathbf{y}, \tag{7.61}$$

$$G_{ij}(\mathbf{x}, \mathbf{y}) \sim \frac{\delta_{ij}}{|\mathbf{x} - \mathbf{y}|} + \frac{(x_i - y_i)(x_j - y_j)}{|\mathbf{x} - \mathbf{y}|^3} \quad \text{as } \mathbf{x} \to \mathbf{y}, \tag{7.62}$$

$$T_{ijk}(\mathbf{x}, \mathbf{y}) \sim -\frac{6(x_i - y_i)(x_j - y_j)(x_k - y_k)}{|\mathbf{x} - \mathbf{y}|^5} \quad \text{as } \mathbf{x} \to \mathbf{y}. \tag{7.63}$$

In addition, in any direction where the liquid domain happens to be unbounded, one further requires the symptotics

$$|P^c_j(\mathbf{x}, \mathbf{y})| \leq O(d^{-2}), \quad |G^c_{ij}(\mathbf{x}, \mathbf{y})| \leq O(d^{-1}) \text{ as } d = |\mathbf{x} - \mathbf{y}| \to \infty, \tag{7.64}$$

$$|T^c_{ijk}(\mathbf{x}, \mathbf{y})| \leq O(d^{-2}) \text{ as } d = |\mathbf{x} - \mathbf{y}| \to \infty. \tag{7.65}$$

Before closing this subsection, let us mention that Jones[38] and Sellier and Pasol[39] also established the Green tensor $\mathbf{G}^c$ for a liquid domain bounded by two motionless and parallel planes (for two solid ones in Ref. 38 and for a thin film in Ref. 39).

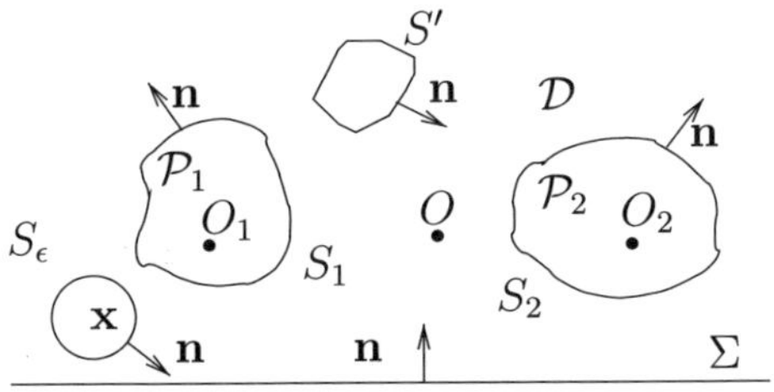

Fig. 7.2. A point $\mathbf{x}$ in the liquid domain $\mathcal{D}$ and the spherical surface S_ϵ centred at $\mathbf{x}$. A closed surface S', introduced before (7.77), is also drawn.

7.4.2. *General integral representation for the velocity field*

Let us go back to the Stokes flow $(\mathbf{u}, p)$ about $N \geq 1$ particles $\mathcal{P}_n$ with surface S_n (see Fig. 7.1). This flow, with stress tensor $\boldsymbol{\sigma}$, obeys (7.13) and exerts a traction $\mathbf{f} = \boldsymbol{\sigma}.\mathbf{n} = f_i \mathbf{e}_i$ on the liquid boundary $\partial\mathcal{D} = \cup_{n=1}^{N} S_n \cup \Sigma$. For a point $\mathbf{x}$ in $\mathcal{D}$ and an arbitrary vector $\mathbf{t} = t_j \mathbf{e}_j$, we introduce the singularity flow $(\mathbf{u}', p')$ induced by a point force with intensity $\mathbf{t}$ located at the pole $\mathbf{x}$. For this flow, the velocity and the pressure at any point $\mathbf{y} \neq \mathbf{x}$ in the liquid thus read

$$\mathbf{u}'(\mathbf{y}) = G_{ij}(\mathbf{y}, \mathbf{x}) t_j \mathbf{e}_i / [8\pi\mu], \quad p'(\mathbf{y}) = P_j(\mathbf{y}, \mathbf{x}) t_j / [8\pi]. \tag{7.66}$$

For $\epsilon > 0$ and small enough, we introduce, as seen in Fig. 7.2, a spherical surface S_ϵ with radius ϵ, center $\mathbf{x}$ and unit normal vector $\mathbf{n}$ directed into the liquid. Moreover, the domain enclosed by S_ϵ lies entirely in $\mathcal{D}$. Exploiting, if needed, the far-field behaviours for $(\mathbf{u}, p)$, with stress tensor $\boldsymbol{\sigma}$, the solution to (7.13) (see Property 2 in Section 7.3.1) and (7.64)–(7.65), the reciprocal identity (7.26) then immediately yields

$$t_j \left\{ \int_{\partial\mathcal{D}} [\mu u_i(\mathbf{y}) T_{ijk}(\mathbf{y}, \mathbf{x}) n_k(\mathbf{y}) - G_{ij}(\mathbf{y}, \mathbf{x}) f_i(\mathbf{y})] dS(\mathbf{y}) \right.$$
$$\left. + \int_{S_\epsilon} [\mu u_i(\mathbf{y}) T_{ijk}(\mathbf{y}, \mathbf{x}) n_k(\mathbf{y}) - G_{ij}(\mathbf{y}, \mathbf{x}) f_i(\mathbf{y})] dS(\mathbf{y}) \right\} = 0. \tag{7.67}$$

From Section 7.3.1 we know that $\mathbf{u}$ is regular near $\mathbf{x}$. Under the behaviors (7.62)–(7.63), the integral over S_ϵ in (7.67) is thus easily seen to tend to $-8\pi\mu u_j(\mathbf{x})$ as ϵ vanishes. Since each t_j is arbitrary, one thus ends up with the following key integral representation of the velocity field $\mathbf{u}$ in the entire liquid domain

$$u_j(\mathbf{x}) = -\frac{1}{8\pi\mu}\int_{\partial\mathcal{D}} G_{ij}(\mathbf{y},\mathbf{x})f_i(\mathbf{y})dS(\mathbf{y}) + \frac{1}{8\pi}\int_{\partial\mathcal{D}} u_i(\mathbf{y})T_{ijk}$$
$$\times(\mathbf{y},\mathbf{x})n_k(\mathbf{y})dS(\mathbf{y}) \quad \text{for } j = 1,2,3 \text{ and } \mathbf{x} \text{ in } \mathcal{D}. \qquad (7.68)$$

The identity (7.68) clearly shows that it is possible to obtain the velocity field $\mathbf{u}$ in the liquid from the knowledge of solely two physical quantities on the entire boundary $\partial\mathcal{D}$ of the liquid domain: the velocity $\mathbf{u}$ and the surface traction $\mathbf{f} = \boldsymbol{\sigma}.\mathbf{n}$ arising in the second double-layer type or the first single-layer type integrations in (7.69), respectively. The integral representation (7.69) is general, since it applies to any selected Green velocity and associated stress tensors $\mathbf{G}$ and $\mathbf{T}$. These tensors are not unique when the liquid domain is bounded by a surface Σ, and two cases occur:

(i) Standard Case: One employs the Oseen–Burgers Green solution (7.48)–(7.49). This is done when the liquid is not externally bounded with $\partial\mathcal{D} = \cup_{n=1}^{N} S_n$ but also in the presence of a bounding surface Σ.
(ii) Special Case: One resorts to another Green solution $(\mathbf{P}^c, \mathbf{G}^c, \mathbf{T}^c)$ for a bounding surface Σ. For example, if there is a no-slip condition on the boundary Σ, the properties $\mathbf{u}(\mathbf{y}) = \mathbf{0}$ and $\mathbf{G}^c(\mathbf{y},\mathbf{x}) = \mathbf{0}$ for $\mathbf{y}$ located on Σ make it possible to replace, in (7.68), the surface $\partial\mathcal{D} = S \cup \Sigma$ with the cluster's boundary $S = \cup_{n=1}^{N} S_n$. For a numerical implementation of (7.68), such a trick has the main merit of involving a smaller surface. Indeed, the boundary Σ is often either unbounded or of a size so large that meshing it adequately would require to many nodal points.

7.4.3. *Integral representation for the pressure field*

The integral representation (7.68) for the velocity suggests deriving a similar identity for the pressure field p. Since p is harmonic in the fluid domain, the usual second Green identity provides p in terms of the values of both p and $\nabla p.\mathbf{n}$ on the entire surface $\partial\mathcal{D}$. Unfortunately, in practice there is no condition regarding the pressure supplementing (7.13) (note that, for instance, the pressure does not appear in the condition (7.6) required at a spherical bubble surface) and one is therefore *unable* to exploit the second Green identity for the harmonic function. Another treatment is thus required. It actually exploits (7.68) and, for the sake of clarity, we again distinguish the previous standard and specific cases (i) and (ii).

(i) Standard Case: The basic property $G^{\infty}_{ij}(\mathbf{y},\mathbf{x}) = G^{\infty}_{ji}(\mathbf{x},\mathbf{y})$ shows that the first term on the right-hand side of (7.68) is a distribution of Green singularities located on the surface $\partial\mathcal{D}$. In addition, recalling (7.51) also shows that the second term in (7.68) is a steady Stokes flow inducing a pressure contribution obtained by combining, this time, pressure contributions $\Pi^{\infty}_{ik}(\mathbf{y},\mathbf{x})$ (see (7.51)). Accordingly, the resulting pressure integral representation in the liquid domain reads

$$p(\mathbf{x}) = -\frac{1}{8\pi}\int_{\partial\mathcal{D}} P^{\infty}_i(\mathbf{x},\mathbf{y})f_i(\mathbf{y})dS(\mathbf{y}) + \frac{\mu}{8\pi}\int_{\partial\mathcal{D}} u_i(\mathbf{y})\Pi^{\infty}_{ik}(\mathbf{x},\mathbf{y})n_k(\mathbf{y})dS(\mathbf{y}) \quad \text{for } \mathbf{x} \text{ in } \mathcal{D}. \tag{7.69}$$

(ii) Specific Case: For that case, a bounding surface Σ occurs. As earlier mentioned, the Green solution $(\mathbf{P}^c, \mathbf{G}^c, \mathbf{T}^c)$ also obeys the property $G^c_{ij}(\mathbf{y},\mathbf{x}) = G^c_{ji}(\mathbf{x},\mathbf{y})$ and admits another pressure tensor with components $\Pi^c_{ik}(\mathbf{x},\mathbf{y})$ as soon as one satisfies $[G^c_{il}T_{ijk}](\mathbf{x},\mathbf{y})n_k(\mathbf{x}) = 0$ for $\mathbf{x}$ on Σ. Under such assumptions, it is clear that (7.69) then holds with superscripts ∞ replaced with c for the same entire boundary $\partial\mathcal{D} = S \cup \Sigma$. By contrast to the velocity representation, the boundary Σ still arises in the pressure representation. Fortunately, for a solid and motionless boundary Σ, one has to replace, in (7.68) and thus also in (7.69), the surface $\partial\mathcal{D}$ with the cluster's boundary S. This approach, however, requires that the quantities Π^c_{ik} be determined, which may be very involved. In addition, one should finally note that for other types of bounding surfaces Σ, one can at least resort to (7.69).

In summary, one builds a suitable pressure integral representation for any encountered problem by carefully inspecting the adopted velocity integral representation (7.68).

7.4.4. *Relevant boundary-integral equations*

As highlighted in Section 1.4.3, the flow $(\mathbf{u}, p)$ with stress tensor $\boldsymbol{\sigma}$, governed by (7.13), is obtained from the velocity $\mathbf{u}$ and the surface traction $\mathbf{f} = \boldsymbol{\sigma}.\mathbf{n}$ on the boundary $\partial\mathcal{D}$. These quantities are calculated by inverting a few relevant boundary integral equations established in this section.

7.4.4.1. *Basic steps and key preliminary results*

For a general problem (7.13), one successively faces the following steps:

(1) Step 1: The problem (7.13) is first supplemented with boundary conditions. One requires $(\mathbf{u}, p)$ to vanish far from the cluster in any direction of unbounded liquid. Such a condition is already ensured by the integral representations adopted for $(\mathbf{u}, p)$ (see, for instance, (7.68)–(7.69)). Additional boundary conditions are prescribed on the cluster's surface $S = \cup_{n=1}^{N} S_n$ and, if it exists, on the bounding surface Σ. As seen in (7.4) and (7.5)–(7.6), these conditions depend upon the nature of the particles and Σ. As announced in the introduction, an external flow $(\mathbf{u}_e, p_e)$ might be imposed on the liquid. This flow is assumed to be a steady Stokes flow in the *entire* domain $\mathcal{D}'$ made of both $\mathcal{D}$ and the particles $\mathcal{P}_n (n = 1, \ldots, N)$. In other words, one assumes that[e]

$$\nabla.\mathbf{u}_e = 0 \quad \text{and} \quad \mu\nabla^2\mathbf{u}_e = \nabla p_e - \rho\mathbf{g} \text{ in } \mathcal{D}' = \mathcal{D} \cup \mathcal{P}_1 \cup \cdots \cup \mathcal{P}_N. \tag{7.70}$$

If the particle $\mathcal{P}_n$ is a solid body, with attached point O_n such that $\mathbf{x}_n = \mathbf{OO_n}$, which moves at the translational velocity $\mathbf{U}^{(n)}$ and the angular velocity $\mathbf{\Omega}^{(n)}$ the required no-slip condition reads

$$\mathbf{u}(\mathbf{x}) = -\mathbf{u}_e(\mathbf{x}) + \mathbf{U}^{(n)} + \mathbf{\Omega}^{(n)} \wedge (\mathbf{x} - \mathbf{x}_n) \quad \text{for } \mathbf{x} \text{ on } S_n. \tag{7.71}$$

whereas if $\mathcal{P}_n$ is a spherical bubble with transational velocity $\mathbf{U}^{(n)}$ one requires that

$$\mathbf{u}.\mathbf{n} = [\mathbf{U}^{(n)} - \mathbf{u}_e].\mathbf{n} \text{ on } S_n, \tag{7.72}$$

$$\boldsymbol{\sigma}.\mathbf{n} - [\mathbf{n}.\boldsymbol{\sigma}.\mathbf{n}]\mathbf{n} = [\mathbf{n}.\boldsymbol{\sigma}_e.\mathbf{n}]\mathbf{n} - \boldsymbol{\sigma}_e.\mathbf{n} \text{ on } S_n. \tag{7.73}$$

For a liquid bounded by another boundary Σ, one adds other conditions for $(\mathbf{u}, p)$ and $(\mathbf{u}_e, p_e)$ on Σ. The most encountered case is a solid and motionless boundary Σ for which the no-slip conditions write

$$\mathbf{u}(\mathbf{x}) = \mathbf{u}_e(\mathbf{x}) = \mathbf{0} \quad \text{for } \mathbf{x} \text{ on } \Sigma. \tag{7.74}$$

Although tractable in price of minor changes by the same approach, surface Σ of different nature will be, for the sake of conciseness, further discarded in this chapter.

[e]There is no required condition for $(\mathbf{u}_e, p_e)$ far from the cluster where, for instance, $\mathbf{u}_e$ might be unbounded (think about the case of external shear or Poiseuille steady flows).

(2) Step 2: One then establishes relevant boundary integral equations for the pair $(\mathbf{u}, \boldsymbol{\sigma}.\mathbf{n})$ on the surface $\partial\mathcal{D} = S \cup \Sigma$ by enforcing the boundary conditions (7.71)–(7.74). This is achieved by letting the point $\mathbf{x}$ tend to $\partial\mathcal{D}$ in (7.68). Such a limit process, however, exhibits quite different behaviours for the single-layer or double-layer contributions $S_j(\mathbf{x})$ or $D_j(\mathbf{x})$ arising on the right-hand side of (7.68), i.e for the terms[f]

$$S_j(\mathbf{x}) = -\frac{1}{8\pi\mu}\int_{\partial\mathcal{D}} G_{ij}(\mathbf{y},\mathbf{x})f_i(\mathbf{y})dS(\mathbf{y}), \tag{7.75}$$

$$D_j(\mathbf{x}) = \frac{1}{8\pi}\int_{\partial\mathcal{D}} u_i(\mathbf{y})T_{ijk}(\mathbf{y},\mathbf{x})n_k(\mathbf{y})dS(\mathbf{y}). \tag{7.76}$$

Indeed, the behaviours (7.62)–(7.63) show that $S_j(\mathbf{x})$ experiences a regular limit whereas $D_j(\mathbf{x})$ admits a weakly singular limit requiring careful attention. How to adequately handle the limit of the double-layer term $D_j(\mathbf{x})$ when the point $\mathbf{x}$ approaches the boundary $\partial\mathcal{D}$ is a key point detailed below.

Recall that $\partial\mathcal{D} = S \cup \Sigma$ with $S = \cup_{n=1}^{N} S_n$ the cluster's surface. For a boundary condition on S_n of type (7.71), it is possible to get rid of the contribution of this surface to $D_j(\mathbf{x})$ (see details in Section 7.4.4.2). For a boundary S_n of different nature such as case of a bubble or clusters made of drops, this is not possible anymore and one needs to cope with the limit of the double-layer term $D_j(\mathbf{x})$ as $\mathbf{x}$ approaches the boundary $\partial\mathcal{D}$. Several methods are available to deal with this basic issue. In this chapter, we propose a regularization approach which is general, simple and quite suitable for a numerical implementation because it solely involves regular integrals and thus permits a very good numerical accuracy. This procedure, worked out in Section 7.4.4.2, exploits a few preliminary properties established below.

We recall that $\mathcal{D}'$ is the liquid domain in the absence of particles, i.e that $\mathcal{D}' = \mathcal{D} \cup \mathcal{P}_1 \cup \cdots \cup \mathcal{P}_N$. Let us consider a closed and simply-connected surface S' drawn in the domain $\mathcal{D}'$ that has a unit outward normal vector $\mathbf{n} = n_i\mathbf{e}_i$ directed *into* $\mathcal{D}'$ (see Fig. 7.2). For instance, S' may be the surface S_n of the particle $\mathcal{P}_n$. The following property then holds

$$\int_{S'} G_{ij}(\mathbf{y},\mathbf{x})n_i(\mathbf{y})dS(\mathbf{y}) = 0 \quad \text{for } j = 1,2,3 \quad \text{and} \quad \mathbf{x} \text{ in } \mathcal{D}'. \tag{7.77}$$

[f]As soon as (7.74), holds one should replace $\partial\mathcal{D}$ with the cluster's surface S in (7.76).

Note that $\mathbf{x}$ in $\mathcal{D}'$ means that $\mathbf{x}$ may lie either *on* S' or *inside or outside* the domain $\mathcal{V}'$ enclosed by S'. For $\mathbf{x}$ outside $\mathcal{V}'$, the velocity $\mathbf{u}'$ defined by (7.66) is divergence-free inside $\mathcal{V}'$ and using the divergence theorem then proves (7.77). *By continuity*, (7.77) also holds for $\mathbf{x}$ on S' and using (7.53) with $\partial R_{ij}(\mathbf{y},\mathbf{x})/\partial y_i = 0$ easily proves (7.77) for $\mathbf{x}$ inside $\mathcal{V}'$.

Another basic property is obtained for any steady Stokes flow $(\mathbf{u}'', p'')$ with stress tensor $\boldsymbol{\sigma}''$ regular inside $\mathcal{V}'$. For $\mathbf{x}$ *outside* $\mathcal{V}'$, the flow $(\mathbf{u}', p')$ given by (7.66) is also regular inside $\mathcal{V}'$ and the reciprocal identity (7.26) then holds. Since the vector $\mathbf{t} = t_j \mathbf{e}_j$ is arbitrary, one thus arrives, for $j = 1, 2, 3$, at the identities

$$\int_{S'} G_{ij}(\mathbf{y},\mathbf{x})[\mathbf{e}_i.\boldsymbol{\sigma}''.\mathbf{n}](\mathbf{y})dS(\mathbf{y})$$
$$= \mu \int_{S'} [\mathbf{u}''.\mathbf{e}_i](\mathbf{y})T_{ikj}(\mathbf{y},\mathbf{x})n_k(\mathbf{y})dS(\mathbf{y}) \quad \text{for } \mathbf{x} \text{ outside } \mathcal{V}'. \quad (7.78)$$

The relations (7.77)–(7.78) are of the utmost importance for the next subsections.

7.4.4.2. *Relevant general boundary integral equations*

The cluster's boundary is $S = \cup_{n=1}^{N} S_n$. Any flow with uniform velocity $\mathbf{a}$ and zero pressure is a steady Stokes flow inside the particle $\mathcal{P}_n$. By virtue of (7.77)–(7.78), if $\mathbf{x}$ lies *outside* $\mathcal{P}_n$ one gets

$$\int_{S_n} T_{ikj}(\mathbf{y},\mathbf{x})n_k(\mathbf{y})dS(\mathbf{y}) = 0 \quad \text{for any pair } (i,j) \text{ and } \mathbf{x} \text{ in } \mathcal{D} \setminus \mathcal{P}_n. \quad (7.79)$$

The relation (7.79) permits one to cast the integral representation (7.68) into another form. Exploiting (7.74), one thus gets for $\mathbf{x}$ in $\mathcal{D}$ and close to $\mathcal{P}_n$,

$$u_j(\mathbf{x}) = -\frac{1}{8\pi\mu}\int_{\partial\mathcal{D}} G_{ij}(\mathbf{y},\mathbf{x})f_i(\mathbf{y})dS(\mathbf{y})$$
$$+\frac{1}{8\pi}\int_{S\setminus S_n} u_i(\mathbf{y})T_{ijk}(\mathbf{y},\mathbf{x})n_k(\mathbf{y})dS(\mathbf{y})$$
$$+\frac{1}{8\pi}\int_{S_n} [u_i(\mathbf{y}) - u_i(\mathbf{x})]T_{ijk}(\mathbf{y},\mathbf{x})n_k(\mathbf{y})dS(\mathbf{y}). \quad (7.80)$$

Each term on the right-hand side of (7.80) admits a regular limit as $\mathbf{x}$ approaches S_n. Thus (7.80) *is still valid* for $\mathbf{x}$ on S_n and one obtains the following set of $3N$ coupled boundary integral equations for the desired velocity $\mathbf{u}$ and traction $\mathbf{f} = \boldsymbol{\sigma}.\mathbf{n}$ on the boundary $\partial\mathcal{D}$

$$\begin{aligned} u_j(\mathbf{x}) = \frac{1}{8\pi}\left[\int_{\mathcal{S}\backslash S_n} u_i(\mathbf{y})T_{ijk}(\mathbf{y},\mathbf{x})n_k(\mathbf{y})dS(\mathbf{y}) \right. \\ \left. - \mu \int_{\partial\mathcal{D}} G_{ij}(\mathbf{y},\mathbf{x})f_i(\mathbf{y})dS(\mathbf{y})\right] + \frac{1}{8\pi}\int_{S_n} [u_i(\mathbf{y}) - u_i(\mathbf{x})] \\ \times T_{ijk}(\mathbf{y},\mathbf{x})n_k(\mathbf{y})dS(\mathbf{y}), \quad \mathbf{x} \text{ on } S_n; n = 1,\ldots,N \end{aligned} \tag{7.81}$$

which must be supplemented with three additional boundary-integral equations whenever Σ exists (see (7.74))

$$\begin{aligned} 0 = -\frac{1}{8\pi\mu}\int_{S\cup\Sigma} G_{ij}(\mathbf{y},\mathbf{x})f_i(\mathbf{y})dS(\mathbf{y}) \\ + \frac{1}{8\pi}\int_S u_i(\mathbf{y})T_{ijk}(\mathbf{y},\mathbf{x})n_k(\mathbf{y})dS(\mathbf{y}) \ \mathbf{x} \text{ on } \Sigma \text{ and } j = 1,2,3. \end{aligned} \tag{7.82}$$

The regularized integral equations (7.81) (or (7.81)–(7.82)) hold, whatever the boundary conditions on the cluster's surface S. One can thus employ these integral equations for clusters made of solid bodies and/or spherical bubbles. Invoking (7.77), furthermore, shows that $\mathbf{f} = \boldsymbol{\sigma}.\mathbf{n}$ is defined on each particle's surface S_n up to a constant multiple of the normal vector $\mathbf{n}$. This means that (7.81) (or (7.81)–(7.82)) does (do) not have a unique solution in general. However, for two different solutions, the induced velocity $\mathbf{u}$ (given by (7.68) or (7.81) is the same in the liquid. Again, we note that when Σ exists, (7.81)–(7.82) hold for any Green solution $(\mathbf{P},\mathbf{G},\mathbf{T})$. Selecting the solution $(\mathbf{P}^c,\mathbf{G}^c,\mathbf{T}^c)$ for which (see (7.74)) $G^c(\mathbf{y},\mathbf{x})$ vanishes for $\mathbf{y}$ on Σ, however, makes it possible to also replace $S\cup\Sigma$ with S in (7.82).

7.4.4.3. *Case of a $N-$bubble cluster*

We consider a cluster made of $N \geq 1$ spherical bubbles and enforce (7.72)–(7.73) on each surface S_n. The traction $\mathbf{f}$ is unknown whereas, on S, one ends up with given data $(\mathbf{d}, d)$ and unknown quantities $(\mathbf{a}, a)$

defined as

$$\mathbf{d} = \mathbf{f} - (\mathbf{f}.\mathbf{n})\mathbf{n}, \quad d = \mathbf{u}.\mathbf{n}, \quad \mathbf{a} = \mathbf{u} - (\mathbf{u}.\mathbf{n})\mathbf{n}, \quad a = \mathbf{f}.\mathbf{n}. \tag{7.83}$$

The decomposition (7.83) satisfies, by definition, the compatibility relation $\mathbf{a}.\mathbf{n} = 0$. Using (7.81) then shows that $(\mathbf{a}, a)$ is governed by the following compatibility relations and coupled boundary integral equations

$$\mathbf{a}.\mathbf{n} = 0 \text{ on } S_n, \tag{7.84}$$

$$\begin{aligned} 8\pi a_j(\mathbf{x}) &- \int_{\mathcal{S}\setminus S_n} a_i(\mathbf{y}) T_{ijk}(\mathbf{y}, \mathbf{x}) n_k(\mathbf{y}) dS(\mathbf{y}) \\ &- \int_{S_n} [a_i(\mathbf{y}) - a_i(\mathbf{x})] T_{ijk}(\mathbf{y}, \mathbf{x}) n_k(\mathbf{y}) dS(\mathbf{y}) \\ &+ \mu \int_{\partial\mathcal{D}} G_{ij}(\mathbf{y}, \mathbf{x}) [a n_i](\mathbf{y}) dS(\mathbf{y}) = s_j^n[\mathbf{d}, d](\mathbf{x}) \text{ on } S_n, \end{aligned} \tag{7.85}$$

if the linear operator $s_j^n[\mathbf{d}, d](\mathbf{x})$ is defined for $\mathbf{x}$ on S_n as

$$\begin{aligned} s_j^n[\mathbf{d}, d](\mathbf{x}) = -8\pi[d n_j](\mathbf{x}) &+ \int_{\mathcal{S}\setminus S_n} [d n_i](\mathbf{y}) T_{ijk}(\mathbf{y}, \mathbf{x}) n_k(\mathbf{y}) dS(\mathbf{y}) \\ &+ \int_{S_n} \{[d n_i](\mathbf{y}) - [d n_i](\mathbf{x})\} T_{ijk}(\mathbf{y}, \mathbf{x}) n_k(\mathbf{y}) dS(\mathbf{y}) \\ &- \mu \int_{\partial\mathcal{D}} G_{ij}(\mathbf{y}, \mathbf{x}) d_i(\mathbf{y}) dS(\mathbf{y}). \end{aligned} \tag{7.86}$$

When a bounding surface Σ is present, one adds a similar counter-part of the equations (7.82).

7.4.4.4. *Case of a $N-$particle cluster involving at least a solid particle*

Let us consider a cluster involving at least one solid particle. On the surface S_m of a *solid* particle $\mathcal{P}_m$, one enforces (7.71) for $n = m$. Since any rigid-body motion and $(\mathbf{u}_e, p_e)$ with stress tensor $\boldsymbol{\sigma}_e$ are steady Stokes flows inside $\mathcal{P}_m$, the identities (7.78) and (7.77) give, for $\mathbf{x}$ in the liquid,

$$\begin{aligned} &- \int_{S_m} G_{ij}(\mathbf{y}, \mathbf{x}) [\mathbf{e}_i.\boldsymbol{\sigma}_e.\mathbf{n}](\mathbf{y}) dS(\mathbf{y}) \\ &= \int_{S_m} u_i(\mathbf{y})_i T_{ikj}(\mathbf{y}, \mathbf{x}) n_k(\mathbf{y}) dS(\mathbf{y}) \text{ for } \mathbf{x} \text{ in } \mathcal{D}. \end{aligned} \tag{7.87}$$

If one denotes by S_s the *solid* subset of the cluster's boundary S, this relation suggests introducing on S_s the auxiliary unknown traction $\mathbf{f}' = \mathbf{f} + \boldsymbol{\sigma}_e.\mathbf{n}$. Under the boundary condition (7.74), the integral representation (7.68) then becomes

$$u_j(\mathbf{x}) = -\frac{1}{8\pi\mu}\left[\int_{\partial\mathcal{D}\backslash S_s} G_{ij}(\mathbf{y},\mathbf{x})f_i(\mathbf{y})dS(\mathbf{y}) + \int_{S_s} G_{ij}(\mathbf{y},\mathbf{x})f_i'(\mathbf{y})dS(\mathbf{y})\right] + \frac{1}{8\pi}\int_{S\backslash S_s} u_i(\mathbf{y})T_{ijk}(\mathbf{y},\mathbf{x})n_k(\mathbf{y})dS(\mathbf{y}) \text{ for } \mathbf{x} \text{ in } \mathcal{D}. \quad (7.88)$$

It shows that the double-layer contribution on the *solid* surface S_s does not appear any more (its influence has been, however, taken into account by replacing the traction $\mathbf{f}$ with the shifted one $\mathbf{f}'$ on this surface). Using again the identity (7.79) for each non-solid surface S_n, the integral representation (7.88) then provides the following boundary integral equations

$$u_j(\mathbf{x}) = -\frac{1}{8\pi\mu}\left[\int_{\partial\mathcal{D}\backslash S_s} G_{ij}(\mathbf{y},\mathbf{x})f_i(\mathbf{y})dS(\mathbf{y}) + \int_{S_s} G_{ij}(\mathbf{y},\mathbf{x})f_i'(\mathbf{y})dS(\mathbf{y})\right] + \frac{1}{8\pi}\int_{S\backslash S_s} u_i(\mathbf{y})T_{ijk}(\mathbf{y},\mathbf{x})n_k(\mathbf{y})dS(\mathbf{y}) \text{ for } \mathbf{x} \text{ on } S_s, \quad (7.89)$$

$$u_j(\mathbf{x}) = -\frac{1}{8\pi\mu}\left[\int_{\partial\mathcal{D}\backslash S_s} G_{ij}(\mathbf{y},\mathbf{x})f_i(\mathbf{y})dS(\mathbf{y}) + \int_{S_s} G_{ij}(\mathbf{y},\mathbf{x})f_i'(\mathbf{y})dS(\mathbf{y})\right] + \frac{1}{8\pi}\int_{S\backslash S_s} [u_i(\mathbf{y}) - u_i(\mathbf{x})]T_{ijk}(\mathbf{y},\mathbf{x})n_k(\mathbf{y})dS(\mathbf{y}) \text{ for } \mathbf{x} \text{ on } S\backslash S_s. \quad (7.90)$$

The left hand-side of (7.89) is given by the boundary conditions (7.71) on each surface S_m of S_s, whereas additional manipulations are required (as for the previous example of an $N-$bubble cluster) to rewrite the equalities (7.90) using the prescribed boundary conditions on the surface $S\backslash S_s$. If a bounding surface Σ, takes place we furthermore impose the following counterpart of (7.82)

$$0 = -\frac{1}{8\pi\mu}\left[\int_{\partial\mathcal{D}\backslash S_s} G_{ij}(\mathbf{y},\mathbf{x})f_i(\mathbf{y})dS(\mathbf{y}) + \int_{S_s} G_{ij}(\mathbf{y},\mathbf{x})f_i'(\mathbf{y})dS(\mathbf{y})\right] + \frac{1}{8\pi}\int_{S\backslash S_s} u_i(\mathbf{y})T_{ijk}(\mathbf{y},\mathbf{x})n_k(\mathbf{y})dS(\mathbf{y}) \text{ for } \mathbf{x} \text{ on } \Sigma. \quad (7.91)$$

A straightforward application is the widely-encountered case of a cluster made of N solid particles. In such circumstances, $S = S_s$ and therefore (7.88), (7.89)–(7.90) and (7.91) readily adopt the simple forms

$$u_j(\mathbf{x}) = -\frac{1}{8\pi\mu}\int_{\Sigma} G_{ij}(\mathbf{y},\mathbf{x})f_i(\mathbf{y})dS(\mathbf{y}) - \frac{1}{8\pi\mu}\int_{S} G_{ij}(\mathbf{y},\mathbf{x})f_i'(\mathbf{y})dS(\mathbf{y}) \text{ for } \mathbf{x} \text{ in } \mathcal{D}\cup S\cup\Sigma. \quad (7.92)$$

For a Green solution, such that $G_{ij}(\mathbf{y},\mathbf{x}) = 0$ for $\mathbf{y}$ on Σ, one can furthermore omit the first integral on the right-hand side of (7.92).

7.5. Numerical Implementation

As seen in the previous section, the flow $(\mathbf{u}, p)$ about a given $N-$particle cluster may be evaluated in the liquid by appealing to velocity and pressure *boundary integral representations* on a prescribed surface (the cluster's boundary S or the surface $S \cup \Sigma$) where it is required to know the velocity $\mathbf{u}$ and the surface traction $\mathbf{f} = \boldsymbol{\sigma}.\mathbf{n}$. Those key surface quantities have been found to satisfy boundary integral equations on the involved surface. For arbitrary clusters (shape, nature and location of each particles) it is not possible to analytically employ or invert those boundary integral representations or equations and a numerical treatment is requested. This section presents a possible and so-called collocation method to achieve such a basic task. This technique was first empoyed to solve potential problems and has since been applied in many other fields, such as elasticity or acoustics. Since it has become standard, we briefly describe the employed steps but, however, pay special attention to the accuracy issues. For additional details, the reader is directed to the available textbooks.[40–42]

7.5.1. *Mesh and boundary-elements*

Let us consider a surface $\mathcal{S}$ which may be either closed or not. This surface is *numerically approximated* by *another* surface $\mathcal{S}_a$. This is obtained by using a mesh consisting of N_n nodal points $\mathbf{y}(n)$ with $n = 1, \ldots, N_n$ located on $\mathcal{S}$ *and* N_e boundary elements Δ_e for $e = 1, \ldots, N_e$, such that $\mathcal{S}_a = \cup_{e=1}^{N_e}\Delta_e$ with $\Delta_e \cap \Delta_{e'} = \emptyset$ for $e \neq e'$. Each element Δ_e is obtained by mapping a standard $Q-$ node *plane* pattern Δ onto Δ_e using interpolation functions. As sketched in Fig. 7.3, the pattern Δ is drawn in the $\xi_1 - \xi_2$ plane and characterized by Q points $\xi_{(q)}$ with coordinates $(\xi_1(q), \xi_2(q))$ for

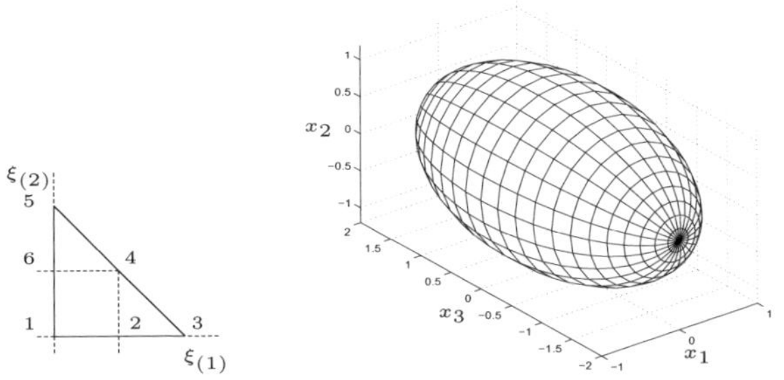

(a) (b)

Fig. 7.3. (a) Standard 6-node triangle Δ in the $\xi_1 - \xi_2$ plane. (b) Example of a mesh for a solid ellipsoid using different *flat* boundary elements Δ_e associated with 3−node triangle (near the poles) and 4−node square standard patterns Δ.

$q = 1, \ldots, Q$. For a given boundary element Δ_e, the image on the surface $\mathcal{S}$ of the point $\xi_{(q)} = (\xi_1(q), \xi_2(q))$ is denoted by $\mathbf{y}(e, q)$ and any point $\mathbf{y}$ of Δ_e is further described by its intrinsic coordinates $\xi = (\xi_1, \xi_2)$ obtained by interpolating the locations of the nodes $\mathbf{y}(e, q)$. This interpolation is defined by providing Q so-called *smooth* shape functions $M_q(q = 1, \ldots, Q)$ satisfying the properties

$$M_q(\xi_{(p)}) = \delta_{pq} \text{ for } p, q \text{ in } \{1, \ldots, Q\}; \quad \sum_{q=1}^{Q} M_q(\xi) = 1 \text{ for } \xi \text{ on } \Delta \tag{7.93}$$

and reads

$$\mathbf{y}(\xi) = \sum_{q=1}^{Q} M_q(\xi)\mathbf{y}(e, q), \quad \xi = (\xi_1, \xi_2). \tag{7.94}$$

Note that under the conditions (7.93), one indeed retrieves the definitions $\mathbf{y}(e, q) = \mathbf{y}(\xi_{(q)})$. In practice, many pairs (Δ, Q) are possible and this key choice dictates the accuracy at which the retained numerical surface $\mathcal{S}_a$ approximates the *exact* surface $\mathcal{S}$. Actually, two types of boundary elements exist:

(i) *Flat boundary elements.* In this case, illustrated in Fig. 7.3(b), each boundary element Δ_e is plane. For instance, one can resort to the following pairs (Δ, Q) :

   (a) Δ is the standard triangle $\mathcal{T}$ with inequations $0 \leq \xi_1 \leq 1, 0 \leq \xi_2 \leq 1, \xi_1 + \xi_2 \leq 1$ and $Q = 3$. The points $\xi_{(q)}$ are the vortices of

$\mathcal{T}$ and each boundary element Δ_e is a flat triangle with corners located on the surface $\mathcal{S}$.

(b) Δ is a unit square $\mathcal{SQ}$ and $Q = 4$ with points $\xi_{(q)}$ at the corners. Each element Δ_e is a flat square with corners on $\mathcal{S}$.

(ii) *Curved boundary elements.* Such high-order and curvilinear elements Δ_e permit a better approximation of $\mathcal{S}$. The usual counterparts of the previous flat elements are the following:

(a) Δ is the standard triangle $\mathcal{T}$ and one uses six nodes ($Q = 6$): the corners and the midsides. This choice is illustrated in Fig 7.3(a).

(b) Δ is the standard square $\mathcal{SQ}$ and one employs nine nodes ($Q = 9$): the center, the corners and the midsides.

Henceforth, we illustrate the method for 6-node curvilinear triangular boundary elements (choice of $\Delta = \mathcal{T}$ and $Q = 6$). Setting $\xi_3 = 1-\xi_1-\xi_2$ and exploiting the relations (7.93) immediately provides the following quadratic shape functions

$$M_1(\xi) = (2\xi_3 - 1)\xi_3, \quad M_2(\xi) = 4\xi_1\xi_3, \quad M_3(\xi) = (2\xi_1 - 1)\xi_1, \tag{7.95}$$

$$M_4(\xi) = 4\xi_1\xi_2, \quad M_5(\xi) = (2\xi_2 - 1)\xi_2, \quad M_6(\xi) = 4\xi_2\xi_3. \tag{7.96}$$

Since a node $\mathbf{y}(n)$ belongs to several boundary elements Δ_e, it is also useful to know, for each node $\mathbf{y}(e, q)$ on a given boundary element Δ_e, its label n in the entire mesh array $(\mathbf{y}_1, \dots, \mathbf{y}_{N_n})$. This is given by the so-called connectivity table or matrix $table(e, q)$ where

$$n = table(e, q) \text{ means that } \mathbf{y}(n) = \mathbf{y}(e, q) \text{ on } \Delta_e. \tag{7.97}$$

In summary, one should be aware that a given surface $\mathcal{S}$ is approximated by a numerical surface $\mathcal{S}_a$ defined by nodal points and boundary elements characterized by the following data:

(1) A given set of N_n nodal points $(\mathbf{y}(1), \dots, \mathbf{y}(N_n))$ located on the *exact* surface $\mathcal{S}$.

(2) The number N_e of boundary elements producing the numerical surface $\mathcal{S}_a$, the pair (Δ, Q) and Q associated shape functions M_q.

(3) The table of connectivity array $table(e, q)$ which is an $N_e \times Q$ matrix.

7.5.2. *Associated functions*

For prescribed mesh and boundary elements, one obtains a surface $\mathcal{S}_a$ which approximates the exact boundary $\mathcal{S}$. For further purposes it is also necessary to approximate the unit ouward normal vector $\mathbf{n}$ and the area element $dS(\mathbf{y})$ on $\mathcal{S}$. Those quantities are obtained by computing two vectors $\mathbf{t}_1$ and $\mathbf{t}_2$ tangent to the boundary element Δ_e. Setting $f_{,\xi_i} = \partial f/\partial \xi_i$, one actually deduces from (7.94)

$$d\mathbf{y} = \sum_{q=1}^{Q} \mathbf{y}(e,q)[M_{q,\xi_1}(\xi)d\xi_1 + M_{q,\xi_2}(\xi)d\xi_2] = \mathbf{t}_1 d\xi_1 + \mathbf{t}_2 d\xi_2. \tag{7.98}$$

The above vectors $\mathbf{t}_1$ and $\mathbf{t}_2$ are tangent to the boundary element on which the area element $dS(\mathbf{y})$ is given by

$$dS(\mathbf{y}) = |\mathbf{t}_1 \wedge \mathbf{t}_2| d\xi_1 d\xi_2 = J(\xi) d\xi_1 d\xi_2 \tag{7.99}$$

with $J(\xi)$ being the Jacobian of the mapping from Cartesian coordinates $y_i(\xi) = \mathbf{y}(\xi).\mathbf{e}_i$ to intrinsic coordinates (ξ_1, ξ_2). This Jacobian is non-zero and using a suitable orientation provides $J > 0$. Accordingly, one gets

$$\mathbf{t}_1 = \sum_{q=1}^{Q} M_{q,\xi_1}(\xi)\mathbf{y}(e,q), \quad \mathbf{t}_2 = \sum_{q=1}^{Q} M_{q,\xi_2}(\xi)\mathbf{y}(e,q), \tag{7.100}$$

$$J(\xi) = \{|\mathbf{t}_1|^2 |\mathbf{t}_2|^2 - (\mathbf{t}_1.\mathbf{t}_2)^2\}^{1/2} > 0. \tag{7.101}$$

For 6-node curvilinear triangular boundary elements, the reader can check, using (7.96), that

$$M_{1,\xi_1}(\xi) = M_{1,\xi_2}(\xi) = (1 - 4\xi_3), \quad M_{2,\xi_1}(\xi) = 4(\xi_3 - \xi_1), \tag{7.102}$$

$$M_{2,\xi_2}(\xi) = -4\xi_1, \quad M_{3,\xi_1}(\xi) = 4\xi_1 - 1, \quad M_{3,\xi_2}(\xi) = 0, \tag{7.103}$$

$$M_{4,\xi_1}(\xi) = 4\xi_2, \quad M_{4,\xi_2}(\xi) = 4\xi_1, \quad M_{5,\xi_1}(\xi) = 0, \tag{7.104}$$

$$M_{5,\xi_2}(\xi) = 4\xi_2 - 1, \quad M_{6,\xi_1}(\xi) = -4\xi_2, \quad M_{6,\xi_2}(\xi) = 4(\xi_3 - \xi_1). \tag{7.105}$$

For other choices (Δ, Q), the reader is directed to the previously quoted textbooks.[40–42] In a similar fashion, the unit normal vector $\mathbf{n}$ pointing outside the surface $\mathcal{S}$ is given (judiciously labelling the tangential vectors $\mathbf{t}_1$ and $\mathbf{t}_2$ in order to select the *outward* normal) by

$$\mathbf{n} = [\mathbf{t}_1 \wedge \mathbf{t}_2]/J(\xi). \tag{7.106}$$

In practice, the approximation of $\mathcal{S}$ is more accurate than the one of $\mathbf{n}$ if one resorts to the formula (7.103). Whenever the exact boundary is known in analytical form, it is *highly preferable* to calculate the *exact* vector $\mathbf{n}$ at each nodal point $\mathbf{y}(n)$ of the selected mesh and then to interpolate this vector on each boundary element Δ_e by mimicking (7.94), i.e. by using the relation

$$\mathbf{n}[\mathbf{y}(\xi)] = \sum_{q=1}^{Q} M_q(\xi)\mathbf{n}[\mathbf{y}(e,q)] \text{ for } \xi = (\xi_1, \xi_2) \text{ and } \mathbf{y}(\xi) \text{ on } \Delta_e. \quad (7.107)$$

7.5.3. *Computation of regular and weakly singular boundary integrals: advocated scheme*

As seen in (7.68)–(7.69) and (7.81)–(7.82), one needs to approximate integrals over the surface $\mathcal{S}$. Of interest is the general integral

$$I(\mathbf{x}) = \int_{\mathcal{S}} v(\mathbf{y})K(\mathbf{y},\mathbf{x})dS(\mathbf{y}) \quad (7.108)$$

for $\mathbf{x}$ lying outside or on $\mathcal{S}, v$ a *regular* function and K a regular or weakly singular kernel experiencing, if singular, the behaviour $K(\mathbf{y},\mathbf{x}) = O(|\mathbf{x} - \mathbf{y}|^{-1}$ as $\mathbf{y}$ approaches $\mathbf{x}$. For given mesh and boundary elements, one gets

$$I(\mathbf{x}) = \sum_{e=1}^{N_e} I^{(e)}(\mathbf{x}), \quad I^{(e)}(\mathbf{x}) = \int_{\Delta} v[\mathbf{y}(\xi)]K[\mathbf{y}(\xi),\mathbf{x}]J(\xi)d\xi_1 d\xi_2. \quad (7.109)$$

The quantity $v[\mathbf{y}(\xi)]$ is approximated by using the same interpolation as (7.94) (case of isoparametric elements). In other words, one writes

$$v[\mathbf{y}(\xi)] = \sum_{q=1}^{Q} M_q(\xi)v(e,q) \quad \text{with } v(e,q) = v[\mathbf{y}(e,q)] \quad (7.110)$$

and thus ends up with the decomposition

$$I(\mathbf{x}) = \sum_{e=1}^{N_e}\sum_{q=1}^{Q} v(e,q)I^{(e,q)}(\mathbf{x}), \quad (7.111)$$

$$I^{(e,q)}(\mathbf{x}) = \int_{\Delta} M_q(\xi)K[\mathbf{y}(\xi),\mathbf{x}]J(\xi)d\xi_1 d\xi_2. \quad (7.112)$$

Accordingly, the task reduces to the numerical evaluation of each integral $I^{(e,q)}(\mathbf{x})$ when the point $\mathbf{x}$ lies either outside or on the boundary element Δ_e. Two cases then immediately occur:

(1) Case of a *regular* integral. This is when the kernel $K[\mathbf{y}(\xi),\mathbf{x}]$ is regular on the boundary element Δ_e for the prescribed location of $\mathbf{x}$. It is then possible to evaluate the regular *integral* $I^{(e,q)}(\mathbf{x})$ by appealing

to standard Gaussian quadrature formulae.[42] Such quadratures are available in the literature both for the standard square and triangle.[44] For a patter Δ and a smooth function u, a $P-$order quadrature approximation reads

$$\int_\Delta u(\xi_1, \xi_2) d\xi_1 d\xi_2 \sim \sum_{i=1}^{P} w_i u[\xi_1(i), \xi_2(i)] \tag{7.113}$$

with P weights w_i and Gaussian points $(\xi_1(i), \xi_2(i))$. Such weights and points are given in[44] for a triangle $\Delta = \mathcal{T}$ and $P = 3, 6, 7, 16, 19, 28$.

(2) Case of a *weakly singular* integral. This happens when the kernel is weakly singular when $\mathbf{y}$ approaches $\mathbf{x}$ which lies on the element Δ_e. More precisely, for a collocation approach, $\mathbf{x}$ turns out to be *one of the nodal points* $\mathbf{y}(e, q)$ of the selected element Δ_e with associated intrinsic coordinates $\eta = (\eta_1, \eta_2)$. In order to remove by hand the weak singularity of the kernel in the vicinity of $\mathbf{x}$, one resorts on the pattern Δ to polar coordinates (ρ, α) centered at (η_1, η_2) with

$$(\xi_1, \xi_2) = (\eta_1, \eta_2) + (\rho \cos\alpha, \rho \sin\alpha), \quad \alpha_1 < \alpha < \alpha_2, \ \ 0 \leq \rho \leq d(\alpha). \tag{7.114}$$

As detailed below, this permits one to express, for any function F, the difference $F(\mathbf{y}) - F(\mathbf{x})$ on the boundary element Δ_e in terms of the distance ρ. Using the interpolation (7.94) for F and exploiting, since $\mathbf{x}$ lies on Δ_e, the approximation (7.93) for $F(\mathbf{x})$ yields

$$F(\mathbf{y}) - F(\mathbf{x}) = \sum_{q=1}^{Q} [M_q(\xi) - M_q(\eta)] u(e, q). \tag{7.115}$$

In practice each smooth shape function M_q is polynomial and its Taylor expansion therefore gives

$$M_q(\xi) - M_q(\eta) = (\xi_1 - \eta_1) P_q^{(1)}(\rho, \alpha, \eta) + (\xi_2 - \eta_2) P_q^{(2)}(\rho, \alpha, \eta) \tag{7.116}$$

with known and regular functions $P_q^{(1)}$ and $P_q^{(2)}$. By virtue of (7.114), it follows that

$$M_q(\xi) - M_q(\eta) = \rho \hat{M}_q(\rho, \alpha, \eta), \quad \hat{M}_q = \cos\alpha P_q^{(1)} + \sin\alpha P_q^{(1)}. \tag{7.117}$$

For example, when $(\Delta, Q) = (\mathcal{T}, 6)$ starting from (7.95)–(7.96) easily provides, under the notation $\eta_3 = 1 - \eta_1 - \eta_2$, the functions

$$\hat{M}_1 = (1 - 4\eta_3)(\cos\alpha + \sin\alpha) + 2\rho(\cos\alpha + \sin\alpha)^2, \tag{7.118}$$

$$\hat{M}_2 = 4[\eta_3 \cos\alpha - \eta_1(\cos\alpha + \sin\alpha) - \rho\cos\alpha(\cos\alpha + \sin\alpha)], \tag{7.119}$$

$$\hat{M}_3 = (4\eta_1 - 1)\cos\alpha + 2\rho\cos\alpha^2, \tag{7.120}$$

$$\hat{M}_4 = 4(\eta_2\cos\alpha + \eta_1\sin\alpha) + 4\rho\cos\alpha\sin\alpha), \tag{7.121}$$

$$\hat{M}_5 = (4\eta_2 - 1)\cos\alpha + 2\rho\sin\alpha^2, \tag{7.122}$$

$$\hat{M}_6 = 4[\eta_3 \sin\alpha - \eta_2(\cos\alpha + \sin\alpha) - \rho\cos\alpha(\cos\alpha + \sin\alpha)]. \tag{7.123}$$

Accordingly, one obtains for $\mathbf{y}$ and $\mathbf{x}$ on Δ_e the key relations

$$F(\mathbf{y}) - F(\mathbf{x}) = \rho\sum_{q=1}^{Q}\hat{M}_q u(e,q), \quad \mathbf{y} - \mathbf{x} = \rho\hat{d}, \quad \hat{d} = \sum_{q=1}^{Q}\hat{M}_q\mathbf{y}(e,q). \tag{7.124}$$

Using (7.114) and (7.124) shows that there exists a regular function $\hat{K}$, such that for $\mathbf{x} = \mathbf{x}(\eta)$ and $\mathbf{y}$ located at the element Δ_e, then

$$K[\mathbf{y}(\xi), \mathbf{x}] = \hat{K}[\rho, \alpha, \eta]/\rho, \quad d\xi_1 d\xi_2 = \rho d\rho d\alpha. \tag{7.125}$$

Hence, the weakly singular integral $I^{(e,q)}(\mathbf{x})$ is written

$$I^{(e,q)}(\mathbf{x}) = \int_{\alpha_1}^{\alpha_2}\left\{\int_0^{d(\alpha)} M_q(\rho,\alpha,\eta)\hat{K}[\rho,\alpha,\eta]d\rho\right\}d\alpha. \tag{7.126}$$

It admits a regular integrand and is thus evaluated by using a standard Gaussian quadrature.

7.5.4. *Computation of regular and weakly singular boundary integrals: accuracy issues*

As seen in Section 7.5.3, each (regular or weakly singular) integral $I^{(e,q)}$ is finally evaluated by employing P−point Gaussian quadratures (in (7.113) or (7.126)) one divides the domain Δ into several subdomains). The choice of P *strongly* depends upon the desired accuracy *and* the location of $\mathbf{x}$ with respect to the selected boundary element Δ_e. For instance, for a regular integral, the value of P that is sufficient to ascertain a prescribed accuracy is small or large if $\mathbf{x}$ lies far from or close to Δ_e, respectively. Thus, a point *of the utmost importance* is the way one selects P versus the pair $(\Delta_e, \mathbf{x})$. Unfortunately, there is no systematic and theoretical answer to this key issue. A nice and carefully-tested guide has been, however, provided in Ref. 45 and the reader is directed to this basic paper for further details.

7.5.5. *Implementation for a N−particle cluster*

This section gives additional details for an $N-$particle cluster with surface $S = \cup_{n=1}^{N} S_n$ immersed in a liquid eventually bounded by the surface Σ. The numerical implementation is achieved through the following steps:

(i) One approximates $S \cup \Sigma$ by proceeding for S_n or Σ as explained in Section 7.5.1 and ends up with an N_t-node mesh and N_{et} boundary elements Δ_e described, on each surface S_n or Σ, by a pattern and smooth shape functions M_q.

(ii) At each node $\mathbf{y}(p)$ on $S \cup \Sigma$ with $p = 1, \ldots, N_t$, three unknown quantities are needed to determine $\mathbf{u}$ and the traction $\mathbf{f}$. Those quantities are governed by a discretized boundary integral equation (see, for instance, (7.81)–(7.82), (7.85)–(7.86) and (7.89)–(7.91)). The resulting $3N_t-$unknown array X obeys a linear system $AX = Y$.

(iii) Since the $3N_t \times 3N_t$ square matrix A is fully populated and unsymmetric, one solves the linear system $AX = Y$ by Gaussian elimination or iterative methods in the case of a large number of particles. A might be singular (see Section 7.3.4.2 for the unicity of the solution) but in practice[30,46] the computed matrix A is non-singular unless very fine meshes are used.

7.6. Illustrating Comparisons and Results

The ability and accuracy of the boundary element approach are illustrated in this section by presenting a few numerical comparisons for a single and solid particle and a fluid bounded by a motionless, solid plane wall or spherical cavity. The associated Green tensors $\mathbf{G}^c$ (see formulae (7.55) and (7.60)) are employed together with the usual free-space Green tensor $\mathbf{G}^\infty$ for the cavity. Moreover, for the sake of accuracy, each surface is meshed using 6-node curved and triangular boundary elements.

7.6.1. *Sedimentation of a solid spheroid near a plane wall*

Let us consider a solid spheroid, with uniform density ρ_s, immersed in a liquid with uniform density ρ and viscosity μ bounded by the $x_3 = 0$ plane, solid and motionless surface Σ. The spheroid has center O and inequation

$x_1^2 + x_2^2 + \epsilon^{-2}(x_3 - H)^2/ \leq a^2$ with $h = H/(\epsilon a) > 1$ being the separation ratio. It, moreover, settles under a prescribed uniform gravity field $\mathbf{g}$. Far from the wall ($h = \infty$), it is known[1] to translate at the velocity $ga^2(\rho_s - \rho)v(\epsilon)/\mu\mathbf{e}_3$ if $\mathbf{g} = g\mathbf{e}_3$ with $v(1) = 2/9$ for a sphere and for *oblate* spheroids

$$v(\epsilon) = \{p(p^2+3)\arctan(1/p) - p^2\}/12, \quad p = \epsilon/(1-\epsilon^2)^{1/2}, \ \epsilon < 1. \tag{7.127}$$

Near the wall, the settling spheroid translates and rotates at the velocity $\mathbf{U}$ and $\mathbf{\Omega}$, respectively. It also experiences net hydrodynamic force $\mathbf{F}$ and torque $\mathbf{C}$ (about O) given by (7.33) (7.34) for $\mathbf{u}_c = \mathbf{0}$, i.e. (see the end of Section 7.3.4.3) the linear relation $(\mathbf{F}, \mathbf{C}) = -\mu\mathbf{M}_s.(\mathbf{U}, \mathbf{\Omega})$. Since the spheroid and the liquid are homogeneous, one gets $(\mathbf{U}, \mathbf{\Omega})$ by enforcing the condition $\mathbf{C} = \mathbf{0}$ and $\mathbf{F} = (\rho - \rho_s)\mathcal{V}\mathbf{g}$ with $\mathcal{V}$ being the particle's volume. Symmetries and linearity further permit one to restrict the attention to $\mathbf{g} = g\mathbf{e}_1$ and $\mathbf{g} = -g\mathbf{e}_3$ with $g > 0$. Assuming that $\rho_s \neq \rho$, the non-zero normalized Cartesian velocity components then read

$$u_1 = \frac{\mu a^{-2}\mathbf{U}.\mathbf{e}_1}{g(\rho_s - \rho)v(\epsilon)}, \quad w_2 = \frac{\mu a^{-3}\mathbf{\Omega}.\mathbf{e}_2}{g(\rho_s - \rho)v(\epsilon)} \text{ if } \mathbf{g} = g\mathbf{e}_1, \tag{7.128}$$

$$u_3 = \frac{\mu a^{-2}\mathbf{U}.\mathbf{e}_3}{g(\rho - \rho_s)v(\epsilon)} \text{ if } \mathbf{g} = -g\mathbf{e}_3 \tag{7.129}$$

and depend upon (ϵ, h). Those quantities have been calculated in Ref. 47 by a collocation method. Here, we employ the boundary integral equation (7.92) with the Blake Green tensor that vanishes on Σ (see (7.33)). Our computations are compared with the bipolar coordinates method[47,48] for a sphere ($\epsilon = 1$) and with[47] for the oblate spheroid $\epsilon = 1/2$ in Table 7.1.

Table 7.1. Computed normalized velocities u_1, w_2 and u_3 (see (7.128)–(7.129)) for a solid sphere (a) and oblate spheroid with $\epsilon = 1/2$ (b) and different N_t−node meshes.

N_1	h	$u_1(a)$	$w_2(a)$	$u_3(a)$	$u_1(b)$	$w_2(b)$	$u_3(b)$
74	1.1	0.4463	0.0245	0.1087	0.6433	−0.0534	0.246
242	1.1	0.4424	0.0259	0.0886	0.6413	−0.0538	0.244
1,058	1.1	0.4430	0.0270	0.0871	0.6411	−0.0538	0.244
[10, 3]	1.1	0.4430	0.0270	0.0873	0.6464	−0.0522	0.241
74	2.0	0.7256	0.0034	0.4726	0.7910	−0.0250	0.473
242	2.0	0.7235	0.0035	0.4707	0.7890	−0.0252	0.472
1,058	2.0	0.7232	0.0035	0.4705	0.7888	−0.0252	0.472
[10, 3]	2.0	0.7232	0.0035	0.4705	0.7892	−0.0252	0.477

As seen in this table, the agreement is excellent for the sphere and very good for the oblate spheroid. Actually,[47,49] kept the extra weakly-singular double-layer integral although it vanishes on theoretical grounds and this might explain the small observed discrepancies for $\epsilon = 1/2$.

7.6.2. *Solid particle in a spherical cavity*

Recently[37] addressed the case of a solid and arbitrarily-shaped particle with surface S immersed in a liquid confined by a solid, motionless and spherical cavity Σ with center O' and radius R. This has been achieved by using the free-space Green tensor $\mathbf{G}^\infty$ (therefore putting Stokeslets on the entire surface $S \cup \Sigma$) or the specific Green tensor $\mathbf{G}^c$ given by (7.60) (then putting Stokeslets solely on S). Here, we report accurate comparisons for a sphere with center O and radius a and direct the reader to the quoted paper for results obtained for solid ellipsoids. Setting $\mathbf{O'O} = d\mathbf{e}_3$ with $0 \leq d < R - a$, reasons of symmetry confine the analysis to four different sphere's rigid-body motion for which we give the net force $\mathbf{F}$ and torque $\mathbf{\Gamma}$ (about O) exerted on the sphere. These selected cases introduce five dimensionless friction coefficients, c_1, c_3, s, t_1 and t_3, which solely depend upon $(d/a, R/a)$ and are defined as follows:

(i) The sphere translates at the velocity $\mathbf{e_1}$. Then $\mathbf{F} = -6\pi\mu a c_1 \mathbf{e}_1$ and $\mathbf{\Gamma} = 8\pi\mu a^2 s \mathbf{e}_2$.
(ii) The sphere translates at the velocity $\mathbf{e_3}$. Then $\mathbf{F} = -6\pi\mu a c_3 \mathbf{e}_3$ and $\mathbf{\Gamma} = \mathbf{0}$.
(iii) The sphere rotates at the velocity $\mathbf{e_1}$. Then $\mathbf{F} = -8\pi\mu a^2 s \mathbf{e}_2$ and $\mathbf{\Gamma} = -8\pi\mu a^3 t_1 \mathbf{e}_1$.
(iv) The sphere rotates at the velocity $\mathbf{e_3}$. Then $\mathbf{F} = \mathbf{0}$ and $\mathbf{\Gamma} = -8\pi\mu a^3 t_3 \mathbf{e}_3$.

We present, below, comparisons for the computed coefficients c_1, c_3, s, t_1, t_3 against analytical or semi-analytical results obtained when the sphere is located at the cavity center or not, respectively. More precisely, when $d = 0$, one gets $s = 0$ and $c := c_1 = c_3$ with the *exact*[1] value

$$c(\beta) = \frac{1 - \beta^5}{1 - \frac{9\beta}{4} + \frac{5\beta^3}{2} - \frac{9\beta^5}{4} + \beta^6}, \quad \beta = a/R < 1. \tag{7.130}$$

We denote by c_∞ or c_c the value of c computed using the Green tensor $\mathbf{G}^\infty$ on $S \cup \Sigma$ (with 1058 collocation points on Σ) or the Green tensor $\mathbf{G}^c$ on S, respectively. The quantities c_∞, c_c and the associated relative

Table 7.2. Computed values $c_\infty, \Delta c_\infty, c_c$ and Δc_c versus the number N of nodal points on the sphere for $R/a = 1.1, 2$ and 5. Note that 1058 nodes have been spread Σ for the evaluation of $(c_s, \Delta c_s)$.

N	R/a	c_∞	Δc_∞	c_c	Δc_c
74	1.1	3,258.137	1.00613	2,097.155	0.29128
242	1.1	2,124.983	0.30842	1,949.547	0.01030
1,058	1.1	1,777.331	0.09436	1,676.260	0.00353
exact	1.1	1,624.089	0	1,624.089	0
74	2.	7.223525	0.00968	7.218993	0.01030
242	2.	7.289179	0.00068	7.284937	0.00126
1,058	2.	7.297493	0.00046	7.293273	0.00012
exact	2.	7.294118	0	7.294118	0
74	5.	1.749799	0.00344	1.749640	0.00353
242	5	1.755232	0.00035	1.755073	0.00044
1,058	5.	1.755937	0.00005	1.755777	0.00004
exact	5.	1.755845	0	1.755845	0

errors $\Delta c_c = |c_c - c|/c$ and $\Delta c_\infty = |c_\infty - c|/c$ are displayed in Table 7.2 versus the number N of nodal points on the sphere's surface S for $R/a = 1.1, 2$ and 5. Inspecting this table reveals that, in addition to a very good agreement with the formula (7.130), the use of the Green tensor $\mathbf{G}^c$ yields more accurate results for the small sphere-cavity gap $R - a = 0.1a$.

When the sphere is not located at the spherical cavity center (for instance if $d > 0$), very accurate semi-analytical results have been recently reported in Ref. 50 using the method of bipolar coordinates (previously worked out in Refs. 21 and 22). For $d > 0$ we denote the sphere-cavity gap by $\eta = (R - d - a)/a$ and compare the results obtained by[50] with two boundary integral approaches obtained when selecting the Green tensors $\mathbf{G}^\infty$ or $\mathbf{G}^c$. More precisely, the coefficients c_1, c_3, s, t_1, t_3 are displayed in Table 7.3 if $\mathbf{G}^\infty$ and Table 7.4 if $\mathbf{G}^c$ for $R = 4a$, a few values of η and several $N-$node meshes on the sphere. For a small sphere-cavity gap it has been necessary to use, on the boundary Σ, more collocation points than for the case of a sphere at the center of the cavity when using the free-space Green tensor $\mathbf{G}^\infty$ (see Table 7.3) and putting 4098 nodes on Σ enables a very good agreement with Jones results for $\eta = 0.5$ (medium gap) and $\eta = 0.1$ (small gap). As seen in Table 7.4, using the Green tensor $\mathbf{G}^c$ yields more accurate results, especially for the small gap $\eta = 0.1$.

Table 7.3. Computed friction coefficients c_1, c_3, t_1, t_3 and s versus the number N of collocation points on the particle surface S (with $N_1 = 74, N_2 = 242, N_3 = 1058$) for $R = 4a$ and two dimensionless sphere-cavity gaps η using the free-space Green tensor $\mathbf{G}^\infty$ and 4098 collocation points on cavity Σ. The results kindly provided by R.B. Jones are labelled J.

N	η	c_1	c_3	t_1	t_3	s
N_1	0.5	2.6330	4.6730	1.1640	1.0789	0.11870
N_2	0.5	2.6473	4.7107	1.1639	1.0755	0.11927
N_3	0.5	2.6488	4.7144	1.1639	1.0755	0.11938
J	0.5	2.6487	4.7131	1.1639	1.0755	0.11933
N_1	0.1	3.9016	15.552	1.6065	1.1960	0.20206
N_2	0.1	3.9273	18.886	1.6145	1.1939	0.19108
N_3	0.1	3.9159	18.832	1.6171	1.1945	0.18494
J	0.1	3.9121	18.674	1.6163	1.1945	0.18344

Table 7.4. Computed friction coefficients c_1, c_3, t_1, t_3 and s versus the number N of collocation points on S (with $N_1 = 74, N_2 = 242, N_3 = 1058$) for $R = 4a$ two dimensionless sphere-cavity gaps η using the Green tensor $\mathbf{G}^c$ (second approach). The results kindly provided by R.B. Jones are labelled J.

N	η	c_1	c_3	t_1	t_3	s
N_1	0.5	2.6327	4.6714	1.1639	1.0789	0.11861
N_2	0.5	2.6471	4.7090	1.1639	1.0755	0.11920
N_3	0.5	2.6486	4.7127	1.1639	1.0755	0.11932
J	0.5	2.6487	4.7131	1.1639	1.0755	0.11933
N_1	0.1	3.9009	15.413	1.6052	1.1960	0.20138
N_2	0.1	3.9237	18.636	1.6134	1.1938	0.19001
N_3	0.1	3.9121	18.711	1.6160	1.1945	0.18353
J	0.1	3.9121	18.674	1.6163	1.1945	0.18344

7.7. Concluding Remarks

The boundary element technique introduced in this chapter admits very closely-related applications in other fields appealing to linear differential equations for which a Green function or tensor exists: linear elasticity, linear acoustics (Helmholtz equation), Brinkman flows. In each case, the required steps are the ones introduced in the previous sections (counterpart of the reciprocal identity and energy dissipation, unicity issue, integral representation of the field in the domain, resulting boundary integral equations on the domain boundary, numerical implementation by resorting to boundary elements and the reader is directed for details to textbooks.[30,40–42]

The comparisons presented in Section 7.6 highlight the very good accuracy of the boundary integral method. For the sake of conciseness, these benchmark tests solely deal with one solid particle but additional comparisons for 2-particle clusters either settling or experiencing a so-called elecrophoretic migration are however available in Refs. 51 and 52 respectively.

Since the finite element method (FEM) is nowadays widely employed and implemented in most of software it is worth comparing its merits and drawbacks with the introduced boundary element method (BEM), at least in the selected field of Low-Reynolds-Number (also named creeping) steady flows. Briefly, one can list the following features for each method:

(1) The FEM is very easy to implement by dividing the liquid domain in cells and using low-order quadrature formulae is sufficient. However, it is necessary to adequately truncate the domain when the liquid is not confined by a cavity resulting in huge linear system to invert.
(2) The BEM solely requires meshing the boundary $S \cup \Sigma$ of the liquid and therefore yields linear systems of reasonable size. It is quite suitable for external problems when the liquid is solely bounded by the particles (no additional bounding surface Σ). In such circumstances, one can employ the free-space tensor $\mathbf{G}^{\infty}$ of very simple and analytical Cartesian components given by (7.48). Moreover, as illustrated in Section 7.6, the method also produces very accurate results for externally bounded liquid domains once one is able to accurately compute (if possible, at a reasonable cpu time cost) both the weakly singular Green tensor $\mathbf{G}^c$ (for regular integrals) and the regular difference $\mathbf{G}^c - \mathbf{G}^{\infty}$ for the weakly singular integrals.
(3) As emphasized in Section 7.5.4, the achieved accuracy of a numerical implementation of the BEM is dictated by the choice of the order of the employed Gaussian quadrature formula when discretizing the boundary integral equations. The nice guide proposed in Ref. 45 is issued from numerical experiments, and any other theroretically-established scheme would therefore be of great interest to select the Gaussian quadratures versus a prescribed accuracy. Such a challenging and cumbersome issue fully deserves attention in the future.

References

1. J. Happel and H. Brenner, *Low Reynolds Number Hydrodynamics* (Martinus Nijhoff, The Hague, 1973).
2. S. Kim and S. J. Karrila, *Microhydrodynamics: Principles and Selected Applications* (Butterworth, Boston, 1991).
3. A. Oberbeck, Uber stationare flussigkeitsbewegungen mit berucksichtigung der inneren reibung, *J. Reine. Angew. Math.* **81**, 62–80, (1876).
4. B. A. Edwardes, Steady motion of a viscous liquid in which an ellipsoid is constrained to rotate about a principal axis, *Quart. J. Math.* **26**, 70–78, (1892).
5. G. B. Jeffery, The motion of ellipsoidal particles immersed in a viscous fluid, *Proc. Roy. Soc. Lond. A* **102**, 161–179, (1922).
6. F. Perrin, Mouvement brownien d'un ellipsoide (i). dispersion diélectrique pour des molécules ellipsoidales, *J. Phys. Radium* **5**, 497–519, (1934).
7. C. W. Oseen, *Neure Methoden und Ergebnisse in der Hydrodynamik.* (Akademische Verlagsgesellschat M. B. H., 1927).
8. H. Lamb, *Hydrodynamics* 6th edition (Cambridge University Press, Cambridge, 1932).
9. G. Hetsroni and S. Haber, Flow in and around a droplet or bubble submerged in an unbound arbitrary velocity field, *Rheol. Acta.* **9**, 488–496, (1970).
10. G. Hetsroni and S. Haber, The hydrodynamic resistance of a fluid sphere submerged in stokes flows, *Z. Angew. Math. Mech.* **15**, 45–50, (1971).
11. G. B. Jeffery, On a form of the solution of laplace's equation suitable for problems relating to two spheres, *Proc R Soc, Lond, A.* **87**, 109–120, (1912).
12. M. Stimson and G. B. Jeffery, The motion of two spheres in a viscous fluid, *Proc R Soc, Lond, A.* **111**, 110–116, (1926).
13. A. J. Goldman, R. G. Cox, and H. Brenner, The slow motion of two identical arbitrarily oriented sphere, through a viscous fluid, *Chemical Engineering Science* **21**, 1151–1170, (1966).
14. M. D. A. Cooley and M. E. O'Neill, On the slow motion of two spheres in contact along their lines of centers through a viscous fluid, *Proc Camb Phil Soc.* **66**, 407–415, (1969).
15. M. D. A. Cooley and M. E. O'Neill, On the slow motion generated in a viscous fluid by the approach of a sphere to a plane wall or stationary sphere, *Mathematika* **16**, 37–49, (1969).
16. M. E. O'Neill, On asymmetrical slow viscous flows caused by the motion of two equal spheres almost in contact, *Proc. Camb. Phil. Soc.* **65**, 543–555, (1969).
17. M. E. O'Neill, Asymmetrical slow viscous fluid motions caused by the translation or rotation of two spheres, *Z.A.M.P.* **21**, 164–179, (1970).
18. H. Brenner, The slow motion of a sphere through a viscous fluid towards a plane surface, *Chemical Engineering Science* **16**, 242–251, (1961).
19. W. R. Dean and M. E. O'Neill, A slow rotation of viscous liquid caused by the rotation of a solid sphere, *Mathematika* **10**, 13–24, (1963).

20. M. E. O'Neill, A slow motion of viscous liquid caused by a slowly moving solid sphere, *Mathematika* **11**, 67–74, (1964).
21. M. E. O'Neill and S. R. Majumdar, Asymmetrical slow viscous motions caused by translation or rotation of two spheres. part i. the determination of exact solutions for any values of the radii ratio and separation parameters, *Z.A.M.P.* **21**, 164–179, (1970).
22. M. E. O'Neill and S. R. Majumdar, Axisymmetric slow viscous motions caused by translation or rotation of two spheres. part ii. asymptotic forms of the solutions when the minimum clearance approaches zero, *Journal of Applied Mathematics Physics (ZAMP)* **21**, 180–187, (1970).
23. S. H. Lee and L. G. Leal, Motion of a sphere in the presence of a plane interface. Part 2. An exact solution in bipolar coordinates, *J. Fluid Mech.* **98**(1), 193–224, (1980).
24. T. A. Dabros, Singularity method for calculating hydrodynamic forces and particle velocities in low-reynolds-number flows, *J. Fluid Mech.* **156**, 1–21, (1985).
25. A. T. Chwang and T. Y. T. Wu, Hydrodynamics of low-reynolds-number flow. part 2, singularity methods, *J. Fluid Mech.* **67**, 787–815, (1975).
26. M. J. Gluckman, R. Pfeffer, and R. Weinbaum, A new technique for treating multiparticle slow viscous flow: Axisymmetric flow past spheres and spheroids, *J. Fluid Mech.* **50**, 705–740, (1971).
27. W. H. Liao and D. A. Krueger, Multipole expansion calculation of slow viscous flow about spheroids of different sizes, *J Fluid Mech.* **197**, 223–241, (1980).
28. Q. Hassonjee, P. Ganatos, and R. Pfeffer, A strong-interaction theory for the motion of arbitrary three-dimensional clusters of spherical particles at low reynolds number, *J. Fluid Mech.* **197**, 1–37, (1988).
29. S. Weinbaum, P. Ganatos, and Z. Y. Yan, Numerical multipole and boundary integral equation technique in stokes flow, *Ann. Rev. Fluid Mech.* **22**, 275–316, (1990).
30. C. Pozrikidis, *Boundary Integral and Singularity Methods for Linearized Viscous Flow.* (Cambridge University Press, 1992).
31. H. A. Lorentz, Ein allgemeiner satz, die bewegung einer reibenden flussigkeit betreffend, nebst einigen anwendungen desselben, *Abhand. Theor. Phys.* **1**, 23–42, (1907).
32. G. K. Youngren and A. Acrivos, Stokes flow past a particle of arbitrary shape: A numerical method of solution, *J. Fluid Mech.* **69**, 377–403, (1975).
33. I. Imai. 2nd inter. jsme symp. fluid machinary and fluidics. pp. 15–23, Tokyo, (1972).
34. O. A. Ladyzhenskaya, *The Mathematical Theory of Viscous Incompressible Flow.* (Gordon & Breach, New-York-London, 1969).
35. J. R. Blake, A note on the image system for a stokeslet in a no-slip boundary, *Proc. Camb. Phil. Soc.* **70**, 303–310, (1971).
36. M. S. E. Gavze, Particles in a shear flow near a solid wall: effect of nonsphericity on forces and velocities, *Int. J. Multiphase Flow* **23**(1), 155–182, (1997).

37. A. Sellier, Slow viscous motion of a solid particle in a spherical cavity, *CMES* **25**(3), 165–180, (2008).
38. R. B. Jones, Spherical particle in Poiseuille flow between planar walls, *Journal of Chemical Physics* **121**, 483–500, (2004).
39. A. Sellier and L. Pasol, Sedimentation of a Solid Particle Immersed in a Fluid Film, *CMES* **16**(3), 187–196, (2006).
40. C. A. Brebbia, J. C. Telles, and L. C. Wrobel, *Boundary Element Techniques, Theory and Applications in Engineering* (Springer-Verlag, Berlin, 1984).
41. D. E. Beskos, *Introduction to Boundary Element Methods* (Elsevier Science Publishers, New-York, 1987).
42. M. Bonnet, *Boundary Integral Equation Methods for Solids and Fluids* (John Wiley & Sons Ltd, New-York, 1999).
43. A. H. Stroud and D. Secrest, *Gaussian Quadrature Formulas* (Prentice-Hall, 1966).
44. J. N. Lyness and D. Jespersen, Moderate degree symmetric quadrature rules for the triangle, *J. Inst. Maths Applics.* **15**, 19–32, (1975).
45. M. Rezayat, D. J. Shippy, and F. J. Rizzo, On time-harmonic elactic-wave analysis by the boundary element method for moderate to high frequencies, *Comp. Meth. in Appl. Mech. Engng.* **55**, 349–367, (1986).
46. T. Tran-Cong and N. Phan-Thien, Stokes problems of multiparticle systems: A numerical method for arbitrary flows, *Phys. Fluids.* **A**(1), 453–460, (1989).
47. R. Hsu and P. Ganatos, Gravitational and zero-drag motion of a spheroid adjacent to an inclined plane at low reynolds number, *J. Fluid Mech.* **268**, 267–292, (1994).
48. M. Chaoui and F. Feuillebois, Creeping flow around a sphere in shear flow close to a wall, *Q. J. Mech. Appl. Math.* **56**, 381–410, (2003).
49. R. Hsu and P. Ganatos, The motion of a rigid body in viscous fluid bounded by a plane wall, *J. Fluid Mech.* **207**, 29–72, (1989).
50. R. B. Jones, Dynamics of a colloid in a spherical cavity, *To appear in Methods of creeping flows. Editors: F.Feuillebois and & A. Sellier.* (2008).
51. A. Sellier, On the slow gravity-driven migration of arbitrary clusters of small solid particles, *Comptes-Rendus Mécanique.* **332**(12), 987–992, (2004).
52. A. Sellier, Electrophoretic motion of two solid particles embedded in an unbounded and viscous electrolyte, *Computational Mechanics* **28**, 202–211, (2002).

Chapter 8

BIT FOR FREE SURFACE FLOWS

G. Baker
Department of Mathematics
The Ohio State University
231 W. 18th Ave, Columbus OH 43210, USA
baker@math.ohio-state.edu

Boundary integral techniques (BIT) offer several advantages for tracking free surfaces in inviscid fluids. They reduce the spatial dimension by one in that only information on the surface is needed to advance the location of the surface. For irrotational flow, the fluid velocity is given by the gradient of a velocity potential which satisfies Laplace's equation. Formal solutions can be expressed in terms of Green's functions for Laplace's equation and these solutions lead to elegant formulations for free surface flows. In particular, dipole distributions along the surface provide a perfect representation for the potential since the normal derivatives are continuous across the surface, one of the required boundary conditions. Lagrangian markers on the surface move with a weighted average of the fluid velocities, the weighting based on the difference in densities across the surface. This choice provides a natural adaption in the representation of the surface in time. Finally, Bernoulli's equation can be used to satisfy the dynamic boundary condition at the surface, leading to an integral equation for the dipole strength. This formulation, provided in this chapter, has been used successfully in the study of water waves, the Rayleigh–Taylor instability and the rise of bubbles. Modifications are presented for the case of multiple surfaces and the presence of solid boundaries.

8.1. The Nature of Free Surface Flows

Free surfaces abound everywhere. The most common example is the interface between water and air. We see these surfaces as rain drops, ocean waves, a glass of water, etc. They appear over a vast range of length scales and they encompass multiple phenomena: jets of water that break into drops, waves that crash on the shore; mixtures of oil and water in porous rock (oil reservoirs); rivulets of water flowing down a window pane.

The list goes on and on. The fascination we find in water surfaces has naturally attracted scientists to develop mathematical models to describe their behaviour. By far, the most common studies assume continuum models with a sharp interface. In particular, the Euler equations or the Navier-Stokes equations are used for the fluid flow and jump conditions that satisfy kinematic and dynamic conditions are imposed at the surface. These models have been thoroughly tested over the last few centuries, and they are now well accepted as good models for free surface flow.

Despite the great success of the standard models for free surface flow, there remain aspects of their behaviour far less well understood, in particular when free surfaces undergo topological changes as in the break-up of a liquid jet into droplets. The separation of a droplet from the jet happens almost instantaneously on a molecular level that makes it extraordinarily difficult to express in a mathematical model, especially one that is a continuum model. Further difficulties in mathematical models arise at the point of contact between a free surface and a solid boundary, the so-called contact-line problem. These difficulties will not be addressed here.

Even when the surface remains smooth without topological changes, the full equations of motion are nonlinear and must be solved in a changing geometry. Not surprisingly, many studies of free surface flow have employed various approximations of the basic equations or restricted the nature of the free surface in some way to make the mathematical problems tractable. A vast body of knowledge has been gained this way and has served the scientific community well. Over the last century, with the advent of high-speed computing, several numerical methods have been developed to track free surface motion. Some of the popular ones include level set methodology, front-tracking techniques and finite volume methods. The overriding factor that emerges from these approaches is the importance of highly accurate calculations of the free surface velocities. Otherwise, the numerical evolution of the surface quickly becomes inaccurate and can lead to unphysical behaviour. The challenge, then, is the continual improvement of numerical techniques that ensure accurate and reliable calculations of free surface flow.

There are two classes of free surface flows where boundary integrals have a natural advantage over other methods. They are characterized by very large or very small Reynolds numbers. The Reynolds number is the ratio of inertial forces to viscous forces. It is small when viscous effects dominate, such as the motion of small drops; it is large when viscous effects are negligible, such as the propagation of water waves. In the former case, the fluid equations can be reduced to the solution of a bi-harmonic equation: this topic is covered thoroughly by Pozrikidis.[1] In the latter case, the fluid equations can be reduced to the solution of Laplace's equation. Here, it is natural to use source or dipole distributions along the surface, and the velocity at the surface can be expressed completely by boundary integrals of these distributions. In essence, only information on the surface is needed to update the surface, thus reducing the effective spatial dimension of the problem by one. Further, the surface may be represented explicitly through a surface parametrization; there is no need to embed the surface in a numerical grid.

In the next section, boundary integrals will be formulated as the solution to Euler's equation of fluid motion with free surfaces. The formulation is for a surface between two fluids of constant but different densities in three-dimensional flow. The restriction to two-dimensional flow is included separately to illustrate its connection to complex analysis, a powerful tool that allows improved numerical methods. The challenge for numerical simulations is reliable and accurate calculation of the boundary integrals which contain singular integrands. This issue is addressed in Section 8.3. In particular, there are some highly accurate methods for surfaces (curves) in two-dimensional flow. These methods are applied to several free surface flow problems in Section 8.4 which contain a single free surface, such as water waves, Rayleigh–Taylor instabilities and rising bubbles. The extension to more free surfaces is straightforward, but an additional boundary integral formulation is needed when solid boundaries are present (described in Section 8.5). Yet, there remain difficulties and the chapter will close in describing some of them.

8.2. Mathematical Formulation

Euler's equations for fluid flow in the absence of viscous effects are:

$$\frac{\partial \rho}{\partial t} + (\mathbf{u} \cdot \nabla)\rho = 0\,, \tag{8.1}$$

$$\frac{\partial \mathbf{u}}{\partial t} + (\mathbf{u} \cdot \nabla)\mathbf{u} = -\frac{1}{\rho}\nabla p + \mathbf{g}\,, \tag{8.2}$$

where $\mathbf{u}$ is the fluid velocity, ρ is the density, p is the hydrodynamic pressure and $\rho\mathbf{g}$ is the gravitational force. For liquids and gases moving slowly, it is usual to assume the flow is incompressible,

$$\nabla \cdot \mathbf{u} = 0. \tag{8.3}$$

Equations (8.1–8.2) are evolution equations for ρ and $\mathbf{u}$, with Eq. (8.3) acting as a constraint that determines p.

Another form of these equations uses the vorticity $\boldsymbol{\omega} = \nabla \times \mathbf{u}$. By taking the curl of Eq. (8.2), an evolution equation for the vorticity is derived.

$$\frac{\partial \boldsymbol{\omega}}{\partial t} + (\mathbf{u} \cdot \nabla)\boldsymbol{\omega} - (\boldsymbol{\omega} \cdot \nabla)\mathbf{u} = \nabla p \times \nabla \left(\frac{1}{\rho}\right). \tag{8.4}$$

Now the velocity must be determined from Eq. (8.3) and the definition of the vorticity. A standard approach is to introduce the vector potential $\mathbf{A}$ by $\mathbf{u} = \nabla \times \mathbf{A}$. This choice automatically satisfies Eq. (8.3), and the definition of the vorticity becomes

$$\nabla^2 \mathbf{A} = -\boldsymbol{\omega}, \tag{8.5}$$

with the additional requirement $\nabla \cdot \mathbf{A} = 0$. Several books[2–4] provide excellent coverage of vorticity and vortex methods, which have direct connections to the methods described in this chapter.

A particular value of the vorticity formulation Eq. (8.4) is that it draws attention to the production of vorticity. The terms on the left-hand side of Eq. (8.4) describe the advection and stretching of vortex lines, and the term on the right-hand side describes how vorticity may be created. In the absence of solid boundaries, vorticity can only be produced from density gradients. For flows where it is reasonable to assume density is uniform, vorticity is conserved. In particular, if there is no initial vorticity, then there will be none in the future.

On the other hand, if there is a sharp interface separating regions of constant but different densities, then vorticity is created at the interface. Unfortunately, the production of vorticity on the interface is a generalized function and difficult to derive from Eq. (8.4) directly. Instead, it is easier to seek solutions in each region separately and then connect them through interfacial conditions at the surface.

In each region, the velocity must be curl-free. Consequently, the velocity may be expressed in terms of a velocity potential $\mathbf{u} = \nabla\phi$. Upon substitution into Eq. (8.2), the equation may be integrated to produce Bernoulli's equation,

$$\frac{\partial \phi}{\partial t} + \frac{1}{2}(\nabla\phi)^2 + \frac{p}{\rho} + gy = C(t), \tag{8.6}$$

where it is now assumed that y is a coordinate in the upward vertical direction. Further, Eq. (8.3) becomes

$$\nabla^2 \phi = 0. \tag{8.7}$$

The way forward is to use Bernoulli's equation, Eq. (8.6), to update the potential at the surface, and then to use the result as a Dirichlet boundary condition for the solution to Laplace's equation, Eq. (8.7). This is where boundary integral techniques come into play.

8.2.1. *Three-dimensional BIT*

Quite clearly, Eq. (8.5) and Eq. (8.7) are candidates for boundary integral methods. To proceed, boundary conditions are needed. Consider a sharp interface between two immiscible fluids of different densities. The interface must move with the fluid: this requires that the normal component of the fluid velocities at the interface match the normal velocity of the interface. Let $\mathbf{n}$ be the unit normal to the surface pointing outwards if the surface is closed and pointing downward if the surface is open and extends to infinity in the horizontal direction. Designate the fluid quantities on either side of the interface by the subscript 1 if the quantity is on the outside (below) the surface and subscript 2 if it is inside (above). Then the kinematic condition is that the normal components of the fluid velocities must match:

$$\mathbf{n} \cdot \mathbf{u}_1 = \mathbf{n} \cdot \mathbf{u}_2. \tag{8.8}$$

Dynamic considerations require that the pressure jump across the interface is balanced by the interfacial force due to surface tension.

$$p_2 - p_1 = T\kappa. \tag{8.9}$$

The surface tension coefficient is T and the curvature is κ. Boundary integrals are needed that satisfy Eqs. (8.8–8.9).

There are many choices for boundary integrals that solve elliptic problems. One in particular offers advantages for free surface flows. Dipole distributions automatically guarantee continuity of the normal derivatives of the potential, in other words, continuity of the normal components of velocity Eq. (8.8). Let μ be a dipole strength distributed along the interface written in parametric form as $\mathbf{x}(\mathbf{p}, t)$, where $\mathbf{p}$ represents two surface coordinates (p_1, p_2).[a] Then the potential generated by the dipole

[a]The dependency on time will no longer be explicitly indicated unless necessary.

distribution along the surface is

$$\phi(\mathbf{x}) = \int \mu(\mathbf{q})\,\mathbf{n}(\mathbf{q}) \cdot \nabla_q G\big(\mathbf{x} - \mathbf{x}(\mathbf{q})\big)\,\mathrm{d}S(\mathbf{q}). \tag{8.10}$$

The subscript q on ∇ indicates that the gradient is with respect to $\mathbf{x}(\mathbf{q})$, i.e. the second argument of G. The Green's function G depends on the dimensions of the elliptic equation. In two dimensions,

$$G(\mathbf{x} - \mathbf{x}(\mathbf{q})) = \frac{1}{2\pi} \ln\big|\mathbf{x} - \mathbf{x}(\mathbf{q})\big|, \tag{8.11}$$

and in three dimensions,

$$G(\mathbf{x} - \mathbf{x}(\mathbf{q})) = -\frac{1}{4\pi} \frac{1}{\big|\mathbf{x} - \mathbf{x}(\mathbf{q})\big|}. \tag{8.12}$$

While the normal derivative of the potential is continuous across the interface, the potential jumps in value by μ. Specifically, as $\mathbf{x} \to \mathbf{x}(\mathbf{p})$ approaches the interface along the normal direction, the potential has the limiting values,[b]

$$\phi_1(\mathbf{p}) = I(\mathbf{p}) - \frac{\mu(\mathbf{p})}{2}, \tag{8.13}$$

$$\phi_2(\mathbf{p}) = I(\mathbf{p}) + \frac{\mu(\mathbf{p})}{2}, \tag{8.14}$$

where

$$I(\mathbf{p}) = \int \mu(\mathbf{q})\,\mathbf{n}(\mathbf{q}) \cdot \nabla_q G\big(\mathbf{x}(\mathbf{p}) - \mathbf{x}(\mathbf{q})\big)\,\mathrm{d}S(\mathbf{q}). \tag{8.15}$$

This integral must be interpreted as a principal-valued integral. It gives the average value of the potentials on either side of the interface, and the dipole strength is the jump in value of the potential across the interface. These statements follow by simply adding or subtracting Eq. (8.13) and Eq. (8.14).

So far, the potential ϕ generated by Eq. (8.10) satisfies Eq. (8.7) and the interfacial condition Eq. (8.8). The other interfacial condition Eq. (8.9) will determine μ. Before describing how, let's determine first how the velocity of the interface can be calculated when μ is considered known. The first issue to face is the choice for the velocity of the interface. While the normal

[b]The limiting behaviour depends on the choice of direction of the normal in Eq. (8.10), so the sign of the jump in the potential may be different from other derivations.

component is continuous, the tangential components may jump. A flexible choice is to take a weighted average,

$$\frac{\partial \mathbf{x}}{\partial t} = \mathbf{u}_I \equiv \frac{1}{2}\big[(1+\alpha)\mathbf{u}_1 + (1-\alpha)\mathbf{u}_2\big], \tag{8.16}$$

where the parameter α controls the weighting. For example, when $\alpha = 1$, the velocity of the interface is the velocity of the outer (lower) fluid. The time derivative of the location of the interface is written with partial derivatives to emphasize that the surface location $\mathbf{p}$ is held fixed. In other words, the motion of the interface is Lagrangian and $\mathbf{p}$ represents the Lagrangian label.

In line with the nature of the potential at the surface, define the average velocity and the jump in velocity as

$$\mathbf{U} = \frac{1}{2}\big(\mathbf{u}_1 + \mathbf{u}_2\big), \qquad \mathbf{m} = \mathbf{u}_2 - \mathbf{u}_1. \tag{8.17}$$

Consequently, the interfacial velocity is

$$\mathbf{u}_I = \mathbf{U} - \frac{\alpha}{2}\mathbf{m}. \tag{8.18}$$

The goal now is to determine $\mathbf{U}$ and $\mathbf{m}$ by using information on the surface only.

The tangential velocity components can be determined directly from the tangential derivatives of the potential evaluated along the surface. Specifically, let $\phi(p_1, p_2) = \phi\big(\mathbf{x}(\mathbf{p})\big)$ represent either the potential above or below evaluated on the surface and differentiate with respect to p_1 and p_2 separately.

$$\frac{\partial \phi}{\partial p_1} = \frac{\partial \mathbf{x}}{\partial p_1} \cdot \nabla\phi, \quad \text{and} \quad \frac{\partial \phi}{\partial p_2} = \frac{\partial \mathbf{x}}{\partial p_2} \cdot \nabla\phi. \tag{8.19}$$

Since

$$\mathbf{t}_1 = \frac{\partial \mathbf{x}}{\partial p_1}, \qquad \mathbf{t}_2 = \frac{\partial \mathbf{x}}{\partial p_2}, \tag{8.20}$$

are tangent vectors (not necessarily unit vectors), the tangential components of the velocities at the interface give

$$\mathbf{t}_1 \cdot \mathbf{U} = \frac{\partial I}{\partial p_1}, \qquad \mathbf{t}_2 \cdot \mathbf{U} = \frac{\partial I}{\partial p_2}, \tag{8.21}$$

$$\mathbf{t}_1 \cdot \mathbf{m} = \frac{\partial \mu}{\partial p_1}, \qquad \mathbf{t}_2 \cdot \mathbf{m} = \frac{\partial \mu}{\partial p_2}. \tag{8.22}$$

The tangent vectors may be used to determine the unit normal to the interface,

$$\big|\mathbf{t}_1 \times \mathbf{t}_2\big|\,\mathbf{n} = \mathbf{t}_1 \times \mathbf{t}_2. \tag{8.23}$$

The assumption is that the orientation of the tangent vectors are such as to make the normal vector point outwards (below). The next step is to find a way to determine the normal velocity component $\mathbf{n}\cdot\nabla\phi$. One way forward is to use Green's theorem to formulate a boundary integral equation for the normal derivative of ϕ, given ϕ on the surface. This approach is often used when either ρ_1 or ρ_2 is zero, but it may be possible to use it in the general case.

An alternative approach is based on tangential derivatives of the vector potential. Fortunately, the vector potential can be determined by a boundary integral of the dipole distribution. Start with $\nabla\phi = \nabla\times\mathbf{A}$, and use the vector identity,

$$\begin{aligned}\nabla_p\Big(\mathbf{n}(q)\cdot\nabla_q G\big(\mathbf{x}(\mathbf{p})-\mathbf{x}(\mathbf{q})\big)\Big) &= \Big(\mathbf{n}(q)\cdot\nabla_p\Big)\Big(\nabla_q G\big(\mathbf{x}(\mathbf{p})-\mathbf{x}(\mathbf{q})\big)\Big)\\ &= -\nabla_p\times\Big(\mathbf{n}(q)\times\nabla_q G\big(\mathbf{x}(\mathbf{p})-\mathbf{x}(\mathbf{q})\big)\Big),\end{aligned}$$

to derive the expression for the vector potential,

$$\mathbf{A}(\mathbf{p}) = -\int \mu(\mathbf{q})\,\mathbf{n}(\mathbf{q})\times\nabla_q G\big(\mathbf{x}(\mathbf{p})-\mathbf{x}(\mathbf{q})\big)\,\mathrm{d}S(\mathbf{q}), \tag{8.24}$$

where

$$\mathrm{d}S(q) = \big|\mathbf{t}_1\times\mathbf{t}_2\big|\,\mathrm{d}p_1\,\mathrm{d}p_2.$$

The vector potential is continuous across the interface so its tangential derivatives do not jump in value. Apply the identity

$$\iint \nabla\times\mathbf{A}\cdot\mathbf{n}\,\mathrm{d}S = \int \mathbf{A}\cdot\mathrm{d}\mathbf{l} \tag{8.25}$$

to a small closed region lying on the interface. The result gives

$$\big(\mathbf{t}_1\times\mathbf{t}_2\big)\cdot\mathbf{U} = \frac{\partial A_2}{\partial p_1} - \frac{\partial A_1}{\partial p_2}, \tag{8.26}$$

where

$$A_1 = \mathbf{t}_1\cdot\mathbf{A}, \quad A_2 = \mathbf{t}_2\cdot\mathbf{A}. \tag{8.27}$$

The quantities A_1 and A_2 are integrals that result from applying the dot product to Eq. (8.24).

Equations (8.20–8.21, and 8.26) determine the velocity at the interface. They constitute a set of linear equations for $\mathbf{U}$ and $\mathbf{m}$ which may be solved in closed form to give

$$|\mathbf{t}_1 \times \mathbf{t}_2|\, \mathbf{U} = \mathbf{t}_2 \times \mathbf{n}\, \frac{\partial I}{\partial p_1} - \mathbf{t}_1 \times \mathbf{n}\, \frac{\partial I}{\partial p_2} + \mathbf{n} \left[\frac{\partial A_2}{\partial p_1} - \frac{\partial A_1}{\partial p_2} \right], \tag{8.28}$$

$$|\mathbf{t}_1 \times \mathbf{t}_2|\, \mathbf{m} = \mathbf{t}_2 \times \mathbf{n}\, \frac{\partial \mu}{\partial p_1} - \mathbf{t}_1 \times \mathbf{n}\, \frac{\partial \mu}{\partial p_2}. \tag{8.29}$$

The result shows that $\mathbf{U}$ can be determined by the surface derivatives of three boundary integrals, I Eq. (8.15) and A_1 and A_2 Eq. (8.27). The two results, Eqs. (8.28–8.29), determine $\mathbf{u}_I$ Eq. (8.15) which completes the specification of the velocity of a Lagrangian marker on the interface Eq. (8.16). So far, no information other than surface quantities need to be known.

The last step is the derivation of an evolution equation for the dipole strength. This derivation must use Bernoulli's equation Eq. (8.6) and the dynamic boundary condition Eq. (8.9), but Bernoulli's equation must be transformed to account for the Lagrangian motion of the surface markers (fixed $\mathbf{p}$). The change in potential following the Lagrangian motion on either side of the interface is

$$\rho_1 \left[\frac{\partial \phi_1}{\partial t} - \mathbf{u}_I \cdot \mathbf{u}_1 + \frac{1}{2} |\mathbf{u}_1|^2 + gy \right] + p_1 = 0, \tag{8.30}$$

$$\rho_2 \left[\frac{\partial \phi_2}{\partial t} - \mathbf{u}_I \cdot \mathbf{u}_2 + \frac{1}{2} |\mathbf{u}_2|^2 + gy \right] + p_2 = 0. \tag{8.31}$$

By subtracting Eq. (8.31) from Eq. (8.30) and using the definitions of the surface velocity Eq. (8.18), one finds after some lengthy algebra,

$$\frac{1}{2} \frac{\partial \mu}{\partial t} - A \frac{\partial I}{\partial t} = -\frac{\alpha}{4} |\mathbf{m}|^2 - A \left[\frac{1}{2} |\mathbf{U}|^2 - \frac{\alpha}{2}\, \mathbf{m} \cdot \mathbf{U} - \frac{1}{8} |\mathbf{m}|^2 - gy \right] - \frac{T}{\rho_1 + \rho_2}\, \kappa. \tag{8.32}$$

The Atwood number measures the jump in the density and is defined as

$$A = \frac{\rho_1 - \rho_2}{\rho_1 + \rho_2}. \tag{8.33}$$

The time derivative of I can be split into two parts:

$$\begin{aligned}\frac{\partial I}{\partial t} &= \int \frac{\partial \mu(\mathbf{q})}{\partial t}\, \mathbf{n}(\mathbf{q}) \cdot \nabla_q G(\mathbf{x}(\mathbf{p}) - \mathbf{x}(\mathbf{q}))\, |\mathbf{t}_1 \times \mathbf{t}_2|\, dp_1\, dp_2 \\ &\quad + \int \mu(\mathbf{q}) \frac{\partial}{\partial t}\Big\{\mathbf{n}(\mathbf{q}) \cdot \nabla_q G(\mathbf{x}(\mathbf{p}) - \mathbf{x}(\mathbf{q}))\, |\mathbf{t}_1 \times \mathbf{t}_2|\Big\}\, dp_1\, dp_2.\end{aligned} \tag{8.34}$$

The time derivative in the second integral must be taken carefully since all the quantities in the braces depend on time through their dependency on $\mathbf{x}(\mathbf{p}, t)$. The time derivative of $\mathbf{x}(\mathbf{p}, t)$ is $\mathbf{u}_I$ which depends only on μ and the location of the interface and may be calculated as already described above. Consequently, the second integral and the right-hand side of Eq. (8.32) may be calculated provided only that μ is known. The first integral in Eq. (8.34) coupled with the first term in Eq. (8.32) forms a Fredholm integral equation of the second kind for the rate of change of the dipole strength, and its solution provides a way to update μ.

In summary, Eqs. (8.16, 8.18, 8.28, 8.29 and 8.32) constitute a set of evolution equations for $\mathbf{x}$ and μ. Knowing $\mathbf{x}$ and μ at some time t, the tangent vectors Eq. (8.20) and the normal Eq. (8.23) may be evaluated, and then the boundary integrals Eqs. (8.15 and 8.27) may be evaluated to determine the average velocity $\mathbf{U}$ and the velocity difference $\mathbf{m}$ from Eqs. (8.28 and 8.29). This gives the interfacial velocity $\mathbf{u}_I$ defined in Eq. (8.18). Finally, the right-hand side terms of Eq. (8.32) may be evaluated, including the second integral in Eq. (8.34), and the integral equation Eq. (8.32) may be solved to obtain the rate of change of the dipole strength. Consequently, the surface location and the dipole strength may be updated in time.

The dipole strength has a direct connection to a special distribution of vorticity on the interface. By integrating by parts, Eq. (8.24) may be written as

$$\mathbf{A} = -\int \boldsymbol{\omega}\, (\mathbf{q})\, G(\mathbf{x}(\mathbf{p}) - \mathbf{x}(\mathbf{q}))\, dS(\mathbf{q}), \tag{8.35}$$

where

$$\begin{aligned}\boldsymbol{\omega}\, (\mathbf{p}) &= \boldsymbol{\Gamma}\, (\mathbf{p})\, \delta(n) \\ &= \frac{1}{|\mathbf{t}_1 \times \mathbf{t}_2|}\left(\frac{\partial \mu}{\partial p_1}\mathbf{t}_2 - \frac{\partial \mu}{\partial p_2}\mathbf{t}_1\right)\delta(n).\end{aligned} \tag{8.36}$$

Equation (8.35) is a formal solution to Eq. (8.5) where $\boldsymbol{\omega}$ is a vorticity distribution. It is a delta distribution on the interface and its direction lies in the tangent plane. The quantity $\mathbf{\Gamma}$ is called the vortex sheet strength and it measures the jump in tangential velocity across the interface. The creation of the dipole strength Eq. (8.32) when $A \neq 0$ or $T \neq 0$ corresponds directly to the creation of a vortex sheet at the interface where the density jumps in value and/or when surface tension effects are present. While it is not easy to derive the equation for the generation of the vortex sheet strength from Eq. (8.4), it is relatively straightforward to derive an evolution equation for $\mathbf{\Gamma}$ from its definition in Eq. (8.36) and the evolution equation for the dipole strength Eq. (8.32), but the result is a complicated expression with no apparent advantage over the dipole formulation.

A specific feature of the particular formulation of free surface flow presented in this chapter is its generality since it allows fluids of different densities on either side of the interface. A similar approach[5] adopted to study water waves chooses the motion of the markers to be purely vertical, which imposes some limitations on the geometry of the surface (breaking waves must be excluded). Both these approaches are the natural extension of the derivation in two-dimensions (one-dimensional surface)[6] which follows the pioneering work of Birkhoff.[7,8]

In contrast, most derivations of boundary integrals for free surface flow have picked one of the fluid densities to be zero, corresponding to one of the choices $A = \pm 1$. Bernoulli's equation Eq. (8.6) is used to update the potential on the free surface and various boundary integral techniques are used to solve Laplace's equation for the potential. The first derivation following this approach[9] uses a dipole representation in three dimensions, but only the application to axisymmetric flows is calculated numerically. Another technique,[10] developed first for two-dimensional flow,[11] uses Green's third identity to obtain a Fredholm integral equation of the first kind for the normal component of the velocity. Numerically, a full system of equations for the discrete approximation must be solved at each time level. In contrast, a Fredholm integral equation of the second kind for the normal component of the velocity given the potential along the interface[12] may be solved iteratively. It is also possible to prove numerical stability when a regularized Green's function is used in the numerical approximations, a technique developed first in two dimensions.[13] A different approach[14] uses a special surface parametrization to take advantage of the Riesz transform to ensure numerical stability of the boundary integral formulation based on dipole distributions. Finally, there are some special techniques, in particular those based on conformal mapping[15] or analytic continuation,[16,17] that can only be used in two-dimensional flows.

Perhaps the greatest strength of the formulation based on dipole distributions is that the integral equation Eq. (8.32) may be solved by modern iterative techniques, such as multigrid and GMRES with a suitable preconditioner. At each stage of the iteration, the cost in calculating the integral may be reduced with multipole or tree-code algorithms.[18,19] Moreover, the integrals for each $\mathbf{p}$ can be performed on separate processors leading to a natural procedure on parallel computer architectures. At the same time, the choice of α allows control on adaptivity. The physical choice $\alpha = A$ takes the density weighted average of the velocities at the interface as the interfacial velocity and proves to be an effective choice in many cases.

Assumptions so far include the presence of only a single interface, either closed or in open periodic geometry where the fluid is at rest far from the interface. If there is an additional external flow, it may be added directly in Eq. (8.18). There are other modifications and extensions for the presence of multiple interfaces and the presence of solid boundaries. These modifications and extensions will be described later for the simpler case of two-dimensional flow. Indeed, a substantial amount of study has concentrated on two-dimensional flows both because of the simplification in the formulation and because of the much lower cost in computational time. It is appropriate, therefore, to state the formulation specifically for two-dimensional flow and describe the various modifications when additional free surfaces and rigid boundaries are present. The approach will be equally valid in three-dimensional flow.

8.2.2. *Two-dimensional BIT*

Introduce a right-handed coordinate system with unit vectors $\mathbf{i}, \mathbf{k}$ lying in the horizontal plane and $\mathbf{j}$ pointing vertically upwards. Gravity will be assumed to act vertically downwards. The flow is now assumed to lie in the $\mathbf{i}, \mathbf{j}$ only. A general parametric form for the interface location is

$$\mathbf{x} = x(p_1)\,\mathbf{i} + y(p_1)\,\mathbf{j} + p_2\,\mathbf{k}, \tag{8.37}$$

and the dipole distribution depends on $p_1 = p$ only,

$$\mu(\mathbf{p}) = \mu(p) \tag{8.38}$$

This choice ensures that the fluid velocity lies in the xy-plane only.

The tangent vectors Eq. (8.20) and the unit normal vector Eq. (8.23) are (with $p_1 = p$)

$$\mathbf{t}_1 = x_p\,\mathbf{i} + y_p\,\mathbf{j}, \quad \mathbf{t}_2 = \mathbf{k}, \quad \mathbf{n} = \frac{y_p}{s_p}\,\mathbf{i} - \frac{x_p}{s_p}\,\mathbf{j}, \tag{8.39}$$

where p is taken to run anti-clockwise (closed) or left to right (open), and

$$s_p^2 = x_p^2 + y_p^2 \tag{8.40}$$

gives the square of the derivative of the arclength: subscript p refers to differentiation.

The boundary integral Eq. (8.15) that determines the average potential at the interface becomes

$$I(p) = -\frac{1}{2\pi}\int \mu(q)\,\frac{y_q(q)\big(x(p)-x(q)\big) - x_q(q)\big(y(p)-y(q)\big)}{\big(x(p)-x(q)\big)^2 + \big(y(p)-y(q)\big)^2}\,\mathrm{d}q. \tag{8.41}$$

The vector potential has only one component $\mathbf{A} = \psi\,\mathbf{k}$; ψ is called the streamfunction. From Eq. (8.27), $A_1 = 0$ and $A_2 = \psi$, and from Eq. (8.24),

$$\psi(p) = \frac{1}{2\pi}\int \mu(q)\,\frac{x_q(q)\big(x(p)-x(q)\big) + y_q(q)\big(y(p)-y(q)\big)}{\big(x(p)-x(q)\big)^2 + \big(y(p)-y(q)\big)^2}\,\mathrm{d}q. \tag{8.42}$$

The velocity of the interface is determined by Eqs. (8.28 and 8.29).

$$s_p^2\mathbf{U} = \big(x_p\,I_p + y_p\,\psi_p\big)\,\mathbf{i} + \big(y_p\,I_p - x_p\,\psi_p\big)\,\mathbf{j}, \tag{8.43}$$

$$s_p^2\mathbf{m} = x_p\,\mu_p\,\mathbf{i} + y_p\,\mu_p\,\mathbf{j}. \tag{8.44}$$

Let $\mathbf{U} = u\,\mathbf{i} + v\,\mathbf{j}$, then u, v are easily determined by Eq. (8.43) and the velocity components of the interface motion Eq. (8.16) are

$$\frac{\partial x}{\partial t} = u_I = u - \frac{\alpha}{2}\frac{\mu_p x_p}{s_p^2}, \tag{8.45}$$

$$\frac{\partial y}{\partial t} = v_I = v - \frac{\alpha}{2}\frac{\mu_p y_p}{s_p^2}. \tag{8.46}$$

The evolution of the dipole strength is determined by Eq. (8.32).

$$\begin{aligned}\frac{\partial \mu}{\partial t} - 2A\frac{\partial I}{\partial t} = {} & -\frac{\alpha}{2}\frac{\mu_p^2}{s_p^2} - A\left[u^2 + v^2 - \alpha\,\frac{\big(u x_p + v y_p\big)\,\mu_p}{s_p^2} - \frac{1}{4}\frac{\mu_p^2}{s_p^2} - 2gy\right] \\ & - \frac{2T}{\rho_1 + \rho_2}\,\frac{x_p y_{pp} - y_p x_{pp}}{s_p^3}.\end{aligned} \tag{8.47}$$

Unfortunately, the expression for the time derivative of I is lengthy, so it is convenient to introduce a more compact notation. Let $z(p) = x(p) + \mathrm{i}y(p)$ be the location of the interface in the complex plane. There is a natural connection between potential theory and complex variables and restatement of the formulation in terms of complex variables leads to other insights in the application of boundary integral methods to free surface flows in two dimensions.

Introduce the complex potential $\phi + \mathrm{i}\psi$. In particular, let the average complex potential at the interface be denoted $\Phi = I + \mathrm{i}\psi$. Then Eq. (8.41) and Eq. (8.42) may be combined into one complex integral,

$$\Phi(p) = \frac{1}{2\pi \mathrm{i}} \int \mu(q) \, \frac{z_q(q)}{z(q) - z(p)} \, \mathrm{d}q. \tag{8.48}$$

Of course, this integral must be taken in the principal-valued sense. It is clearly the parametric form of the Cauchy integral.

The complex velocity $w = u + \mathrm{i}v$ is related to the derivative of the complex potential. The average complex velocity Eq. (8.43) at the interface will be

$$w^*(p) = \frac{\Phi_p(p)}{z_p(p)}, \tag{8.49}$$

where the superscript $*$ refers to complex conjugation and the subscripts p indicate differentiation as before. The motion of the interface Eqs. (8.45 and 8.46) becomes the complex weighted velocity

$$W^* = w^* - \frac{\alpha}{2} \frac{\mu_p}{z_p}, \tag{8.50}$$

$$\frac{\partial z}{\partial t} = W. \tag{8.51}$$

Equation (8.47) may be rewritten as

$$\frac{\partial \mu}{\partial t} - 2A\Re\left\{ \frac{1}{2\pi \mathrm{i}} \int \frac{\partial \mu}{\partial t}(q) \, \frac{z_q(q)}{z(q) - z(p)} \, \mathrm{d}q \right\} = R(p) \tag{8.52}$$

where

$$R(p) = 2A\Re\left\{ \frac{1}{2\pi \mathrm{i}} \int \mu(q) \, \frac{W_q(q)}{z(q) - z(p)} \, \mathrm{d}q \right.$$
$$\left. - \frac{1}{2\pi \mathrm{i}} \int \mu(q) \, \frac{z_q(q)\left(W(q) - W(p)\right)}{\left(z(q) - z(p)\right)^2} \, \mathrm{d}q \right\} + \left(\frac{A}{4} - \frac{\alpha}{2} \right) \frac{\mu_p^2}{s_p^2}$$

$$- A\left[w^* w - \alpha\,\mu_p\,\Re\left\{\frac{w}{z_p}\right\} - 2g\Im\{z\}\right] - \frac{2T}{\rho_1 + \rho_2}\frac{1}{s_p}\Im\left\{\frac{z_{pp}}{z_p}\right\}. \tag{8.53}$$

Written this way, it is clear that Eq. (8.52) constitutes a Fredholm integral equation of the second kind for the rate of change of the dipole strength.

The connection between a dipole distribution and a vortex sheet representation is very easy to establish in this new notation. By integrating by parts, the complex velocity can be written as

$$\begin{aligned} w^* = \frac{\Phi_p}{z_p} &= -\frac{1}{2\pi \mathrm{i}}\int \mu(q)\,\frac{\mathrm{d}}{\mathrm{d}q}\left(\frac{1}{z(q) - z(p)}\right)\mathrm{d}q \\ &= \frac{1}{2\pi \mathrm{i}}\int \frac{\mu_q(q)}{z(q) - z(p)}\,\mathrm{d}q, \end{aligned} \tag{8.54}$$

which is the complex form for the Birkhoff-Rott integral that gives the motion of a vortex sheet. The vortex sheet strength is $\Gamma = \mu_p/s_p$; see Eq. (8.36). It is convenient, and common, to express μ_p as the unnormalized vortex sheet strength γ. The evolution equation for γ follows simply from differentiating Eq. (8.47) with respect to p. If there is no jump in density $A = 0$ and if surface tension may be neglected $T = 0$, the choice $\alpha = 0$ (the motion of the interface is the average velocity) means γ remains a constant in time. Thus,

$$\frac{\partial z^*}{\partial t} = \frac{1}{2\pi \mathrm{i}}\int \frac{\gamma(q)}{z(q) - z(p)}\,\mathrm{d}q \tag{8.55}$$

is the equation of motion for a vortex sheet, a well-studied equation as the model for a thin shear layer.[20,21]

There is an important difference between a dipole and vortex sheet representation. The mean value of the dipole strength is dynamically unimportant and the far-field motion is at rest. If the vortex sheet strength has a mean value then there is a net circulation in the region containing the interface and the far-field motion will reflect its presence. In particular, if the vortex sheet is in open, periodic geometry, then the fluids above and below the sheet flow past each other in a shearing motion. In this case, the mean value of γ is the jump in the velocities of the two fluids far from the sheet. In other words, if there is a mean shear flow at the interface, it is more appropriate to use a vortex sheet representation.

This completes the basic derivation of boundary integral methods for free surface flows. However, there is a useful modification for open, periodic geometries which is needed for studies of water waves and the Rayleigh–Taylor instability.

8.2.3. *Open, periodic geometry*

The Green's function, Eq. (8.11) or Eq. (8.12), used in the derivation of the boundary integrals for free surface flow is the free space Green's function. If there are geometrical constraints on the motion, there may be other Green's functions more appropriate for the boundary integrals. An obvious example is open, periodic geometry where the interface and its motion remain periodic in the horizontal plane. A periodic Green's function may be obtained by the method of images which then guarantees periodic motion.

For two-dimensional motion, the assumption will be that $z(p + L) = L + z(p)$ and $\mu(p + L) = \mu(p)$. Then

$$\begin{aligned}\Phi(p) &= \frac{1}{2\pi \mathrm{i}} \int_{-\infty}^{\infty} \mu(q) \, \frac{z_q(q)}{z(q) - z(p)} \, \mathrm{d}q \\ &= \frac{1}{2\pi \mathrm{i}} \int_0^L \mu(q) \, z_q(q) \sum_{n=-\infty}^{\infty} \frac{1}{z(q) - z(p) - nL} \, \mathrm{d}q \\ &= \frac{1}{2L\mathrm{i}} \int_0^L \mu(q) \, z_q(q) \, \cot\left\{ \frac{\pi}{L} \left(z(q) - z(p) \right) \right\} \mathrm{d}q. \qquad (8.56)\end{aligned}$$

The formula for the sum can be found in Reference 22. The standard choice is to scale z with $L/2\pi$ or equivalently set $L = 2\pi$. Besides the obvious replacement of the Green's function in the integrals in Eq. (8.52) and Eq. (8.53), the equations of motion for the interface remain unchanged. The value of Eq. (8.56) is that the range of integration is finite, and the integrand is periodic allowing highly accurate spectral methods to be used numerically.

Unfortunately, the periodic Green's function in three dimensions requires a double sum that does not even converge. Fortunately, the sum can be modified so that it is convergent without destroying the nature of the Green's function; only the mean level of the Green's function is affected. Since only derivatives of the Green's function are needed in the boundary integrals, they remain unchanged. Even so, the modified sum has no known closed form and converges very slowly. Instead, the sum can be converted to rapidly converging Ewald sums.[9]

The boundary integral equations for free surface flow are nonlinear and only a few special solutions are known. A simple example is the perfectly flat interface which is stationary and the dipole strength is constant. Instead, the solutions to the equations must be constructed numerically. Section 8.3 will present the standard numerical methods for the case of open, periodic geometry.

8.2.4. *Linear stability analysis*

Before designing a numerical method, it is wise to have some understanding of the basic mathematical properties of the solution to the equations. A good starting point is to consider the stability of a flat interface with small perturbations. A linear analysis can establish the stability of the solutions and give insight into the general behavior of the interfacial motion. However, it is best to consider as general a case as possible, which means the inclusion of a mean value γ_0 to the vortex sheet strength to allow for a mean shearing motion.

Because the Lagrangian markers may move along the interface, the perturbations are written in the form,

$$z(p) = p + \frac{\alpha}{2}\gamma_0 t + a\mathrm{e}^{imp} + b\mathrm{e}^{-imp}, \tag{8.57}$$

$$\gamma(p) = \gamma_0 + c\mathrm{e}^{imp} + c^*\mathrm{e}^{-imp}, \tag{8.58}$$

where the coefficients a, b, c are assumed small. Note that the form of the perturbation in γ is designed to ensure that it is a real function: c^* is the complex conjugate of c. For convenience, only one mode m is selected. Once the results are known, a linear superposition of all integer modes will produce in effect a complete Fourier series allowing for arbitrarily small perturbations.

The evolution equations for the coefficients can be obtained by substituting Eqs. (8.57 and 8.58) into Eqs. (8.51 and 8.52), and retaining only those terms that are linear in the coefficients a, b, c;

$$\frac{\mathrm{d}a}{\mathrm{d}t} = -\frac{1+\alpha}{2}\left(c - im\gamma_0 b^*\right), \tag{8.59}$$

$$\frac{\mathrm{d}b^*}{\mathrm{d}t} = \frac{1-\alpha}{2}\left(c - im\gamma_0 a\right), \tag{8.60}$$

$$\frac{\mathrm{d}c}{\mathrm{d}t} = (A-\alpha)\mathrm{i}m\gamma_0 c + \frac{A-\alpha}{2}\gamma_0^2 m^2\left(a + b^*\right) + \Omega\left(a - b^*\right), \tag{8.61}$$

where

$$\Omega = Agm + \frac{Tm^3}{\rho_1 + \rho_2}. \tag{8.62}$$

The general solution to this system of ordinary differential equations is

$$a \;= \varepsilon_1 + \varepsilon_2 a_+ \mathrm{e}^{\sigma_+ t} + \varepsilon_3 a_- \mathrm{e}^{\sigma_- t}, \tag{8.63}$$

$$b^* = \varepsilon_1 + \varepsilon_2 b_+^* \mathrm{e}^{\sigma_+ t} + \varepsilon_3 b_-^* \mathrm{e}^{\sigma_- t}, \tag{8.64}$$

$$c \;= \varepsilon_1 i m \gamma_0 + \varepsilon_2 c_+ \mathrm{e}^{\sigma_+ t} + \varepsilon_3 c_- \mathrm{e}^{\sigma_- t}, \tag{8.65}$$

where

$$\sigma_\pm = \frac{A - \alpha}{2} i m \gamma_0 \pm \left(\frac{1 - A^2}{4} \gamma_0^2 m^2 - \Omega \right)^{1/2}, \tag{8.66}$$

and

$$a_\pm = -\frac{1+\alpha}{2}\left(\sigma_\pm - \frac{1-\alpha}{2} i m \gamma_0\right),$$

$$b_\pm^* = \frac{1-\alpha}{2}\left(\sigma_\pm + \frac{1+\alpha}{2} i m \gamma_0\right),$$

$$c_\pm = \sigma_\pm^2 - \frac{1-\alpha^2}{4}\gamma_0^2 k^2.$$

The constants ε_j are determined by the initial conditions.

The constant ε_1 reflects a choice of initial condition in which the interface remains perfectly flat, while the Lagrangian markers are placed along the interface so that $\mu = \gamma_0 x$ remains unchanged. Normally, the choice of ε_1 is made to start with an initial concentration of Lagrangian markers where resolution is needed later in the calculation.

The more important behavior of the linear solution is the nature of the growth rates $\sigma_\pm$. Note first that the first term in Eq. (8.66) controls the tangential motion of the markers. It suggests that the choice $\alpha = A$ is a natural one; this choice states that the average velocity at the interface should be weighted with the ratio of densities.

The physically relevant part of the behavior of the linear solution is determined by the term with the square root in Eq. (8.66). Consideration of a few special cases helps to shed light on the general case. First, consider $A = 1$. Physically, the density about the interface is negligible, as commonly assumed in the study of water waves, for example. Indeed,

$$\sigma_{\pm} = \pm \mathrm{i}\left(gm + \frac{Tm^3}{\rho_1}\right)^{1/2}, \tag{8.67}$$

which is just the standard dispersion relation for water waves.

Second, consider $A = -1$. Now the density below the interface is negligible and the surface is unstable (the Rayleigh–Taylor instability). The growth rates are

$$\sigma_{\pm} = \pm\left(gm - \frac{Tm^3}{\rho_2}\right)^{1/2}. \tag{8.68}$$

The instability is stabilized by surface tension for large wavenumbers.

The final choice is $A = 0$. There is no density difference across the interface as might occur for two immiscible fluids of equal density. If, in addition, $T = 0$, then the growth rates are

$$\sigma_{\pm} = \pm\frac{\gamma_0 m}{2}. \tag{8.69}$$

The motion is linearly ill-posed;[23] the modes with the smallest length scales (largest m) grow the fastest. This choice has received considerable attention for the last three decades since it describes the motion of a vortex sheet Eq. (8.55), the standard long wave model for a thin layer of vorticity. At first, hopes were that nonlinear effects would restore well-posedness, but these hopes have been dashed.[24,25] Indeed, there is strong evidence in Cowley et al. (and the references cited therein)[26] that vortex sheets develop curvature singularities in finite time. Surprisingly, the inclusion of surface tension effects, which stabilizes the highest modes according to linear theory, does not prevent singularity formation,[27] although the nature of the singularity has changed.

The above linear stability results have more applications than might be expected. By restricting attention to regions of an interface where it is locally flat and allowing the gravity vector to compensate for the orientation of the interface, it is possible to use the linear results to predict the stability under more general circumstances.[28] A good example is the late time development of the Rayleigh–Taylor instability ($-1 < A < 0$) when heavier fluid falls in long spikes into the lower fluid. Along the sides of the spike, a local γ_0 grows slowly in time as the heavier fluid rushes by the lighter fluid and triggers the onset of a curvature singularity[29] and the subsequent roll up of the interface into a plume. In contrast, the case $0 < A < 1$ with no initial γ_0, does not appear to form a curvature singularity in finite time unless a breaking wave develops (a topological singularity of a different nature then occurs). Presumably, there are no regions where a mean vortex sheet strength arises for sufficient time to trigger curvature singularities. Singularity formation in interfacial flow remains an active area of research and is one of the remaining challenges for boundary integral methods.

For completeness and subsequent use, the linear results for the dipole strength are included when $\gamma_0 = 0$. Let

$$\mu(p) = d\mathrm{e}^{\mathrm{i}mp} + d^*\mathrm{e}^{-\mathrm{i}mp}. \tag{8.70}$$

Then the linear part of the potential Eq. (8.56) becomes

$$\begin{aligned}\Phi(p) &= \frac{1}{4\pi\mathrm{i}}\int_0^{2\pi}\left(d\mathrm{e}^{\mathrm{i}mq} + d^*\mathrm{e}^{-\mathrm{i}mq}\right)\cot\left(\frac{q-p}{2}\right)\mathrm{d}q \\ &= \frac{d}{2}\mathrm{e}^{\mathrm{i}mp} - \frac{d^*}{2}\mathrm{e}^{-\mathrm{i}mp}.\end{aligned} \tag{8.71}$$

The results Eqs. (8.63, 8.64 and 8.66) still stand, and $d_\pm = c_\pm/(\mathrm{i}m)$.

In summary, care must be taken in the design of numerical methods for interfacial flow because of the possibility of ill-posedness and subsequent formation of curvature singularities.

8.3. Numerical Approximation

The method of lines is a natural choice for the numerical treatment of the evolution of the interface. Markers along the interface are distributed in the Lagrangian variable p with an associated dipole strength. Then, the velocity and the rate of change of the dipole strength are calculated and used to update the marker's location and dipole strength through a standard method, such as the 4th-order Runge-Kutta method. Details will be provided here for free surface motion based on a dipole representation in open, 2π-periodic geometry, such as water waves. It is relatively straightforward to adapt the methods to other circumstances, such as the vortex sheet representation.

From now on, N markers are assumed to be evenly-spaced in p. Note that the definition of p is done in conjunction with the specification of the initial conditions and can be done to improve resolution in the subsequent motion. Subsequent examples with water wave motion and the Rayleigh–Taylor instability will illustrate this idea. Let the location of the markers be $x_j = x(jh)$, $y_j = y(jh)$ with $h = 2\pi/N$. Similarly, let $\mu_j = \mu(jh)$. Since the interface is taken as 2π-periodic, the data may be expressed in terms of a discrete Fourier series. Let $f(p)$ stand for any 2π-periodic function in p (note that $x(p) - p$ is 2π-periodic). Then,

$$f_j = F_0^e + F_{N/2}^e(-1)^j + \sum_{m=1}^{N/2-1} \Big(F_m^e \cos(mjh) + F_m^o \sin(mjh) \Big). \qquad (8.72)$$

The odd/even Fourier coefficients F^o, F^e can be obtained through an application of the fast Fourier transform.

The first step in the numerical procedure is the calculation of the derivatives x_p, y_p and μ_p. Spectrally accurate results can be obtained by differentiating their Fourier series. The derivatives will be denoted as $(\mathcal{D}f)_j = f_p(jh)$, etc. Thus,

$$(\mathcal{D}f)_j = \sum_{m=1}^{N/2-1} mS(m)\Big(F_m^o \cos(mjh) - F_m^e \sin(mjh) \Big). \qquad (8.73)$$

Knowing the Fourier coefficients, the Fourier series in Eq. (8.72) can be evaluated through the fast Fourier transform. The inclusion of the factor $S(m)$ is to help suppress the ill-posed effects of differentiation. At the very least, it is important to suppress the effects of round-off errors by removing all Fourier coefficients that are close to machine precision. Otherwise, coefficients in the tail of the spectrum that might contain only round-off errors will be multiplied by a large m and exacerbate the effects of round-off in the derivatives.

The next step is the evaluation of Eq. (8.56). The difficulty in using a numerical approximation is that it is a principal-valued integral. Fortunately, the result

$$\frac{1}{4\pi \mathrm{i}} \int_0^{2\pi} z_q(q) \cot\left\{ \frac{z(q) - z(p)}{2} \right\} \mathrm{d}q = 0 \qquad (8.74)$$

may be used to rewrite Eq. (8.56) without the pole singularity;

$$\Phi(p) = \frac{1}{4\pi \mathrm{i}} \int_0^{2\pi} \big(\mu(q) - \mu(p)\big)\, z_q(q) \cot\left\{ \frac{z(q) - z(p)}{2} \right\} \mathrm{d}q. \qquad (8.75)$$

The integrand is now in a suitable form for the trapezoidal rule except for the indeterminate form when $q = p$. The limit is easily calculated, however, and the approximation to the integral becomes

$$\Phi_j = \frac{1}{2N\mathrm{i}} \sum_{\substack{k \neq j \\ k=0}}^{N-1} \big(\mu_k - \mu_j\big)\, (\mathcal{D}z)_k \cot\left\{ \frac{z_k - z_j}{2} \right\} + \frac{(\mathcal{D}\mu)_j}{N\mathrm{i}}. \tag{8.76}$$

It is also possible to apply the trapezoidal rule at alternate points that skip over the point $q = p$:

$$\Phi_j = \frac{1}{N\mathrm{i}} \sum_{\substack{k+j-\mathrm{odd} \\ k=0}}^{N-1} \big(\mu_k - \mu_j\big)\, (\mathcal{D}z)_k \cot\left\{ \frac{z_k - z_j}{2} \right\}. \tag{8.77}$$

Both approximations are spectrally accurate, but Eq. (8.77) has a lower cost since only half the points are used. It also has the advantage that it is easier to programme on vector/parallel computers. It is still important to include the pole subtraction in Eq. (8.77) since it reduces the effects of round-off errors.

Another approach to the numerical treatment of the principal-valued integrals is to regularize the Green's function through convolution with a suitable smoothing function.[30] There are many possible choices for smoothing functions[31] and the specific choice does not seem to be crucial. A popular choice is the Krasny vortex blob method,[33] where

$$\cot\left\{ \frac{z(q) - z(p)}{2} \right\} = \frac{\sin\big(x(q) - x(p)\big) - \mathrm{i}\sinh\big(y(q) - y(p)\big)}{\cosh\big(y(q) - y(p)\big) - \cos\big(x(q) - x(p)\big)} \tag{8.78}$$

is replaced by

$$\frac{\sin\big(x(q) - x(p)\big) - \mathrm{i}\sinh\big(y(q) - y(p)\big)}{\cosh\big(y(q) - y(p)\big) - \cos\big(x(q) - x(p)\big) + \delta^2}. \tag{8.79}$$

The integrand is no longer singular when $q = p$, and the pole singularity has been smoothed over a distance of $\mathcal{O}(\delta)$.

The original motivation for the regularization Eq. (8.79) was to ensure that vortex sheet motion ($A = \alpha = T = 0$) would exist globally in time. Indeed, appropriate regularizations[32] of the integrand guarantee solutions for all time and the limit of $\delta \to 0$ converges to a weak solution beyond the time of singularity formation. With $\delta > 0$, the vortex sheet rolls up into a spiral where the curvature singularity would otherwise form.[33] Only partial success is achieved when $-1 < A < 0$.[30] On the other hand, the regularized kernel does provide accurate simulations of water waves

($A = 1$) especially when δ is made proportional to s_p.[30] In other words, the blob size adjusts to the spacing of the Lagrangian markers. Incidentally, Eq. (8.78) or Eq. (8.79) can be evaluated quickly by computing $\cos(x_j)$, $\sin(x_j)$, $\cosh(y_j)$ and $\sinh(y_j)$ first and using the expansion formulas to compute these functions with arguments that contain a difference.

The next step is to use the results of the numerical integration of the boundary integrals to determine the interfacial velocities Eqs. (8.50 and 8.51) by simply evaluating $(\mathcal{D}\Phi)_j$ through Eq. (8.73). Alternatively, the velocity may be determined directly from Eq. (8.54).

The last step requires a solution to Eq. (8.52). The discrete form is

$$\frac{\mathrm{d}\mu_j}{\mathrm{d}t} - 2A\Re\left\{\frac{1}{2N\mathrm{i}}\sum_{\substack{k\neq j\\k=0}}^{N-1}\left(\frac{\mathrm{d}\mu_k}{\mathrm{d}t} - \frac{\mathrm{d}\mu_j}{\mathrm{d}t}\right)(\mathcal{D}z)_k \cot\left\{\frac{z_k - z_j}{2}\right\}\right.$$
$$\left. + \frac{1}{N\mathrm{i}}\left(\mathcal{D}\frac{\mathrm{d}\mu}{\mathrm{d}t}\right)_j\right\} = R_j. \tag{8.80}$$

Note that R_j contains two integrals (see Eq. (8.53)) and it is advisable to remove the pole singularity in these integrals at $q = p$ by replacing $\mu(q)$ with $\mu(q) - \mu(p)$. Further, if the trapezoidal rule is to be applied as in Eq. (8.76), then the limiting values of the indeterminate forms in these integrals must be included. They require additional derivatives, which can be determined as in Eq. (8.73). The inclusion is not needed if the alternate point trapezoidal rule Eq. (8.77) is used.

Equation (8.80) is a system of linear equations for the rate of change of the dipole strength at the Lagrangian points. There are two main ways this system can be solved: direct methods based on an LU-decomposition, or iterative methods. Which choice is preferable depends on several factors, but generally iterative methods have the most advantages. The simplest iterative procedure is:

$$\frac{\partial\mu_j}{\partial t}^{(\nu+1)} = 2A\Re\left\{\frac{1}{2N\mathrm{i}}\sum_{\substack{k\neq j\\k=0}}^{N-1}\left(\frac{\mathrm{d}\mu_k}{\mathrm{d}t}^{(\nu)} - \frac{\mathrm{d}\mu_j}{\mathrm{d}t}^{(\nu)}\right)(\mathcal{D}z)_k\right.$$
$$\left. \times \cot\left\{\frac{z_k - z_j}{2}\right\} + \frac{1}{N\mathrm{i}}\left(\mathcal{D}\frac{\mathrm{d}\mu}{\mathrm{d}t}\right)_j^{(\nu)}\right\} + R_j. \tag{8.81}$$

Convergence[6] of the iteration has been established for $|A| \leq 1$. More importantly, when the interface is nearly horizontal, as in the propagation of water waves, the rate of convergence is very large so that only a few iterations are needed for even very high accuracy.

Several refinements are available. Each iteration in Eq. (8.81) takes $\mathcal{O}(N^2)$ operations to evaluate all the sums. Multipole expansions and tree codes[19] can reduce the count to $\mathcal{O}\big(N\log_2(N)\big)$ but with large setup costs so that N must be much larger than about 1,000 before significant savings are realized. Since the eigenvalue structure of the iteration is much like the typical structure of iteration operators for elliptic problems, multigrid strategies can be useful in reducing the effective number of iterations to just a few, and obviously the two methods can be combined. Finally, the first guess for the iterative solution can be improved since the dipole strength is evolving in time. For example, if a fourth-order predictor-corrector routine is used, it is simple to approximate the rate of change of the dipole strength by a cubic over the last four time steps and extrapolate (predict) the new value as the first guess for the iteration. Significant speed-up can be obtained this way.[6]

8.3.1. *Numerical stability*

Numerical stability is difficult to establish in general because of the differing nature of the solutions with different choices of A. But it is possible to conduct a linear stability analysis along the lines of Section 8.2.4. The difference is that the analysis depends on the specifics of the numerical approximation. It is advisable to break up the analysis into different parts to assess their influence on the stability. The starting point is to write

$$z_j = jh + \hat{z}_j, \quad \mu_j = \hat{\mu}_j, \tag{8.82}$$

where quantities with 'hats' will be assumed small. The linear version of Eq. (8.76) is

$$\Phi_j = \frac{1}{2N\mathrm{i}} \sum_{\substack{k\neq j\\ k=0}}^{N-1} \big(\hat{\mu}_k - \hat{\mu}_j\big) \cot\left\{\frac{kh - jh}{2}\right\} + \frac{(\mathcal{D}\hat{\mu})_j}{N\mathrm{i}}. \tag{8.83}$$

In line with Eq. (8.70), assume

$$\hat{\mu}_j = d\,\mathrm{e}^{\mathrm{i}mjh} + d^*\,\mathrm{e}^{-\mathrm{i}mjh}. \tag{8.84}$$

Equation (8.83) may be evaluated exactly by using the sum[34]

$$\frac{1}{2N\mathrm{i}}\sum_{\substack{k\neq j\\k=0}}^{N-1}\mathrm{e}^{\pm \mathrm{i}mkh}\cot\left\{\frac{kh-jh}{2}\right\}=\pm\frac{1}{2}\mathrm{e}^{\pm \mathrm{i}mjh}\left(1-\frac{mh}{\pi}\right), \tag{8.85}$$

valid for $0 < mh \leq \pi$, to obtain

$$\Phi_j=\frac{d}{2}\,\mathrm{e}^{\mathrm{i}mjh}R(m)-\frac{d^*}{2}\,\mathrm{e}^{-\mathrm{i}mjh}R(m), \tag{8.86}$$

where

$$R(m)=1+\frac{mh}{\pi}\left(S(m)-1\right).$$

The presence of $S(m)$ in $R(m)$ arises from the contribution from $D\hat{\mu}_j$. By comparison with Eq. (8.71), $R(m)-1$ is a measure of the relative error. For most spectral filters, $S(m) = 1$ for small m, and the approximation is exact. For mh near π, $S(m) \approx 0$, and the relative error is approximately 1. The error arises because the derivative of the high modes has been damped with $S(m)$ and the correction no longer balances the presence of μ_j in the sum. A simple remedy is to take the Fourier transform of μ_j, apply the filter $S(m)$ to the amplitudes and reconstitute a filtered version of μ to use in the sum. The result is that $R(m) = S(m)$; the potential has been effectively filtered. Remember to adopt the same procedure for the numerical approximation to the second and third integrals in Eq. (8.53).

The error for the alternate point quadrature Eq. (8.77) can be calculated in the same way by using the sum,

$$\frac{1}{N\mathrm{i}}\sum_{\substack{k+j=\mathrm{odd}\\k=0}}^{N-1}\mathrm{e}^{\pm \mathrm{i}mkh}\cot\left\{\frac{kh-jh}{2}\right\}=\begin{cases}\pm\frac{1}{2}\mathrm{e}^{\pm \mathrm{i}mjh}, & 0<mh<\pi,\\ 0, & mh=0,\pi,\end{cases} \tag{8.87}$$

to obtain Eq. (8.86) except that $R(m) = 1$ for $0 < mh < \pi$, and $R(m) = 0$ for $mh = 0, \pi$. In other words, alternate point quadrature gives the exact results except for the Nyquist freguency $m = N/2$.

The rest of the analysis is straightforward and, with the appropriate expression for $R(m)$ depending on the choice of quadrature, leads to

$$\frac{\mathrm{d}a}{\mathrm{d}t} = -\frac{R(m)+\alpha}{2} imd, \tag{8.88}$$

$$\frac{\mathrm{d}b^*}{\mathrm{d}t} = \frac{R(m)-\alpha}{2} imd, \tag{8.89}$$

$$\frac{\mathrm{d}d}{\mathrm{d}t} = \frac{\tilde{\Omega}}{im}\left(a - b^*\right), \tag{8.90}$$

where

$$\tilde{\Omega} = Agm + \frac{Tm^3}{\rho_1+\rho_2} S(m), \tag{8.91}$$

which is the numerical version of Eqs. (8.59–8.62), except that $c = imd$ since we are using the dipole formulation and $\gamma_0 = 0$. The key difference is that the dispersion relation Eq. (8.66) is replaced by

$$\tilde{\sigma}_{\pm} = \pm \mathrm{i}\sqrt{\tilde{\Omega}\, R(m)\, S(m)}\,. \tag{8.92}$$

This numerical dispersion relationship is very close to the exact relationship, the difference being only the influence of the Fourier filter $S(m)$. Note, in particular, that $m = N/2$ gives $\tilde{\sigma}_{\pm} = 0$, and this may lead to nonlinear resonances.[35] In most methods,[36] this "sawtooth" mode is problematic, and remedies include its suppression,[36] regularization of the integrals,[30] interpolation of additional integration points[34,37] and polynomial smoothing.[11] For $0 < A \leq 1$, the modes are purely oscillatory and a suitable time-stepping method must be used, especially if N is large (the problem becomes 'oscillatory stiff'). Provided the time step is chosen small enough, the standard Runge-Kutta method is appropriate. The fourth-order Adams-Moulton method has also been used.[6]

8.4. Applications with a Single Surface

Given the importance of water in human existence, it is not surprising that the free surface flow of water and air occupies a large body of scientific study. Some important examples include the nature of water waves, the Rayleigh–Taylor instability, rising bubbles and falling drops. The development of boundary integrals methods for free surface flow has led to improved understanding of these flows through direct numerical simulation and mathematical analysis. The following three canonical examples will serve to illustrate the power of boundary integral methods. There are many more applications but the list is too long to review.

8.4.1. *Waves on deep water*

A wave travelling to the right on the surface of water whose mean height is zero can be constructed from the results of the linear analysis Eqs. (8.63, 8.64 and 8.70). Specifically, $A = \alpha = 1$, and for waves with long wavelengths, $T = 0$. Set $\varepsilon_1 = \varepsilon_2 = 0$ and $\varepsilon_3 = \varepsilon/\sqrt{gm}$ where ε is real and gives the amplitude of the wave. The linear analysis predicts the motion to be

$$z(p) = p - \varepsilon \sin\big(mp - \sqrt{gm}\, t\big) + \mathrm{i}\varepsilon \cos\big(mp - \sqrt{gm}\, t\big), \tag{8.93}$$

$$\mu(p) = -2\varepsilon \sqrt{\frac{g}{m}} \sin\big(mp - \sqrt{gm}\, t\big). \tag{8.94}$$

An initial condition for numerical simulation is easily obtained by setting $t = 0$. Tests with ε small provide a good way to check the code. For moderate values of ε, around $\varepsilon = 0.3$, the waves steepen and break into little spillers. Larger choices for ε demonstrate breaking plungers as illustrated in Fig. 8.1. The physical scales are set with the choice $g = m = 1$, and $\varepsilon = 0.5$. The numerical parameters are $N = 4{,}096$, a time step of 0.0002 in the standard 4th-order Runge-Kutta method, and

$$S(m) = \frac{1}{2}\left\{1 - \tanh\left[40\left(\frac{2m}{N} - \frac{1}{4}\right)\right]\right\}. \tag{8.95}$$

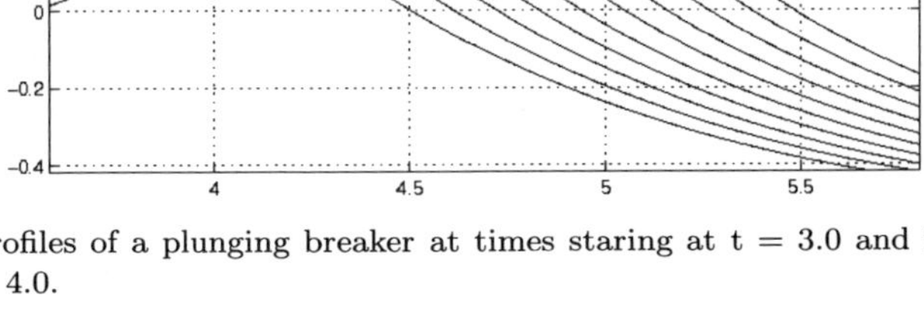

Fig. 8.1. Profiles of a plunging breaker at times staring at t = 3.0 and increasing by 0.1 until t = 4.0.

This filter suppresses the top half of the discrete Fourier spectrum, in essence de-aliasing the spectral method and ensuring stability. The integral equation is solved by iteration until the iterates don't change by more than 10^{-10}.

What is striking about the results is the very high curvature at the tip of the plunging breaker. It reaches a value of 3.7×10^3. This level of accuracy is almost impossible to reach with other methods and highlights the impressive advantages that boundary integral methods can exhibit. It is also possible to confirm that the very large curvature is associated with a square root singularity in the complex p-plane.[17] The singularity approaches the real axis of p but does not seem to reach it in finite time.

The plunging tip of the breaker has strong similarities to the falling spike that occurs during the Rayleigh–Taylor instability. The classical Rayleigh–Taylor instability occurs when heavy fluid falls into a vacuum, $A = -1$, $\alpha = -1$. For a perturbation with a single mode, a pattern of falling spikes and rising bubbles[38] quickly emerges. The spikes fall freely under gravity and the bubbles rise with constant speed. Eventually, the tip of the spike will develop a curvature large enough for surface tension effects to become important.[39,40] Then a drop forms at the tip which will subsequently detach, but this process cannot be described by the current boundary integral formulation. The nature of the Rayleigh–Taylor instability for a fluid layer is discussed in Section 8.5 and provides a typical example of a falling spike.

8.4.2. *Rising bubbles*

When a circular bubble of air is released, it rises and deforms into the shape of a hemispherical cap. Under these circumstances, $A = \alpha = 1$. The mathematical nature of the underlying elliptic problem reveals a deficiency in the dipole representation of the interfacial motion. The elliptic problem is an exterior one where the solution in the far field must be specified either as a constant or a logarithmic variation. A constant has no dynamic significance and may be neglected if there is no fluid motion in the far field. On the other hand, if the bubble expands as it rises, then there must be a net outflow of fluid in the far field to maintain conservation of mass. The outward flux F of fluid on the boundary of a large circle $r = R$ that encases the bubble at its centre will be

$$F = \int_0^{2\pi} \frac{\partial \phi}{\partial r}(R, \theta)\, R d\theta. \tag{8.96}$$

Since the far-field behavior of a dipole is $1/|z|$, a dipole distribution cannot give the correct far-field behavior. Instead, a source term must be added to complete the representation of the potential; Eq. (8.48) is replaced with

$$\Phi_E(p) = \Phi + \frac{F(t)}{2\pi} \ln\bigl(z(p)\bigr), \tag{8.97}$$

where F is real-valued. As a consequence, Eq. (8.49) must be replaced with

$$w^*(p) = \frac{\Phi_p(p)}{z_p(p)} + \frac{F}{2\pi z(p)}. \tag{8.98}$$

The only other changes that are necessary occur in Eq. (8.52). A term must be included to account for the difference between the pressure inside the bubble P_i and the pressure in the far-field P_o. Also, the change in outward flux must be included. As a consequence, Eq. (8.52) becomes (recalling that $A = \alpha = 1$, so $\rho_1 = \rho$ and $\rho_2 = 0$),

$$\begin{aligned} &\frac{\partial \mu}{\partial t}(p) - 2\Re\left\{ \frac{1}{2\pi \mathrm{i}} \int \frac{\partial \mu}{\partial t}(q) \, \frac{z_q(q)}{z(q) - z(p)} \, \mathrm{d}q \right\} \\ &\quad = R(p) + \frac{1}{\pi} \frac{\mathrm{d}F}{\mathrm{d}t} \ln|z(p)| + \frac{F}{\pi} \Re\left\{ \frac{w(p)}{z(p)} \right\} + 2\, \frac{P_o - P_i}{\rho}. \end{aligned} \tag{8.99}$$

Setting aside for the moment that the rate of change of F is unknown, the integral equation for the rate of change of the dipole distribution is singular. A uniform rate of change of dipole distribution is a homogeneous solution to Eq. (8.99). Solutions to Eq. (8.99) exist only if the Fredholm alternative is satisfied.[41]

Let τ satisfy the homogeneous adjoint equation,

$$\tau(p) + 2\Re\left\{ \frac{z_p(p)}{2\pi \mathrm{i}} \int \frac{\tau(q)}{z(q) - z(p)} \, \mathrm{d}q \right\} = 0. \tag{8.100}$$

Then the application of the Fredholm alternative produces the relation

$$\begin{aligned} \frac{1}{2\pi} \frac{\mathrm{d}F}{\mathrm{d}t} \int \tau(p) \, \ln|z(p)| \, \mathrm{d}p &= \int \left[\frac{P_i - P_o}{\rho} - R(p) \right] \tau(p) \, \mathrm{d}p \\ &\quad - \frac{F}{2\pi} \int \Re\left\{ \frac{w(p)}{z(p)} \right\} \tau(p) \, \mathrm{d}p, \end{aligned} \tag{8.101}$$

which provides an evolution equation for F. Equation (8.101) provides a clear connection between the outward flux and the pressure difference between the interior of the bubble and the far-field, modified by hydrodynamic and hydrostatic effects that appear in the other terms. There are several ways this connection can be exploited. For example, set $P_i = P_o$, and let the bubble rise under the influence of buoyancy effects. As the bubble rises, the hydrostatic pressure will lead to changes in $R(p)$, causing the bubble to expand ($F(t)$ will increase). The inside of the bubble may be regarded as a gas and P_i will change according to the equation of state as the volume of the bubble increases. All of these effects can be accommodated in Eq. (8.101).

Two examples of this formulation have been considered before. In one,[42] the initial rise and distortion of the bubble is simulated by both the dipole and vortex sheet methods for comparison purposes. Here, the expansion of the bubble is neglected, so $F(t) = 0$ and P_i is allowed to adjust to satisfy Eq. (8.101) (not explicitly calculated). Equation (8.99) may still be solved by iteration as long as the undetermined homogeneous solution is specified. Since it is merely a uniform constant, it has no dynamic consequences. The iterations converge with the dipole held constant in time at the top of the bubble. An initially circular bubble of unit radius is released in the presence of a unit vertical gravity field. The rise and distortion of the bubble are shown in Fig. 8.2. The bottom of the bubble rises faster than the top, forming an inward plume. The plume broadens and its sides approach the sides of the bubble. The appearance of the bubble is now that of a hemispherical cap with two attached lobes at its sides.

What is also noticeable in these results is the close approach of different parts of the surface to each other. Indeed, At $t = 4.0$ the code begins to fail as the neck of the two side lobes begin to pinch off. From the stand point of the numerical approximation, difficulties are arising because the denominator in the integrand in Eq. (8.48) is becoming very small for values of q not close to p leading to large spikes in the integrand. This matter will be revisited as a future challenge in Section 8.6.

The other example where this formulation is used occurs when a high external pressure is used to attempt to collapse the bubble. Buoyancy is neglected because the outside pressure is large enough to force the bubble to collapse quickly not allowing the bubble enough time to rise. The inside pressure is allowed to increase according to the equation of state for gases, and it grows large enough to reverse the pressure gradient at a later time causing the surface to undergo Rayleigh–Taylor instability. For the numerical results presented in Ref. 9, the bubble is assumed

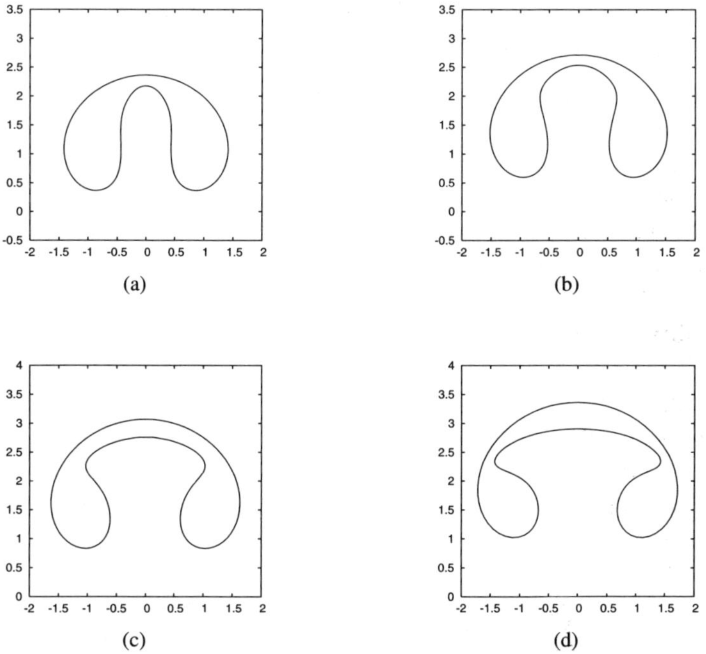

Fig. 8.2. Profiles of a rising bubble at times (a) $t = 2.5$, (b) $t = 3.0$ (c) $t = 3.5$ (d) $t = 4.0$.

axisymmetric but the derivation of the equations is essentially the same as the derivation for two-dimensional flow. In either case, Eq. (8.101) or its axisymmetric version must be used. See Reference 9 for further details.

8.5. Applications with Two Surfaces

In principle, the addition of more interfaces means simply the addition of more surface integrals. A good example is the study of triadic resonances between water waves and internal waves.[6] But there can be new features. The acceleration of thin liquid layers in gases requires modifications to the basic formulation similar to the formulation of external flow outside a single interface given in Sec. 8.4.

New considerations arise if one of the surfaces is a rigid boundary, as in shallow water or in the motion of a submerged body. Two specific examples will illustrate both of the necessary modifications.

8.5.1. *Rayleigh–Taylor instability of a liquid layer of finite thickness*

A liquid layer of density ρ lies between two horizontal surfaces with gas of negligible density above and below. If the surfaces are perfectly flat, then a pressure gradient $(P_1 - P_2)/(\rho g H)$ can hold the layer in place against the force of gravity. Here, P_1 and P_2 are the pressures below and above the layer respectively, and H is the mean thickness of the layer. This layer is unstable to perturbations, an example of the Rayleigh–Taylor instability. Also, the pressure difference may cause the layer to accelerate, a possibility that is allowed in the following description of the equations of motion.[43,44]

Let all quantities on the lower and upper surface be designated with a subscript 1 and 2 respectively. From the assumed form of the layer, $A_1 = \alpha_1 = -1$ and $A_2 = \alpha_2 = 1$. The flow is assumed 2π-periodic and $\rho = g = 1$. The average complex potential at the surfaces induced by the dipole distributions is

$$\Phi_j(p) = \frac{1}{4\pi \mathrm{i}} \sum_{k=1}^{2} \int_0^{2\pi} \mu_k(q)\, z_{k,q}(q)\, \cot\left\{ \frac{z_k(q) - z_j(p)}{2} \right\} \mathrm{d}q. \tag{8.102}$$

Since the layer may move vertically with speed $V(t)$, a behavior not represented by dipole distributions, a term $-\mathrm{i}Vz_j$ must be added to Eq. (8.102). The complex velocity becomes

$$w_j^* = \frac{\Phi_{j,p}}{z_{j,p}} - \mathrm{i}V, \tag{8.103}$$

$$\frac{\partial z_j^*}{\partial t} = W_j^* = w_j^* - \alpha_j \frac{\mu_{j,p}}{z_{j,p}}. \tag{8.104}$$

The evolution equation for the dipole distributions are modifications of Eq. (8.52) as done in Eq. (8.99).

$$\frac{\partial \mu_j}{\partial t}(p) - 2A_j \sum_{k=1}^{2} \Re\left\{ \frac{1}{4\pi \mathrm{i}} \int_0^{2\pi} \frac{\partial \mu_k}{\partial t}(q)\, z_{k,q}(q) \cot\left\{ \frac{z_k(q) - z_j(p)}{2} \right\} \mathrm{d}q \right\}$$
$$= R_j(p) + 2A_j \frac{\mathrm{d}V}{\mathrm{d}t} y_j + 2A_j V v_j(p) + 2A_j \frac{P_j}{\rho}. \tag{8.105}$$

This coupled system of Fredholm integral equations of the second kind is singular and solutions exist provided the Fredholm alternative is satisfied.

$$\frac{\mathrm{d}V}{\mathrm{d}t}\sum_{j=1}^{2} A_j \int_0^{2\pi} y_j(p)\,\tau_j(p)\,\mathrm{d}p = -\sum_{j=1}^{2} A_j \int_0^{2\pi}\left[\frac{P_j}{\rho} + V v_j(p)\right]\tau_j(p)\,\mathrm{d}p - \sum_{j=1}^{2}\int_0^{2\pi} R_j(p)\,\tau(p)\,\mathrm{d}p, \tag{8.106}$$

where τ_j is a nontrivial solution to the homogeneous adjoint equations,

$$\tau_j(p) + 2A_j \sum_{k=1}^{2} \Re\left\{\frac{z_{k,p}(p)}{4\pi\mathrm{i}}\int_0^{2\pi} \tau_k(q)\,\cot\left\{\frac{z_k(q)-z_j(p)}{2}\right\}\mathrm{d}q\right\} = 0. \tag{8.107}$$

The procedure to solve the adjoint equations numerically follow the methods already described in this chapter. Assume z_j and μ_j are known at some moment. They can be updated as follows. First, the integrals in Eq. (8.102) are calculated by either Eq. (8.76) or Eq. (8.77) after applying the standard treatment for the principal-valued integrals. Second, the complex velocities are calculated, Eq. (8.103) and Eq. (8.104). Third, the eigenvector τ_j of the adjoint problem Eq. (8.107) is found by iteration. Fourth, the acceleration of the layer is calculated from Eq. (8.106) and finally the dipole equations Eq. (8.105) are solved by iteration. The iterative solution of the integral equations is made slightly complicated by the presence of two eigenvalues for the iteration matrix, one being $\lambda = 1$ which corresponds to the singular nature of the integral equations and another being $\lambda = -1$ which prevents convergence of the iteration. All other eigenvalues are less than 1 in magnitude. By shifting the eigenvalues and by specifying the unimportant homogeneous solutions to the dipole equations, the iterations can be made to converge.[41,43]

Results are shown in Fig. 8.3 for a layer located initially at

$$z_1(p) = P + \mathrm{i}\varepsilon\cos(P),$$
$$z_2(p) = P + \mathrm{i}H,$$

with

$$P = p - \pi + \frac{1}{2}\sin(p).$$

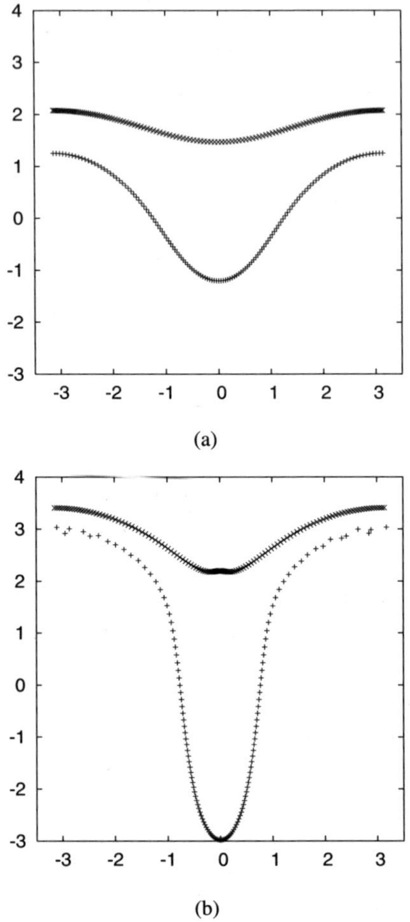

Fig. 8.3. Profiles of the layer at (a) $t = 2.0$, (b) $t = 3.0$.

The mapping $P(p)$ is a good example of the way to cluster the Lagrangian points initially for good effect later in the calculation. They are concentrated in the region which will later become the rising bubble since the markers there are stretched apart as the bubble rises. The layer is assumed at rest initially, so $\mu_1 = \mu_2 = 0$. The fourth-order Adams-Bashforth-Moulton predictor-corrector is used with a time step of 0.005 and 128 markers. The layer, if perfectly flat, would be in hydrostatic balance, $P_1 = P_2 + \rho g H$. Physical scales are set with $\rho = g = 1$, $P_2 = 0$, $H = \pi/2$ and $\varepsilon = \pi/10$. Rather than drawing the location of the surfaces as curves passing through the Lagrangian markers, only the positions of the Lagrangian markers are shown in Fig. 8.3 to highlight the formation of numerical instability on the lower surface at late times.

The profiles in Fig. 8.3 show the drainage of fluid into falling spikes, while bubbles rise causing thin regions in the layer. Visible at $t = 3.0$ is the development of numerical instabilities on the lower surface at the top of the bubble. This is an example of the loss of accuracy when two different surfaces come close together causing the denominator $z_1(q) - z_2(p)$ in the integrals to become very small. This limitation prevents the numerical simulation from proceeding and being able to track this thinning region as it rises. This limitation is unfortunate since it prevents the calculations from establishing how quickly the region thins and eventually snaps apart under the influence of van der Waal forces.

8.5.2. *Water waves in finite depth*

There are many circumstances where fluid interfaces occur in the presence of rigid boundaries. A good example is the propagation of water waves over bottom topography. Laplace's equation for the velocity potential Eq. (8.7) must now include a boundary condition at the rigid surface. Physically, the normal velocity of the fluid must match the normal velocity of the rigid surface. Let the rigid surface by represented in parametric form by $z_b(p)$ and its complex velocity $W_b(p)$. The downward normal velocity is

$$\frac{u_b y_{b,p} - v_b x_{b,p}}{s_{b,p}} = \Im\left\{ \frac{W_b^* z_{b,p}}{s_{b,p}} \right\} = \frac{\psi_{b,p}}{s_{b,p}}. \tag{8.108}$$

Even if the surface is moving, there is usually no flux through the surface. This implies that

$$\Im\left\{ \int W_b^* z_{b,p} \, \mathrm{d}p \right\} = \int \psi_{b,p} \, \mathrm{d}p = 0. \tag{8.109}$$

Since there is no mean value to $\psi_{b,p}$, it has a Fourier series of the form Eq. (8.72), except that $F_0^e = 0$. The series may be integrated term-by-term and the result remains periodic. After integration, ψ_b has an arbitrary constant which may be taken as zero. Note in particular that if $W_b = 0$, the surface is stationary, and $\psi_b = 0$ is the obvious result.

Since ψ is the harmonic conjugate to ϕ, it too satisfies Laplace's equation. The Dirichlet boundary condition $\psi = \psi_b$ at the rigid surface can also be solved by a dipole distribution. For a 2π-periodic layer between a free surface $z_f(p)$ and the rigid boundary z_b, the complex potential may be written as

$$\phi + \mathrm{i}\psi = \frac{1}{4\pi \mathrm{i}} \int_0^{2\pi} \mu_f(q)\, z_{f,q}(q) \cot\left\{ \frac{z_f(q) - z}{2} \right\} dq + \frac{1}{4\pi} \int_0^{2\pi} \mu_b(q)\, z_{b,q}(q) \cot\left\{ \frac{z_b(q) - z}{2} \right\} dq. \qquad (8.110)$$

The application of the boundary condition $\psi = \psi_b$ at $z = z_b$ leads to a Fredholm integral equation also of the second kind for μ_b.

$$\mu_b(p) + 2\Im\left\{ \frac{1}{4\pi \mathrm{i}} \int_0^{2\pi} \mu_f(q)\, z_{f,q}(q) \cot\left\{ \frac{z_f(q) - z_b(p)}{2} \right\} dq + \frac{1}{4\pi} \int_0^{2\pi} \mu_b(q)\, z_{b,q}(q) \cot\left\{ \frac{z_b(q) - z_b(p)}{2} \right\} dq \right\} = \psi_b. \qquad (8.111)$$

As before, the second integral in Eq. (8.111) must be interpreted in the principal-valued sense. The evolution for μ_f follows the standard derivation for the dipole distribution at a free surface Eq. (8.52), except that the second integral in Eq. (8.110) must be included and leads to a time derivative of μ_b. To complete the set of evolution equations, the time derivative of Eq. (8.111) must be included, leading to a set of coupled Fredholm integral equations of the second kind for the time derivatives of both dipole distributions.[6] These equations have been used to study water waves in shallow depth[6] and the downward acceleration of a rigid bottom that produces Rayleigh–Taylor instability on the falling free surface.[45]

A different application of this method calculates the motion of a rigid body beneath a water surface.[46] Here, the flow is not 2π-periodic. Instead, an infinite wave train is generated. A finite computational domain is chosen with damping layers attached to absorb the water waves as they leave the computational domain.

8.6. Some Challenges and Improvements

Spectral accuracy proves very desirable for tracking free surfaces in two-dimensional flow. The smooth calculation of the surface velocities prevents unwanted oscillations or instabilities from appearing during the motion. Unfortunately, in three-dimensional flow the accuracy of the numerical quadrature is limited, typically to third-order, and can result in numerical instability. Fortunately, there are accurate ways to treat axisymmetric geometry[47,48] through special properties of the elliptic functions which appear in the integrand. The hope is that there might be similar approaches for the full three-dimensional integrands.

Even for spectral methods in two-dimensional flow, significant accuracy is lost when two parts of the same surface or two different surfaces approach each other closely. Good examples are the pinch-off region in the rising bubble in Fig. 8.2 or the thinning region at the top of the rising bubble in Fig. 8.3. The culprit for the loss of accuracy is the rapid variation in the integrand which is nearly singular when the denominator in the integrand becomes small. To be specific, consider the integral,

$$I(p) = \frac{1}{2\pi \mathrm{i}} \int \frac{\mu_1(q)\, z_{1,q}(q)}{z_1(q) - z_2(p)}\, \mathrm{d}q. \tag{8.112}$$

When $z_1(q)$ is very close to $z_2(p)$, the denominator has large spikes of opposite sign on either side of the point of closest approach. Unless integration points are made available to resolve these spikes, large errors can arise in the numerical calculation of the integral.[41] One way to treat this difficulty is through interpolation by increasing the local resolution.[41] Another method uses the blob regularization and correction terms.[49]

A new method currently under development is the removal of the presence of the nearby pole singularity in the complex q plane of the integrand.[50] Seek the complex point Q such that $z_1(Q) = z_2(p)$. This means that Q is a function of p. Then rewrite

$$I(p) = \frac{1}{2\pi \mathrm{i}} \int \frac{\Big(\mu_1(q) - \mu_1(Q)\Big)\, z_{1,q}(q)}{z_1(q) - z_2(p)}\, \mathrm{d}q \pm \frac{\mu_1(Q)}{2}. \tag{8.113}$$

The sign of the additional term is determined by whether $z_1(q)$ lies below $z_2(p)$ $(+)$ or above $(-)$. Tests on a thin annulus reveal that spectral accuracy is restored more or less independently of the spacing between the surfaces. Unfortunately, the method relies on the ability to analytically continue both $z_1(q)$ and $\mu_1(q)$ into the complex q plane. Analytic continuation by means of the Fourier series is possible,[50] but other methods are currently being tested. Ultimately, the method must demonstrate success for free surface flows such as shown in Fig. 8.3.

There remain challenges for BITs when surfaces reconnect, for example when the tip of the plunging breaker shown in Fig. 8.1 reaches the surface below, or when the lobes of air pinch-off from the sides of the bubble in Fig. 8.2. There is no mathematical reason why these surfaces can't meet in finite time, but when they do, the boundary integral formulation becomes invalid. One might expect the surfaces to reconnect, but how? It is not just a matter of mathematics but also of physics since the processes by which reconnection occurs are not fully understood and probably occur on microscopic scales. There are methods, contour surgery in front tracking or level set methods which allow reconnection. There might be ways to use those techniques to design reconnection methods in boundary integral techniques.

Acknowledgments

I am indebted to two of my graduate students, Chao Xie and Jeong-Sook Im, for checking the results reported here. Chao Xie also provided me with the data for Fig. 8.1.

References

1. C. Pozrikidis, *Boundary Integral and Singularity Methods for Linearized Viscous Flow* (Cambridge University Press, Cambridge, 1992).
2. G. Cottet and P. Koumoutsakos, *Vortex Methods* (Cambridge University Press, Cambridge, 2000).
3. A. Majda and A. Bertozzi, *Vorticity and Incompressible Flow* (Cambridge University Press, Cambridge, 2001).
4. P. Saffman, *Vortex Dynamics* (Cambridge University Press, Cambridge, 1992).
5. D. Haroldsen and D. Meiron, Numerical calculation of three dimensional interfacial potential flows using the point vortex method, *SIAM J. Sci. Comp.* **20**, 648–683, (1998).
6. G. Baker, D. Meiron, and S. Orszag, Generalized vortex methods for free-surface flow problems, *J. Fluid. Mech.* **123**, 477–501, (1982).
7. G. Birkhoff, *Los Alamos Sci. Lab. Report LA-1862* (1954).
8. G. Birkhoff, Helmholtz and Taylor instability, *Proc Symp Appl Maths Soc* **13**, 55–76, (1962).
9. G. Baker, D. Meiron, and S. Orszag, Boundary integral methods for axisymmetric and three-dimensional Rayleigh–Taylor instability problems, *Physica D* **12**, 19–31, (1984).
10. J. Broeze, E. van Daalen, and P. Zandbergen, A three-dimensional panel method for nonlinear free surface waves on vector computers, *Comput. Mech.* **13**, 12–28, (1993).
11. M. Longuet-Higgins and E. Cokelet, The deformation of steep surface waves on water I. A numerical method of computation, *Proc Roy Soc Lond.* **A350**, 1–26, (1976).
12. J. Beale, A convergent boundary integral method for three-dimensional water waves. *Math. Comp.* **70**, 977–1029, (2001).

13. J. Beale, T. Hou, and J. Lowengrub, Convergence of a boundary integral method for water waves, *SIAM J. Numer. Anal.* **33**, 1797–1843, (1996).
14. T. Hou and P. Zhang, Convergence of a boundary integral method for 3-D water waves, *Disc. Cont. Dynam. Sys.-Ser. B* **2**, 1–34, (2002).
15. R. Menikoff and C. Zemach, Rayleigh–Taylor instability and the use of conformal maps for ideal fluid flow, *J. Comput. Phys.* **51**, 28–64, (1983).
16. S. Tanveer, Singularities in water waves and Rayleigh–Taylor instability, *Proc. Roy. Soc. Lond.* **435**, 137–158, (1991).
17. S. Tanveer, Singularities in the classical Rayleigh–Taylor flow: Formation and subsequent motion, *Proc. Roy. Soc. Lond.* **441**, 501–525, (1993).
18. J. Hamilton and G. Majda, On the Rokhlin-Greengard method with vortex blobs for problems posed in all space or periodic in one direction, *J. Comput. Phys.* **121**, 29, (1995).
19. K. Lindsay and R. Krasny, A particle method and adaptive treecode for vortex sheet motion in three-dimensional flow, *J. Comput. Phys.* **172**, 879–907, (2001).
20. D. Moore, The equations of motion of a vortex layer of small thickness, *Stud. Appl. Maths* **58**, 119, (1978).
21. M. Dhanak, Equation of motion of a diffusing vortex sheet, *J. Fluid Mech.* **269**, 265–281, (1994).
22. G. Carrier, M. Crook, and C. Pearson, *Functions of a Complex Variable.* McGraw-Hill, New York (1966).
23. P. Saffman and G. Baker, Vortex Interactions, *Ann. Rev. Fluid Mech.* **11**, 95–122, (1979).
24. R. Caflisch and O. Orellana, Singular solutions and ill-posedness for the evolution of vortex sheets, *SIAM J. Math. Anal.* **20**, 293–307, (1989).
25. J. Duchon and O. Roberts, Global vortex-sheet solutions of Euler equations in the plane, *J. Diff. Equat.* **73**, 215–224, (1988).
26. S. Cowley, G. Baker, and S. Tanveer, On the formation of Moore curvature singularities in vortex sheets, *J. Fluid Mech.* **378**, 233–267, (1999).
27. T. Hou, J. Lowengrub and M. Shelley, The long-time motion of a vortex sheets with surface tension, *Phys. Fluids* **9**, 1933–1954, (1997).
28. D. Joseph and T. Liao, Kelvin-Helmholtz mechanism for side branching in the displacement of light with heavy fluid under gravity, *Euro. J. Mech. B/Fluids* **11**, 253–264, (1992).
29. G. Baker, R. Caflisch, and M. Siegel, Singularity formation during Rayleigh–Taylor instability, *J. Fluid Mech.* **252**, 51–78, (1993).
30. G. Baker and J. Beale, Vortex blob methods applied to interfacial motion, *J. Comput. Phys.* **196**, 233–258, (2003).
31. G. Baker and L. Pham, A comparison of blob methods for vortex sheet roll-up, *J. Fluid Mech.* **547**, 297–316, (2006)
32. J.-G. Liu and Z. Xin, Convergence of vortex methods for weak solutions to the 2D Euler equations with vortex sheet data, *Comm. Pure Appl. Maths* **48**, 611–628, (1995).
33. R. Krasny, Desingularization of periodic vortex sheet roll-up, *J. Comput. Phys.* **65**, 292, (1986).

34. G. Baker and A. Nachbin, Stable methods for vortex sheet motion in the presence of surface tension, *SIAM J. Sci. Comput.* **19**, 1737–1766, (1998).
35. D. Moore, A point vortex method applied to interfacial waves, In: H. Hornung and E.-A. Müller (eds.), *Vortex Motion* (Vieweg and Sohn, Braunschweig, Weisbaden, 1982).
36. J. Beale, T. Hou, and J. Lowengrub, On the well-posedness of two fluid interfacial flows with surface tension, In: R. Caflisch and G. Papanicolaou (eds.), *NATO Adv. Sci. Inst. Ser. A*, 11–38, (1993).
37. A. Roberts, A stable and accurate numerical method to calculate the motion of a sharp interface between fluids, *IMA J. Appl. Math.* **31**, 13–35, (1983).
38. G. Baker, D. Meiron, and S. Orszag, Vortex simulations of the Rayleigh–Taylor instability, *Phys. Fluids* **23**, 1485–1490, (1980).
39. D. Pullin, Numerical studies of surface-tension effects in nonlinear Kelvin-Helmholtz and Rayleigh–Taylor instabilities, *J. Fluid Mech.* **119**, 507–532, (1982).
40. Y. Yang, The initial value problem of a rising bubble in a two-dimensional vertical channel, *Phys. Fluids A* **4**, 913–920, (1992).
41. G. Baker and M. Shelley, Boundary integral techniques for multi-connected domains, *J. Comput. Phys.* **64**, 112–132, (1986).
42. G. Baker and D. Moore, The rise and distortion of a two-dimensional gas bubble in an inviscid liquid, *Phys. Fluids A* **1**, 1451–1459, (1989).
43. G. Baker, R. McCrory, C. Verdon, ans S. Orszag, Rayleigh–Taylor instability of fluid layers, *J. Fluid Mech.* **178**, 161–175, (1987).
44. G. Baker and Q. Nie, The asymptotic motion of an accelerating, thick layer of inviscid liquid, *Phys. Fluids* **10**, 101–112, (1998).
45. G. Baker, The asymptotic motion of a tall column of liquid in a tube, in *Proc. 22nd Symp. Numer. Math.*, SANUM, Cape Town, RSA, (1998).
46. G. Baker, D. Meiron, and S. Orszag, Generalized vortex methods for free surface flow problems II: Radiating waves, *J. Sci. Computing* **4**, 237, (1990).
47. M. Nitsche, Axisymmetric vortex sheet motion: Accurate evaluation of the principal-value integral, *SIAM J. Sci. Computing* **21**, 1066–1084, (1999).
48. Q. Nie, The nonlinear evolution of vortex sheets with surface tension in axisymmetric flows, *J. Comput. Phys.* **174**, 438–459, (2001).
49. J. Beale and M.-C. Lai, A method for computing nearly singular integrals, *SIAM J. Numer. Anal.* **38**, 1902–1925, (2001).
50. G. Baker and N. Golubeva, Improved accuracy for points near a surface in BIT, In: V. Minutolo, M. Aliabadi (eds.), *Advances in Boundary Element Techniques VIII* (EC Ltd, UK, 2007).

Chapter 9

SIMULATION OF CAVITATING AND FREE SURFACE FLOWS USING BEM

Spyros A. Kinnas*
Ocean Engineering Group, ECJ 8.604
Civil, Architectural and Environmental Engineering Department-EWRE
Austin, TX 78712-0280, U.S.A.
kinnas@mail.utexas.edu[†]

Boundary Element Methods (BEMs) have been applied extensively in the past for the prediction of steady or unsteady flows around or inside bodies, including the modelling of cavities and/or free-surfaces. The major advantage of the BEM is that they are computationally very efficient, and thus are well suited for design. On the other hand, their major disadvantage is that they ignore the effects of viscosity. However, in the case where the effects of viscosity are limited in a thin boundary layer close to the body, the BEM *coupled* with boundary layer analysis can still provide a viable alternative to viscous flow solvers. The latter include Reynolds Averaged Navier–Stokes (RANS) solvers, Large-Eddy Simulations (LES), or Detached Eddy Simulations (DES), and most of them implement finite volume methods for the numerical solution of the Navier–Stokes equations in the whole flow domain, and use some type of turbulence model, given the fact that Direct Navier–Stokes (DNS) solvers still require enormous computer resources (especially when applied in three dimensions). In this chapter, the author will present recent applications of BEM in three cases: (a) cavitating flows around hydrofoils and propeller blades, (b) free-surface flows and (c) flows around hydrofoils or propeller blades entering (or exiting) a free-surface, which can be considered as a combination of cases (a) and (b). In all these cases there are flow boundaries of *unknown* shape (cavities, free-surfaces and ventilated surfaces) on which the *pressure is a known constant*, which is either equal to the vapour pressure[a] for case (a), or equal to the atmospheric pressure for cases (b) and (c). All the mentioned

*Professor, Department of Civil, Architectural and Environmental Engineering.
†The University of Texas at Austin.
[a]In the case of ventilated hydrofoils or propellers the pressure inside the *cavity* (or ventilated surface) is also constant and equal to the pressure of the air injected through slits on the body surface.

problems are inherently *non-linear* since they involve boundaries of which the shape must also be determined as part of the solution, and thus require the application of an iterative process. Linearized formulations have been developed in the past, but will not be addressed in this chapter.

9.1. Introduction

Sheet cavitation, the type of cavitation which is characterized by a 'continuous' liquid/vapour interface which is 'attached' to the body, can often occur during the operation of hydraulic or hydrodynamic devices, some of which are shown in Fig. 9.1. Despite its undesirable nature, some sheet (or other types of) cavitation often has to be accepted, especially in the case of high-speed and/or high displacement hull applications, so that the efficiency of these devices is not excessively low. The prediction of cavitation therefore becomes a very crucial aspect during the hydrodynamic design and assessment stage.

Experiments on models of these devices that simulate the conditions of sheet cavitation are also very common despite the fact that scale effects can

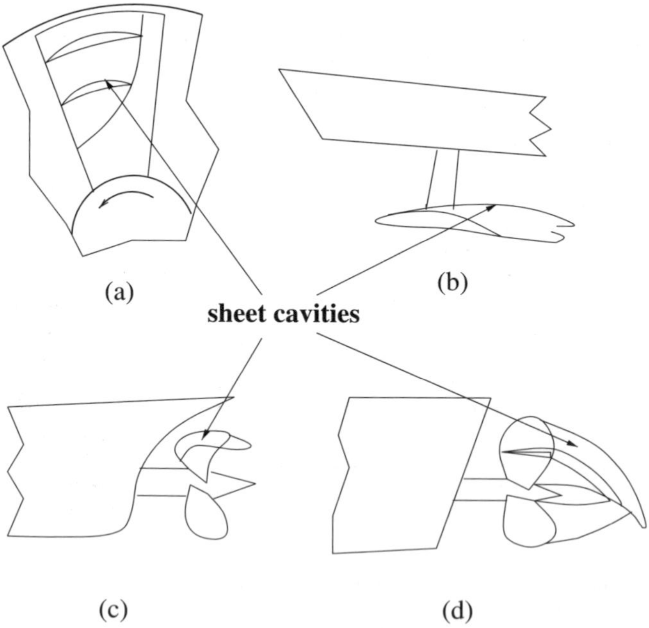

Fig. 9.1. Schematic of some applications in which sheet cavitation often occurs: (a) liquid pump, (b) hydrofoil, (c) marine propeller, and (d) supercavitating propeller.

seriously affect the cavitation inception stage. Some photographs from sheet cavitation experiments which were performed at MIT's variable pressure Marine Hydrodynamics Water Tunnel are shown in Fig. 9.2. Other types of cavitation on a marine propeller are shown schematically in Fig. 9.3. A general description of the different types of cavitation may be found in Kato.[1] In general, sheet cavitation is of an unsteady nature, i.e. the cavity extent and volume varies with time, and this is due to either inflow non-uniformity (in the case of marine propellers) or unsteady motion of the body (in the case of hydrofoil boats). In the laboratory steady sheet cavitation can be produced, although due to the turbulence in the inflow, small variations

Fig. 9.2. Photos from experiments at MIT's Marine Hydrodynamics Water Tunnel showing a sheet cavity on top of a hydrofoil (top), from;[6] and a propeller with sheet cavities cavities[8] at blade angles (from left to right) −30°, 6°, +30° (bottom). Flow goes from left to right.

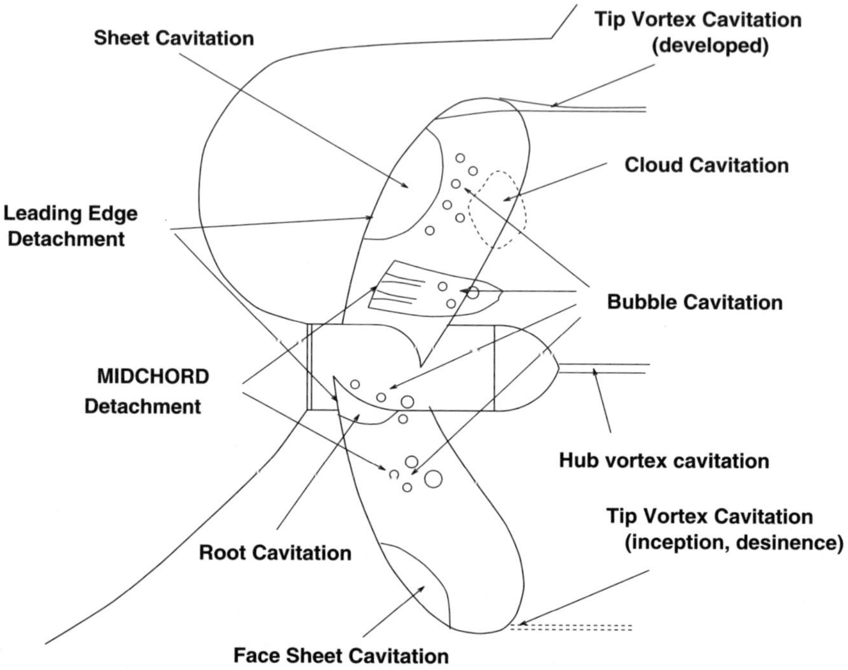

Fig. 9.3. Different types of cavitation on a marine propeller.

in the cavity shape are always present. Applications of sheet cavitation on hydrofoils and propellers will be emphasized in this paper, even though the formulation and the involved methods are general enough to deal with arbitrary geometries (e.g. bluff bodies or bodies of revolution).

According to observations and measurements,[2–6] the flow around a 2-D hydrofoil with a sheet cavity is depicted in the top portion of Fig. 9.4 with the corresponding pressure distribution on the suction side shown on the bottom portion of the figure. The following things are worth noting: (a) the cavity pressure is constant and equal to the vapour pressure over a large part of the cavity, (b) the cavity detaches at a point upstream of which pressures can be *smaller* than vapour pressure, (c) there is a pressure recovery region in the vicinity of the cavity end and (d) there is a viscous wake region downstream of the cavity end.

Cavitating or free-streamline flows were first addressed in non-linear theory via the hodograph technique as introduced by Helmholtz, Kirchoff and Levi–Civita.[7] The cavity surface in steady flow was taken

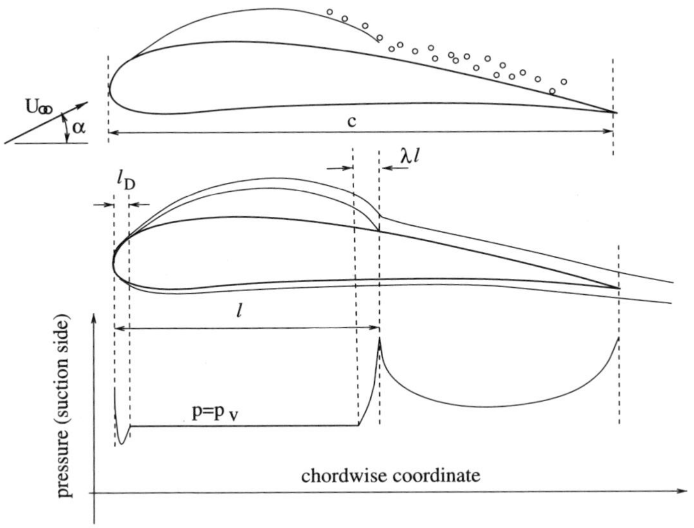

Fig. 9.4. Depiction of the actual sheet cavity (top) and the utilized model with its main parameters (bottom). Adapted from Kato.[1]

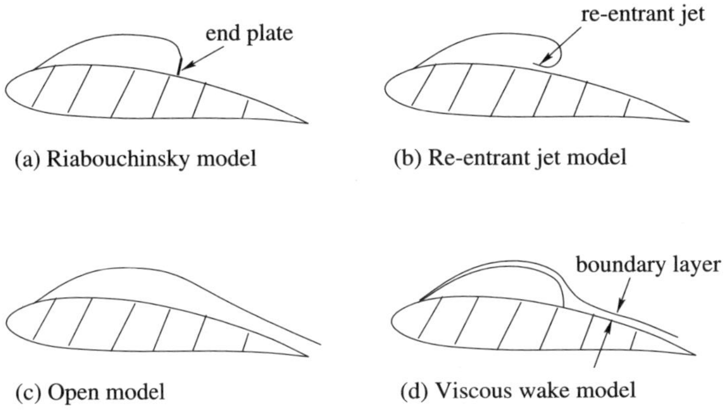

Fig. 9.5. Some of the existing cavity termination models.

as a *free* streamline with constant pressure (thus, constant velocity). The formulation of the cavitating flow around bodies created a lot of diversity on the cavity termination models. Some of the first known models were the Riabouchinsky end-plate model and the re-entrant jet model. Some of the existing cavity termination models are shown in Fig. 9.5.

Due to the difficulty of the hodograph technique in dealing with general body shapes, very few cases have been treated analytically.

The linearized cavity theory was introduced by Tulin[9] and became quickly very popular, as proven by the vast number of publications[b] which made use of it. Linearized cavity theory is actually an extension of the classical hydrofoil theory, where the hydrofoil (and cavity) is replaced with a zero thickness segment, essentially a 'branch cut', on which the boundary conditions are applied.

Unfortunately, based on the results of non-linear cavity theories, the linearized theory for partially cavitating hydrofoils predicts that, by increasing the thickness of a hydrofoil, the extent and size of the cavity, for constant flow conditions, also increases, which is *opposite* to the trend predicted by non-linear cavity theories. In addition, the linear cavity theory overpredicts the extent and amount of cavitation.[12]

A non-linear numerical method was employed to analyse cavitating hydrofoils by using surface vorticity techniques and by applying the exact boundary conditions on the cavity and on the hydrofoil.[13,14] An end-plate cavity termination model was implemented. A reduction in the size of the cavity as the foil thickness increased was predicted. A surface vorticity technique to deal with thick foil sections which employed an open cavity model was developed by Yamaguchi & Kato.[15] Similar BEM techniques were developed by Lemonnier & Rowe[16] and by Rowe & Blottiaux.[17] Potential-based BEMs were finally applied by Kinnas & Fine[12,18] and by Lee *et al.*[19]

Numerical BEMs within non-linear cavity theory were naturally extended to treat super-cavitating 3-D hydrofoils[20] and 3-D hydrofoils with partial cavities[12] or cavities with mixed (partial and super-cavities) planforms.[21] Similar methods were also developed by Kim *et al.*[22] and Pellone & Peallat.[23]

Non-linear potential-based BEMs were first applied to cavitating propellers in non-uniform flows by Kinnas & Fine,[24] Fine & Kinnas,[25] and then by Kim & Lee.[26]

The inviscid cavity flow method was coupled with a boundary layer solver in the case of partial and super-cavitating 2-D hydrofoils by Kinnas *et al.*[27] and Brewer & Kinnas.[6] This allowed for the inclusion of the viscous boundary layer in the wake of the cavity and for determining the cavity detachment point based on the viscous flow upstream of the cavity.

[b]An extended list of which may be found in Tulin & Hsu[10] or Kinnas.[11]

RANS solvers have also been applied for the prediction of attached sheet cavitation on 2-D hydrofoils[28,29] and, more recently on propellers by Rhee *et al.*,[30] where a two-phase model has been implemented.

In the present chapter, non-linear methods for the prediction of steady and unsteady sheet cavitation will be summarized, and some comparisons with experiments will be presented.

9.2. 2-D Hydrofoil

9.2.1. *Formulation*

Consider the 2-D partially cavitating hydrofoil,[c] as shown in Fig. 9.6. Assuming inviscid and irrotational flow, the governing equation everywhere inside the fluid region is given by[d]:

$$\nabla^2 \phi = 0 \tag{9.1}$$

where ϕ is the the *perturbation potential*, defined from:

$$\mathbf{q} = \mathbf{U}_\infty + \nabla \phi \tag{9.2}$$

where $\mathbf{q}$ is the total velocity vector in the flow. In order to uniquely determine ϕ, the following boundary conditions are imposed:

- On the foil surface, the following kinematic boundary condition is applied, which requires the fluid flow to be tangent to the surface of

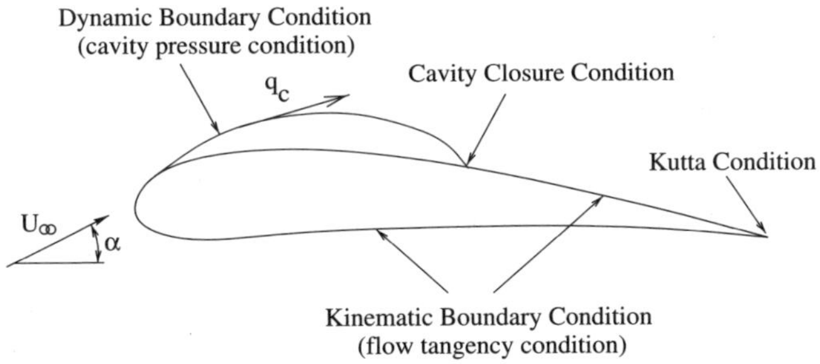

Fig. 9.6. Formulation of the inviscid cavitating flow problem

[c]The flow around partially cavitating hydrofoils will be addressed in this work. The extension to super-cavitating hydrofoils is straight-forward and the reader can find more details in the sited references.

[d]The cavity is assumed to start at a known location on the foil. A criterion for determining this location will be discussed in a later section.

the foil. Therefore,

$$\frac{\partial \phi}{\partial n} = -\mathbf{U}_\infty \cdot \mathbf{n} \tag{9.3}$$

where $\mathbf{n}$ is the surface unit normal vector.

- At infinity, the perturbation velocities should go to zero:

$$\nabla \phi \rightarrow 0 \tag{9.4}$$

- The Kutta condition requires finite velocities at the trailing edge of the foil:

$$\nabla \phi = \text{finite} \tag{9.5}$$

- The dynamic boundary condition specifies constant pressure on the cavity, or (via the Bernoulli equation applied far upstream and on the cavity) constant cavity velocity q_c:

$$q_c = U_\infty \sqrt{1+\sigma} \tag{9.6}$$

where the cavitation number, σ, is defined as:

$$\sigma = \frac{p_\infty - p_v}{\frac{\rho}{2} U_\infty^2} \tag{9.7}$$

where p_∞ is the pressure corresponding to a point in the free-stream and p_v is the cavity pressure (vapour pressure of water).

- Kinnas & Fine[18] assumed a viscous wake model as shown in part (d) of Fig. 9.4. Under this model, the following conditions apply:

  (1) The cavity closes at its trailing edge. The inclusion of the viscous wake downstream of the cavity will be addressed in a later section.

  (2) Near the trailing edge of the cavity, the pressure recovery termination model renders for the cavity velocity, q_{tr}, over a region λl:

$$q_{tr} = U_\infty \sqrt{1+\sigma}\,[1 - f(x)]\,, \tag{9.8}$$

  where $f(x)$ is an algebraic function defined in Kinnas & Fine.[12]

The problem of finding the cavitation number for given cavity extent l will be addressed first. A method for determining the cavity extent for a given cavitation number will be described in Sec. 9.2.3.

9.2.2. *The boundary element method*

Uhlman[13,14] applied a numerical non-linear BEM based on the surface vorticity formulation. Equations (9.1) to (9.6) were satisfied numerically, while the Riabouchinsky cavity termination model was employed. The shape of the cavity was determined in an iterative manner. The relatively slow convergence of the method with number of iterations made it prohibitively expensive to apply in three dimensions.

A potential-based boundary element method was applied[18] for the analysis of partially and super-cavitating hydrofoils. This method was applied first for the analysis of non-cavitating flows.[31] The perturbation potential on the combined foil and cavity surface satisfies the following integral equation (Green's third identity):

$$\pi\phi_p = \int_{S_B}\left[\frac{\partial\phi}{\partial n}\ln R - \phi\frac{\partial \ln R}{\partial n}\right]dS - \int_{S_W}\triangle\phi_W\frac{\partial \ln R}{\partial n}dS \tag{9.9}$$

where $\mathbf{n}$ is a unit vector normal to the foil or cavity surface, S_B is the foil and cavity surface, S_W is the trailing wake surface, and R is the distance between a point, P, and the point of integration over the foil or wake surface.

The foil and cavity surface are discretized into flat panels, as shown in Fig. 9.7. The source and dipole strengths are assumed to be constant over each panel. On the foil surface, the source strengths, which are proportional

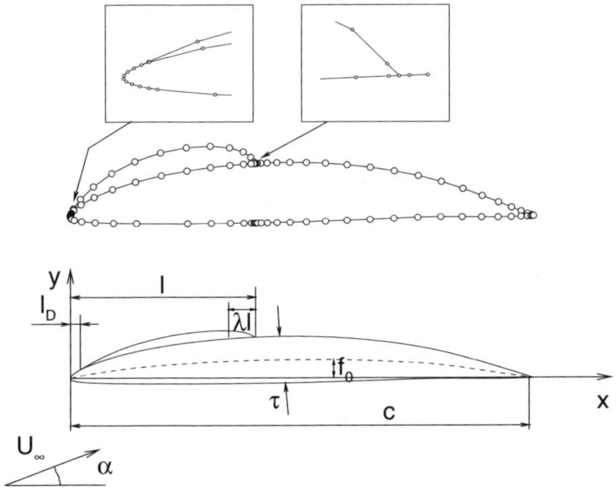

Fig. 9.7. Panel arrangement on the cavity and foil in numerical non-linear cavity theory. Details show dense panelling at the foil leading edge and cavity end.

to $\partial\phi/\partial n$, are given by the kinematic boundary condition, equation (9.3). On the cavity, the dipole strengths, which are proportional to ϕ, are determined by first expressing the dynamic boundary condition (9.6) and (9.8). For simplicity λ is zero for the rest of this paper; the complete derivation is given in Kinnas & Fine.[12] It should be noted that the length of the transition region λ has been found to affect the results only locally. In the case of $\lambda = 0$ and when a large number of panels is used the beginning of a re-entrant jet structure can be captured, as shown in Krishnaswamy *et al.* (2001)[32]:

$$q_c = \mathbf{U}_\infty \cdot \mathbf{s} + \frac{\partial \phi}{\partial s} = U_\infty \sqrt{1+\sigma} \tag{9.10}$$

where s is the arclength along the cavity surface (measured from the cavity detachment point), and $\mathbf{s}$ is the unit vector tangent to the cavity surface. Integration of equation (9.10) renders the potential on the cavity surface:

$$\phi(s) = \phi(0) - \mathbf{U}_\infty \cdot \mathbf{s} + sU_\infty\sqrt{1+\sigma}, \tag{9.11}$$

where $\phi(0)$ is the potential at the leading edge of the cavity. In the numerical scheme, $\phi(0)$ is expressed in terms of the (unknown) potentials on the wetted part of the foil in front of the cavity.

The Kutta condition is enforced via the Morino condition[31]:

$$\Delta\phi_w = \phi_T^+ - \phi_T^- \tag{9.12}$$

where $\Delta\phi_w$ is the potential jump in the wake of the foil and cavity and ϕ_T^+ and ϕ_T^- are the potentials at the upper and lower trailing edge panels, respectively. In two dimensions, the effect of the trailing wake surface (in the case of partial cavitation) is equivalent to the effect of a concentrated vortex at the foil trailing edge with strength equal to $\Delta\phi_w$.

The cavity closure condition requires the thickness at the cavity trailing edge, $h(s_L)$, to be equal to zero. This is enforced by the following equation:

$$h(s_L) = \frac{1}{q_c}\int_0^{s_L}\left[\mathbf{U}_\infty \cdot \mathbf{n} + \frac{\partial\phi}{\partial n}\right] ds = 0 \tag{9.13}$$

where s_L is the total arclength along the cavity surface.

The integral equation (9.9) is applied at the panel midpoints, together with equations (9.12) and (9.13). The resulting linear system of equations is inverted in order to provide the unknowns: (a) ϕ on the wetted foil, (b) $\partial\phi/\partial n$ on the cavity, and (c) the corresponding cavitation number σ.

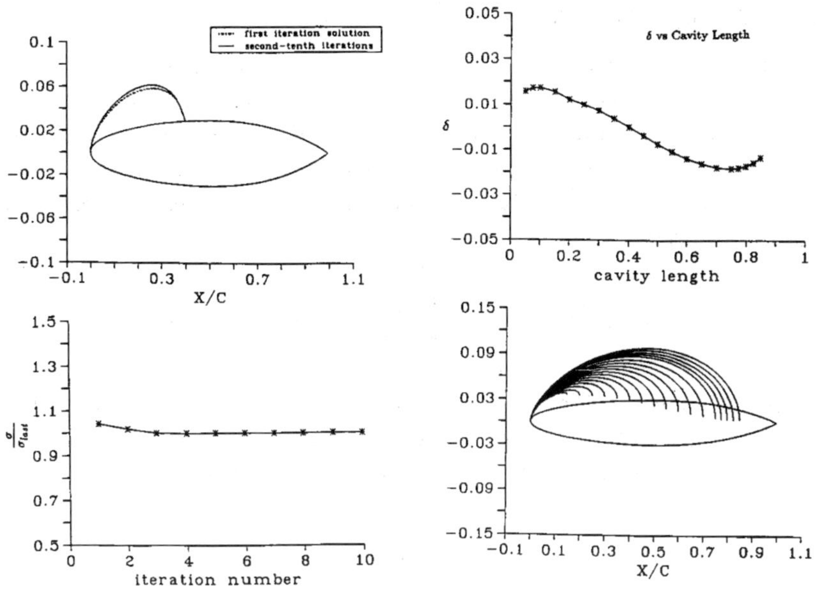

Fig. 9.8. Convergence of predicted cavity shape (top left) and cavitation number with number of iterations (bottom right), at *given cavity lengths*. Predicted cavity trailing edge thickness (top right) and cavity shape for various cavity lengths (bottom right) at *given* $\sigma = 1.097$ (corresponding to $l = 0.4$). From Kinnas & Fine.[12]

The cavity shape is determined in an iterative manner. In the first iteration the panels representing the cavity are placed on the foil surface directly under the cavity. In subsequent iterations the cavity shape is updated by an amount $h(s)$ (applied normal to the cavity surface) which is determined by integrating the following ordinary differential equation:

$$q_c \frac{dh}{ds} = \mathbf{U}_\infty \cdot \mathbf{n} + \frac{\partial \phi}{\partial n} \tag{9.14}$$

The history of predicted cavity shapes and cavitation numbers with number of iterations is shown in Fig. 9.8. Notice the remarkable convergence with number of iterations. In fact, even the predicted shape and cavitation number from the first iteration (when the cavity panels are placed on the foil surface) are very close to the converged values. This particular feature of the presented method makes it very attractive for 3-D applications, where carrying more than one iterations would increase the computation time substantially. The extension of the method in three dimensions is presented in the next sections.

9.2.3. *Cavity extent for given cavitation number*

In the previous sections the cavity length is assumed to be known. For *given (fixed)* cavitation numbers the BEM can still be applied for various cavity lengths. In this case though, due to the fact that the cavitation number is given (instead of being determined), the cavity closure condition, equation (9.13), will not be satisfied. Instead the thickness at the cavity end, δ, will be given from the expression:

$$\delta(l;\sigma) \equiv h(s_L) = \frac{1}{q_c} \int_0^{s_L} \left[\mathbf{U}_\infty \cdot \mathbf{n} + \frac{\partial \phi}{\partial n} \right] ds \tag{9.15}$$

This is shown in Fig. 9.8 where the predicted δ and the cavity shapes for fixed σ and different values of cavity length are shown. An iterative scheme has been developed[12] for determining the cavity extent for given σ by searching for the cavity length, l, for which[e]:

$$\delta(l;\sigma) = 0 \tag{9.16}$$

In Fig. 9.9, the cavity shapes are predicted, for the same cavitation number, from applying the fully converged method versus the method in which the first iteration has only been applied, i.e. when the panels representing the cavity are placed on the foil under the cavity. The latter method is also called the *thin cavity theory.* It is essential to see that the differences are not significant. In the extensions of this method to 3-D hydrofoils and propellers (to be addressed later), the thin cavity theory is applied.

9.2.4. *Effects of flow unsteadiness*

In order to understand the importance of the flow unsteadiness on the boundary conditions and the cavity dynamics, an unsteady two-dimensional model of a partially cavitating hydrofoil is studied, wherein the cavitation number is allowed to vary harmonically around a mean value, i.e. $\sigma(t) = \sigma_o + \sigma_g \cos(\omega t)$, while the inflow is steady and uniform.[f] This model allows us

[e]An open cavity model can be readily implemented within this method by requiring the right-hand side of equation (9.16) to be equal to a specified thickness at the cavity end.
[f]This is equivalent to applying a small sinusoidal disturbance either on the ambient pressure or on the vapour pressure. In the case of propellers, to be addressed later, the unsteadiness is due the variation of the inflow to which the blade is subject to, when in the presence of a non-axisymmetric ship wake.

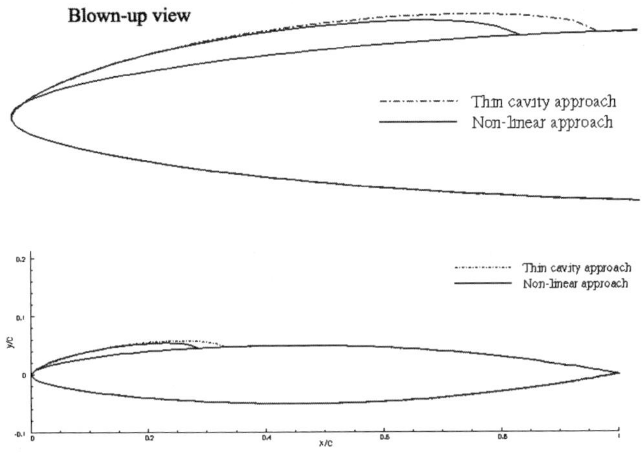

Fig. 9.9. Comparison of cavity shapes predicted from thin cavity theories (in which the cavity panels are placed on the hydrofoil surface, underneath the cavity), and fully non-linear cavity theories, at the *same cavitation number*. From Sun.[33]

to see the effects of the unsteady terms on the solution. Bernoulli's equation will render, in this case:

$$p_\infty + \rho\frac{U_\infty^2}{2} = \rho\frac{\partial\phi}{\partial t} + p_v + \rho\frac{q^2}{2} \tag{9.17}$$

or equivalently,

$$q_c(s,t) = \mathbf{U}_\infty \cdot \mathbf{s} + \frac{\partial\phi}{\partial s} = \sqrt{[1+\sigma(t)]U_\infty^2 - 2\frac{\partial\phi}{\partial t}} \tag{9.18}$$

where $q_c(s,t)$ is the cavity velocity which, instead of being constant (as in the case of steady flow), now varies with time and location along the cavity surface. As in the case of steady cavity flow, equation (9.18) renders:

$$\frac{\partial\phi}{\partial s} = -\mathbf{U}_\infty \cdot \mathbf{s} + \sqrt{[1+\sigma(t)]U_\infty^2 - 2\frac{\partial\phi}{\partial t}} \tag{9.19}$$

which, when integrated along s, provides an expression for $\phi(s,t)$ along the cavity. Note that the right-hand side of equation (9.19) now involves $\partial\phi/\partial t$ which is not known and may be determined via a fourth order accurate backward finite difference scheme.[24] The cavity height, $h(s,t)$, is determined by inverting the following partial differential equation[34]:

$$q_c(s,t)\frac{\partial h}{\partial s} + \frac{\partial h}{\partial t} = \mathbf{U}_\infty \cdot \mathbf{n} + \frac{\partial\phi}{\partial n} \tag{9.20}$$

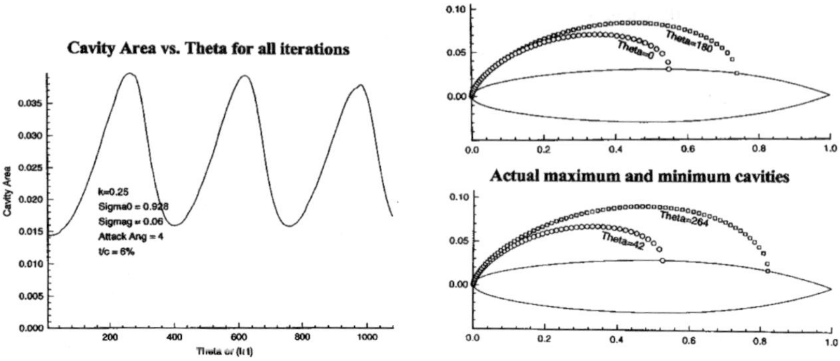

Fig. 9.10. Predicted unsteady cavity area (left), maximum and minimum cavity shapes from quasi-steady (top right) and fully unsteady solution (bottom right); $\sigma_o = 0.928$, $\sigma_g = 0.06$, and $k = 0.25$.

Equations (9.9) and (9.12) are applied at each time step. Due to the smallness of the introduced disturbance, the effect of the shed vorticity from the foil trailing edge on the solution is ignored, i.e. the wake dipole strength, $\Delta\phi_w$, is assumed to only change with time and not with location along the wake. In addition, a cavity closure condition is applied at the instantaneous cavity trailing edge:

$$h(s_L, t) = 0 \tag{9.21}$$

The time history of the cavity length on an NACA16006 hydrofoil operating at an angle of attack of 4° and with a time-dependent cavitation number equal to $\sigma(t) = 0.928 + 0.06 \cos\omega t$ at a reduced frequency of $k = \omega c/(2U_\infty) = 0.25$ was computed. Fig. 9.10 shows the time history of the cavity volume (area) with time, and the predicted cavity shapes from a 'quasi-steady' and the fully unsteady solution. As expected, the fully unsteady solution is shifted to the right, i.e. the growth stage lasts longer than the collapse stage, a well known characteristic of unsteady cavitation. Note the substantial change in the cavity characteristics due to a small change in the cavitation number. In experiments, the unsteadiness in the inflow (due to turbulence) causes fluctuations of the cavity extent, leading to the well known instability in the flow as the cavities transition from partial to supercavitation.

9.2.5. *Effects of viscosity*

The effects of viscosity on the inviscid cavitating solution can be included via coupling with a boundary layer solver,[35] as shown in Fig. 9.11. This

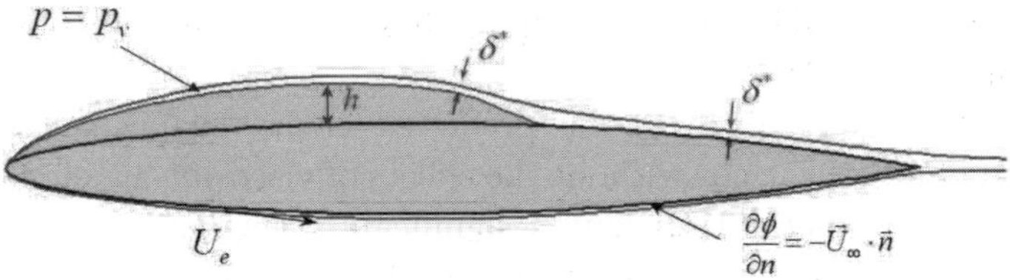

Fig. 9.11. The boundary layer displacement thickness (δ^*) is shown together with the cavity and the hydrofoil.

is accomplished by adding 'blowing' sources to the inviscid cavity flow problem,[6,36] in the right hand side of the Green's third identity, equation (9.9), as follows:

$$\pi\phi_p = \int_{S_B}\left[\frac{\partial\phi}{\partial n}\ln R - \phi\frac{\partial \ln R}{\partial n}\right]dS - \int_{S_W}\Delta\phi_W\frac{\partial \ln R}{\partial n}dS + \int_{S_B\cup S_W}\hat{\sigma}\ln R dS \tag{9.22}$$

The strength of these sources is given as:

$$\hat{\sigma} = \frac{d(U_e\delta^*)}{ds}, \tag{9.23}$$

where s is the arclength along the foil, U_e is the velocity at the edge of the boundary layer, and δ^* is the displacement thickness.

The terms in equation (9.22) can be rearranged to produce the following equation:

$$U_e = U_e^{inv} + \mathcal{E}\{U_e\delta^*\} \tag{9.24}$$

where U_e^{inv} is the value of the inviscid velocity on the hydrofoil surface.

Equation (9.24) gives the edge velocity at each panel in terms of the inviscid edge velocity and a mass defect term, $m = U_e\delta^*$. $\mathcal{E}$ is a geometry-dependent operator, the discretized version of which is given in Hufford *et al.*[37] and Drela.[35]

The boundary layer equations[35] are solved first with the edge velocity distribution given from the inviscid cavity theory. Once δ^* is found, U_e is updated via equation (9.24). This process continues until convergence is achieved. It should be noted that a zero friction condition is applied on the cavity surface.[6] In addition, an iterative method has been developed to ensure that the *viscous* pressure distribution over the cavity is constant.[33,36]

This viscous/inviscid interactive coupling has also been extended in the case of propeller blades.[33,38]

At the top of Fig. 9.12, the predicted cavity shapes for the same cavitation number, with or without the effects of viscosity, are shown. It can be seen that the method predicts a significantly smaller cavity, when the effects of viscosity are included. To verify the current method, a hydrofoil and the predicted cavity (with the effects of viscosity included in the

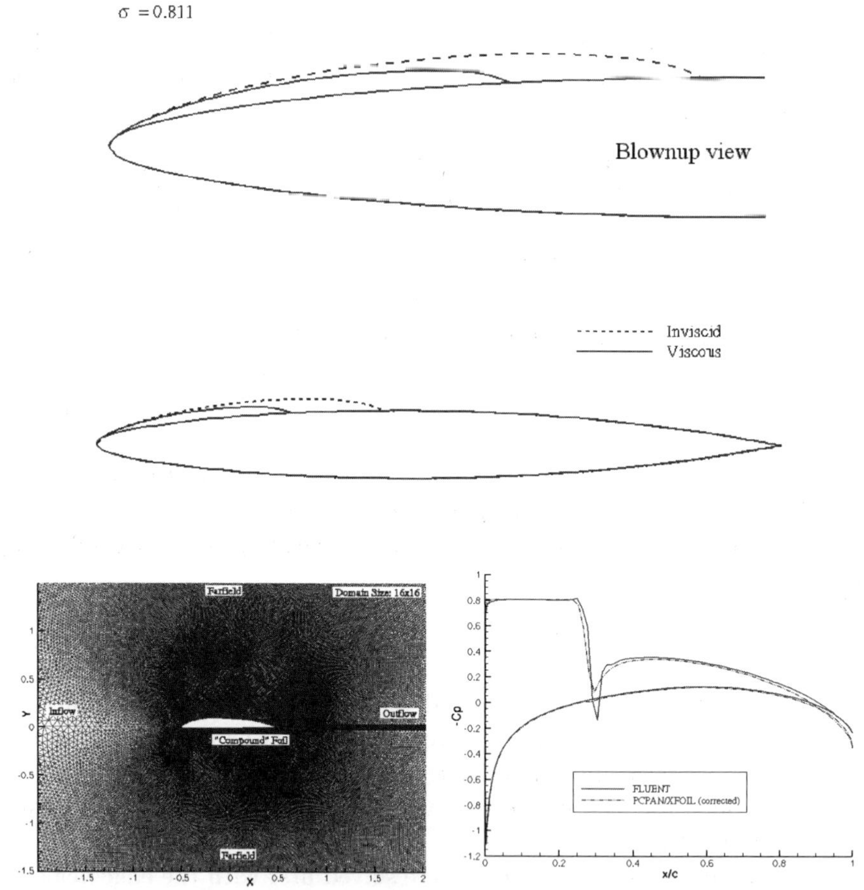

Fig. 9.12. The cavity for *given* cavitation number predicted by using the present BEM with and without the effects of viscosity included (top). The cells in the finite volume method (Fluent) used to solve for the viscous flow applied on the cavity, as predicted by the current BEM, *after* coupling with the boundary layer solver (bottom left), and the resulting pressures from the BEM method and Fluent (bottom right). The Reynolds number based on chord is 10^6. From Sun.[33]

BEM) are analyzed via the commercial RANS solver, FLUENT, and the predicted pressures from the BEM and FLUENT are shown together at the bottom right of Fig. 9.12. The required cell arrangement in FLUENT is shown on the bottom left of Fig. 9.12. The pressures distributions from the two methods are remarkably close to each other. It should be noted that FLUENT required about 200,000 cells and about 8 hours of CPU on a SUN BLADE 2,000 (2×1.2 GHz, 8G RAM) workstation. To produce the same answer with the BEM takes only 1–2 minutes!

9.2.6. *The cavity detachment point*

The cavity detachment location can be determined using the smooth detachment criterion (or the Villat-Brillouin condition) which is equivalent to requiring the following conditions: (a) the cavity has non-negative thickness at its leading edge, and (b) the pressure in front of the cavity is *no* smaller than the vapour pressure. The detachment criterion has been applied in the case of 2-D or 3-D hydrofoils as well as propeller blades in Young & Kinnas,[39] where cavities are searched for on both sides of the blade. The effect of the cavity detachment location on the predicted cavity extent and shape is shown in Fig. 9.13. It should be noted that experimental evidence[2,40] has shown that the cavity detachment occurs behind a point of laminar separation. This criterion has been implemented in the inviscid/viscous cavity formulation.[6]

Figure 9.13 demonstrates the significance of determining the cavity detachment points through an iterative procedure, as opposed to simply picking the points where the non-cavitating pressure is equal to vapour pressure.

9.3. 3-D Hydrofoil

Consider now a 3-D hydrofoil which is subject to a uniform inflow $\mathbf{U}_\infty$ as shown in Fig. 9.14. The cavity surface is denoted by S_C, the wetted hydrofoil surface by S_{WS}, and the trailing wake surface by S_W. The total flow velocity field $\mathbf{q}(x, y, z)$ can be written in terms of the perturbation potential $\phi(x, y, z)$, as follows:

$$\mathbf{q}(x, y, z) = \mathbf{U}_\infty + \nabla\phi(x, y, z). \tag{9.25}$$

In the next four sections the necessary equations and conditions for determining $\phi(x, y, z)$, as well as the cavity planform and shape are outlined.

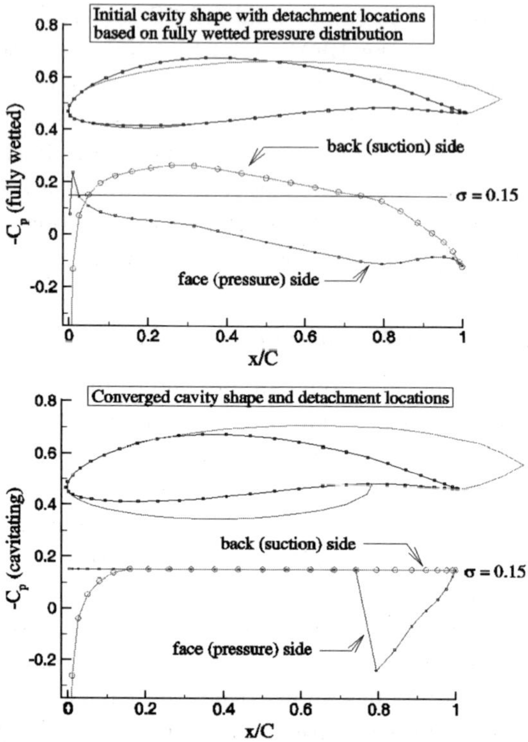

Fig. 9.13. Effect of the detachment location on the predicted cavity extent and shape via BEM. The cavity on the left has been predicted by assuming that the cavity on either side of the hydrofoil starts at the points where the wetted pressure distribution (shown) becomes equal to the vapour pressure (e.g. $-C_{P,wetted} = \sigma$). Note that the predicted cavity on the upper ('back') side has negative thickness (a mathematically correct solution has produced a non-physical cavity!), due to the 'wrong' location of the cavity detachment. The cavity on the right, on the other hand, has been predicted by determining, in an iterative manner,[39] the location of the cavity detachment on either side of the hydrofoil based on the smooth detachment condition. Note that the predicted cavities on the right have positive thickness and that the pressures on the wetted part of the hydrofoil (ahead of the cavity detachment points) are larger than vapour pressure (e.g. $-C_P < \sigma$). From Kinnas *et al.*[41]

9.3.1. *The green's third identity*

As in the case of 2-D hydrofoil Green's third identity renders the following integral equation for $\phi(x, y, z)$ as

$$2\pi\phi = \int_{S_{WS}\cup S_C} \left[\phi\frac{\partial G}{\partial n} - G\frac{\partial \phi}{\partial n}\right] dS + \int_{S_W} \Delta\phi_W \frac{\partial G}{\partial n} dS \tag{9.26}$$

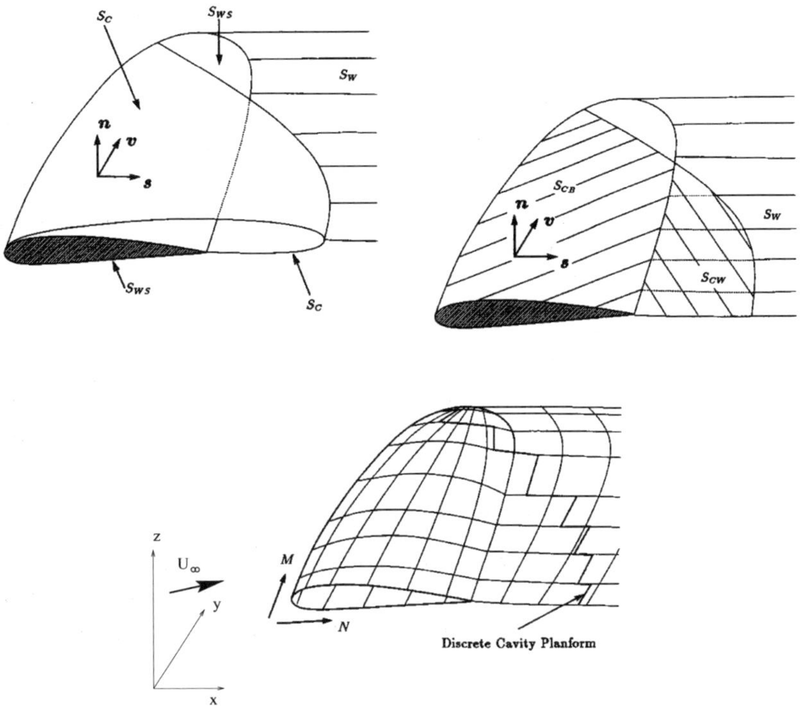

Fig. 9.14. Definition of the 'exact' (top left), the approximate (top right), and the discretized 3-D cavity and foil surface (bottom right).

where **n** is the unit vector normal to the foil wetted surface, the cavity surface or the wake surface; $\Delta\phi$ is the potential jump across the wake sheet and $G = 1/R$ is the Green's function, where R is the distance between a point P and the point of integration along the foil and cavity surface.

Equation (9.26) should be applied on the 'exact' cavity surface as shown in Fig. 9.14. Based on the fast convergence of the BEM with number of iterations in 2-D (as described in the previous section), only the first iteration non-linear solution is carried out. This is equivalent to applying Green's formula on an approximate surface, as shown in Fig. 9.14. This surface is comprises the foil surface and the trailing wake surface. The treatment of a supercavity within this scheme is discussed in detail in Fine & Kinnas.[25] For simplicity, only the partially cavitating 3-D hydrofoil problem will be described here. The approximate surface is discretized as shown at the bottom part of Fig. 9.14. Constant strength dipoles and sources are distributed on each of the wetted or cavitating flow panels. The strength of the source distribution on the wetted foil surface is expressed via the kinematic boundary condition:

$$\frac{\partial \phi}{\partial n} = -\mathbf{U}_\infty \cdot \mathbf{n}; \quad \text{on } S_{WS} \tag{9.27}$$

In the case of partial cavitation the trailing wake, S_W, is treated the same way as in the case of fully wetted flows.[31] In other words, $\Delta\phi_W$ is only a function of the spanwise location, y, given as:

$$\Delta\phi_W(y) = \phi_T^+(y) - \phi_T^-(y) = \Gamma(y) \tag{9.28}$$

where $\phi_T^+(y)$ and $\phi_T^-(y)$ are the values of the potential at the upper (suction side) and lower (pressure side) foil trailing edge, respectively. The difference in those potentials is also equal to the circulation Γ at a spanwise location y.

9.3.2. *The dynamic boundary condition*

The dynamic boundary condition (DBC) requires that the pressure everywhere inside and on the cavity be constant and equal to the known cavity pressure, p_v. As in the case of 2-D hydrofoil, Bernoulli's equation will give:

$$q_c = U_\infty \sqrt{1+\sigma}, \tag{9.29}$$

where q_c is the magnitude of the cavity velocity $\mathbf{q}_c$. Note that for simplicity the hydrostatic terms, which would be important in the case of a vertical 3-D foil, have not been included. The cavity velocity vector, $\mathbf{q}_c$, may also be expressed as follows[12]:

$$\mathbf{q}_c = \frac{V_s\,[\mathbf{s} - (\mathbf{s}\cdot\mathbf{v})\mathbf{v}] + V_v\,[\mathbf{v} - (\mathbf{s}\cdot\mathbf{v})\mathbf{s}]}{\|\mathbf{s}\times\mathbf{v}\|^2} \tag{9.30}$$

where s and v are the curvilinear coordinates[g] along the cavity surface (as shown in Fig. 9.14); $\mathbf{s}$ and $\mathbf{v}$ are the corresponding unit vectors. V_s and V_v are given as follows:

$$V_s = \frac{\partial \phi}{\partial s} + \mathbf{U}_\infty \cdot \mathbf{s}; \quad V_v = \frac{\partial \phi}{\partial v} + \mathbf{U}_\infty \cdot \mathbf{v} \tag{9.31}$$

Equations (9.29) and (9.30) may then be combined to form an equation which is quadratic in the unknown chordwise perturbation velocity, $\partial\phi/\partial s$.

[g] In general non-orthogonal.

The solution to this quadratic[h] renders $\partial\phi/\partial s$ in terms of the cavitation number, the inflow velocity, and the unknown crossflow $\partial\phi/\partial v$:

$$\frac{\partial \phi}{\partial s} = -\mathbf{U}_\infty \cdot \mathbf{s} + V_v \cos\theta + \sin\theta\sqrt{q_c^2 - V_v^2}, \tag{9.32}$$

with θ is the angle between s and v, and q_c is given by equation (9.29). Equation (9.32) is integrated once to form a Dirichlet boundary condition on ϕ:

$$\phi(s) = \phi(0) + \int_0^s [\text{right-hand side of equation (9.32)}]ds \tag{9.33}$$

The value of $\phi(0)$ in equation (9.33) at each strip is determined (as in the case of 2-D hydrofoil) via a cubic extrapolation in terms of the unknown potentials on the wetted panels on the same strip in front of the cavity.

The crossflow term, $\partial\phi/\partial v$, in equation (9.33) is included in an iterative sense.[12]

9.3.3. *The cavity thickness distribution*

The kinematic boundary condition on the cavity requires that the velocity normal to the cavity is zero (in the case of steady flows) or, more generally (also valid in the case of unsteady flows), that the following substantial derivative is zero:

$$\frac{D}{Dt}(n-h) = \left(\frac{\partial}{\partial t} + \mathbf{q}_c \cdot \nabla\right)(n-h) = 0 \tag{9.34}$$

where n is the coordinate normal to the foil surface under the cavity (with unit vector $\mathbf{n}$) and $h(s,v,t)$ is the thickness of the cavity normal to the foil surface at the point (s,v) at time t. Expressing the gradient in terms of the local directional derivatives

$$\nabla = \frac{[\mathbf{s} - (\mathbf{s}\cdot\mathbf{v})\mathbf{v}]\frac{\partial}{\partial s} + [\mathbf{v} - (\mathbf{s}\cdot\mathbf{v})\mathbf{s}]\frac{\partial}{\partial v}}{\|\mathbf{s}\times\mathbf{v}\|^2} + \mathbf{n}\frac{\partial}{\partial n}, \tag{9.35}$$

and performing the dot product with $\mathbf{q}_c$ (as defined in equation (9.30)) and finally substituting the result in equation (9.34) yields the following partial

[h]The root which corresponds to cavity velocity vectors pointing downstream is selected.

differential equation for the cavity thickness:

$$\frac{\partial h}{\partial s}\left[V_s - \cos\theta V_v\right] + \frac{\partial h}{\partial v}\left[V_v - \cos\theta V_s\right] = \sin^2\theta\left(V_n - \frac{\partial h}{\partial t}\right) \qquad (9.36)$$

where

$$V_n = \frac{\partial \phi}{\partial n} + \mathbf{U}_\infty \cdot \mathbf{n} \qquad (9.37)$$

The partial differential equation (9.36) can be integrated over the entire cavity planform in order to provide the cavity thickness. In the case of steady flow (as considered in the beginning of this section), $\partial h/\partial t = 0$.

9.3.4. *The cavity planform*

The extent (planform) of the 3-D cavity is not known and has to be determined as a part of the solution. This is accomplished by finding the appropriate $l(y)$ (cavity length at each spanwise location y) which satisfies the cavity closure condition for the given cavitation number, σ:

$$\delta(y;\sigma) = h(l(y), y) = 0 \qquad (9.38)$$

9.3.5. *Numerical aspects*

The objective of the numerical analysis is to invert equation (9.26) subject to the conditions (9.27), (9.33) and (9.28). The numerical implementation is described in detail in Kinnas & Fine[12,18] and Fine & Kinnas.[21] In brief, for a given cavity planform, equation (9.26) is solved with respect to the unknown ϕ on the wetted foil and for the unknown $\partial\phi/\partial n$ on the cavity. The cavity height at the trailing edge of the used cavity planform is then determined by integrating equation (9.36). The cavity planform is adjusted accordingly and the solution is carried over again until the corresponding heights at the cavity end are equal to zero within some given tolerance. The numerics of the method have been extensively validated in Kinnas & Fine[12] and Fine & Kinnas.[21]

9.3.6. *The split panel technique*

A very crucial issue in the numerical implementation was found to be related to the treatment of panels which were intersected by the cavity trailing edge. In order to avoid recomputing influence coefficients a technique was devised, namely the *split* panel technique,[12,34] in which the intersected panel is

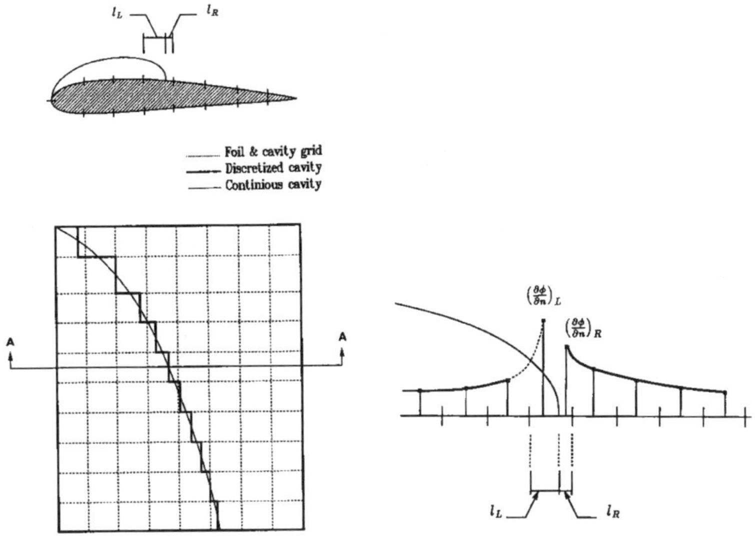

Fig. 9.15. The split panel technique applied to the cavity trailing edge in three dimensions. The extrapolated values for $\partial\phi/\partial n$ into the two parts of the split panel are also shown.

treated as one panel with each of the ϕ and $\partial\phi/\partial n$ being determined as the weighted average of the values on the wetted and the cavitating part of the panel. This technique, as depicted in Fig. 9.15, provided substantial savings on computer time since the same panel discretization can handle arbitrary cavity planforms.

9.3.7. *Multiplicity of solutions*

Equation (9.38) may accept more than one solutions, i.e. produce more than one cavity planform for a given value of cavitation number. This is a very well known fact in two dimensions, where for some cavitation numbers there are three solutions (two partial cavities and one supercavity). The present method has also been found to predict multiple solutions in three dimensions,[21,24] as can be seen in Fig. 9.16. Note that for $\sigma = 0.85$, two cavity planforms are predicted: one partial cavity and one mixed cavity (slightly supercavity at midspan). The partial cavity was produced when the initial guess was a partial cavity and the mixed cavity when the initial guess was a supercavity. The cavity length at midchord vs α/σ is also shown at the bottom part of Fig. 9.16. Note the striking similarity of this curve to the well known characteristic curve for a two-dimensional cavitating flat

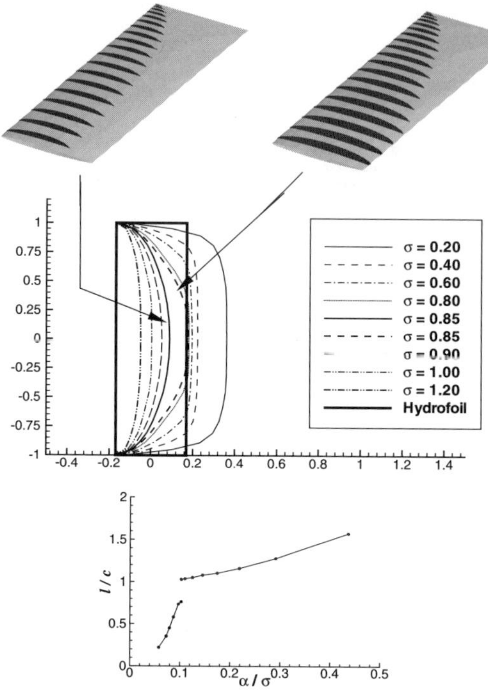

Fig. 9.16. Multiple solutions in 3-D for $\sigma = 0.85$. The predicted cavity shapes are shown over half of the span at the top and the l_{max}/c vs. α/σ curve is shown at the bottom.

plate (not shown in the Figure). This multiplicity of solutions in 3-D can also be confirmed from the observed instability on the cavity extent during experiments on cavitating 3-D foils, as the cavity transitions from partial to super-cavitation.

9.4. Propellers

Consider a cavitating propeller subject to a non-uniform wake inflow, $\mathbf{U}_W(x_s, r_s, \theta_s)$, as shown in Fig. 9.17. The wake inflow is referenced with respect to the ship fixed system of coordinates (x_s, y_s, z_s) or (x_s, r_s, θ_s) and is assumed to include the propeller/inflow interaction, i.e. it is the *effective* wake. The propeller is often immersed in the boundary layer of the ship's hull, and thus, its inflow is non-axisymmetric and carries strong vorticity in it. The actual (or effective) propeller inflow can be determined by coupling the presented BEM, in an iterative manner, with a Euler solver or a RANS solver which models the global flow around the propeller and hull. The

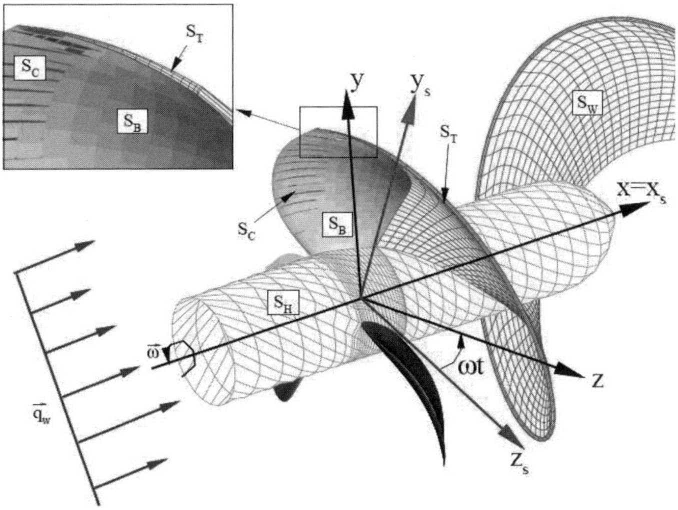

Fig. 9.17. Cavitating propeller subject to non-axisymmetric wake inflow. The surface of a developed tip vortex cavity, S_T, is also shown.

propeller is represented with body forces (also called source terms) within the global flow Euler or RANS solver.[42,43]

The propeller flow is considered with respect to the propeller fixed system $[(x, y, z)$ or $(x, r, \theta)]$, which rotates with an angular velocity $\boldsymbol{\omega}$ with respect to the ship fixed system. The inflow, with respect to the propeller rotating system will thus be[i]:

$$\mathbf{U}_{in}(x, y, z, t) = \mathbf{U}_W(x, r, \theta - \omega t) + \boldsymbol{\omega} \times \mathbf{x} \tag{9.39}$$

Equation (9.26) still applies in the time domain while the perturbation potential $\phi(x, y, z, t)$ is defined from:

$$\mathbf{q} = \mathbf{U}_{in} + \nabla\phi \tag{9.40}$$

where $\mathbf{q}$ is the total velocity vector of the flow with respect to the propeller fixed system.

The treatment of the time-dependent sheet cavity $S_C(t)$ on the propeller blades is very similar to that in the case of 3-D hydrofoils, described in the previous section. Equations (9.26), (9.27), (9.28), (9.33),

[i]Details of the application of the BEM on cavitating propellers may be found in Kinnas & Fine.[24]

(9.36) and (9.38) now apply at every time step in the numerical solution. The main differences, according to Kinnas & Fine[24] and Fine,[34] consist of:

- Replacing $\mathbf{U}_\infty$ with $\mathbf{U}_{in}$
- Replacing q_c in the right-hand side of equation (9.32) with:

$$q_c = \sqrt{n^2 D^2 \sigma_n + |\mathbf{U}_{in}|^2 - 2gy_s - 2\frac{\partial \phi}{\partial t}}, \tag{9.41}$$

where D is the propeller diameter, n the revolutions per second, g the acceleration of gravity and σ_n the cavitation number defined as:

$$\sigma_n = \frac{p_{sh} - p_v}{(\rho/2) n^2 D^2} \tag{9.42}$$

with p_{sh} being the hydrostatic pressure at the centre of the shaft of the propeller. Note that the hydrostatic terms $(-2gy_s)$ are included in the pressure calculation.

The $\frac{\partial \phi}{\partial t}$ term in equation (9.41) is not known, and is determined in an iterative manner by using the value when the blade is at the same angle in the previous revolution (or period).[24]

Other important issues, which were not addressed in this chapter, are (a) the location of the trailing wake sheet, which is *not known a priori* and must be determined from applying a *force-free* condition in the wake, (b) the values of the dipoles in the wake are known from previous time-steps, and (c) the sectional shape of the *developed* tip vortex cavity, which is modelled with panels in the presented approach, is not known and can be determined from applying a constant pressure condition (dynamic condition) on the tip vortex cavity surface. The reader is referred to Kinnas & Hsin[44] for (b), Lee & Kinnas[45] for (a), and Lee & Kinnas[46] for (c).

Finally, when comparing with experiments, which often take place inside confined spaces, the tunnel walls must also be modelled in the BEM.[47] A similar approach was recently applied for the prediction of the cavitating flow through a water-jet pump.[38]

9.4.1. *Super-cavitating propellers*

The method has also been extended in the case of super-cavitating hydrofoils and propellers, of which the blade sections have a sharp leading edge and a non-zero thickness trailing edge.[48] It should be noted that in predicting the performance of these propellers correctly, it is essential

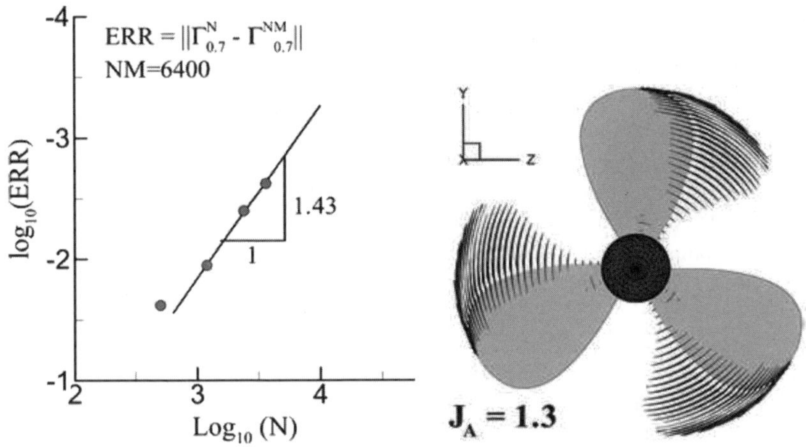

Fig. 9.18. Rate of convergence of predicted circulation distribution on a super-cavitating propeller, of which the predicted cavity shapes are shown on the right; from Young & Kinnas.[48]

that cavities be allowed on both sides of the blade, and that the cavity detachment is searched for. Fig. 9.18 shows the rate of convergence of the results of the BEM and the predicted cavities in the case of a super-cavitating propeller.

9.5. Comparisons with Experiments

Several comparisons of results from the presented BEM, applied on 3-D hydrofoils or propellers, with experimental measurements or observations are provided in Figs. 9.19 to 9.22.

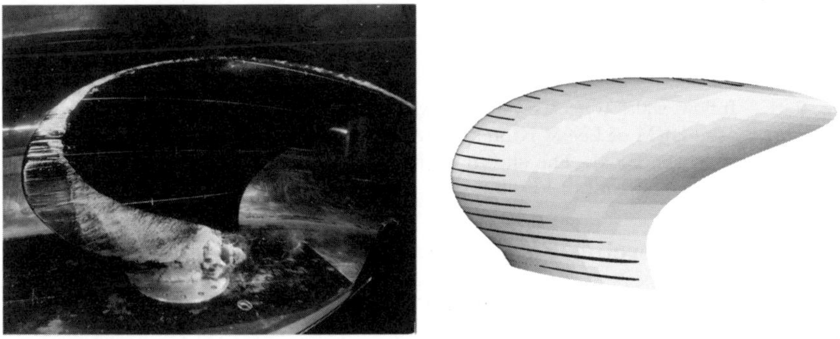

Fig. 9.19. 4990 propeller hydrofoil observed in experiment performed at the 36 inch cavitation tunnel at NSWC-CD (left), and predicted (right). The effects of the tunnel walls have been modelled in the BEM.

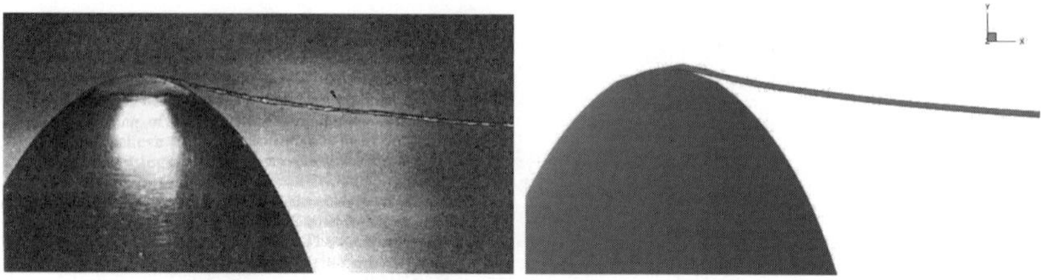

Fig. 9.20. Comparison of observed developed tip vortex cavity in experiment (left)[49] and predicted (right) by the BEM of Lee & Kinnas.[46]

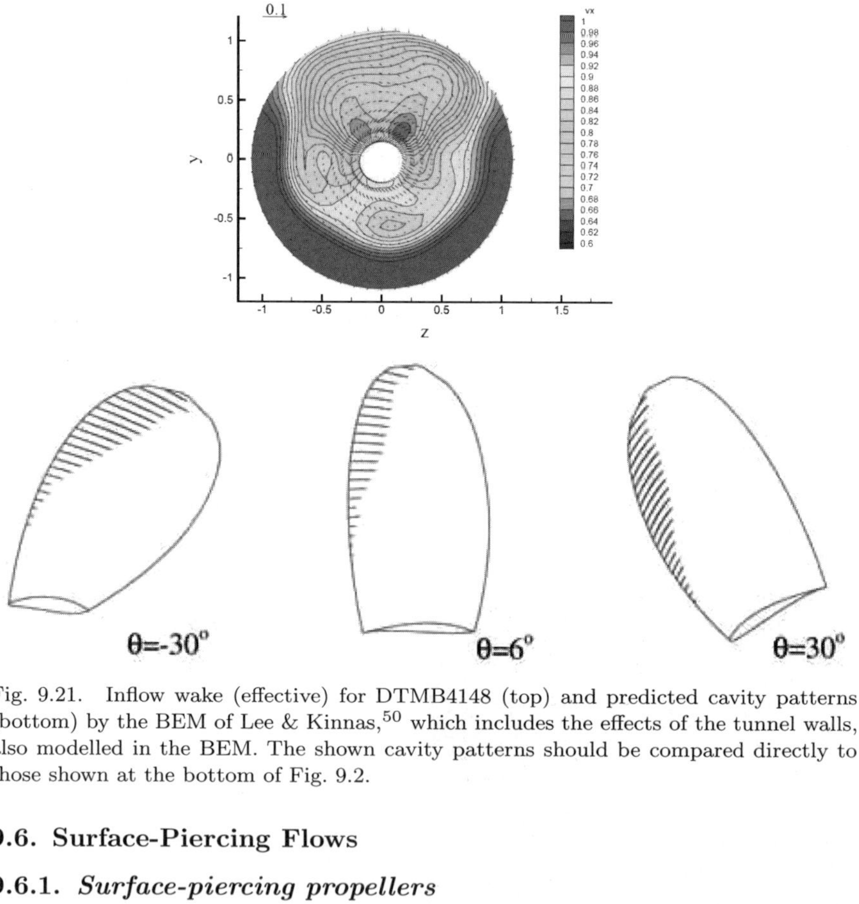

Fig. 9.21. Inflow wake (effective) for DTMB4148 (top) and predicted cavity patterns (bottom) by the BEM of Lee & Kinnas,[50] which includes the effects of the tunnel walls, also modelled in the BEM. The shown cavity patterns should be compared directly to those shown at the bottom of Fig. 9.2.

9.6. Surface-Piercing Flows

9.6.1. *Surface-piercing propellers*

The effect of freesurface on *submerged* cavitating hydrofoils or propellers may be determined by modelling the free-surface via the BEM, as was done in Bal *et al.*,[51] Bal & Kinnas.[52]

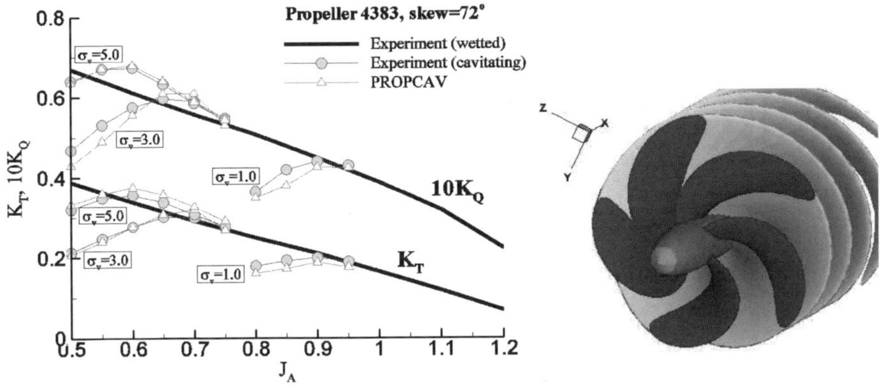

Fig. 9.22. Predicted by the BEM of Kinnas *et al.*[41] and measured thrust and torque on propeller DTMB4383 (panelled geometry shown on the right) versus advance ratio J_A ($J_A = V_A/nD$; V_A is the uniform inflow velocity), at various cavitation numbers. Note significant loss of the thrust and torque at lower J_A, due to the presence of supercavitation. PROPCAV is the name of the corresponding source code.

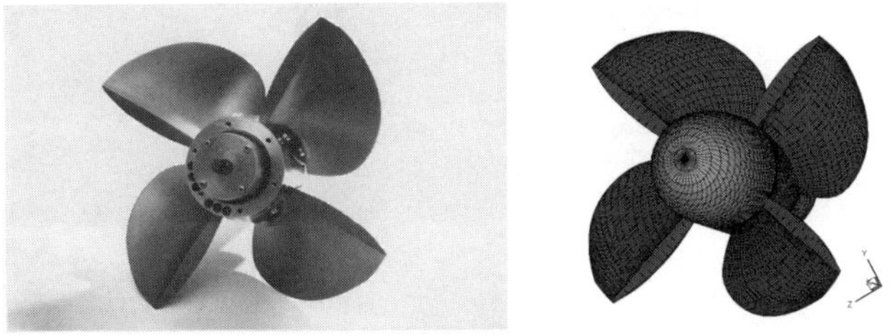

Fig. 9.23. Propeller model 841-B from Olofsson[56] with the corresponding BEM model on the right (from Young & Kinnas[54]).

In the case of surface-piercing (also called partially submerged) propellers, special consideration must be given to the treatment of the flow as the blade enters into or exits from the free-surface. The actual geometry of a surface-piercing propeller is shown in Fig. 9.23, together with the geometry used within a BEM model.

Young[53] and Young & Kinnas[54] modified the BEM for submerged propellers (as presented in the previous sections) by making the fundamental assumption that the free-surface is a flat surface, thus neglecting the influence of the free-surface jets and spray formed as a part of the ventilated surface. The free-surface boundary conditions are linearized and their validity is ensured via the *negative image* method, as depicted in Fig. 9.24. The model is based on that developed for surface-piercing

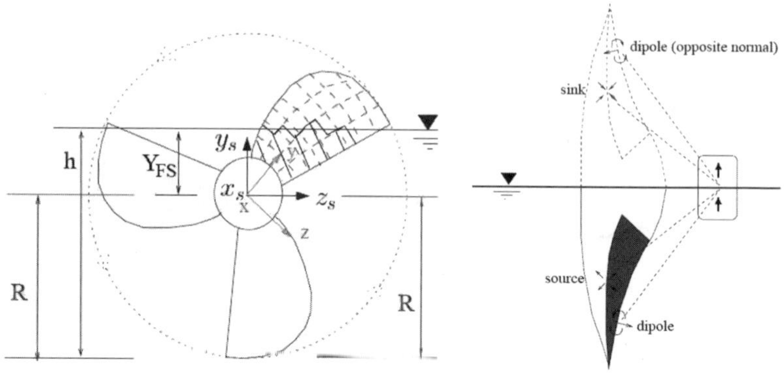

Fig. 9.24. Surface-piercing propeller model using a linearized free-surface condition: discretized propeller geometry in BEM (left), and depiction of the utilized 'negative' image model (right).

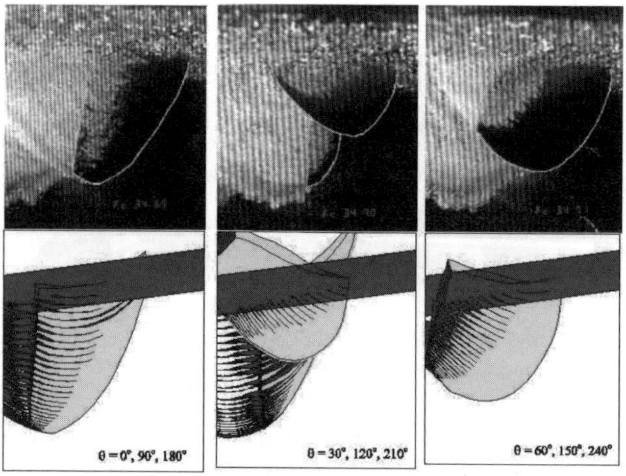

Fig. 9.25. Ventilated patterns on surface-piercing propeller: observed (top) and predicted by BEM (bottom).

hydrofoils,[55] and applies for large Froude numbers, $F_{nD} = \frac{V}{\sqrt{gD}} > 4$, where V is the ship speed and D is the propeller diameter.

Comparisons of the predicted ventilated surfaces with those observed in Olofsson,[56] are presented in Fig. 9.25. As reported in Young[53] and Young & Kinnas,[54] even though the mean performance of a surface-piercing propeller can be predicted within reasonable accuracy, the time history of the blade forces at the entry and exit stages seem to be inaccurate, and that has been attributed to the lack of the modelling of the jets during those stages. A BEM to model the effects of the jets is presented in the next section.

Finally, as shown in the measurements of Olofsson,[56] the effects of *hydro-elasticity* can be significant, and should be included by coupling the present BEM with a method for the structural analysis of the blade, by using a similar approach to that presented by Young & Liu[57] in the case of super-cavitating propellers.

9.6.2. *Surface-piercing 2-D hydrofoils: entry stage*

This section presents a BEM to model the water entry of a surface-piercing hydrofoil by employing the *non-linear* free-surface boundary conditions. The BEM scheme used here was developed initially in the context of the roll-motion of Floating-Production-Storage-Offloading vessel hull-sections, as described in Vinayan *et al.*[58] and Vinayan & Kinnas.[59] Only a brief summary of the mathematical and numerical formulation is presented here and the complete details of the scheme can be found in Vinayan & Kinnas.[60] The approach combines the mixed Eulerian-Lagrangian method of Longuet-Higgins & Cokelet,[61] for tracking the free-surface, with the solution of a Boundary Integral Equation (BIE) at each time-step, and is similar to to that presented by Young[53] and Young & Kinnas,[62] and to that developed by Zhao & Faltinsen[63] for the study of the vertical entry of two-dimensional wedge shaped hulls at small angles of deadrise.

9.6.2.1. *Formulation*

Consider a rigid, 2-D hydrofoil entering an initially calm domain with a constant velocity $\vec{V}$ and an angle of attack α, as shown in Fig. 9.26. An ideal fluid is considered and the flow is assumed to be irrotational. A *fixed* (non-rotating) Cartesian coordinate system is chosen to represent the flow with its origin at the undisturbed water level. The flow is represented in terms of a harmonic function $\phi(\mathbf{x}, t)$ commonly referred to as the *velocity potential.* Also, the local fluid velocity is given as $\vec{q}(\mathbf{x}, t) = \nabla\phi = (\phi_x, \phi_y) = (u, v)$. Here, $\mathbf{x} = (x, y)$ represents the spatial location with respect to the fixed coordinate system, with x being the horizontal measure and y the vertical measure positive upward. The fluid domain and the corresponding boundary surfaces are shown in Fig. 9.26. $S_{\mathbb{WB}}(t)$ represents the '*wetted*' part of the hydrofoil surface, $S_{\mathbb{F}}(t)$ is the free-surface that also includes a part of the ventilated surface on the suction side of the

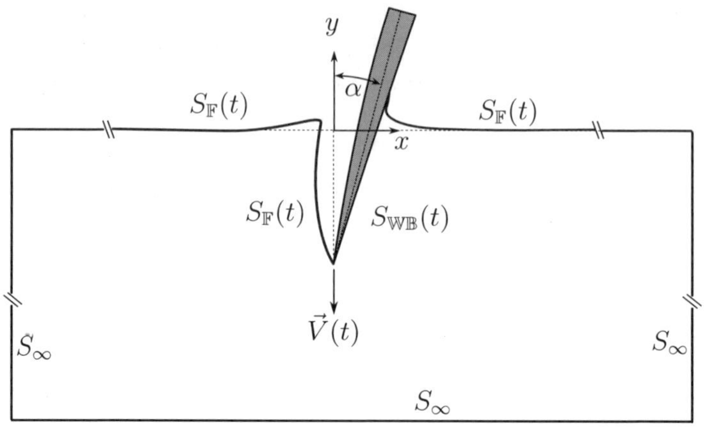

Fig. 9.26. Entry of a surface-piercing hydrofoil: fluid domain and boundaries.

hydrofoil and S_∞ is the far-field boundary placed *far* enough to minimize interference.

A BIE is solved at each time-step of a higher-order time-stepping scheme in order to obtain the velocity potential. Once the solution is obtained for a particular time, the time-dependent boundary conditions are updated and the solution scheme progresses onto the next one.

The boundary value problem for the velocity potential is converted into a BIE by introducing a two-dimensional Green's function $G(\mathbf{p},\mathbf{q}) = -\frac{1}{2\pi}\ln r_{\mathbf{pq}}$ (satisfying the Laplace equation), where $r_{\mathbf{pq}} = |\mathbf{p}-\mathbf{q}|$, $\mathbf{p} \equiv \mathbf{p}(\mathbf{x})$ is the field point and $\mathbf{q} \equiv \mathbf{q}(\mathbf{x})$ is the source point. The BIE obtained by applying Green's third identity to $\phi(\mathbf{x},t)$ and $G(\mathbf{p},\mathbf{q})$ is

$$\alpha(\mathbf{p})\phi(\mathbf{p}) + \int_\Gamma \phi(\mathbf{q})G_n(\mathbf{p},\mathbf{q})\, d\Gamma_{\mathbf{q}} = \int_\Gamma G(\mathbf{p},\mathbf{q})\phi_n(\mathbf{q})\, d\Gamma_{\mathbf{q}}, \tag{9.43}$$

where $2\pi\alpha(\mathbf{p})$ is the internal angle formed at the boundaries. $G_n(\mathbf{p},\mathbf{q}) = \nabla G(\mathbf{p},\mathbf{q}) \cdot \vec{n}_{\mathbf{q}}$ and $\phi_n(\mathbf{p},\mathbf{q}) = \nabla\phi(\mathbf{p},\mathbf{q}) \cdot \vec{n}_{\mathbf{q}}$ and $\vec{n}_{\mathbf{q}}$ is the normal vector at $\mathbf{q}$, positive out of the fluid.

The above BIE is solved via a BEM which employs linear elements (straight segments) with *linear* dipole and source distributions over each element. The numerical implementation of this BEM is described in more detail in Vinayan & Kinnas,[59] Vinayan & Kinnas[60] and Vinayan.[64]

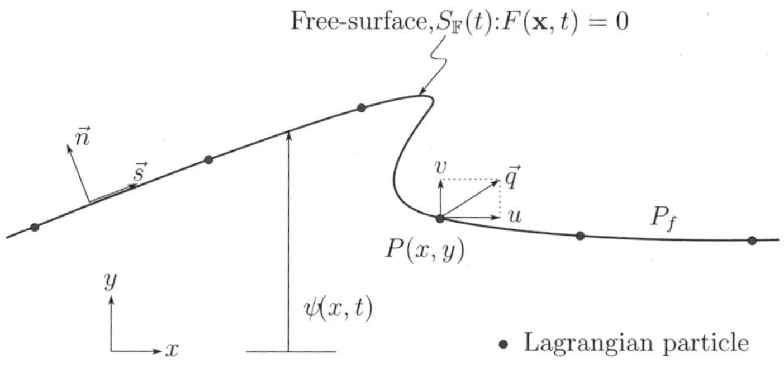

Fig. 9.27. Free-surface schematic

• *Kinematic Boundary Condition (KBC) on* $S_{\mathbb{F}}(t)$*:*
The KBC is obtained by assuming $S_{\mathbb{F}}(t)$ to be a bounding surface, i.e. no material passes across the free-surface.[65] Based on the schematic shown in Fig. 9.27, if we represent the free-surface as $F(\mathbf{x},t) = y - \eta(x,t) = 0$, the KBC on the free-surface is given as[65,66]

$$\frac{D}{Dt}F(\mathbf{x},t) = 0, \tag{9.44}$$

where $\frac{D}{Dt} = \frac{\partial}{\partial t} + \nabla\phi \cdot \nabla$ is the material derivative. $\vec{q} = \nabla\phi = (\phi_x, \phi_y)$ is the fluid velocity on the free-surface, and $y = \eta(x,t)$ is the free-surface elevation.

For any particle $P(x,y)$ on the free-surface, the KBC is given as[61,65,66]

$$\frac{D\mathbf{x}}{Dt} = \nabla\phi = \vec{q} \quad \text{or} \quad \left\{ \begin{array}{l} \dfrac{Dx}{Dt} = u = \phi_x \\ \dfrac{Dy}{Dt} = v = \phi_y \end{array} \right\} \quad \mathbf{x} \in S_{\mathbb{F}}(t) \tag{9.45}$$

The DBC is obtained from the Bernoulli's equation and assuming the pressure to be continuous across the free-surface. It is assumed that the wavelength of the free-surface elevation is long enough to neglect the effects of surface tension. Thus, the pressure underneath the free-surface must equal the atmospheric pressure above, giving the most general form of the free-surface DBC

$$\frac{\partial\phi}{\partial t} + \frac{1}{2}|\nabla\phi|^2 + g\eta + \frac{P_f}{\rho} = 0, \quad \mathbf{x} \in S_{\mathbb{F}}(t) \tag{9.46}$$

where g is the acceleration due to gravity. The common form of the DBC is obtained by expressing the pressure as gage pressure, in which case the pressure on the free-surface, $P_f = P - P_{atm} = 0$. For a Lagrangian particle $P(x, y)$, the DBC can be rewritten as

$$\frac{D\phi}{Dt} = \frac{1}{2}|\nabla\phi|^2 - g\eta = \frac{1}{2}q^2 - g\eta, \quad \mathbf{x} \in S_{\mathbb{F}}(t) \tag{9.47}$$

where $q = |\vec{q}| = |\nabla\phi|$

• *Boundary Condition on Hydrofoil* $S_{\mathbb{WB}}(t)$

On the 'wetted' part of the hydrofoil surface $S_{\mathbb{WB}}(t)$,

$$\nabla\phi \cdot \vec{n} = \vec{V}(t) \cdot \vec{n}, \quad \mathbf{x} \in S_{\mathbb{WB}}(t) \tag{9.48}$$

where $\vec{V}(t)$ is the prescribed velocity of the hydrofoil.

• *Boundary Condition on Far-field Boundary* S_∞

The far-field boundary S_∞ is assumed to be a no-flux surface with

$$\nabla\phi \cdot \vec{n} = 0; \quad \mathbf{x} \in S_\infty \tag{9.49}$$

Special attention is paid in placing the boundary far away from the body to avoid reflection of the waves generated by its motion.

9.6.2.2. *Algorithm of marching free-surface in time*

A fourth-order Runge-Kutta scheme is employed in order to integrate equations (9.45) and (9.47) in time, and determine the free-surface location, $\vec{x}(t)$ and the corresponding potential $\phi(t)$. The following steps start after the end of time step t and end at the next time step $t + \Delta t$. Please note that each time the free-surface location is updated, the BEM, equation (9.43), is solved.

Step 1: Solve the BEM on $\vec{x}(t)$ with known $\phi(t)$ to find $\left(\frac{\partial\phi}{\partial n}\right)_1$, and[j] $\vec{q}_1$.

$$\vec{x}_1 = \vec{x}(t) + \vec{q}_1 \frac{\Delta t}{2} \tag{9.50}$$

$$\phi_1 = \phi(t) + \left[\frac{1}{2}q_1^2 - g\eta(t)\right] \frac{\Delta t}{2} \tag{9.51}$$

[j]The derivative $\left(\frac{\partial\phi}{\partial s}\right)$ is determined by using $\phi(t)$ and the arc-length s along the free surface.

Step 2: Solve the BEM on $\vec{x}_1$ with known ϕ_1 to find $\left(\frac{\partial \phi}{\partial n}\right)_2$, and $\vec{q}_2$.

$$\vec{x}_2 = \vec{x}(t) + \vec{q}_2 \frac{\Delta t}{2} \tag{9.52}$$

$$\phi_2 = \phi(t) + \left[\frac{1}{2}q_2^2 - g\eta_1\right]\frac{\Delta t}{2} \tag{9.53}$$

Step 3: Solve the BEM on $\vec{x}_2$ with known ϕ_2 to find $\left(\frac{\partial \phi}{\partial n}\right)_3$, and $\vec{q}_3$.

$$\vec{x}_3 = \vec{x}(t) + \vec{q}_3 \Delta t \tag{9.54}$$

$$\phi_3 = \phi(t) + \left[\frac{1}{2}q_3^2 - g\eta_2\right]\Delta t \tag{9.55}$$

Step 4: Solve the BEM on $\vec{x}_3$ with known ϕ_3 to find $\left(\frac{\partial \phi}{\partial n}\right)_4$, and $\vec{q}_4$. Then determine the location of the free surface at $t + \Delta t$ and the corresponding potential by using the following equations:

$$\begin{aligned}
\vec{x}(t+\Delta t) &= \vec{x}(t) + \frac{\Delta t}{6}\left[\vec{q}_1 + 2\vec{q}_2 + 2\vec{q}_3 + \vec{q}_4\right] \\
\phi(t+\Delta t) &= \phi(t) + \frac{\Delta t}{12}\left[q_1^2 + 2q_2^2 + 2q_3^2 + q_4^2\right] \\
&\quad - g\frac{\Delta t}{6}\left[\eta(t) + 2\eta_1 + 2\eta_2 + \eta_3\right]
\end{aligned} \tag{9.56}$$

9.6.2.3. *Results*

The detailed numerical implementation, including the initial conditions and the treatment of the free-surface jets, is given in Vinayan & Kinnas.[60]

Some results from the present BEM are shown in Fig. 9.28, together with those from using the commercial RANS solver, FLUENT, and those observed in the experiment of Cox.[67] The specifics of the FLUENT run are given in Vinayan & Kinnas.[60]

It is worth noting that the ventilated surfaces and the jets predicted from the BEM and from FLUENT are very close to each other, and also quite close to those observed. The total CPU time for the BEM on a single processor is about 16 minutes. In comparison, the FLUENT simulations take about 36 hours (wall-time), run parallel on six processors (processor: 64-bit 1.6GHz AMD Opteron).

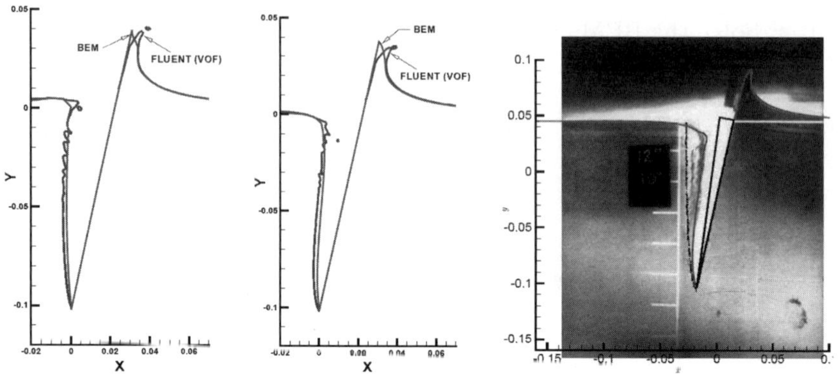

Fig. 9.28. Predicted ventilated surface free-surface elevations for a surface-piercing hydrofoil by using a BEM[60] and a commercial viscous flow solver, Fluent, with the effects of gravity excluded (left) or included (middle). The observed ventilated surface, free-surface elevations, and hydrofoil[67] are shown on the right, with the surfaces predicted by Fluent, superimposed. The wedge-shaped 6-inch long hydrofoil enters the free-surface at an angle of 10° with respect to the vertical, and with the vertical speed of 2.45 m/sec.

9.7. Concluding Remarks

BEMs, as applied in the case of steady and unsteady cavitating flows around hydrofoils and propellers, were presented. Only the modelling of sheet and developed tip vortex cavitation was addressed in this chapter. The effects of viscosity can be incorporated by coupling the BEM with a boundary layer solver in the case the flows are attached. The interaction with the viscous inflow and adjacent boundaries (hulls, pods, ducts, rudders), was only briefly addressed in this chapter, and can be accounted for via coupling of the BEM with RANS solvers.[68–71] Coupling techniques for isolated bubbles (modelled via a BEM) with the viscous inflow have also been developed by Chahine[72] and Hsiao & Chahine.[73]

Finally, the effects of the free-surface can also be modelled via BEM, by solving for the location of the free-surface. In the case of surface-piercing hydrofoils, BEM can predict the flow as accurately as viscous flow solvers, by using limited computer resources and by requiring only a very small fraction of the required CPU for the viscous flow calculations.

9.8. Acknowledgements

The related research was funded by the Consortium on Cavitation Performance of High Speed Propulsors with current members: American

Bureau of Shipping, USA; Daewoo Shipbuilding and Marine Engineering Co. Ltd., Korea; Kawasaki Heavy Industries Ltd., Japan; Naval Surface Warfare Center Carderock Division and Office of Naval Research (contracts N00014-04-1-0287 and N00014-07-1-0616), USA; Rolls-Royce Marine AS, Norway; Rolls-Royce Marine AB, Sweden; Samsung Heavy Industries Co. Ltd., Korea; SSPA AB, Sweden; Andritz Hydro, Germany; Wärtsilä Propulsion AS, Norway; Wärtsilä Propulsion BV, The Netherlands; Wärtsilä Lips Defense SAS., France and Wärtsilä Propulsion CME Zhenjiang Propeller Co. Ltd., PRC. The more recent work on surface-piercing propellers was supported by the US Office of Naval Research (contract N00014-07-1-0932). Finally, I would like to thank Dr. Vimal Vinayan of the Ocean Engineering Group for his great help with this chapter.

References

1. H. Kato, *Cavitation.* In: Advances in Marine Hydrodynamics, Ohkusu M. Chapter 5, Computational Mechanics Publications, pp. 233–277 (1996).
2. J. Franc and J. Michel, Attached cavitation and the boundary layer: Experimental investigation and numerical treatment, *Journal of Fluid Mechanics* **154**, pp 63–90 (1985).
3. Q. Le, J. Franc, and J. Michel, Partial cavities: Global behavior and mean pressure distribution, *Journal of Fluids Engineering* **115**, pp. 243–248 (1993).
4. Q. Le, J. Franc, and J. Michel, Partial cavities: Pressure pulse distribution around cavity closure, *Journal of Fluids Engineering* **115**, pp. 249–254 (1993).
5. H. Kato, Recent advances in cavitating foil research, In: *International Conference on Hydrodynamics*, Wuxi, China, pp. 80–89, (1994).
6. W. Brewer and S. Kinnas, Experiment and viscous flow analysis on a partially cavitating hydrofoil, *Journal of Ship Research* **41**(3), pp. 161–171 (1997).
7. G. Birkhoff and E. Zarantonello, *Jets, Wakes and Cavities.* (Academic Press Inc., New York, 1957).
8. S. Mishima, S. Kinnas, and D. Egnor, The CAvitating PRopeller EXperiment (CAPREX), Phases I & II. Technical report, Department of Ocean Engineering, MIT (August, 1995).
9. M. Tulin, Steady two-dimensional cavity flows about slender bodies, Technical Report 834, DTMB (May, 1953).
10. M. Tulin and C. Hsu, New applications of cavity flow theory, In: *13th Symposium on Naval Hydrodynamics*, Tokyo, Japan, (1980).
11. S. Kinnas, Leading-edge corrections to the linear theory of partially cavitating hydrofoils, *Journal of Ship Research* **35**(1), pp. 15–27 (1991).
12. S. Kinnas and N. Fine, A numerical nonlinear analysis of the flow around two- and three-dimensional partially cavitating hydrofoils, *Journal of Fluid Mechanics* **254**, 151–181 (1993).

13. J. Uhlman, The surface singularity method applied to partially cavitating hydrofoils, *Journal of Ship Research* **31**(2), pp. 107–124 (1987).
14. J. Uhlman, The surface singularity or boundary integral method applied to supercavitating hydrofoils, *Journal of Ship Research* **33**(1), pp. 16–20 (1989).
15. H. Yamaguchi and H. Kato, On application of nonlinear cavity flow theory to thick foil sections, In: *Second International Conference on Cavitation*, pp. 167–174. IMechE, (1983).
16. H. Lemonnier and A. Rowe, Another approach in modelling cavitating flows, *Journal of Fluid Mechanics* **195**, (1988).
17. A. Rowe and O. Blottiaux, Aspects of modeling partially cavitating hydrofoils, *Journal of Ship Research* **37**(1), pp. 34–48, (1993).
18. S. Kinnas and N. Fine, Non-Linear Analysis of the Flow Around Partially or Super-Cavitating Hydrofoils by a Potential Based Panel Method, *Boundary Integral Methods-Theory and Applications, Proceedings of the IABEM-90 Symposium, Rome, Italy, October 15–19, 1990*, pp. 289–300, (Heidelberg, Springer-Verlag, 1991).
19. C.-S. Lee, Y.-G. Kim, and J.-T. Lee, A potential-based panel method for the analysis of a two-dimensional super- or partially- cavitating hydrofoil, *Journal of Ship Research* **36**(2), 168–181 (1992).
20. C. Pellone and A. Rowe. Supercavitating hydrofoils in non-linear theory. *Third International Conference on Numerical Ship Hydrodynamics*, Paris, France (June, 1981). Basin d'essais des Carènes.
21. N. Fine and S. Kinnas, A boundary element method for the analysis of the flow around 3-D cavitating hydrofoils, *Journal of Ship Research* **37**, 213–224 (1993).
22. Y.-G. Kim, C.-S. Lee, and J.-C. Suh. Surface panel method for prediction of flow around a 3-D steady or unsteady cavitating hydrofoil, *Second International Symposium on Cavitation*, pp. 113–120, Tokyo, Japan (April 5–7, 1994).
23. C. Pellone and J. Peallat, Non-linear analysis of three-dimensional partially cavitating hydrofoil, *CAV'95 International Symposium on Cavitation*, pp. 433–440, Deauville, France (May 2-5, 1995).
24. S. Kinnas and N. Fine, A nonlinear boundary element method for the analysis of unsteady propeller sheet cavitation, *Nineteenth Symposium on Naval Hydrodynamics*, pp. 717–737, Seoul, Korea (August, 1992).
25. N. Fine and S. Kinnas, The nonlinear numerical prediction of unsteady sheet cavitation for propellers of extreme geometry, *Sixth International Conference On Numerical Ship Hydrodynamics*, pp. 531–544, University of Iowa, Iowa (August, 1993).
26. Y.-G. Kim and C.-S. Lee, Prediction of unsteady performance of marine propellers with cavitation using surface-panel method, *Twenty-first Symposium on Naval Hydrodynamics*, Trondheim, Norway (June, 1996).
27. S. Kinnas, S. Mishima, and W. Brewer, Nonlinear analysis of viscous flow around cavitating hydrofoils, *Twentieth Symposium on Naval Hydrodynamics*, pp. 446–465, University of California, Santa Barbara (August, 1994).

28. A. Kubota, H. Kato, and H. Yamaguchi, Finite difference analysis of unsteady cavitation on a two-dimensional hydrofoil, *Fifth International Conference on Numerical Ship Hydrodynamics*, pp. 667–683, Hiroshima, Japan (September, 1989).
29. M. Deshpande, J. Feng, and C. Merkle, Navier–Stokes analysis of 2-D cavity flow, *Cavitation and Multiphase Flow Forum*, pp. 149–155. FED-154, ASME, (1993).
30. S. H. Rhee, T. Kawamura, and H. Y. Li, Propeller cavitation study using an unstructured grid based Navier–Stokes solver, *Journal of Fluids Engineering.* **127**, 986–994, (2005).
31. L. Morino and C.-C. Kuo, Subsonic Potential Aerodynamic for Complex Configurations: A General Theory, *AIAA Journal.* **12**(2), pp 191–197 (February, 1974).
32. P. Krishnaswamy, P. Andersen, and S. Kinnas, Re-entrant jet modeling for partially cavitating two-dimensional hydrofoisl, *Fourth International Symposium on Cavitation*, Pasadena, CA (June 20–23, 2001).
33. H. Sun, Performance Prediction of Cavitating Propulsors using a Viscous/Inviscid Interaction Method, PhD thesis, Ocean Engineering Group, Department of Civil, Architectural Environmental Engineering, The University of Texas at Austin (August, 2008).
34. N. E. Fine, Nonlinear Analysis of Cavitating Propellers in Nonuniform Flow, PhD thesis, Department of Ocean Engineering, MIT, (October, 1992).
35. M. Drela, Integral boundary layer formulation for blunt trailing edges, *AIAA Paper.* **89-2200**, (1989).
36. H. Sun and S. A. Kinnas, Simulation of sheet cavitation on propulsor blades using a viscous/inviscid interactive method, *CAV2006: Sixth International Symposium on Cavitation*, Wageningen, The Netherlands (September, 2006).
37. G. Hufford, M. Drela, and J. Kerwin, Viscous flow around marine propellers using boundary-layer strip theory, *Journal of Ship Research.* **38**(1), pp. 52–62 (1994).
38. H. Sun and S. A. Kinnas, Performance prediction of cavitating water-jet propulsors using a viscous/inviscid interactive method, *Transactions SNAME.* **116**, (2008).
39. Y. L. Young and S. A. Kinnas, A bem for the prediction of unsteady mid-chord face and/or back propeller cavitation, *Journal of Fluids Engineering.* **123**, 311–319 (2001).
40. H. Arakeri, Viscous effects on the position of cavitation separation from smooth bodies, *Journal of Fluid Mechanics.* **68**(4), 779–799 (1975).
41. S. A. Kinnas, J.-K. Choi, H. S. Lee, Y. L. Young, H. Gu, K. Kakar, and S. Natarajan, Prediction of cavitation performance of single or multi-component propulsors and their interaction with the hull, *Transactions SNAME.* **110**, 215–244 (2002).
42. C. L. Warren, T. E. Taylor, and J. E. Kerwin. A coupled viscous/potential-flow method for the prediction of propulsor-induced maneuvering forces. *Propellers/Shafting '00 Symposium*, Virginia Beach, VA (September, 2000). The Society of Naval Architects & Marine Engineers.

43. J.-K. Choi and S. A. Kinnas, Prediction of unsteady effective wake by an euler solver/vortex-lattice coupled method, *Journal of Ship Research* **47**(2), 131–144 (2003).
44. S. Kinnas and C.-Y. Hsin, A boundary element method for the analysis of the unsteady flow around extreme propeller geometries, *AIAA Journal* **30**(3), 688–696 (1992).
45. H. S. Lee and S. A. Kinnas, Fully unsteady wake alignment for propellers in non-axisymmetric flows, *Journal of Ship Research* **49**(3), 176–190 (2005).
46. H. S. Lee and S. A. Kinnas, Application of BEM in the prediction of unsteady blade sheet and developed tip vortex cavitation on marine propellers, *Journal of Ship Research* **48**(1), 15–30 (2004).
47. J.-K. Choi and S. Kinnas, Numerical model of a cavitating propeller inside of a tunnel, *Journal of Fluids Engineering* **121**, 297–304 (1999).
48. Y. L. Young and S. A. Kinnas, Analysis of supercavitating and surface-piercing propeller flows via bem, *Journal of Computational Mechanics* **32**(4–6), 269–280 (2003).
49. R. Arndt, V. Arakeri, and H. Higuchi, Some observations of tip-vortex cavitation, *Journal of Fluid Mechanics* **229**, 269–289, (1991).
50. H. S. Lee and S. A. Kinnas, A BEM for the modeling of unsteady propeller sheet cavitation inside a cavitation tunnel, *Journal of Computational Mechanics* **37**(1), 41–51, (2005).
51. S. Bal, S. A. Kinnas, and H. S. Lee, Numerical analysis for 2-D and 3-D cavitating hydrofoils under a free surface, *Journal of Ship Research.* **45**, 34–49, (2001).
52. S. Bal and S. A. Kinnas, A BEM for the prediction of free surface effects on cavitating hydrofoils, *Journal of Computational Mechanics* **28**(3–4), 260–274 (2002).
53. Y. L. Young, Numerical modeling of super-cavitating and surface-piercing propellers, PhD thesis, Department of Civil Engineering, The University of Texas at Austin (May, 2002).
54. Y. L. Young and S. A. Kinnas, Performance prediction of surface-piercing propellers, *Journal of Ship Research* **48**(4), 288–304 (2004).
55. C. Savineau and S. Kinnas, A numerical formulation applicable to surface piercing hydrofoils and propellers, *24th American Towing Tank Conference*, (Texas A&M University, College Station, TX, 1995).
56. N. Olofsson. Force and Flow Characteristics of a Partially Submerged Propeller, PhD thesis, Department of Naval Architecture and Ocean Engineering, Chalmers University of Technology, Göteborg, Sweden (February, 1996).
57. Y. Young and Z. Liu, Performance prediction of newton-rader propellers, *Journal of Ship Research* **52**(2), 124–145 (June, 2008).
58. V. Vinayan, S. A. Kinnas, and Y.-H. Yu, Modeling of the flow around fpso hull sections subject to roll motion: Effect of nonlinear boundary conditions, *Proceedings 24th Int. Conference of Offshore Mech. and Arct. Eng. (OMAE2005)*, Halkidiki, Greece **3**, 805–815, (2005).

59. V. Vinayan and S. A. Kinnas, A BEM for the propagation of nonlinear planar free-surface waves, *Electronic Journal of Boundary Elements* **5**(1), pp 17–40, (2007).
60. V. Vinayan and S. A. Kinnas, Numerical modeling of surface piercing hydrofoils and propellers, *27th Symposium on Naval Hydrodynamics, Seoul, Korea* (5–10 October, 2008).
61. M. S. Longuet-Higgins and E. D. Cokelet, The deformation of steep surface waves on water. I. A numerical method of computation, *Proceedings of the Royal Society of London. Series A, Mathematical and Physical Sciences* **350** (1660), 1–26 (July 30, 1976).
62. Y. L. Young and S. A. Kinnas, A bem technique for the modeling of supercavitating and surface-piercing propeller flows, *Twenty fourth Symposium on Naval Hydrodynamics*, Fukuoka, Japan (July, 2002).
63. R. Zhao and O. Faltinsen, Water entry of two-dimensional bodies, *Journal of Fluid Mechanics* **246**, pp. 593–612 (1993).
64. V. Vinayan, A boundary element method for the strongly nonlinear analysis of ventilating water-entry and wave-body interaction problems, PhD thesis, Ocean Engineering Group, Department of Civil Engineering, Architectural and Environmental Engineering (University of Texas at Austin, Austin, Texas, 2009).
65. E. B. Dussan V., On the difference between a bounding surface and a material surface, *Journal of Fluid Mechanics* **75**(4), pp. 609–623 (1976).
66. J. V. Wehausen and E. V. Laitone, Surface waves, *Handbuch der Physik* **9**, 446–778, (1960).
67. B. Cox, Hydrofoil theory for vertical water entry. PhD thesis, Department of Naval Architecture, MIT (1971).
68. S. A. Kinnas, H. S. Lee, H. Gu, and Y. Deng, Prediction of performance and design via optimization of ducted propellers subject to non-axisymmetric inflows, *Transactions SNAME.* **113**, 99–121 (2005).
69. J. E. Kerwin, T. J. Michael, and S. K. Neely, Improved algorithms for the design/analysis of multi-component complex propulsors, *Eleventh propeller/Shafting Symposium, Society of Naval Architects and Marine Engineers*, Williamsburg, Virginia, USA (September, 2006).
70. S. A. Kinnas, H. S. Lee, B. Mishra, L. He, S.-H. Rhee, and B. Sasunapuri, *Eleventh Propellers/Shafting Symposium (PROPS), Society of Naval Architects and Marine Engineers*, Williamsburg, Virginia, USA (September, 2006).
71. S. A. Kinnas, H. S. Lee, H. Gu, and S. Natarajan, Prediction of sheet cavitation on a rudder subject to propeller flow, *Journal of Ship Research* **51**, 65–75 (March, 2007).
72. G. Chahine, Numerical simulation of bubble flow interactions, Paper presented at WIMRC 2nd International Cavitation Forum, Turbo-machinery and Medical Applications (Warwick University, UK, 2008).
73. C.-T. Hsiao and G. Chahine, Numerical study of cavitation inception due to vortex/vortex interaction in a ducted propulsor, *Journal of Ship Research* (2008), in publication.

[illegible] (2007).
77. [illegible] presented at AVMFC 2nd International Conference [illegible] and Medical Applications (Warwick University, UK, 2008).
78. C. J. Wang and C. Chandler, [illegible] interaction in a ducted propeller [illegible] (2009), in publication.

Chapter 10

CONDITION NUMBERS AND LOCAL ERRORS IN THE BOUNDARY ELEMENT METHOD

W. Dijkstra

Eindhoven University of Technology, Department of Mathematics and Computing Science, P.O. Box 513, 5600 MB Eindhoven, The Netherlands,

G. Kakuba and R.M.M. Mattheij

Mecal, P.O. Box 375, 5500 AJ Veldhoven, The Netherlands,
willemdijkstra1979@gmail.com, g.a.kakuba@tue.nl, r.m.m.mattheij@tue.nl

In this chapter we investigate local errors and condition numbers in the BEM. The results of these investigations are important in guiding adaptive meshing strategies and the solvability of linear systems in the BEM. We show that the local error for the BEM with constant or linear elements decreases quadratically with the boundary element mesh size. We also investigate better ways of treating boundary conditions to reduce the local errors. The results of our numerical experiments confirm the theory. The values of the condition numbers of the matrices that appear in the BEM depend on the shape and size of the domain on which a problem is defined. For certain critical domains, these condition numbers can even become infinitely large. We show that this holds for several classes of boundary value problems and propose a number of strategies to guarantee moderate condition numbers.

10.1. Introduction

The subject of errors in the boundary element method (BEM) is still a very interesting one and some aspects have not yet been as explored as they have been in other numerical methods like the finite element method (FEM) and the finite difference method. Errors in BEM solutions may be due to discretization or to inaccuracies in the solver that involves the use of BEM matrices with high condition numbers. For a given discretization, there are several ways to implement the BEM because of the choice in collocation and nodal points. These will all influence the resulting error in the solution. In most cases where error measurement has been performed, like in adaptive refinement, the main focus has been a measure of the error rather than a fundamental analysis of the error and its behaviour. This chapter presents recent results on such fundamental analysis on local errors in BEM solutions. The results presented will not only be helpful in choosing an implementation strategy but also in guiding adaptive refinement techniques.

Several techniques have been used to measure BEM errors in the area of adaptive refinement. For instance, the discretization error is estimated by the difference between two solutions obtained using different collocation points but the same discretization.[1] Another technique uses the first (singular integral equation) and the second (hypersingular integral equation) kinds of formulation to provide an error estimate.[2] Data from the first kind of equation is substituted into the second kind to obtain a residual which is then used to estimate the error. Some authors have used a posteriori error estimation in FEM as a guiding tool to develop error estimates for the BEM.[3] Unfortunately, such techniques are usually restricted to the Galerkin BEM. In our case, we would like to start from the basics of the boundary integral equation (BIE) discretization to develop error analyses for collocation BEM for potential problems. Though the ideas could easily be adapted to 3-D, we will discuss 2D problems and the Laplace equation in particular.

It is well known that the Laplace equation in differential form with either Dirichlet or mixed boundary conditions has a unique solution. However, when the Laplace equation is transformed to a BIE it is not straightforward that this carries over to the BIE. It is noted that the BIE for the Dirichlet Laplace equation does not always have a unique solution.[4–7] Certain domains can be distinguished on which the BIE becomes singular and a non-trivial solution of the homogeneous equations can be found. A multiple of this solution can be added to the solution of the non-homogeneous equations, which is then no longer unique. For each domain there exists exactly one rescaled version of this domain for which the BIE becomes singular. This introduces an extraordinary phenomenon for the BIEs; uniqueness of the solutions depends on the scale of the domain.

After discretization of the boundary of the domain, the BIE transforms into a linear system of equations. When the BIE is singular, one may expect that the linear system is also singular, or at least ill-conditioned. This is reflected by the condition number of the system matrix. If this condition number is infinitely large, then the linear system is singular. As a consequence, the linear system does not have a unique solution. If the condition number is not infinitely large, but still large, the linear system is ill-conditioned and difficult to solve. Hence the condition number also greatly affects the local error in the BEM solution. Note that it is a matter of taste to decide what 'large' means in the context of condition numbers.

The domains on which the BIE does not have a unique solution are related to the so-called *logarithmic capacity*. The logarithmic capacity is a real positive number being a function of the domain. This concept originates from the field of measure theory, but it also appears in potential theory. The concept of a capacity applied to a single domain may be a bit confusing, as usually the *electrical* capacity is defined as a charge difference between two conducting objects. The logarithmic capacity, however, is related to a single domain.

In potential theory, it is shown that when the logarithmic capacity of a domain is equal to one, then the homogeneous BIE for the Dirichlet Laplace equation on the boundary of that domain has a non-trivial solution.[8–10] This allows us to *a priori* detect whether a BIE will become singular on a certain domain. Namely, we have to compute the logarithmic capacity and verify whether it is equal to one. Additionally, the logarithmic capacity also enables us to modify the BIE such that it does not become singular. We can scale the domain in such a way that its logarithmic capacity is not equal to one. The BIE on the corresponding boundary will then be nonsingular.

The BIE for the Laplace equation with mixed boundary conditions did not receive much attention until now.[11] However, a similar phenomenon as for the Dirichlet case takes place for mixed conditions. For each domain, there exists exactly one rescaled version of this domain for which the BIE becomes singular. In this chapter, we investigate both the BIE for the Laplace equation with Dirichlet conditions and mixed conditions and the related linear systems. We also extend the research to flow problems in particular to BEM solutions of the Stokes equations.

This chapter is outlined as follows. Section 10.2 gives a recapitulation of the BEM formulation. An outline of the integral equation formulation for a Laplace problem and its discretization is presented. In Section 10.3, we present a discussion on local errors in the BEM and give the theory for the expected convergence rates of the local error for both constant and linear elements. Numerical examples and results for a Dirichlet and mixed problem are presented to illustrate the theory. Then a general insight into the various ways to implement the BEM on a circle is presented.

In Section 10.4, we give a number of illustrative examples of BEM matrices with large condition numbers. This motivates us to study the condition numbers of BEM matrices in potential problems in Section 10.5.

It is shown that these condition numbers may become infinitely large under certain conditions. This is also true for the condition numbers of BEM matrices in flow problems, which is the topic of Section 10.6. The chapter is concluded with a discussion of the results presented.

10.2. BEM Formulation for Potential Problems

In this section, we briefly present a BEM formulation. This, we do here only for potential problems but we will generalize it to Stokes equations in a later section. The BEM is used to approximate solutions of boundary value problems that can be formulated as integral equations. For the sake of clarity, we consider potential problems governed by the Laplace equation on a simply connected domain Ω in 2D with boundary Γ. That is

$$\nabla^2 u(\boldsymbol{r}) = 0, \quad \boldsymbol{r} \in \Omega. \tag{10.1}$$

We will consider problems for which either the function u (Dirichlet) or its outward normal derivative $q := \partial u/\partial n$ (Neumann) is given on Γ, or the mixed problem where on one part of the boundary u is given and on the other q is given. The integral equation formulation of (10.1) reads

$$cu + \mathcal{K}^d u = \mathcal{K}^s q, \quad \boldsymbol{r} \in \Gamma \tag{10.2}$$

Here, $\mathcal{K}^s$ and $\mathcal{K}^d$ are the *single* and *double layer operator*, defined as

$$\begin{aligned}(\mathcal{K}^s q)(\boldsymbol{r}) :&= \int_\Gamma q(\boldsymbol{r}')G(\boldsymbol{r};\boldsymbol{r}')d\Gamma_{r'},\\ (\mathcal{K}^d u)(\boldsymbol{r}) :&= \int_\Gamma u(\boldsymbol{r}')\frac{\partial}{\partial n'}G(\boldsymbol{r};\boldsymbol{r}')d\Gamma_{r'},\end{aligned} \tag{10.3}$$

where $\boldsymbol{r}$ and $\boldsymbol{r}'$ are points at the boundary and n' is the unit outward normal at $\boldsymbol{r}'$ at Γ. The function G,

$$G(\boldsymbol{r};\boldsymbol{r}') := \frac{1}{2\pi}\log\frac{1}{||\boldsymbol{r}-\boldsymbol{r}'||},$$

is the fundamental solution for the Laplace equation in 2D. The constant $c(\boldsymbol{r}) = 1/2$ if Γ is a smooth boundary at $\boldsymbol{r}$. Thus (10.2) expresses the potential at any point $\boldsymbol{r}$ in terms of its values $u(\boldsymbol{r}')$ and the values of its outward normal derivatives $q(\boldsymbol{r}')$ on the boundary. Some of this information will be given as boundary conditions and the rest has to be solved for. We note that the most important step of BEM is to approximate the missing

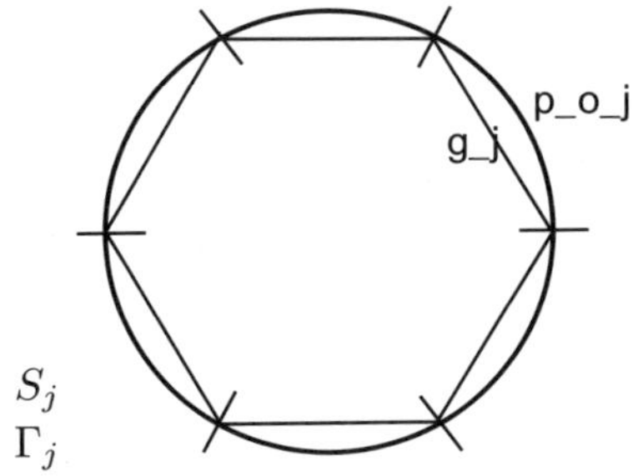

Fig. 10.1. A BEM discretization in which a polygon of rectilinear elements is used to represent a circular boundary.

boundary data as accurately as possible since the final step of computing the unknown function in Ω is merely a case of post-processing.

In a BEM discretization, the boundary Γ is divided into N partitions Γ_j such that $\cup_{j=1}^{N}\Gamma_j = \Gamma$. Each Γ_j is then represented by a numerical boundary S_j (see Fig. 10.1 for a two-dimensional case of rectilinear elements). Then on each S_j, the functions $u(\boldsymbol{r}')$ and $q(\boldsymbol{r}')$ are assumed to vary as the so-called shape functions. Let us denote these shape functions by $f_u(\xi)$ and $f_q(\xi)$, respectively, where ξ is a local coordinate on S_j. For instance, $f_u(\xi)$ and $f_q(\xi)$ are constant functions in the case of constant elements and linear functions in the case of linear elements. Thus, the discretized integral equation is

$$c(\boldsymbol{r}_i)u(\boldsymbol{r}_i) + \sum_{j=1}^{N} \int_{S_j} f_u(\boldsymbol{r}') \frac{\partial G}{\partial n'}(\boldsymbol{r}_i; \boldsymbol{r}')\, dS = \sum_{j=1}^{N} \int_{S_j} G(\boldsymbol{r}_i; \boldsymbol{r}') f_q(\boldsymbol{r}')\, dS. \tag{10.4}$$

where $\boldsymbol{r}_i$ is a node on the i-th element.

At this stage we would like to note two important sources of numerical error. First, replacing the physical boundary by a numerical boundary and second, representing the unknown functions by shape functions. The local error should be a measure of how well the solution satisfies the original integral equation on each element. In some geometries or discretization $S_j \equiv \Gamma_j$, thus eliminating one inherent source of error. The size of the local error will be small depending on how well $f_u(\xi)$ and $f_q(\xi)$ represent the original functions. Errors may also be due to integration and discontinuities at element boundaries but these can be reduced to negligible amounts by using suitable techniques. The integral equation (10.4) is written for N collocation points $\boldsymbol{r}_i$ and boundary conditions are appropriately applied to obtain a linear system of equations

$$\mathbf{A}\mathbf{x} = \mathbf{b}, \tag{10.5}$$

where $\mathbf{x}$ is a vector of the unknown values of either u or q at the boundary.

10.3. Local Errors in Potential Problems

Consider a Dirichlet problem given by,

$$\begin{cases} \nabla^2 u(\boldsymbol{r}) = 0, & \boldsymbol{r} \in \Omega, \\ u(\boldsymbol{r}) = g(\boldsymbol{r}), & \boldsymbol{r} \in \Gamma. \end{cases} \tag{10.6}$$

The unknown in this case is the outward normal derivative $q(\boldsymbol{r}')$. Thus, what we want to solve for using BEM are the values of the normal flux at the boundary Γ; That is, $\mathbf{x} = \mathbf{q} = (q_1 \; q_2 \; \cdots \; q_N)^T$, a vector of the unknown values of $q(\boldsymbol{r}')$ at the boundary. Here we have introduced the notation

$$u(\boldsymbol{r}'_i) =: u_i \quad \text{and} \quad q(\boldsymbol{r}'_i) =: q_i.$$

Let $\tilde{\mathbf{q}} := (\tilde{q}_1 \; \tilde{q}_2 \; \cdots \; \tilde{q}_N)^T$ be the corresponding BEM solution and let $\mathbf{q}^* := (q_1^* \; q_2^* \; \cdots \; q_N^*)^T$ denote the exact solution at the corresponding nodes. The global error $\mathbf{e}$ is defined as

$$\mathbf{e} := \mathbf{q}^* - \tilde{\mathbf{q}}, \tag{10.7}$$

that is, pointwise,

$$e_j = q_j^* - \tilde{q}_j.$$

In order to advance our error investigations, let us define the exact values of the integrals on the j-th element from source node i as

$$I_{ij}^u := \int_{\Gamma_j} u(\boldsymbol{r}') \frac{\partial G}{\partial n'}(\boldsymbol{r}_i; \boldsymbol{r}') \, d\Gamma, \tag{10.8a}$$

$$I_{ij}^q := \int_{\Gamma_j} G(\boldsymbol{r}_i; \boldsymbol{r}') q(\boldsymbol{r}') \, d\Gamma. \tag{10.8b}$$

Let us also define the corresponding numerical estimates of (10.8) in the BEM as

$$\tilde{I}_{ij}^u := \int_{S_j} f_u(\boldsymbol{r}') \frac{\partial G}{\partial n'}(\boldsymbol{r}_i; \boldsymbol{r}') \, dS, \tag{10.9a}$$

$$\tilde{I}_{ij}^q := \int_{S_j} G(\boldsymbol{r}_i; \boldsymbol{r}') f_q(\boldsymbol{r}') \, dS. \tag{10.9b}$$

We then define the two contributions to the local error on element j due to source node i as

$$d_{ij}^q := I_{ij}^q - \tilde{I}_{ij}^q, \tag{10.10a}$$

$$d_{ij}^u := I_{ij}^u - \tilde{I}_{ij}^u. \tag{10.10b}$$

The total local error on element j due to source node i is

$$d_{ij} := d_{ij}^q + d_{ij}^u. \tag{10.11}$$

The cumulative local error for the i-th equation due to contributions from all the elements is therefore given by

$$d_i := -\sum_{j=1}^{N} d_{ij}^u + \sum_{j=1}^{N} d_{ij}^q. \tag{10.12}$$

Consider the right-hand side of the system (10.5); if we were to compute the vector $\mathbf{b}$ exactly, then the i-th component would be

$$b_i^E := -c_i u(\boldsymbol{r}_i') - \sum_{j=1}^{N} \int_{\Gamma_j} u(\boldsymbol{r}') \frac{\partial G}{\partial n}(\boldsymbol{r}_i'; \boldsymbol{r}')\, d\Gamma.$$

The one we actually use in the BEM is

$$b_i = -c_i u(\boldsymbol{r}_i') - \sum_{j=1}^{N} \int_{S_j} f_u(\boldsymbol{r}') \frac{\partial G}{\partial n}(\boldsymbol{r}_i'; \boldsymbol{r}')\, dS.$$

Let us define

$$\begin{aligned} \delta b_i := b_i^E - b_i &= -\sum_{j=1}^{N} \int_{\Gamma_j} \frac{\partial G}{\partial n}(\boldsymbol{r}_i; \boldsymbol{r}') u(\boldsymbol{r}') d\Gamma \\ &\quad + \sum_{j=1}^{N} \int_{S_j} f_u(\boldsymbol{r}') \frac{\partial G}{\partial n}(\boldsymbol{r}_i'; \boldsymbol{r}')\, dS = -\sum_{j=1}^{N} d_{ij}^u. \end{aligned} \tag{10.13}$$

The local error in (10.12) can therefore be expressed as

$$d_i = \delta b_i + \sum_{j=1}^{N} d_{ij}^q. \tag{10.14}$$

If the right-hand side is evaluated exactly, then $\delta b_i = 0$ and the error (10.12) would be

$$d_i = \sum_{j=1}^{N} d_{ij}^q. \tag{10.15}$$

Now, we will asses further the behaviour of these local errors.

Theorem 10.1. *The error in constant elements BEM is second order with respect to grid size.*

Proof. We need to show that the local error is third order in grid size. Take the case of an element with a Dirichlet boundary condition where the unknown is $q(\boldsymbol{r}')$. Consider the integral (10.8b), that is

$$I_{ij}^q = \int_{\Gamma_j} G(\boldsymbol{r}_i; \boldsymbol{r}') q(\boldsymbol{r}')\, d\Gamma. \tag{10.16}$$

Let l_j be the length of S_j and ξ be a local coordinate on S_j such that

$$-l_j/2 \leq \xi \leq l_j/2.$$

Then (10.16) is transformed and evaluated in terms of ξ. Let ξ_j be the midpoint of S_j. Suppose we have a Taylor series expansion of $q(\xi)$ about ξ_j, that is,

$$q(\xi) = q(\xi_j) + q'(\xi_j)(\xi - \xi_j) + \frac{q''(\xi_j)}{2}(\xi - \xi_j)^2 + \frac{q^{(3)}(\xi_j)}{6}(\xi - \xi_j)^3 + \cdots. \tag{10.17}$$

Then, if we use (10.17) in (10.16), we have

$$\begin{aligned} 2\pi I_{ij}^q = &\int_{-l_j/2}^{l_j/2} \ln[r(\xi)] q(\xi_j)\, d\xi + \int_{-l_j/2}^{l_j/2} \ln[r(\xi)] q'(\xi_j)(\xi - \xi_j)\, d\xi \\ &+ \int_{-l_j/2}^{l_j/2} \ln[r(\xi)] \frac{q''(\xi_j)}{2}(\xi - \xi_j)^2\, d\xi + \cdots, \end{aligned} \tag{10.18}$$

where

$$r(\xi) := ||\boldsymbol{r}_i - \boldsymbol{r}'(\xi)||.$$

In constant elements BEM, we only use the first term, that is,

$$I_{ij}^q \approx \frac{1}{2\pi} \int_{-l_j/2}^{l_j/2} \ln[r(\xi)] q(\xi_j)\, d\xi. \tag{10.19}$$

So we have a truncation error whose principal term is the second term on the right of (10.18). Let us consider the following integral of this term,

$$I_2 := \int_{-l_j/2}^{l_j/2} q'(\xi_j)\ln[r(\xi)](\xi - \xi_j)\,d\xi. \tag{10.20}$$

We note that

$$r_1 := r(\xi = -l_j/2) \leq r(\xi) \leq r_2 := r(\xi = l_j/2). \tag{10.21}$$

We assume that r_i is far enough such that $r(\xi)$ remains nonzero. The distance $r(\xi)$ can be expanded about the point ξ_j as

$$r(\xi) = r(\xi_j) + r'(\xi_j)(\xi - \xi_j) + \mathrm{O}((\xi - \xi_j)^2). \tag{10.22}$$

We introduce the definitions

$$\begin{aligned}\rho_0 &:= r(\xi_j),\\ \rho_1 &: + r'(\xi_j),\end{aligned}$$

so that we can write (10.22) as

$$r(\xi) = \rho_0\left(1 + \frac{\rho_1}{\rho_0}(\xi - \xi_j)\right) + \mathrm{O}((\xi - \xi_j)^2).$$

Then

$$\ln[r(\xi)] = \ln(\rho_0) + \frac{\rho_1}{\rho_0}(\xi - \xi_j) + \mathrm{O}((\xi - \xi_j)^2).$$

The integral I_2 can now be evaluated as

$$\begin{aligned}I_2 &= \int_{-l_j/2}^{l_j/2} \ln[r(\xi)](\xi - \xi_j)\,d\xi = \int_{-l_j/2}^{l_j/2} \ln(\rho_0)(\xi - \xi_j)\,d\xi\\ &\quad + \int_{-l_j/2}^{l_j/2}\left[\frac{\rho_1}{\rho_0}(\xi - \xi_j) + \mathrm{O}((\xi - \xi_j)^2)\right](\xi - \xi_j)\,d\xi.\end{aligned} \tag{10.23}$$

Since we use the midpoint as ξ_j, the first integral on the right of (10.23) is zero so that we remain with

$$I_2 = \int_{-l_j/2}^{l_j/2} \frac{\rho_1}{\rho_0}(\xi - \xi_j)^2\,d\xi + \mathrm{O}((\xi - \xi_j)^5). \tag{10.24}$$

Then using the mean value theorem,

$$I_2 \approx \frac{\rho_1}{\rho_0}(\zeta - \xi_j)^2 l_j \quad \zeta \in (-l_j/2,\ \ l_j/2). \tag{10.25}$$

For ξ_j a midpoint of S_j

$$|\zeta - \xi_j| \leq l_j/2. \tag{10.26}$$

Putting (10.25) and (10.26) together we have

$$|I_2| \leq \frac{l_j^3}{4}|\rho_1/\rho_0|,$$

which is third order in grid size. This shows that constant elements BEM can be expected to be of second order. □

Theorem 10.2. *The error in linear elements BEM is second order with respect to grid size.*

Proof. Likewise, we need to show that the local error is third order in grid size. In linear elements, the unknown function is assumed to vary linearly on the element, that is,

$$q(\xi) \approx p_1(\xi) := \alpha_0 + \alpha_1 \xi,$$

where α_0 and α_1 are constants and ξ a local coordinate on S_j. The error in this case is due to the error when we interpolate by an order one polynomial which is given by[12]

$$q(\xi) - p_1(\xi) = \frac{q''(\eta)}{2}(\xi - \xi_0)(\xi - \xi_1), \quad \eta \in (\xi_0, \xi_1).$$

Using this result, the local error in BEM will be

$$E_{s_1} = \frac{1}{4\pi}\int_{-l_j/2}^{l_j/2} \ln[r(\xi)]q''(\eta)(\xi - \xi_0)(\xi - \xi_1)\, d\xi, \tag{10.27}$$

where ξ_0 and ξ_1 are the interpolation coordinates. Again using the second mean value theorem, we have

$$\begin{aligned} E_{s_1} &\approx \frac{\ln[r(\beta)]q''(\eta)}{4\pi}\int_{-l_j/2}^{l_j/2} (\xi - \xi_0)(\xi - \xi_1)\, d\xi \\ &= -\frac{l_j^3}{24\pi}\ln[r(\beta)]q''(\eta), \end{aligned} \tag{10.28}$$

where $\xi_0 = -l_j/2$, $\xi_1 = l_j/2$ and β is an intermediate point. This result shows that linear elements BEM can be expected to be of second order. □

10.3.1. *Numerical examples*

In this section we will perform numerical experiments to illustrate the above claims about the error. Results show that indeed we obtain second order convergence of the error. Our experiments are performed on a circle which rules out the errors caused by discontinuities at corners in other geometries like a square. It would be a natural choice to work on a unit circle but in our examples we avoid this choice for reasons that will become clear in later sections.

Example 10.3.1. Consider the boundary value problem

$$\begin{cases} \nabla^2 u(\boldsymbol{r}) = 0, & \boldsymbol{r} \in \Omega := \{\boldsymbol{r} \in \mathbb{R}^2 : ||\boldsymbol{r}|| \leq 1.2\}, \\ u(\boldsymbol{r}) = g(\boldsymbol{r}'), & \boldsymbol{r} \in \Gamma. \end{cases} \tag{10.29}$$

The boundary condition function $g(\boldsymbol{r}')$ is chosen such that the exact solution is

$$q(\boldsymbol{r}') = \frac{(\boldsymbol{r}' - \boldsymbol{r}_s) \cdot \boldsymbol{n}(\boldsymbol{r}')}{||\boldsymbol{r}' - \boldsymbol{r}_s||^2}, \tag{10.30}$$

where the source point $\boldsymbol{r}_s = (0.36, 1.8)$ is a fixed point outside Ω and $\boldsymbol{n}(\boldsymbol{r}')$ is the outward normal at $\boldsymbol{r}'$.

We will use this example to verify the results of theorems 10.1 and 10.2. Since we have the analytic expression for the unknown $q(\boldsymbol{r}')$ we can compute the global error in (10.7). As we see in Tables 10.1 and 10.2, the results show that indeed the error for both constant and linear elements exhibits second order behaviour.

In Table 10.1, we discretize as in Fig. 10.1 using an inscribed regular polygon of N sides. We refine by a factor of three because we are using constant elements with the midpoints as nodes and we would like to compute the pointwise error as shown in the second column. For a better

Table 10.1. Errors to the Dirichlet problem (10.29) using constant elements BEM with rectilinear elements.

N	$e_j = \lvert q_j^* - \tilde{q}_j \rvert$	$\lvert\lvert \mathbf{e} \rvert\rvert_2/\sqrt{N}$	$e_j(N)/e_j(3N)$
5	6.57E-02	2.80E-01	8.66
15	7.58E-03	3.53E-02	9.43
45	8.04E-04	5.40E-03	9.12
135	8.82E-05	6.58E-04	9.04
405	9.76E-06	7.53E-05	9.01
1,215	1.08E-06	8.45E-06	9.00
3,645	1.20E-07	9.42E-07	—

Table 10.2. Errors to the Dirichlet problem (10.29) using linear elements BEM.

N	$e_j = \|q_j^* - \tilde{q}_j\|$	$\|\|\mathbf{e}\|\|_2/\sqrt{N}$	$e_j(N)/e_j(3N)$
5	1.39E-02	1.29E-01	70.1
15	1.98E-04	2.11E-02	12.9
45	1.53E-05	4.89E-03	10.7
135	1.42E-06	6.07E-04	9.6
405	1.47E-07	7.00E-05	9.2
1,215	1.59E-08	7.88E-06	9.2
3,645	1.72E-09	8.79E-07	—

comparison, we use the same refinement factor for linear elements as well. Both the pointwise error and the median 2-norm of the error show the same trend of error convergence. The pointwise error is computed at the point $\boldsymbol{r}' = (-0.370820, -1.141268)$ on thc circle. As we can see from the ratios of consecutive errors, the convergence is second order.

A discretization using arcs instead of rectilinear elements as shown in Fig. 10.2 gives similar results, see Table 10.3.

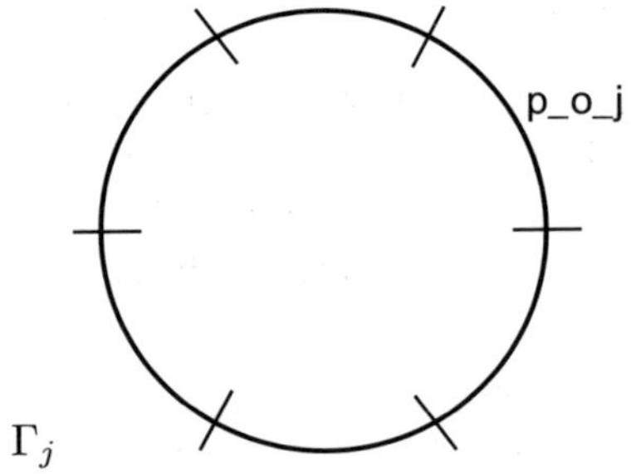

Fig. 10.2. A BEM discretisation of a circle using circular arcs.

Table 10.3. Errors to Dirichlet problem (10.29) when we use constant elements BEM with circular arc elements.

N	$\|e_j\| = \|q_j^* - \tilde{q}_j\|$	$\|\|\mathbf{e}\|\|_2/\sqrt{N}$	$e_j(N)/e_j(3N)$
5	5.54E-03	1.44E-001	13.6
15	4.06E-04	1.25E-002	10.8
45	3.77E-05	2.55E-003	9.6
135	3.94E-06	3.55E-004	9.2
405	4.28E-07	4.21E-005	9.1
1,215	4.72E-08	4.78E-006	9.0
3,645	5.23E-09	5.34E-007	—

Example 10.3.2. Consider the Neumann boundary value problem

$$\begin{cases} \nabla^2 u(\boldsymbol{r}) = 0, & \boldsymbol{r} \in \Omega := \{\boldsymbol{r} \in \mathbb{R}^2 : ||\boldsymbol{r}|| \leq 1.2\}, \\ q(\boldsymbol{r}) = h(\boldsymbol{r}'), & br \in \Gamma. \end{cases} \tag{10.31}$$

The boundary condition function $h(\boldsymbol{r}')$ is chosen such that the exact solution is

$$u(\boldsymbol{r}') = \log\left(||\boldsymbol{r}' - \boldsymbol{r}_s||\right), \tag{10.32}$$

where the source point $\boldsymbol{r}_s = (0.36, 1.8)$ is a fixed point outside Ω and $\boldsymbol{n}(\boldsymbol{r}')$ is the outward normal at $\boldsymbol{r}'$. In the implementation, a Dirichlet boundary condition is prescribed at the last node $j = N$ as a remedy to obtain a well posed problem. So, in actual sense, we solve a mixed boundary problem. We also compute the pointwise and median 2-norm errors and the results are shown in Table 10.4. The convergence behaviour is the same as that for the Dirichlet problem. Both errors show a second order convergence as we can see from the ratios of consecutive errors.

10.3.2. *A detailed study of local errors*

As mentioned earlier, the freedom to choose collocation points for a given discretization suggests that there are several ways to implement the BEM. In this section we survey the possibilities but restrict ourselves to a circular domain.

Suppose we use rectilinear elements to discretize the circle such that the numerical boundary is a polygon like the one shown in Fig. 10.3. Besides, the way we treat the boundary conditions is also very important as we will see in the survey that follows.

In the usual constant elements formulation, the nodes are the midpoints of the elements and also function as the collocation points. For the

Table 10.4. Errors to the mixed (one Neumann boundary node) problem using constant elements BEM.

N	$\|e_j\| = \|u_j^* - \tilde{u}_j\|$	$\|\|\mathbf{e}\|\|_2/\sqrt{N}$	$e_j(N)/e_j(3N)$
5	1.39E-01	4.66E-01	23.5
15	5.92E-03	2.25E-02	13.0
45	4.55E-04	2.88E-03	8.7
135	5.24E-05	3.27E-04	8.9
405	5.89E-06	3.65E-05	9.0
1,215	6.57E-07	4.07E-06	9.0
3,645	7.31E-08	4.52E-07	—

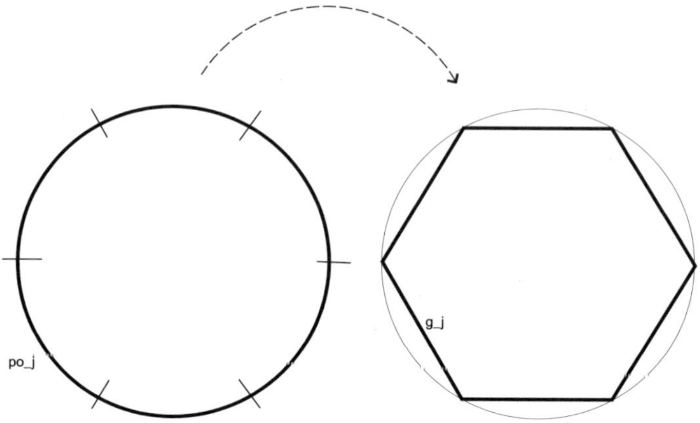

Fig. 10.3. Discretizsation of a circle by a six element polygon.

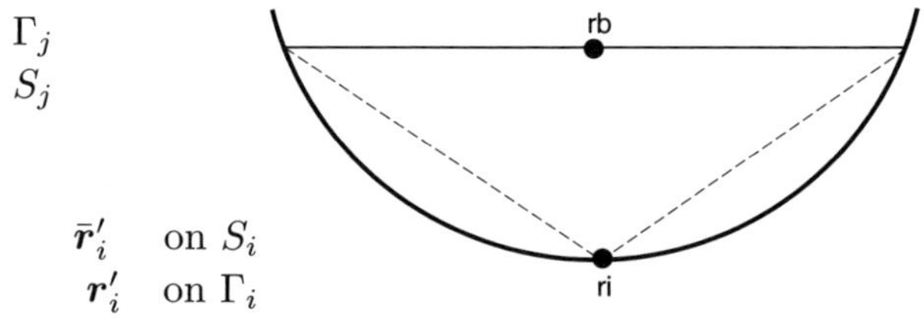

Fig. 10.4. Numerical representation of boundary points; $\boldsymbol{r}'_i$ the exact point and $\bar{\boldsymbol{r}}'_i$ its numerical representation.

discretization shown in Fig. 10.3, the numerical boundary does not coincide with the physical boundary and therefore we can talk of exact points $\boldsymbol{r}'_i$ and their numerical representations $\bar{\boldsymbol{r}}'_i$ as shown in Fig. 10.4. Since our goal is to come up with a proper understanding of (local) errors for the BEM, we would like to explore the various possibilities we have to choose the collocation points and nodes. These possibilities are summarized in Table 10.5.

As mentioned earlier, it is important to differentiate between nodal points (the points used to approximate the functions) and the collocation points (the points where the integral equation is applied). Nevertheless, the boundary condition on S_i shown in Fig. 10.4 is $u(\boldsymbol{r}'_i)$. We consider again the Dirichlet problem, (10.29), and compute errors for some of the numbered cases in the table. Similar to the examples of the previous section, the pointwise error is computed at the point $(-0.370820,\ -1.141267)$.

Table 10.5. Summary of possibilities on implements of constant elements BEM on a circle.

Collocation on S_j	u constant	integration of b on S_j	integration of A on S_j	(1)
			integration of A on Γ_j	
		integration of b on Γ_j	integration of A on S_j	(2)
			integration of A on Γ_j	
	u exact	integration of b on S_j	integration of A on S_j	(3)
			integration of A on Γ_j	
		integration of b on Γ_j	integration of A on S_j	(4)
			integration of A on Γ_j	
Collocation on Γ_j	u constant	integration of b on S_j	—	
			—	
		integration of b on Γ_j	—	
			integration of A on Γ_j	(5)
	u exact	integration of b on S_j	—	
			—	
		integration of b on Γ_j	—	
			integration of A on Γ_j	(6)

Case (1) in Table 10.5 is the traditional constant elements BEM, that is, on S_j, u is constant and given by $\bar{u}_j := u(\boldsymbol{r}'_j)$. For the linear system of equations, we have,

$$b_i = -c_i\bar{u}_i - \sum_{j=1}^{N} \int_{S_j} \bar{u}_j \frac{\partial G}{\partial n'}(\bar{\boldsymbol{r}}'_i; \boldsymbol{r}')\, dS,$$

$$A_{ij} = -\int_{S_j} G(\bar{\boldsymbol{r}}'_i; \boldsymbol{r}')\, dS.$$

In case (3) in the table, we have the 'exact' right-hand side. But it is not yet really exact because we integrate on S_j, and not Γ_j thus the boundary function is evaluated on the numerical boundary. The difference now is that we make no assumptions on the given boundary function u. That is,

$$b_i = -c_i u(\boldsymbol{r}'_i) - \sum_{j=1}^{N} \int_{S_j} u(\boldsymbol{r}') \frac{\partial G}{\partial n'}(\bar{\boldsymbol{r}}'_i; \boldsymbol{r}')\, dS,$$

$$A_{ij} = -\int_{S_j} G(\bar{\boldsymbol{r}}'_i; \boldsymbol{r}')\, dS.$$

Case (5) is like the traditional BEM but with the integrations carried out on the arcs. The fixed point i is on the arc and the functions $u(\boldsymbol{r})$ and

$q(\boldsymbol{r})$ are assumed constant on each element j.

$$b_i = -c_i\bar{u}_i - \sum_{j=1}^{N} \int_{\Gamma_j} \bar{u}_j \frac{\partial G}{\partial n'}(\boldsymbol{r}'_i; \boldsymbol{r}')\, dS,$$

$$A_{ij} = -\int_{\Gamma_j} G(\boldsymbol{r}'_i; \boldsymbol{r}')\, dS.$$

In case (6) we integrate the exact boundary function $u(\boldsymbol{r})$ as given. Thus we use 'the exact right-hand side'

$$b_i = -c_i u(\boldsymbol{r}'_i) - \sum_{j=1}^{N} \int_{\Gamma_j} u(\boldsymbol{r}') \frac{\partial G}{\partial n'}(\boldsymbol{r}'_i; \boldsymbol{r}')\, dS,$$

$$A_{ij} = -\int_{\Gamma_j} G(\boldsymbol{r}'_i; \boldsymbol{r}')\, dS.$$

We expect that the results of case (3) are better than those of case (1) since case (3) should capture the variation of the boundary function better. For the same reasons, the results of case (6) are expected to be better than those of case (5). Since in cases (5) and (6) the numerical boundary coincides with the physical boundary, the corresponding results should be better than those of cases (1) and (3).

Comparing the results in Tables 10.6 and 10.7, we see that indeed we obtain better results in case (3) than in case (1). The gain is not much in this example and this is because of the smooth variation of the boundary condition function on the large part of the boundary. This also explains the small difference between the results in Tables 10.8 and 10.9. As we expected, the results of cases (5) and (6) are better than those of cases (1) and (3) respectively. We see that the pointwise errors in Table 10.8 are at least one order smaller than those in Table 10.6.

Table 10.6. Errors for case (1) in Table 10.5.

N	$\lvert e_j \rvert = \lvert q_j^e - \tilde{q}_j \rvert$	$\lvert\lvert \mathbf{e} \rvert\rvert_2/\sqrt{N}$
5	6.57E-02	2.81E-01
15	7.58E-03	3.53E-02
45	8.04E-04	5.40E-03
135	8.82E-05	6.58E-04
405	9.76E-06	7.53E-05
1,215	1.08E-06	8.45E-06
3,645	1.20E-07	9.42E-07

Table 10.7. Errors for case (3) in Table 10.5.

N	$\|e_j\| = \|q_j^e - \tilde{q}_j\|$	$\|\|\mathbf{e}\|\|_2/\sqrt{N}$
5	3.01E-01	4.57E-01
15	4.57E-03	2.17E-02
45	4.76E-04	3.70E-03
135	5.20E-05	4.73E-04
405	5.74E-06	5.49E-05
1,215	6.37E-07	6.19E-06
3,645	7.07E-08	6.91E-07

Table 10.8. Errors for case (5) in Table 10.5.

N	$\|e_j\| = \|q_j^e - \tilde{q}_j\|$	$\|\|\mathbf{e}\|\|_2/\sqrt{N}$
5	5.54E-03	1.44E-01
15	4.06E-04	1.25E-02
45	3.77E-05	2.55E-03
135	3.94E-06	3.55E-04
405	4.28E-07	4.21E-05
1,215	4.72E-08	4.78E-06
3,645	5.23E-09	5.34E-07

Table 10.9. Errors for case (6) in Table 10.5.

N	$\|e_j\| = \|q_j^e - \tilde{q}_j\|$	$\|\|\mathbf{e}\|\|_2/\sqrt{N}$
5	2.96E-02	1.59E-01
15	4.05E-04	1.24E-02
45	3.62E-05	2.40E-03
135	3.88E-06	3.49E-04
405	4.26E-07	4.19E-05
1,215	4.31E-08	4.77E-06
3,645	1.66E-08	5.34E-07

10.4. Large Condition Numbers

In all practical applications of the BEM, a linear system needs to be solved eventually. In most cases it is assumed that the condition number of the system matrix is moderate, i.e. the linear system has a unique solution and can be solved accurately. However, it is not very clear if this is always true.

It is known that the condition number of the BEM matrix in potential problems with Dirichlet conditions depends on the size and shape of the domain on which the potential problem is defined.[4–7] On most domains the condition number is moderate, but on certain domains the condition number becomes infinitely large.

When the BEM is applied to the Laplace equation with Dirichlet conditions on a circular domain, the condition number of the BEM matrix can be computed analytically.[13,14] Here, rectilinear elements are used and a constant approximation of the functions at each element. It can be shown that the condition number depends on the radius R of the circle and the number of boundary elements N according to

$$\mathrm{cond}(\mathbf{G}) = \frac{\max\left(\frac{1}{2}, |\log R|\right)}{\min\left(\frac{1}{N}, |\log R|\right)}. \tag{10.33}$$

This leads to the observation that the condition number is infinitely large when the radius of the circle is equal to one. Hence, the size of the domain affects the condition number of the BEM matrix and, consequently, it also affects the solvability of the linear system. This introduces a remarkable phenomenon: the solvability of a linear system in the BEM depends on the size of the domain.

For the potential problem with Neumann conditions the corresponding BEM matrix is always infinitely large, since this problem is ill-posed, regardless of the size and shape of the domain. The question then arises whether we may expect infinitely large condition numbers when solving potential problems with mixed boundary conditions. In this case, the BEM matrix is a composite matrix, consisting of a number of columns from the Dirichlet BEM matrix, and a number of columns from the Neumann BEM matrix. The first matrix is ill-conditioned on some domains, the latter is ill-conditioned always. It is not clear how this affects the condition number of the composite matrix.

For the Laplace equation with mixed conditions on a circular domain, it is possible to give an estimate for the BEM matrix.[15] Again, it shows that for the unit circle the condition number is infinitely large. Hence, the mixed problem seems to inherit the solvability problem from the Dirichlet problem.

Figures 10.5 and 10.6 correspond to solving the Laplace equation on a triangular domain and an ellipsoidal domain, respectively. The size of the triangle is characterized by its side length l, whereas the ellipse is characterized by the lengths of the semi-axes a and $a/2$. We show the condition numbers for the Dirichlet BEM matrix (squares) and the mixed BEM matrix (circles). For the mixed problem, we set the number of boundary elements with Dirichlet conditions equal to the number of

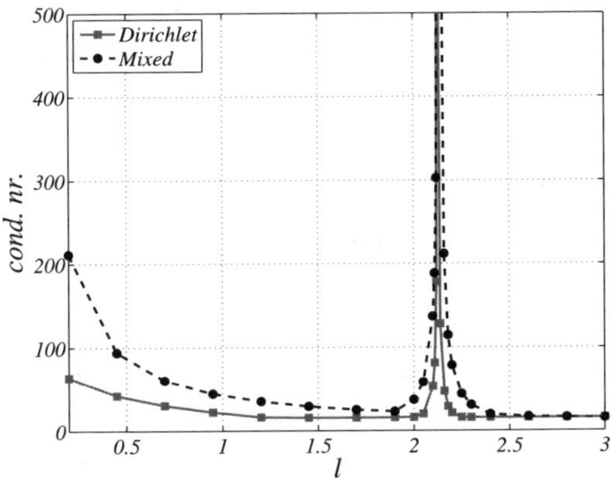

Fig. 10.5. The condition number of the matrices **A** (circles) and **G** (squares), corresponding to the Laplace equation with mixed boundary conditions and Dirichlet boundary conditions, respectively, for a triangular domain.

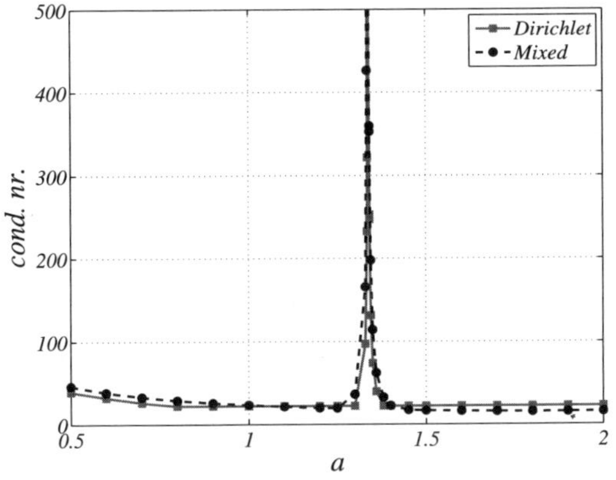

Fig. 10.6. The condition number of the matrices **A** (circles) and **G** (squares), corresponding to the Laplace equation with mixed boundary conditions and Dirichlet boundary conditions, respectively, for an ellipsoidal domain.

boundary elements with Neumann conditions. Again, we see that the condition numbers are infinitely large for certain sizes of the domains. Hence, again the size of the domain influences the solvability of the linear systems in the BEM.

In the next section, we present the theory behind this remarkable phenomenon by studying the boundary integral equations that lay the foundations of the BEM. We show that, regardless of the shape of the domain, there always exists a particular size of that domain for which the boundary integral equation becomes singular. We also propose a number of strategies to avoid these singularities.

10.5. Condition Numbers in Potential Problems

This section investigates the condition numbers of the system matrices that appear in the BEM when applied to potential problems. We analyse potential problems with Dirichlet and mixed boundary conditions and show that these problems do not always have a unique solution. This happens when the domains on which the potential problems are defined have a logarithmic capacity that is equal to one. We will proceed by introducing the concept of logarithmic capacity.

10.5.1. *Logarithmic capacity*

To study the uniqueness properties of the Dirichlet and the mixed problem, we need to introduce the notion of *logarithmic capacity*. We define the energy integral I by a double integral over a contour Γ,

$$I(q) := \int_\Gamma \int_\Gamma \log \frac{1}{\|\boldsymbol{r} - \boldsymbol{r}'\|} q(\boldsymbol{r}) q(\boldsymbol{r}') d\Gamma_r d\Gamma_{r'}, \tag{10.34}$$

and the logarithmic capacity $C_l(\Gamma)$ is related to this integral by

$$-\log C_l(\Gamma) := \inf_q I(q). \tag{10.35}$$

Here, the infimum is taken over all functions q, with the restriction that

$$\int_\Gamma q(\boldsymbol{r}) d\Gamma_r = 1. \tag{10.36}$$

Let us give a physical interpretation of the logarithmic capacity. For simplicity, let the domain Ω be contained in the disc with radius $1/2$. In that case, it can be shown that the integral $I(q)$ is positive. The function q can be seen as a charge distribution over a conducting domain Ω. Faraday demonstrated that this charge will only reside at the exterior boundary of the domain ('cage'), in our case at Γ. We normalize q in such a way that the total amount of charge at Γ is equal to one, cf. condition (10.36). The function

$$\frac{1}{2\pi} \int_\Gamma \log \frac{1}{\|\boldsymbol{r} - \boldsymbol{r}'\|} q(\boldsymbol{r}') d\Gamma_{r'} \equiv (\mathcal{K}^s q)(\boldsymbol{r}) \tag{10.37}$$

is identified as the potential due to the charge distribution q. Note that the integral I can also be written as

$$I(q) = 2\pi \int_{\Gamma} (\mathcal{K}^s q)(\boldsymbol{r}) q(\boldsymbol{r}) d\Gamma_r. \tag{10.38}$$

Hence, I can be seen as the energy of the charge distribution q. The charge will distribute itself over Γ in such a way that the energy I is minimized, so the quantity $-\log C_l(\Gamma)$ is the minimal amount of energy. Hence, the logarithmic capacity $C_l(\Gamma)$ is a measure for the capability of the boundary Γ to support a unit amount of charge.

For most boundaries the logarithmic capacity is not known explicitly. It is only for a few elementary domains that the logarithmic capacity can be calculated analytically[9]; we have listed some in Table 10.10.

There are also some useful properties that help us to determine or estimate the logarithmic capacity.[8,16]

(1) If Γ is the outer boundary of a closed bounded domain Ω, then $C_l(\Gamma) = C_l(\Omega)$. This agrees with the idea of Faraday's cage, mentioned above.
(2) Denote by d_Γ the Euclidean diameter of Ω, then $C_l(\Gamma) \le d_\Gamma$. Hence, the radius of the smallest circle in which Γ is contained is an upper bound for the logarithmic capacity of Γ.
(3) If $\Gamma = \boldsymbol{x} + \alpha\Gamma_1$, then $C_l(\Gamma) = \alpha C_l(\Gamma_1)$. Hence, the logarithmic capacity behaves linearly with respect to scaling and is invariant with respect to translation.
(4) If $\Omega_1 \subset \Omega_2$, then $C_l(\Omega_1) \le C_l(\Omega_2)$.
(5) For a convex domain Ω,

$$C_l(\Omega) \ge \left(\frac{\text{area}(\Omega)}{\pi}\right)^{1/2}. \tag{10.39}$$

Table 10.10. The logarithmic capacity of some domains. Note that $\Gamma(\cdot)$ represents the gamma-function.

Boundary Γ	Logarithmic capacity
circle with radius R	R
square with side L	$\frac{\Gamma(\frac{1}{4})^2}{4\pi^{3/2}} L \approx 0.59017 \cdot L$
ellipse with semi-axes a and b	$(a+b)/2$
interval of length a	$\frac{1}{4}a$
isosceles right-angled triangle side l	$\frac{3^{3/4}\Gamma(1/4)^2}{2^{7/2}\pi^{3/2}}\, l \approx 0.476\, l$

If the properties from the list above do not supply accurate enough estimates, the logarithmic capacity can also be approximated numerically with the help of linear programming[17] or using the BEM-matrices explicitly.[18]

10.5.2. *Dirichlet problem*

We consider the Laplace equation (10.1) with Dirichlet boundary conditions, with corresponding BIE (10.2). For the single layer operator $\mathcal{K}^s$ in this BIE we have the following result.

Theorem 10.3. *There exists a nonzero q_e such that*

$$(\mathcal{K}^s q_e)(\boldsymbol{r}) = -\frac{1}{2\pi} \log C_l(\Gamma), \quad \boldsymbol{r} \in \Gamma. \tag{10.40}$$

Sketch of proof. In the following, we briefly present the major steps in the proof of the theorem.[8,19,20] We observe that the energy integral (10.34) takes values $-\infty < I(q) \leq \infty$. If the infimum of the energy integral is infinitely large, then by definition the logarithmic capacity is equal to zero. Suppose that $C_l(\Gamma) > 0$ and thus $-\infty < I(q) < \infty$. It is proven in Hille[8] (p. 282) that for each boundary Γ, there exists a unique minimizer q_e of $I(q)$, i.e.

$$I(q_e) = \inf_q I(q) = -\log C_l(\Gamma) \text{ with } \int_\Gamma q_e(\boldsymbol{r}) d\Gamma_r = 1. \tag{10.41}$$

For the minimizer q_e, the following result in Hille[8] (p. 287) is proven. Let Γ be the boundary of a closed bounded domain with positive logarithmic capacity and a connected complement. Then $2\pi\mathcal{K}^s q_e \leq -\log C_l(\Gamma)$ in the whole plane and $2\pi\mathcal{K}^s q_e = -\log C_l(\Gamma)$ at Γ, except possibly for a subset which has zero logarithmic capacity. ∎

Theorem 10.3 leads to the following result.

Corollary 10.1. *If $C_l(\Gamma) = 1$, there exists a nonzero q_e such that $\mathcal{K}^s q_e = 0$.*

Thus in the specific case that $C_l(\Gamma) = 1$, the single layer operator $\mathcal{K}^s$ admits an eigenfunction q_e with zero eigenvalue. Hence, $\mathcal{K}^s$ is not positive definite and the Dirichlet problem does not have a unique solution.

If the BIE does not have a unique solution, we may expect that its discrete counterpart, the linear system, also does not have a unique solution.

That is, the linear system is singular, or at least ill-conditioned. This is reflected by the condition number of the system matrix, as shown in Figs. 10.5 and 10.6. Here, the condition number of the BEM matrix is plotted for two problems: the Laplace equation with Dirichlet conditions on a triangular and an ellipsoidal domain. The triangle is an isosceles right-angled triangle with sides of length l. For such a triangle the logarithmic capacity is given by

$$C_l(\text{triangle}) = \frac{3^{3/4}\Gamma^2(1/4)}{2^{7/2}\pi^{3/2}}\, l \approx 0.476\, l. \tag{10.42}$$

This implies that the condition number will be large when the scaling parameter l is close to the critical scaling $l^* := 1/0.476 \approx 2.1$. The ellipse has semi-axes of length a and $a/2$, which has a logarithmic capacity equal to $3a/4$. Hence, we may expect a large condition number when the scaling parameter a is close to $a^* := 4/3$. We observe that the condition number for the Dirichlet case (solid line) is indeed infinitely large for these two critical scalings.

If we rescale the domain such that the Euclidean diameter is smaller than one, then the second property in Section 10.5.1 shows us that the logarithmic capacity will also be smaller than one. In this way we can guarantee a unique solution of the BIE and thus a low condition number.

Recall that the non-trivial solution q_e of the homogeneous BIE $\mathcal{K}^s q = 0$ has a contour integral equal to 1. At the same time we realize that a solution q of $\mathcal{K}^s q = 0$ has to satisfy

$$\int_\Gamma q d\Gamma = \int_\Omega \Delta u d\Omega = 0, \tag{10.43}$$

where we make use of Gauss' theorem. By supplementing this requirement for q to the BIE, we exclude the possibility that q_e is a solution of the homogeneous BIE. This provides a second strategy to ensure unique solutions of the BIE.

A third option to guarantee a unique solution is to adjust the integral operator $\mathcal{K}^s$. Note that the function G_α,

$$G_\alpha(\boldsymbol{r};\boldsymbol{r}') := \frac{1}{2\pi}\log\frac{\alpha}{\|\boldsymbol{r}-\boldsymbol{r}'\|}, \quad \alpha \in \mathbb{R}^+, \tag{10.44}$$

is also a fundamental solution for the Laplace operator. The corresponding single layer potential reads

$$\mathcal{K}^s_\alpha q := \frac{1}{2\pi}\int_\Gamma \log\frac{\alpha}{\|\boldsymbol{r}-\boldsymbol{r}'\|} q(\boldsymbol{r}')d\Gamma_{r'} = \mathcal{K}^s q + \frac{\log\alpha}{2\pi}\int_\Gamma q d\Gamma. \tag{10.45}$$

For the minimizer q_e we get

$$\begin{aligned}\mathcal{K}^s_\alpha q_e &= \mathcal{K}^s q_e + \frac{\log\alpha}{2\pi}\int_\Gamma q_e d\Gamma = -\frac{1}{2\pi}\log C_l(\Gamma) + \frac{1}{2\pi}\log\alpha \\ &= \frac{1}{2\pi}\log\frac{\alpha}{C_l(\Gamma)}. \end{aligned} \tag{10.46}$$

This is only equal to zero if $\alpha = C_l(\Gamma)$. We may choose for α any positive real number unequal to $C_l(\Gamma)$ and obtain $\mathcal{K}^s_\alpha q_e \neq 0$. In that case, q_e is no longer an eigenfunction of the single layer potential operator with zero eigenvalue. Hence, the BIE (10.2) with Dirichlet conditions is uniquely solvable if $\mathcal{K}^s$ is replaced by $\mathcal{K}^s_\alpha$.[19,21] The advantage of this procedure is that we do not need to rescale the domain or add an extra equation. Furthermore, we do not need to know the logarithmic capacity explicitly: a rough estimate of the capacity suffices to choose α such that $\alpha \neq C_l(\Gamma)$.

There are also ways to ensure a unique solution of the BIE for the Laplace equation that can be used without having to know the logarithmic capacity. For instance, adding an extra collocation node at the interior or exterior of the domain can change the BIE in such a way that it is not singular any longer[22]; although this does depend on the location of the extra collocation node. Another option is to use the hypersingular formulation of the BIE.[23] The hypersingular BIE is the normal derivative of the standard BIE and does not involve the single layer operator. As a consequence, the BIE does not become singular at certain domains.

10.5.3. *Mixed problem*

We consider the Laplace equation with mixed boundary conditions,

$$\begin{aligned} \nabla^2 u &= 0, \quad \boldsymbol{r} \in \Omega, \\ u &= \tilde{u}, \quad \boldsymbol{r} \in \Gamma_1, \\ q &= \tilde{q}, \quad \boldsymbol{r} \in \Gamma_2, \end{aligned} \tag{10.47}$$

where $\Gamma = \Gamma_1 \cup \Gamma_2$. To investigate this boundary value problem we have to rewrite the BIE in (10.2). For $i = 1, 2$, we introduce the functions $u_i := u|_{\Gamma_i}$ and $q_i := q|_{\Gamma_i}$ and the boundary integral operators

$$(\mathcal{K}^s_i q)(\boldsymbol{r}) := \int_{\Gamma_i} G(\boldsymbol{r};\boldsymbol{r}')q(\boldsymbol{r}')d\Gamma_{r'}, \quad \boldsymbol{r} \in \Gamma, \tag{10.48a}$$

$$(\mathcal{K}^d_i u)(\boldsymbol{r}) := \int_{\Gamma_i} \frac{\partial}{\partial n_y} G(\boldsymbol{r};\boldsymbol{r}')u(\boldsymbol{r}')d\Gamma_{r'}, \quad \boldsymbol{r} \in \Gamma. \tag{10.48b}$$

Note that the boundary conditions in (10.47) provide $u_1 = \tilde{u}$ and $q_2 = \tilde{q}$. By distinguishing $\boldsymbol{r} \in \Gamma_1$ and $\boldsymbol{r} \in \Gamma_2$, we write (10.2) as a system of two BIEs,

$$\mathcal{K}_2^d u_2 - \mathcal{K}_1^s q_1 = \mathcal{K}_2^s \tilde{q} - \frac{1}{2}\tilde{u} - \mathcal{K}_1^d \tilde{u}, \quad \boldsymbol{r} \in \Gamma_1, \tag{10.49a}$$

$$\frac{1}{2}u_2 + \mathcal{K}_2^d u_2 - \mathcal{K}_1^s q_1 = \mathcal{K}_2^s \tilde{q} - \mathcal{K}_1^d \tilde{u}, \quad \boldsymbol{r} \in \Gamma_2. \tag{10.49b}$$

In this system, all prescribed boundary data are at the right-hand side of the equations.

Theorem 10.4. *If $C_l(\Gamma) = 1$, the homogeneous equations of (10.49a) and (10.49b) have a non-trivial solution pair (q_1, u_2).*

Sketch of proof. We have to find a non-trivial pair of functions (q_1, u_2) such that the left-hand sides of (10.49a) and (10.49b) are equal to zero when $C_l(\Gamma) = 1$. It can be shown that such a pair of functions exists by using information from the Dirichlet problem.[11] ■

Theorem 10.4 tells us that the BIE for the mixed problem does not have a unique solution when $C_l(\Gamma) = 1$, i.e. the BIE is singular. Moreover the division of Γ into a part Γ_1 with Dirichlet conditions and a part Γ_2 with Neumann conditions does not play a role in this. It does not make a difference whether we take Γ_1 very small or very large; the singular BIE relates solely to the whole boundary Γ.

In Figs. 10.5 and 10.6, we observe that indeed the condition number for the mixed problems is infinitely large at (almost) the same scalings as for the Dirichlet problems, agreeing with the theory. The small difference that is present is caused by numerical inaccuracies due to the discretization.

To guarantee a unique solution for the mixed problem, we have the same options as for the Dirichlet problem. The simplest remedy is to rescale the domain, thus avoiding a unit logarithmic capacity. A second option is to demand that the function q have a zero contour integral. Since part of q is already prescribed, this yields the following condition for the unknown part of q,

$$\int_{\Gamma_1} q_1 d\Gamma = -\int_{\Gamma_2} \tilde{q} d\Gamma. \tag{10.50}$$

As a last option to obtain nonsingular BIEs, we can also replace the single layer operator $\mathcal{K}^s$ by $\mathcal{K}_\alpha^s$ (see Section 10.5.2).

10.6. Condition Numbers in Flow Problems

The Laplace equation and the Stokes equations have at least one thing in common: the Laplace operator appears in both equations. As we have seen in the previous section, the Laplace equation may lead to a singular BIE for certain critical domains. The question arises whether this is also the case for the equations in viscous flow problems: can the corresponding BIEs become singular on certain critical domains?

In this section, we study the BIEs following from the Stokes equations. In particular, we focus on the eigenvalues of the integral operators. It is shown that for certain critical domains these integral operators admit zero eigenvalues. Hence, again we find that the BIEs become singular for a number of critical domains.

For the Laplace equation, it is possible to *a priori* determine the critical domains. For a number of simple domains, the logarithmic capacity can be used to exactly compute the critical size. For more involved domains, the logarithmic capacity can be used to estimate the critical size of the domain. Unfortunately, the critical domains for the Stokes equations do not coincide with the critical domains for the Laplace equations. Hence, we cannot use the logarithmic capacity to *a priori* determine the critical domains on which the BIEs for Stokes equations become singular. It is only by numerical experiments that we can distinguish the critical domains.

Let Ω be a two-dimensional simply-connected domain with a piece-wise smooth boundary Γ. The Stokes equations for a viscous flow in Ω read

$$\begin{aligned} \nabla^2 \boldsymbol{v} - \nabla p &= \boldsymbol{0}, \\ \nabla . \boldsymbol{v} &= 0, \end{aligned} \tag{10.51}$$

where $\boldsymbol{v}$ is the velocity field of the fluid and p its pressure. Let Γ be divided into a part Γ_1 on which the velocity $\boldsymbol{v}$ is prescribed, and a part Γ_2 on which the pressure p is prescribed, $\Gamma = \Gamma_1 \bigcup \Gamma_2$. Hence, the Stokes equations are subject to the boundary conditions

$$\begin{aligned} \boldsymbol{v} &= \tilde{\boldsymbol{v}}, \quad \boldsymbol{r} \in \Gamma_1, \\ p &= \tilde{p}, \quad \boldsymbol{r} \in \Gamma_2. \end{aligned} \tag{10.52}$$

Either Γ_1 or Γ_2 can be empty, leading to a purely Neumann or Dirichlet problem, respectively. The Stokes equations in differential form can be transformed to a set of two BIEs:[6,24,25]

$$\frac{1}{2}v_i(\boldsymbol{r}) + \int_\Gamma q_{ij}(\boldsymbol{r},\boldsymbol{r}')v_j(\boldsymbol{r}')d\Gamma_{r'}$$
$$= \int_\Gamma u_{ij}(\boldsymbol{r},\boldsymbol{r}')b_j(\boldsymbol{r}')d\Gamma_{r'}, \quad \boldsymbol{r}\in\Gamma,\ i=1,2. \tag{10.53}$$

Here, a repeated index means summation over all possible values of that index. The vector function $\boldsymbol{b}$ is the normal stress at the fluid boundary,

$$\boldsymbol{b} := \sigma(p,\boldsymbol{v})\boldsymbol{n}, \tag{10.54}$$

with $\boldsymbol{n}$ being the outward unit normal at the boundary and the stress tensor σ being defined by

$$\sigma_{ij}(p,\boldsymbol{v}) := -p\delta_{ij} + \left(\frac{\partial v_i}{\partial x_j} + \frac{\partial v_j}{\partial x_i}\right). \tag{10.55}$$

Hence, the boundary integral formulation involves two variables: the velocity $\boldsymbol{v}$ and the normal stress $\boldsymbol{b}$. In correspondence to (10.52), at each point of the boundary either $\boldsymbol{v}$ or $\boldsymbol{b}$ is prescribed:

$$\begin{aligned} \boldsymbol{v} &= \tilde{\boldsymbol{v}}, & \boldsymbol{r}&\in\Gamma_1,\\ \boldsymbol{b} &= -\tilde{p}\boldsymbol{n}, & \boldsymbol{r}&\in\Gamma_2. \end{aligned} \tag{10.56}$$

The kernels u_{ij} and q_{ij} in the integral operators are defined as

$$\begin{aligned} q_{ij}(\boldsymbol{r},\boldsymbol{r}') &:= \frac{1}{\pi}\frac{(x_i-y_i)(x_j-y_j)(x_k-y_k)n_k}{\|\boldsymbol{r}-\boldsymbol{r}'\|^4},\\ u_{ij}(\boldsymbol{r},\boldsymbol{r}') &:= \frac{1}{4\pi}\left\{\delta_{ij}\log\frac{1}{\|\boldsymbol{r}-\boldsymbol{r}'\|} + \frac{(x_i-y_i)(x_j-y_j)}{\|\boldsymbol{r}-\boldsymbol{r}'\|^2}\right\}, \end{aligned} \tag{10.57}$$

for $i,j=1,2$. We introduce boundary integral operators,

$$\begin{aligned} (\mathcal{G}\varphi)_i(\boldsymbol{r}) &:= \int_\Gamma u_{ij}(\boldsymbol{r},\boldsymbol{r}')\varphi_j(\boldsymbol{r}')d\Gamma_{r'},\\ (\mathcal{H}\psi)_i(\boldsymbol{r}) &:= \int_\Gamma q_{ij}(\boldsymbol{r},\boldsymbol{r}')\psi_j(\boldsymbol{r}')d\Gamma_{r'}, \end{aligned} \tag{10.58}$$

which enables us to write (10.53) as

$$\left(\frac{1}{2}\mathcal{I}+\mathcal{H}\right)\boldsymbol{v} = \mathcal{G}\boldsymbol{b}. \tag{10.59}$$

The operators $\mathcal{G}$ and $\mathcal{H}$ are called the *single and double layer operator* for the Stokes equations. For the Dirichlet problem, the velocity $\boldsymbol{v}$ at the boundary is given ($\Gamma_2 = \emptyset$) and we would like to reconstruct the normal stress $\boldsymbol{b}$ at the boundary. To this end, we need to invert the operator $\mathcal{G}$. This can only be done when all eigenvalues of $\mathcal{G}$ are unequal to zero. In this section we investigate the conditions under which $\mathcal{G}$ admits a zero eigenvalue.

For the mixed problem, the velocity at Γ_1 is prescribed and the normal stress at Γ_2 is prescribed. We would like to reconstruct the unknown velocity at Γ_2 and the unknown normal stress at Γ_1. After rearranging known and unknown terms (see Section 10.6.2) we again need to invert a boundary integral operator. This can only be done when all eigenvalues of the operator are unequal to zero. We will show that zero eigenvalues occur under the same conditions as for the Dirichlet problem.

10.6.1. *Flow problems with dirichlet boundary conditions*

In this subsection, we study the solvability of the BIE (10.59) with Dirichlet conditions on a piece-wise smooth closed boundary Γ. We search for eigenfunctions $\boldsymbol{b}$ of the boundary integral operator $\mathcal{G}$ with zero eigenvalue, hence $\mathcal{G}\boldsymbol{b} = 0$. If such eigenfunctions exist, the boundary integral operator $\mathcal{G}$ is not invertible and the integral equation (10.59) is not uniquely solvable. First, we show that at least one such eigenfunction with zero eigenvalue exists.

Theorem 10.5. *For any smooth boundary Γ, the outward unit normal $\boldsymbol{n}(\boldsymbol{r})$ is an eigenfunction of the boundary integral operator $\mathcal{G}$ with eigenvalue zero.*

Proof. The i-th component of $\mathcal{G}\boldsymbol{n}$ equals

$$\begin{aligned}(\mathcal{G}\boldsymbol{n})_i &= \int_\Gamma u_{ij}(\boldsymbol{r},\boldsymbol{r}')n_j(\boldsymbol{r}')d\Gamma_{r'}\\ &= \frac{1}{4\pi}\int_\Gamma \left[\delta_{ij}\log\frac{1}{\|\boldsymbol{r}-\boldsymbol{r}'\|} + \frac{(x_i-y_i)(x_j-y_j)}{\|\boldsymbol{r}-\boldsymbol{r}'\|^2}\right] n_j(\boldsymbol{r}')d\Gamma_{r'}\\ &= \frac{1}{4\pi}\int_\Omega \frac{\partial}{\partial x_j}\left[\delta_{ij}\log\frac{1}{\|\boldsymbol{r}-\boldsymbol{r}'\|} + \frac{(x_i-y_i)(x_j-y_j)}{\|\boldsymbol{r}-\boldsymbol{r}'\|^2}\right] d\Omega_y\\ &= -\int_\Omega \frac{\partial}{\partial x_j}u_j^i d\Omega_y = -\int_\Omega \nabla.\boldsymbol{u}^i d\Omega = 0. \end{aligned} \tag{10.60}$$

Here, the vector $\boldsymbol{u}^i$ is the velocity due to a Stokeslet,[24] i.e. the velocity field induced by a point force in the $\boldsymbol{e}_i$-direction. This velocity field satisfies the incompressibility condition $\nabla.\boldsymbol{u}^i = 0$. □

In the sequel to this section, we assume that the solutions of the Dirichlet problem (10.59) are sought in a function space that excludes the normal. Hence, the eigenfunctions $\boldsymbol{b}$ of $\mathcal{G}$ we are looking for are perpendicular to $\boldsymbol{n}$.

We now show that for each boundary Γ there exist (at most) two critical scalings of the boundary such that the operator $\mathcal{G}$ in the Dirichlet problem (10.59) is not invertible. This phenomenon has been observed and proven[26] and we will partly present the analysis here. The scaling for which the operator $\mathcal{G}$ is not invertible is called a *critical scaling*, and the corresponding boundary a *critical boundary*. The domain that is enclosed by the critical boundary is referred to as the *critical domain*.

Theorem 10.6. *For all given functions* $\boldsymbol{f}$ *and constant vectors* $\mathbf{d}$ *the system of equations*

$$\begin{cases} \mathcal{G}\boldsymbol{b} + \mathbf{c} = \boldsymbol{f}, \\ \int_\Gamma \boldsymbol{b} d\Gamma = \boldsymbol{d}, \end{cases} \tag{10.61}$$

has a unique solution pair $(\mathbf{b}, \mathbf{c})$*, where* $\mathbf{b}$ *is a function and* $\mathbf{c}$ *is a constant vector.*

Sketch of proof. The main idea is to show that the operator that maps the pair $(\boldsymbol{b}, \boldsymbol{c})$ to the left-hand side of (10.61) is an isomorphism.[26] ■

We proceed by introducing the two unit vectors $\boldsymbol{e}_1 = [1,0]^T$ and $\boldsymbol{e}_2 = [0,1]^T$. Theorem 10.6 guarantees that two pairs $(\boldsymbol{b}^1, \boldsymbol{c}^1)$ and $(\boldsymbol{b}^2, \boldsymbol{c}^2)$ exist that are the unique solutions to the two systems

$$\begin{cases} \mathcal{G}\boldsymbol{b}^1 + \boldsymbol{c}^1 = \mathbf{0}, \\ \int_\Gamma \boldsymbol{b}^1 d\Gamma = \boldsymbol{e}_1, \end{cases} \qquad \begin{cases} \mathcal{G}\boldsymbol{b}^2 + \boldsymbol{c}^2 = \mathbf{0}, \\ \int_\Gamma \boldsymbol{b}^2 d\Gamma = \boldsymbol{e}_2. \end{cases} \tag{10.62}$$

We define the matrix $\mathbf{C}_\Gamma$ as $\mathbf{C}_\Gamma := [\boldsymbol{c}^1 | \boldsymbol{c}^2]$.

Theorem 10.7. *If* $\det(\mathbf{C}_\Gamma) = 0$*, then the operator* $\mathcal{G}$ *is not invertible.*

Proof. Suppose that $\det(\mathbf{C}_\Gamma) = 0$, then the columns $\boldsymbol{c}^1$ and $\boldsymbol{c}^2$ are dependent, say $\boldsymbol{c}^1 = \alpha \boldsymbol{c}^2$ for some $\alpha \in \mathbb{R}$, $\alpha \neq 0$. In that case

$$\begin{aligned} \mathbf{0} &= \left(\mathcal{G}\boldsymbol{b}^1 + \boldsymbol{c}^1\right) - \alpha\left(\mathcal{G}\boldsymbol{b}^2 + \boldsymbol{c}^2\right) \\ &= \mathcal{G}(\boldsymbol{b}^1 - \alpha \boldsymbol{b}^2) + \alpha \boldsymbol{c}^2 - \alpha \boldsymbol{c}^2 \\ &= \mathcal{G}(\boldsymbol{b}^1 - \alpha \boldsymbol{b}^2). \end{aligned} \tag{10.63}$$

The function $\boldsymbol{b}^1 - \alpha\boldsymbol{b}^2$ cannot be equal to zero, since this requires $\int_\Gamma(\boldsymbol{b}^1 - \alpha\boldsymbol{b}^2)d\Gamma$ to be equal to zero, while we have

$$\int_\Gamma (\boldsymbol{b}^1 - \alpha\boldsymbol{b}^2)d\Gamma = \boldsymbol{e}_1 - \alpha\boldsymbol{e}_2 \neq \boldsymbol{0}. \tag{10.64}$$

So $\boldsymbol{b}^1 - \alpha\boldsymbol{b}^2$ is an eigenfunction of $\mathcal{G}$ with zero eigenvalue. This eigenfunction cannot be equal to the normal $\boldsymbol{n}$, since $\boldsymbol{n}$ also requires $\int_\Gamma \boldsymbol{n} d\Gamma = 0$. □

Corollary 10.2. *There are (at most) two critical scalings of the domain Γ for which the operator $\mathcal{G}$ is not invertible.*

Proof. We rescale the domain Γ by a factor a, i.e. $\Gamma \to a\Gamma$. With the definition of the operator $\mathcal{G}$, it can be shown that

$$\mathcal{G}\boldsymbol{b} \to \mathcal{G}^a\boldsymbol{b} := -\frac{1}{4\pi}\int_\Gamma a\log a\ \boldsymbol{b}d\Gamma + a\mathcal{G}\boldsymbol{b}. \tag{10.65}$$

Then the two systems in (10.62) change into

$$\begin{cases} a\mathcal{G}\boldsymbol{b}^j + \boldsymbol{c}^j - \frac{1}{4\pi}\int_\Gamma a\log a\ \boldsymbol{b}^j d\Gamma = \boldsymbol{0}, \\ \qquad a\int_\Gamma \boldsymbol{b}^j d\Gamma = \boldsymbol{e}_j,\ j = 1,2. \end{cases} \tag{10.66}$$

In defining $\boldsymbol{b}_a^j := a\boldsymbol{b}^j$ for $j = 1,2$, we obtain

$$\begin{cases} \mathcal{G}\boldsymbol{b}_a^j + \boldsymbol{c}^j - \frac{1}{4\pi}\int_\Gamma \log a\ \boldsymbol{b}_a^j d\Gamma = \boldsymbol{0}, \\ \qquad \int_\Gamma \boldsymbol{b}_a^j d\Gamma = \boldsymbol{e}_j,\ j = 1,2. \end{cases} \tag{10.67}$$

Substituting the second equation into the first equation, we get

$$\begin{cases} \mathcal{G}\boldsymbol{b}_a^j + \boldsymbol{c}^j - \frac{1}{4\pi}\log a\ \boldsymbol{e}_j = \boldsymbol{0}, \\ \qquad \int_\Gamma \boldsymbol{b}_a^j d\Gamma = \boldsymbol{e}_j,\ j = 1,2. \end{cases} \tag{10.68}$$

These systems have the same form as the original systems in (10.62), except for the change $\boldsymbol{c}^j \to \boldsymbol{c}^j - \frac{1}{4\pi}\log a\ \boldsymbol{e}_j$ for $j = 1,2$. Define the new matrix $\mathbf{C}_{a\Gamma}$ by

$$\mathbf{C}_{a\Gamma} := \mathbf{C}_\Gamma - \frac{1}{4\pi}\log a\ \mathbf{I}_2, \tag{10.69}$$

then $\mathcal{G}^a$ is not invertible when $\det(\mathbf{C}_{a\Gamma}) = 0$. Hence, when $\frac{1}{4\pi}\log a$ is an eigenvalue of $\mathbf{C}_\Gamma$, the operator $\mathcal{G}^a$ is not invertible. This implies that when $\mathbf{C}_\Gamma$ has two distinct eigenvalues, there are two critical scalings a for which $\mathcal{G}^a$ is not invertible. If $\mathbf{C}_\Gamma$ has one eigenvalue with double multiplicity these critical scalings coincide. □

The result of corollary 10.2 shows that the BIEs for the Dirichlet Stokes equations become singular for certain sizes of the domain. As a consequence, the equations are not uniquely solvable. This solvability problem is an artifact of the boundary integral formulation: the Stokes equations in differential form always have a unique solution. In the previous section, a similar phenomenon was observed for the Laplace equation with Dirichlet conditions; in its differential form, the problem is well-posed, while the corresponding BIE is not solvable at critical boundaries.

For the Laplace equation, the critical scaling is related to the logarithmic capacity of the domain. By calculating or estimating the logarithmic capacity one can determine or estimate the critical domains without computing BEM matrices and evaluating their condition numbers. It is this *a priori* information that allows us to modify the standard BEM formulation such that the BIE becomes uniquely solvable.

For the Stokes equations, there does not exist an equivalent to the logarithmic capacity. Hence we cannot *a priori* determine the critical domains. One way to determine the critical domains is by computing the BEM-matrices and evaluating their condition numbers. If the condition number jumps to infinity for a certain domain, then this domain is a critical domain. Hence, this strategy requires the solution of many BEM problems.

Another possibility for determining the critical domains is by solving the systems in (10.62). This yields the matrix $\mathbf{C}_\Gamma$ and, subsequently, the matrix $\mathbf{C}_{a\Gamma}$. By calculating the eigenvalues of the latter matrix, the critical scalings can be found. Again, we have to solve two non-standard BEM problems to compute the critical scalings.

Remark. The BIEs for the Stokes flow in 2D are similar to the equations for plane elasticity. Hence, the BIEs for the latter equations suffer from the same solvability problems as the Stokes equations. A proof of this phenomenon for plane elasticity is found in the literature[27] and is similar to the proof sketched above.

10.6.2. *Flow problems with mixed boundary conditions*

In the previous subsection, we showed that the boundary integral operator $\mathcal{G}$ for the Dirichlet Stokes equations is not invertible for all domains. In this subsection, we will show that this phenomenon extends to the Stokes equations with mixed boundary conditions.

The starting point is again the BIE for the Stokes equations,

$$\frac{1}{2}\boldsymbol{v} + \mathcal{H}\boldsymbol{v} = \mathcal{G}\boldsymbol{b}, \text{ at } \Gamma. \tag{10.70}$$

Suppose that the boundary Γ is split into two parts, $\Gamma = \Gamma_1 \bigcup \Gamma_2$. On Γ_1, we prescribe the velocity $\boldsymbol{v}^1$ while the normal stress $\boldsymbol{b}^1$ is unknown. On Γ_2, we prescribe the normal stress $\boldsymbol{b}^2$ while the velocity $\boldsymbol{v}^2$ is unknown. The boundary integral operators $\mathcal{G}$ and $\mathcal{H}$ are split accordingly,

$$\begin{aligned}
[\mathcal{G}\boldsymbol{b}]_i &= \int_\Gamma u_{ij} b_j d\Gamma = \int_{\Gamma_1} u_{ij} b_j^1 d\Gamma + \int_{\Gamma_2} u_{ij} b_j^2 d\Gamma =: [\mathcal{G}^1\boldsymbol{b}^1]_i + [\mathcal{G}^2\boldsymbol{b}^2]_i, \\
[\mathcal{H}\boldsymbol{v}]_i &= \int_\Gamma q_{ij} v_j d\Gamma = \int_{\Gamma_1} q_{ij} v_j^1 d\Gamma + \int_{\Gamma_2} q_{ij} v_j^2 d\Gamma =: [\mathcal{H}^1\boldsymbol{v}^1]_i + [\mathcal{H}^2\boldsymbol{v}^2]_i.
\end{aligned} \tag{10.71}$$

With these notations, the BIE is written in the following way:

$$\frac{1}{2}\boldsymbol{v}^k + \mathcal{H}^1\boldsymbol{v}^1 + \mathcal{H}^2\boldsymbol{v}^2 = \mathcal{G}^1\boldsymbol{b}^1 + \mathcal{G}^2\boldsymbol{b}^2, \quad \text{at } \Gamma_k,\ k = 1, 2. \tag{10.72}$$

We arrange the terms in such a way that all unknowns are at the left-hand side and all knowns are at the right-hand side,

$$\begin{aligned}
\mathcal{H}^2\boldsymbol{v}^2 - \mathcal{G}^1\boldsymbol{b}^1 &= \mathcal{G}^2\boldsymbol{b}^2 - \mathcal{H}^1\boldsymbol{v}^1 - \frac{1}{2}\boldsymbol{v}^1, \quad \text{at } \Gamma_1, \\
\frac{1}{2}\boldsymbol{v}^2 + \mathcal{H}^2\boldsymbol{v}^2 - \mathcal{G}^1\boldsymbol{b}^1 &= \mathcal{G}^2\boldsymbol{b}^2 - \mathcal{H}^1\boldsymbol{v}^1, \quad \text{at } \Gamma_2.
\end{aligned} \tag{10.73}$$

Now we can define an operator $\mathcal{A}$ that assigns to the pair $(\boldsymbol{b}^1, \boldsymbol{v}^2)$ the two functions at the left-hand side of (10.73),

$$\begin{bmatrix} \boldsymbol{b}^1 \\ \boldsymbol{v}^2 \end{bmatrix} \overset{\mathcal{A}}{\rightarrow} \begin{bmatrix} \mathcal{H}^2\boldsymbol{v}^2 - \mathcal{G}^1\boldsymbol{b}^1 \\ \frac{1}{2}\boldsymbol{v}^2 + \mathcal{H}^2\boldsymbol{v}^2 - \mathcal{G}^1\boldsymbol{b}^1 \end{bmatrix} \tag{10.74}$$

To study the invertibility of this operator, we need to study the homogeneous version of the eqs in (10.73),

$$\begin{aligned}
\mathcal{H}^2\boldsymbol{v}^2 - \mathcal{G}^1\boldsymbol{b}^1 &= 0, \quad \text{at } \Gamma_1, \\
\frac{1}{2}\boldsymbol{v}^2 + \mathcal{H}^2\boldsymbol{v}^2 - \mathcal{G}^1\boldsymbol{b}^1 &= 0, \quad \text{at } \Gamma_2.
\end{aligned} \tag{10.75}$$

Theorem 10.8. *There are (at most) two scalings of Γ such that the homogeneous eqs in (10.75) have a non-trivial solution, i.e. $\mathcal{A}$ is not invertible.*

Sketch of proof. We have to find a non-trivial pair of functions $(\boldsymbol{b}_1, \boldsymbol{v}_2)$ such that the left-hand sides of (10.75) are equal to zero when $C_l(\Gamma) = 1$. Similar to the Laplacian case, it can be shown that such a pair of functions exists by using information from the Dirichlet problem.[28] ■

This result shows that the BIE for the Stokes equations with mixed boundary conditions may also become singular. This happens for the same critical boundaries as for the Stokes equations with Dirichlet boundary conditions. Hence, the mixed problem inherits the singularities from the Dirichlet problem. The division of the boundary into a Dirichlet and a Neumann part does not play a role in this.

Note that the Laplace equation exhibits the same behaviour. The boundary integral equation for the Laplace equation with mixed boundary conditions also inherits the solvability problems from the BIE for the Dirichlet case.[11]

10.6.3. *Numerical examples*

To solve the BIEs (10.59), the boundary Γ is discretized into a set of N linear elements. At each element, the velocity $\boldsymbol{v}$ and normal stress $\boldsymbol{b}$ are approximated linearly. In this way, the BIEs are transformed into a linear system of algebraic equations. (For details about the discretization, refer to any BEM handbook.[29,30]) We introduce vectors $\mathbf{v}$ and $\mathbf{b}$ of length $2N$ containing the coefficients of $\boldsymbol{v}$ and $\boldsymbol{b}$ at the nodal points. Then the system of equations can be written in short-hand notation as

$$\mathbf{G}\mathbf{b} = \left(\frac{1}{2}\mathbf{I} + \mathbf{H}\right)\mathbf{v}. \tag{10.76}$$

Here, $\mathbf{G}$ and $\mathbf{H}$ are the discrete counterparts of the single and double layer operator.

If the boundary integral operator $\mathcal{G}$ is not invertible, then its discrete counterpart, the matrix $\mathbf{G}$, is ill-conditioned. To visualize this, we compute the condition number of $\mathbf{G}$: if the condition number is infinitely large, then the matrix is not invertible. As a consequence, the linear system (10.76) is singular and cannot be solved for arbitrary right-hand side vectors. If the condition number is bounded but very large, then the problem (10.76) is still difficult to solve accurately.

In the following examples, we construct the matrix $\mathbf{G}$ for a certain boundary Γ and compute the condition number of this matrix. Then we rescale the boundary Γ by a factor a, i.e. $\Gamma \rightarrow a\Gamma$. Again, we compute the condition number of the matrix $\mathbf{G}$. We do this for several values of the scaling parameter a. According to the theory in the previous sections, there are two critical scalings for which the integral operator $\mathcal{G}$ is not invertible. For these two scalings the matrix $\mathbf{G}$ is not invertible, or at least very ill-conditioned. Hence, we expect that the condition number of $\mathbf{G}$ will jump to infinity at these two scalings. The scaling for which such large condition numbers appear is called a *critical scaling* and has, ideally, the same value as the critical scaling defined for the BIE in the previous section. However, due to the discretization of the equations, the critical scaling for the discrete problem may be slightly different from the critical scaling for the BIE. In the limit $N \rightarrow \infty$ these differences vanish.

As the results for the Dirichlet and mixed problems are similar, we do not present any examples for the mixed case here.

Example 10.6.1. In Fig. 10.7, we show the condition number as a function of the scale a. We do this for an ellipse (ellipse 1) with aspect ratio 0.4 and for an ellipse (ellipse 2) with aspect ratio 0.7. We observe that for both cases two critical scalings exist for which the condition number goes to infinity. Moreover, these critical scalings differ significantly for the two ellipses. For ellipse 1, we find critical scalings 1.9 and 2.9 approximately, while for ellipse 2 we find 1.8 and 2.1. Hence the shape of the ellipse, i.e. its aspect ratio, greatly influences the values of the critical scalings.

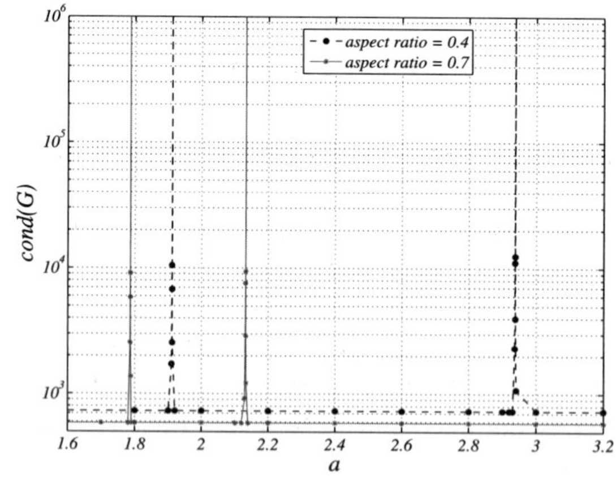

Fig. 10.7. Condition number of $\mathbf{G}$ for an ellipsoidal domain with aspect ratios 0.4 and 0.7 as a function of scaling parameter a.

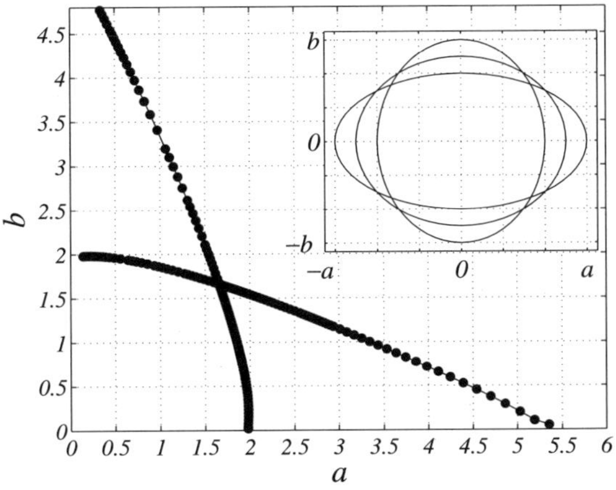

Fig. 10.8. The critical sizes of an ellipse for which the condition number of **G** is very large.

Figure 10.8 visualizes all ellipses for which the condition number of **G** is very large, i.e. all critical ellipses with different aspect ratios. At the horizontal axis is the length a of the horizontal semi-axis of the ellipse, at the vertical axis is the length b of the vertical semi-axis of the ellipse. We compute the condition number of **G** for several values of a and b. We call the values of a and b for which the condition number goes to infinity the *critical sizes* and the corresponding ellipse the critical ellipse. At the *critical sizes* we plot a dot in the (a, b)-plane of Fig. 10.8. We see that the critical sizes lie on two curved lines, which are symmetric around the line $a = b$. It can be concluded that for an ellipse with fixed aspect ratio $d := a/b \neq 1$, two critical sizes exist. For a circle, where $d = 1$, only one critical size exists. The values corresponding to this critical size are approximately $a = b = 1.65$, which agrees with the critical scale $\exp(1/2) \approx 1.649$ that can be derived analytically.[28]

10.7. Discussion

In this chapter, we have looked at the fundamental causes of error in the BEM and provided expressions for local errors. A proper definition of error will be at the core of any adaptive grid refinement strategy. We have shown that the local error for both constant and linear elements is of quadratic order. This is confirmed by the results in Tables 10.1 to 10.3. These results plus those in Table 10.4 show that this error behaviour holds for Dirichlet as well as Neumann problems. These findings provide an informed guide for choosing meshing strategies. Comparing the results of Tables 10.6 and 10.7 and those of Tables 10.8 and 10.9 shows that the exact treatment of boundary conditions as we explained leads to more accurate results. This will be particularly important in problems with localized regions of high variations in the boundary condition where we need to capture the high activity.

The value of the condition number of BEM matrices, apart from the number of boundary elements and problem type, depends on the size and shape of the domain. Critical domains exist for which the condition numbers are infinitely large. For such domains, the corresponding linear systems are singular and any resulting BEM solution is not unique. This phenomenon, whereby the size and shape of the domain affect the solvability, is a typical two-dimensional problem. In three dimensions this does not occur. The core of the phenomenon is the logarithmic term in the 2D fundamental solution for the differential operators that transforms a scaling of the domain into an additive term. In three dimensions, such a logarithmic term does not appear in the fundamental solution.

We have presented a number of strategies to avoid critical domains, which can be implemented easily in existing BEM codes. This will guarantee the uniqueness of the BEM solutions for domains of all sizes and shapes.

For potential problems, one can *a priori* determine or estimate the critical domains on which the condition number of the BEM matrix is infinitely large. This is achieved by exploiting the concept of logarithmic capacity. Unfortunately, we cannot use this strategy to determine or estimate the critical domains for flow problems, since an equivalent concept of the logarithmic capacity for flow problems does not yet exist.

In this chapter, we have shown that critical domains exist for potential problems and flow problems, and at the same time critical domains occur when solving the biharmonic equation with the BEM.[31–33] The question arises whether critical domains exist for a broader class of boundary value problems. To the authors' knowledge this is still an open question.

References

1. M. Guiggiani and F. Lombardi, Self-adaptive boundary elements with *h*-hierachical shape functions, *Advances in Engineering Software, Special issue on error estimates and adaptive meshes for FEM/BEM* **15**, 269–277, (1992).
2. J. T. C. M. T. Liang and S. S. Yang, Error estimation for boundary element method., *Engineering Analysis with Boundary Elements* **23**, 257–265, (1999).

3. C. Carsten and E. P. Stephan, A posteriori error estimates for boundary element methods., *Mathematics of Computation* **64**, 483–500, (1995).
4. G. Hsiao, On the stability of integral equations of the first kind with logarithmic kernels, *Arch. Ration. Mech. Anal.* **94**, 179–192, (1986).
5. M. Jaswon and G. Symm, *Integral Equation Methods in Potential Theory and Elastostatics* (Academic Press, London, 1977).
6. H. Power and L. Wrobel, *Boundary Integral Methods in Fluid Mechanics*, (Computational Mechanics Publications, Southampton, 1995).
7. I. Sloan and A. Spence, The Galerkin method for integral equations of the first kind with logarithmic kernel: Theory, *IMA J. Numer. Anal.* **8**, 105–122, (1988).
8. E. Hille, *Analytic Function Theory: Vol II.* (Ginn, London, 1959).
9. N. Landkof, *Foundations of Modern Potential Theory* (Springer Verlag, Berlin, 1972).
10. M. Tsuji, *Potential Theory in Modern Function Theory* (Maruzen, Tokyo, 1959).
11. W. Dijkstra and R. Mattheij, A relation between the logarithmic capacity and the condition number of the BEM-matrices, *Comm. Numer. Methods Engrg.* **23**(7), 665–680, (2007).
12. M. T. Heath, *Scientific Computing, an Introductory Survey* (McGraw Hill, New York, 2002).
13. S. Christiansen, Condition number of matrices derived from two classes of integral equations, *Mat. Meth. Appl. Sci.* **3**, 364–392, (1981).
14. S. Christiansen, On two methods for elimination of non-unique solutions of an integral equation with logarithmic kernel, *Appl. Anal.* **13**, 1–18, (1982).
15. W. Dijkstra and R. Mattheij, The condition number of the BEM-matrix arising from Laplace's equation, *Electron. J. Bound. Elem.* **4**(2), 67–81, (2006).
16. R. Barnard, K. Pearce, and A. Solynin, Area, width, and logarithmic capacity of convex sets, *Pacific J. Math.* **212**, 13–23, (2003).
17. J. Rostand, Computing logarithmic capacity with linear programming, *Experiment. Math.* **6**, 221–238, (1997).
18. W. Dijkstra and M. Hochstenbach, Numerical approximation of the logarithmic capacity, *Appl. Numer. Math.* (2008). Submitted for publication.
19. W. McLean, *Strongly Elliptic Systems and Boundary Integral Equations* (Cambridge University Press, Cambridge, 2000).
20. Y. Yan and I. Sloan, On integral equations of the first kind with logarithmic kernels, *J. Integral Equations Appl.* **1**, 549–579, (1988).
21. C. Constanda, On the solution of the Dirichlet problem for the two-dimensional Laplace equation, *Proc. Amer. Math. Soc.* **119**(3), 877–884, (1993).
22. J. Chen, C. Lee, I. Chen, and J. Lin, An alternative method for degenerate scale problems in boundary element methods for the two-dimensional Laplace equation, *Eng. Anal. Bound. Elem.* **26**, 559–569, (2002).
23. J. Chen, J. Lin, S. Kuo, and Y. Chiu, Analytical study and numerical experiments for degenerate scale problems in boundary element method

using degenerate kernels and circulants, *Eng. Anal. Bound. Elem.* **25**, 819–828, (2001).
24. O. Ladyzhenskaya, *The Mathematical Theory of Viscous Incompressible Flow.* (Gordon and Beach, New York-London, 1963).
25. C. Pozrikidis, *Boundary Integral and Singularity Methods for Linearized Viscous Flows* (Cambridge University Press, Cambridge, 1992).
26. V. Domínguez and F. Sayas, A BEM-FEM overlapping algorithm for the Stokes equation, *Appl. Math. Comput.* **182**, 691–710, (2006).
27. R. Vodička and V. Mantič, On invertibility of elastic single-layer potential operator, *J. Elasticity.* **74**, 147–173, (2004).
28. W. Dijkstra and R. Mattheij, Condition number of the BEM matrix arising from the Stokes equations in 2D, *Eng. Anal. Bound. Elem.* **32**(9), 736–746, (2008).
29. A. Becker, *The Boundary Element Method in Engineering* (McGraw-Hill, Maidenhead, 1992).
30. C. Brebbia, J. Telles, and L. Wrobel, *The Boundary Element Method in Engineering* (McGraw-Hill Book Company, London, 1984).
31. B. Fuglede, On a direct method of integral equations for solving the biharmonic Dirichlet problem, *ZAMM.* **61**, 449–459, (1981).
32. M. Costabel and M. Dauge, On invertibility of the biharmonic single-layer potential operator, *Integr. Equ. Oper. Theory* **24**, 46–67, (1996).
33. C. Constanda, On the Dirichlet problem for the two-dimensional biharmonic equation, *Mat. Meth. Appl. Sci.* **20**, 885–890, (1997).